The Springer Series on Demographic Methods and Population Analysis

Volume 50

In recent decades, there has been a rapid development of demographic models and methods and an explosive growth in the range of applications of population analysis. This series seeks to provide a publication outlet both for high-quality textual and expository books on modern techniques of demographic analysis and for works that present exemplary applications of such techniques to various aspects of population analysis.

Topics appropriate for the series include:

- General demographic methods
- Techniques of standardization
- Life table models and methods
- Multistate and multiregional life tables, analyses, and projections
- Demographic aspects of biostatistics and epidemiology
- Stable population theory and its extensions
- Methods of indirect estimation, Stochastic population models
- Event history analysis, duration analysis, and hazard regression models
- Demographic projection methods and population forecasts
- Techniques of applied demographic analysis, regional and local population estimates and projections
- Methods of estimation and projection for business and health care applications
- Methods and estimates for unique populations such as schools and students

Volumes in the series are of interest to researchers, professionals, and students in demography, sociology, economics, statistics, geography and regional science, public health and health care management, epidemiology, biostatistics, actuarial science, business, and related fields.

More information about this series at http://www.springer.com/series/6449

Christos H. Skiadas • Charilaos Skiadas
Editors

Demography of Population Health, Aging and Health Expenditures

Springer

Editors
Christos H. Skiadas
ISAST
Athens, Attiki, Greece

Charilaos Skiadas
Mathematics & Computer Science
Hanover College
Hanover, IN, USA

ISSN 1389-6784 ISSN 2215-1990 (electronic)
The Springer Series on Demographic Methods and Population Analysis
ISBN 978-3-030-44694-9 ISBN 978-3-030-44695-6 (eBook)
https://doi.org/10.1007/978-3-030-44695-6

This Springer imprint is published by the registered company Springer Nature Switzerland AG
The registered company address is: Gewerbestrasse 11, 6330 Cham, Switzerland

Preface

This book, *Demography of Population Health, Aging and Health Expenditures*, provides theoretical and applied material for estimating vital parts of demography and health issues including the healthy aging process along with calculating the healthy life years lost to disability. Furthermore, the appropriate methodology for the optimum health expenditure allocation is included. Data analysis, statistical and stochastic methodology and probability approach are used, and important applications are provided. Aging and mortality, birth-death processes, self-perceived age, lifetime and survival are explored as well as pension and labour force. The book addresses a methodological approach to health problems in demography and society including and quantifying important parameters in order to be a valuable guide for researchers, theoreticians and practitioners from various disciplines in their studies and research.

This Book Is Divided into Five Parts

The First Part Includes Five Chapters on Healthy Aging, Healthy Life Years Lost and Health Expenditure Allocation

In Chap. 2, "Relation of the Weibull Shape Parameter with the Healthy Life Years Lost Estimates: Analytical Derivation and Estimation from an Extended Life Table", **Christos H. Skiadas** and **Charilaos Skiadas** analytically derive a more general model of survival and mortality in which a parameter related to the healthy life years lost (HLYL) is estimated, leading to the Weibull model and the corresponding shape parameter as a specific case as Matsushita et al. (1992) had observed. It was also demonstrated that the results found for the general HLYL parameter proposed provide results similar to those provided by the World Health Organization for the healthy life expectancy (HALE) and the corresponding HLYL estimates. An easy-to-apply Excel programme is developed as an extension of the classical life table including four more columns to estimate the cumulative mortality, the average mortality, the person life years lost and finally the HLYL parameter b_x. The latest versions of this programme appear in the Demographics2019 website: http://www.asmda.es/demographics2019.html.

In Chap. 3, "Direct Healthy Life Expectancy Estimates from Life Tables with a Sullivan Extension. Bridging the Gap Between HALE and Eurostat Estimates", **Christos H. Skiadas** and **Charilaos Skiadas** provide an analytic derivation of a more general model of survival and mortality and the estimation of parameter b_x related to the healthy life years lost (HLYL) followed by the formulation of a computer programme as an extension of the classical life table including more columns to estimate the cumulative mortality, the average mortality, the person life years lost and finally the HLYL parameter b_x. Even more, a further extension of the Excel programme based on the Sullivan method provides estimates of the healthy life expectancy at every year of the lifespan for five different types of estimates which are direct, WHO, Eurostat, equal and other. Estimates for several countries are presented. A methodology and a programme to bridge the gap between the World Health Organization (HALE) and Eurostat (HLE) healthy life expectancy estimates is also presented. The latest version of this programme (SKI-6 Program) appears in the Demographics2020 website: http://www.smtda.net/demographics2020.html.

In Chap. 4, "Modeling the Health Expenditure in Japan, 2011. A Healthy Life Years Lost Methodology", **Christos H. Skiadas** and **Charilaos Skiadas** present the main part of the healthy life years lost (HLYL) methodology with more details and illustrations and develop and extend a life table important to estimate the healthy life years lost along with the fitting to the health expenditure in the related case. The application results in Japan are quite promising and important to support decision-makers and health agencies with a powerful tool to improve the health expenditure allocation and the future predictions.

In Chap. 5, "Healthy Ageing in Czechia", **Tomáš Fiala** and **Jitka Langhamrová** explore the implications of the actual values of healthy life years (HLY) and the total life expectancy in Czechia separately for males and females. The limitation prevalence rates will be based on the European Union's Statistics on Income and Living Conditions (EU-SILC) survey and the European Health Interview Survey (EHIS). The comparison and the analysis of the differences between males and females and between healthy and total life expectancies using decomposition method are presented.

In Chap. 6, "Evolution of Systems with Power-Law Memory: Do We Have to Die? (Dedicated to the Memory of Valentin Afraimovich)", **Mark Edelman** explores the hypothesis on the similarity between the human evolution and the evolution of simple discrete nonlinear fractional (with power-law memory) systems, as is the Caputo fractional/fractional difference logistic map. This is a simple discrete system with power-/asymptotically power-law memory and quadratic nonlinearity. The underlying reason for modelling the evolution of humans by fractional systems is the observed power law in human memory and the viscoelastic nature of organ tissues of living species. Models with power-law memory may explain the observed decrease at very large ages of the rate of increase of the force of mortality and imply limited lifespans.

The Second Part on Mortality Modelling and Applications Includes Seven Chapters

In Chap. 7, "Application of Structural Equation Modeling to Infant Mortality Rate in Egypt", **Fatma Abdelkhalek** and **Marianna Bolla** apply structural equation modelling (SEM) to examine the factors that affect infant mortality rate (IMR) over time. They use data for five indicators: gross domestic product (GDP) per capita, current health expenditure as a percentage of the GDP, out-of-pocket health expenditure as a percentage of current health expenditures, "hepatitis B" immunizations and the maternal mortality ratio, all available at the World Bank website. SEM results show the direct, indirect and total effects of each indicator on the IMR. SEM provides important sequential causal relationships that can help policymakers set programme priorities.

In Chap. 8, "Modeling of Mortality in Elderly by Lung Cancer in the Northeast of Brazil", **João Batista Carvalho** and **Neir Antunes Paes** identify sociodemographic and socioeconomic determinants of mortality by lung cancer in older people, applying the confirmatory factor analysis (CFA) in the Northeast of Brazil. A cross-sectional ecological study was adopted using microdata information from the 2010 census and projected population for 2015, linking registered data on cause-specific mortality by lung cancer by sex to the 188 micro-regions of the Northeast of Brazil.

In Chap. 9, "Demographics of the Russian Pension Reform", **Dalkhat M. Ediev** uses recently developed techniques for improving estimates of life expectancy and extending mortality data to old age. He provides the first-ever analysis of the demographic implications of pension reform in Russia in both the cross-sectional and cohort perspectives. A comparison of the Russian case to Sweden is also made.

In Chap. 10, "Using the Developing Countries Mortality Database (DCMD) to Probabilistically Evaluate the Completeness of Death Registration at Old Ages", **Nan Li**, **Hong Mi** and **Xiaotong Tang** propose a probabilistic completeness of the death registration (DR) whose samples are the values of deterministic completeness. When the difference between 1 and the mean of probabilistic completeness is statistically insignificant, the DR is probabilistically complete. Focusing on old age and the level of mortality rather than the number of deaths, the effects of migration and census error are largely reduced in the Developing Countries Mortality Database (DCMD, www.lifetables.org), which is used to provide applications of the probabilistic evaluation in their paper.

In Chap. 11, "Mortality Developments in Greece from the Cohort Perspective", **Konstantinos N. Zafeiris**, **Anastasia Kostaki** and **Byron Kotzamanis** use period life table data for the male and female population in order to obtain 1-year probabilities of death for any birth cohort formed in Greece after the 1950s. Partial life expectancies and the expected years lost between birth and several other ages were calculated for each of them. The results of the analysis are indicative of the mortality transition observed in Greece for previous years and give a clear picture of the existing gender differences.

In Chap. 12, "On Demographic Approach of the BGGM Distribution Parameters on Italy and Sweden", **Panagiotis Andreopoulos** and **Alexandra Tragaki** study a new model of mathematical mortality (**BGGM** distribution) applied on Italian and Swedish data over a 114-year period (1900–2013). The aim of this work is to evaluate the proper adaptation of this distribution, to estimate its values and to identify eventual spatial and temporal differentiations.

In Chap. 13, "Alcohol Consumption in Selected European Countries", **Jana Vrabcová**, **Kornélia Svačinová** and **Markéta Pechholdová** assess alcohol consumption in Europe from the perspective of recorded per capita alcohol consumption and self-reported drinking patterns. The aim was to combine information from multiple data sources to reveal similarities and differences in drinking, including gender, educational and income disparities.

The Third Part Includes Three Chapters on Birth-Death Process, Self-Perceived Age and Gender Differences

In Chap. 14, "Modelling Monthly Births and Deaths Using Seasonal Forecasting Methods as an Input for Population Estimates", **Jorge Miguel Bravo** and **Edviges Coelho** evaluate the forecasting performance of alternative linear and nonlinear time series methods (seasonal ARIMA, Holt-Winters and state-space models) to death and birth monthly forecasting at the subnational level. Additionally, they investigate how well the models perform in terms of predicting the uncertainty of future monthly birth and death counts. They use the series of monthly death and birth data from 2000 to 2018 disaggregated by sex for the 25 Portuguese NUTS3 regions to compare the models' short-term (1-year) forecasting accuracy using a backtesting time series cross-validation approach.

In Chap. 15, "Births by Order and Childlessness in the Post-socialist Countries", **Filip Hon** and **Jitka Langhamrová** contribute to women's fertility research in the Czech Republic and other European countries. Their work focuses on the phenomenon of childlessness and on children by the order of birth. The female population by the number of children ever born is analysed by age groups. Besides the international comparison, this chapter also deals with the projection of the monitored characteristics in the future.

In Chap. 16, "On the Evaluation of 'Self-perceived Age' for Europeans and Americans", **Apostolos Papachristos** and **Georgia Verropoulou** estimate the "self-perceived age" by reference to life tables and evaluate its validity in comparison with actual mortality patterns. They use data from the 6th Wave of the Survey of Health, Ageing and Retirement in Europe (RAND SHARE) and the 12th Wave of Health and Retirement Study (RAND HRS) and life tables from the Human Mortality Database (HMD). They employ regression models for statistical analysis. The results indicate that health status and frequency of physical activities imply similar patterns of "self-perceived age" and actual mortality patterns.

The Fourth Part Includes Seven Chapters on Theoretical Issues and Applications

In Chap. 17, "Spatio-Temporal Aspects of Community Well-Being in Multivariate Functional Data Approach", **Włodzimierz Okrasa**, **Mirosław Krzyśko** and **Waldemar Wołyński** apply a functional data measurement approach – multivariate functional principal component analysis (MFPCA) – in a parallel way (independently) to two types of multidimensional measures characterizing community and individual (residents') levels of quality (development or deprivation) and subjective well-being. Having constructed classifications of both local communities (communes) and their residents for the same years (2004–2016), the spatial perspective is involved in the second part of the presentation.

In Chap. 18, "Properties and Dynamics of the Beta Gompertz Generalized Makeham Distribution", **Panagiotis Andreopoulos**, **Alexandra Tragaki**, **George Antonopoulos** and **Fragkiskos G. Bersimis** investigate the statistical properties of the proposed distribution with six parameters, called Beta Gompertz Generalized Makeham distribution. This includes verifying the probability density function, the cumulative density function and the hazard function.

In Chap. 19, "Increasing Efficiency in the EBT Algorithm", **Tin Nwe Aye** and **Linus Carlsson** deal with the Escalator Boxcar Train (EBT), a commonly used method for solving physiologically structured population models. The main goal was to overcome computational disadvantages of the Escalator Boxcar Train (EBT) method. They prove convergence for a general class of EBT models in which they modify the original EBT formulation, allowing merging of cohorts. They show that this modified EBT method induces a bounded number of cohorts, independent of the number of time steps, improving the numerical algorithm from polynomial to linear time.

In Chap. 20, "Psychometric Validation of Constructs Defined by Ordinal-Valued Items", **Anastasia Charalampi**, **Catherine Michalopoulou** and **Clive Richardson** perform exploratory factor analysis (EFA) on one half-sample in order to assess the construct validity of the scale. Then, the structure suggested by EFA is validated by carrying out a confirmatory factor analysis (CFA) in the second half. Based on the full sample, the psychometric properties of the resulting scales or subscales are assessed. They carry out the investigation and assessment of the 2012 European Social Survey (ESS) short eight-item version of the Center for Epidemiologic Studies Depression (CES-D 8) scale for Italy and Spain where items are considered as ordinal.

In Chap. 21, "Robust Minimal Markov Model for Dengue Virus Type 3", **Jesús E. García** and **V.A. González-López** suggest procedures to select the samples which will be used to establish a minimal Markov model from the whole set of samples applied to nine complete genomic sequences of dengue virus type 3 (DENV3), from the outbreak occurred in Henan, China, in 2013. The final model of the Henan set was built using the most representative sequences. It can be described by 14 basic units (parts).

In Chap. 22, "Determining Influential Factors in Spatio-temporal Models", **Rebecca Nalule Muhumuza, Olha Bodnar, Joseph Nzabanita** and **Rebecca N. Nsubuga** deal with general spatio-temporal models by applying the LOESS predictor for both the spatial interpolation and the temporal prediction. The number of closest neighbouring regions to be used in its construction is determined by cross-validation. They also discuss the computational aspects in the case of large-dimensional data and apply the theoretical findings to real data consisting of the number of influenza cases observed in the south of Germany.

In Chap. 23, "Describing Labour Market Dynamics Through Non-Homogeneous Markov System Theory", **Maria Symeonaki** and **Glykeria Stamatopoulou** apply the non-homogeneous Markov system (NHMS) theory to labour market transitions and provide a cross-national comparison of labour market flows among southern European countries. The theoretical adaptation of the NHMS model to labour market dynamics and its basic parameters are presented. Raw data, drawn from the European Union Labour Force Survey (EU-LFS), is used in order to estimate and compare the distribution of transition probabilities from the labour market state of employment, unemployment and inactiveness and vice versa for the selected European countries.

The Fifth Part Includes Seven Chapters on Lifetime, Survival, Pension, Labour Force and Further Estimates

In Chap. 24, "The Wide Variety of Regression Models for Lifetime Data", **Chrys Caroni** presents and discusses regression models for lifetime data, their differences and in what circumstances each is useful. Some theoretical relations are presented. Proportional hazards, proportional odds, proportional reversed hazards and inverse Gaussian, among others, are included.

In Chap. 25, "Analysing the Risk of Bankruptcy of Firms: Survival Analysis, Competing Risks and Multistate Models", **Francesca Pierri** and **Chrys Caroni**, after a brief review of quantitative methods applied in credit scoring, focus on analysing different causes of failure applied to a data set of 15,184 small business enterprises in Italy from 2008 to 2013. Firstly, the competing risk methodology from survival analysis is applied. Secondly, the occurrence of a second event in addition to the first is taken into account, by applying a multistate model to transition and survival probabilities.

In Chap. 26, "A Bayesian Modeling Approach to Private Preparedness Behavior Against Flood Hazards", **Pedro Araújo**, **Gilvan Guedes** and **Rosangela Loschi** investigate if the hypothesis supporting the Protective Action Decision Model (PADM) is satisfied in a survey involving individuals under risk of river floods in Brazil. Their model (1) is based on a probabilistic sample, with 1164 individuals interviewed in a city with a large share of the population under risk of river floods, (2) introduces a hierarchical Bayesian logistic model relating the probability of adopting protective measures against floods, (3) measures effectiveness and opportunity cost through Bayesian item response theory (IRT) models and (4) includes a random effect reflecting unmeasured individual features to correlate the individual responses to the different protective measures considered.

In Chap. 27, "Assessing Labour Market Mobility in Europe", **Maria Symeonaki** and **Glykeria Stamatopoulou** estimate a labour market mobility index that aims at capturing not only the extent of labour market mobility of individuals but also the quality of their transitions. The proposed methodology is illustrated for the case of young individuals aged between 15 and 29, for the latest at the time available data of the European Union Labour Force Survey (EU-LFS), i.e. the year 2016.

In Chap. 28, "The Implications of Applying Alternative-Supplementary Measures of the Unemployment Rate to Subpopulations and Regions: Evidence from the European Union Labour Force Survey for Southern Europe, 2008–2015", **Aggeliki Yfanti**, **Catherine Michalopoulou** and **Stelios Zachariou** investigate the implications of applying two broader alternative definitions of the unemployment rate to subpopulations and regions of interest for social policy purposes. The analysis is based on the 2008–2015 annual data sets of the EU-LFS for Southern Europe: Greece, Italy, Portugal and Spain. The results indicate the need, especially in recessionary times, for implementing alternative measures of the unemployment rate to the EU-LFS in the tradition of the Current Population Survey.

In Chap. 29, "Reverse Mortgages: Risks and Opportunities", **E. Di Lorenzo**, **G. Piscopo**, **M. Sibillo** and **R. Tizzano** focus on reverse mortgage contracts whose expiry is a function of the contractor's lifespan and whose assets depend on the evolution of real estate market prices. Reverse mortgages (RM) provide an attractive way to increase retirement incomes and to face the needs of health care for elderly people. A neural network procedure is employed in order to include a range of explanatory variables as part of the reverse mortgage evaluation algorithms.

In Chap. 30, "Estimating the Health State at Retirement: A Stochastic Modeling Approach", **Christos H. Skiadas** and **Charilaos Skiadas** calculate the gradual loss of health due to aging by using a quantitative methodology of estimating the declining health state function of a population and the growing deterioration function as an alternative. According to the theory developed, the health state is at level one at birth and gradually declines to zero at death. The development of the health state function and the deterioration function provides quantitative tools for the estimation of the health state of the human capital of a population, thus providing a supporting tool for the estimation of the average exit age from the labour force and the estimation of the optimal retirement age. Applications are also presented.

We thank all the contributors of this book, the authors, the chapter and book reviewers, and of course the Springer team for their help, guidance and support.

Athens, Greece
Hanover, IN, USA
December 2019

Christos H. Skiadas
Charilaos Skiadas

Contents

1 **Preliminary Notes** 1
Christos H. Skiadas and Charilaos Skiadas

Part I Healthy Aging, Healthy Life Years Lost and Health Expenditure Allocation

2 **Relation of the Weibull Shape Parameter with the Healthy Life Years Lost Estimates: Analytical Derivation and Estimation from an Extended Life Table** 9
Christos H. Skiadas and Charilaos Skiadas

3 **Direct Healthy Life Expectancy Estimates from Life Tables with a Sullivan Extension. Bridging the Gap Between HALE and Eurostat Estimates** 25
Christos H. Skiadas and Charilaos Skiadas

4 **Modeling the Health Expenditure in Japan, 2011. A Healthy Life Years Lost Methodology** 43
Christos H. Skiadas and Charilaos Skiadas

5 **Healthy Ageing in Czechia** 53
Tomáš Fiala and Jitka Langhamrová

6 **Evolution of Systems with Power-Law Memory: Do We Have to Die? (Dedicated to the Memory of Valentin Afraimovich)** 65
Mark Edelman

Part II Mortality Modeling and Applications

7 **Application of Structural Equation Modeling to Infant Mortality Rate in Egypt** 89
Fatma Abdelkhalek and Marianna Bolla

8 **Modeling of Mortality in Elderly by Lung Cancer in the Northeast of Brazil** ... 101
João Batista Carvalho and Neir Antunes Paes

9 **Demographics of the Russian Pension Reform** ... 111
Dalkhat M. Ediev

10 **Using the Developing Countries Mortality Database (DCMD) to Probabilistically Evaluate the Completeness of Death Registration at Old Ages** ... 133
Nan Li, Hong Mi, and Xiaotong Tang

11 **Mortality Developments in Greece from the Cohort Perspective** ... 151
Konstantinos N. Zafeiris, Anastasia Kostaki, and Byron Kotzamanis

12 **On Demographic Approach of the BGGM Distribution Parameters on Italy and Sweden** ... 169
Panagiotis Andreopoulos and Alexandra Tragaki

13 **Alcohol Consumption in Selected European Countries** ... 187
Jana Vrabcová, Kornélia Svačinová, and Markéta Pechholdová

Part III Birth-Death Process, Self-Perceived Age and Gender Differences

14 **Modelling Monthly Births and Deaths Using Seasonal Forecasting Methods as an Input for Population Estimates** ... 203
Jorge Miguel Bravo and Edviges Coelho

15 **Births by Order and Childlessness in the Post-socialist Countries** ... 223
Filip Hon and Jitka Langhamrová

16 **On the Evaluation of 'Self-perceived Age' for Europeans and Americans** ... 235
Apostolos Papachristos and Georgia Verropoulou

Part IV Theoretical Issues and Applications

17 **Spatio-Temporal Aspects of Community Well-Being in Multivariate Functional Data Approach** ... 251
Włodzimierz Okrasa, Mirosław Krzyśko, and Waldemar Wołyński

18 **Properties and Dynamics of the Beta Gompertz Generalized Makeham Distribution** ... 275
Panagiotis Andreopoulos, Alexandra Tragaki, George Antonopoulos, and Fragkiskos G. Bersimis

19 **Increasing Efficiency in the EBT Algorithm** 289
Tin Nwe Aye and Linus Carlsson

20 **Psychometric Validation of Constructs Defined by Ordinal-Valued Items** 319
Anastasia Charalampi, Catherine Michalopoulou, and Clive Richardson

21 **Robust Minimal Markov Model for Dengue Virus Type 3** 335
Jesús E. García and V. A. González-López

22 **Determining Influential Factors in Spatio-temporal Models** 347
Rebecca Nalule Muhumuza, Olha Bodnar, Joseph Nzabanita, and Rebecca N. Nsubuga

23 **Describing Labour Market Dynamics Through Non Homogeneous Markov System Theory** 359
Maria Symeonaki and Glykeria Stamatopoulou

Part V Life-Time, Survival, Pension, Labor Force and Further Estimates

24 **The Wide Variety of Regression Models for Lifetime Data** 377
Chrys Caroni

25 **Analysing the Risk of Bankruptcy of Firms: Survival Analysis, Competing Risks and Multistate Models** 385
Francesca Pierri and Chrys Caroni

26 **A Bayesian Modeling Approach to Private Preparedness Behavior Against Flood Hazards** 395
Pedro Araújo, Gilvan Guedes, and Rosangela Loschi

27 **Assessing Labour Market Mobility in Europe** 409
Maria Symeonaki and Glykeria Stamatopoulou

28 **The Implications of Applying Alternative-Supplementary Measures of the Unemployment Rate to Subpopulations and Regions: Evidence from the European Union Labour Force Survey for Southern Europe, 2008–2015** 419
Aggeliki Yfanti, Catherine Michalopoulou, and Stelios Zachariou

29 **Reverse Mortgages: Risks and Opportunities** 435
E. Di Lorenzo, G. Piscopo, M. Sibillo, and R. Tizzano

30 Estimating the Health State at Retirement: A Stochastic Modeling Approach 443
Christos H. Skiadas and Charilaos Skiadas

Correction to: Births by Order and Childlessness in the Post-socialist Countries C1

Chapter 1
Preliminary Notes

Christos H. Skiadas and Charilaos Skiadas

The World Health Organization defines health as: "a state of complete physical, mental, and social well-being, not merely the absence of disease or infirmity".

Further definitions exist depending on the special theoretical and applied needs. Some definitions are important for qualitative studies in demography while others are crucial for quantitative studies. In several of the chapters of this volume a special definition of health is extremely important. Especially when dealing with the stochastic character of the health state of a system or an organism. This is in connection of health state to the deterioration of the organism and the gradual health decline to the end. The data needed could come only from the life tables and the so-called Force of Mortality.

Accordingly, health is a highly probabilistic term related to the status of a complex system, an organism or even mechanical devise. At the individual level, it is impossible to estimate the health course during time due to the probabilistic-stochastic character of the phenomenon. The end may come any time and usually it is modeled by a first exit time stochastic process. However, following our experience, the health state process estimates for a large ensample can be expressed in a rather smooth form. The death probability density of a large population is relatively smooth. It could be modeled by an equation form. At the macro level, in a country, the mean health state curve and the dispersion of points around it are adequate to express the health state process in a population. The dispersion is usually called the variance or the standard deviation of the process. Weitz and Fraser (2001)

C. H. Skiadas (✉)
ISAST, Athens, Attiki, Greece
e-mail: skiadas@cmsim.net

C. Skiadas
Mathematics & Computer Science, Hanover College, Hanover, IN, USA
e-mail: skiadas@hanover.edu

C. H. Skiadas, C. Skiadas (eds.), *Demography of Population Health, Aging and Health Expenditures*, The Springer Series on Demographic Methods and Population Analysis 50, https://doi.org/10.1007/978-3-030-44695-6_1

had applied a simple stochastic model to express the health state of a large ensample of Mediterranean Flies from data collected by James Carey. The health state model they suggested is the co-called Inverse Gaussian already proposed at least from 1915 whereas an earlier form is due to Bachelier (1900).

1.1 The Simplest Stochastic Modeling Case

A decreasing function of concave form and a standard deviation for the health state are assumed for the human population. The standard deviation and the model parameters are estimated by a non-linear regression analysis algorithm by fitting the appropriate model to data. It is also important to decide on a very crucial problem, that is the starting point of the mean health status. Weitz and Fraser had decided the unity as a starting point of the mean health state level by means of the mean organism starting at a 100% health state. This is in accordance to Torrance's (1976) proposal. In the time course, the mean health state decreases gradually to zero level. As a mathematical problem this is equivalent to estimating the first exit time probability density function for a particle crossing a boundary. Weitz and Fraser selected the form $H(x) = 1 - bx$ for the Health State $H(x)$ at age x with b as the deterioration parameter. The health state declines linearly while the deterioration (Det) of the organism grows linearly as $\mathrm{Det}(x) = bx$. Then the first exit time probability density function is $g(x) = \frac{1}{\sigma\sqrt{2\pi x^3}} \exp\left(-\frac{(1-bx)^2}{2\sigma^2 x}\right)$. By fitting this formula to Mediterranean Fly data, the deterioration parameter b and standard deviation σ are estimated. The graphs of the data and of the fit estimates are illustrated in Fig. 1.1a. However, several equation forms could have a good fit to the same data. The most important point is to reproduce the data by applying stochastic simulations. This is presented in Fig. 1.1b where a large number of stochastic paths are generated by using the formula $h(x) = 1 + \sigma W_x$, where W_x is the standard Wiener process. The barrier is set by the deterioration function $\mathrm{Det}(x) = bx$. The result, after a large number of simulations, is presented in Fig. 1.1a verifying the

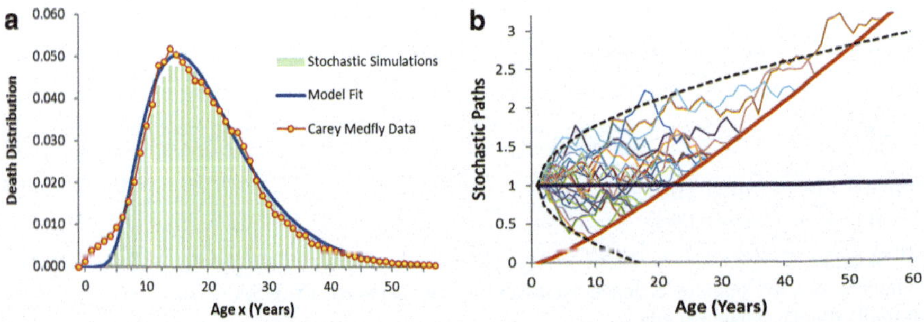

Fig. 1.1 (**a**) Carey Medfly data. Model fit and simulation. (**b**) Carey Medfly data. Stochastic simulations

argument of a successful reproduction of the processes and the good standing of the selected model of a linear deterioration and the associated linear health state process.

1.2 Stochastic Modeling in the Human Population

Though several cases are already known from physics, no specific case was applicable to the human population. This is because an extension of the Inverse Gaussian was needed. Fortunately, some mathematical forms where already developed in the previous decades, with the Jennen and Lerche (1981) approach to be more appropriate to model the Human Populations' case. In our paper (Skiadas and Skiadas 2010) a successful adaptation was performed followed by various applications. The extension was based on a health state form $H(x) = 1 - (bx)^c$, and a deterioration function $Det(x) = (bx)^c$. The stochastic model adopted is an extension of the Inverse Gaussian $g(x) = \frac{1+(c-1)(bx)^c}{\sigma\sqrt{2\pi x^3}} \exp\left(-\frac{(1-(bx)^c)^2}{2\sigma^2 x}\right)$. Soon afterwards, it was possible to reproduce the original data by stochastic simulations following the high fitting performance of the related model. It was evident that the health mean function was a smooth decreasing function associated with an increasing exponential-like deterioration function. Both the health state and the deterioration functions provide characteristic estimates of the mean level of the health or the deterioration of the population. One of the numerous applications verifying the fitting and explanatory performance of the model is illustrated in Fig. 1.2a, b for females in Sweden in 2016, whereas, a comparative application of males' and females' deterioration is also illustrated if Fig. 1.2c. Further to presenting the high significance of the methodology for estimating the deterioration function of a population, this comparative application is extremely useful. It was possible to respond to the important question on the level of health state or the deterioration at retirement and the related differences between males and females.

1.3 Barriers to Stochastic Modeling

Over the years, life table data sets dominated the quantitative demography estimates. Stochastic modeling is mainly based on probability distributions and related data sets. The total probability of a stochastic process is 1. Accordingly, the total probability of a population dying in the time course is 1. Instead of the Life Tables including the related data for dead and alive, they traditionally start from a total population of 100.000 at birth, decreasing to zero over the full life span. However, by dividing by 100.000 we can easily arrive to a probability distribution for people alive and also calculate the probability density function for people dead. From this point of view, the data needed for stochastic calculations are extracted

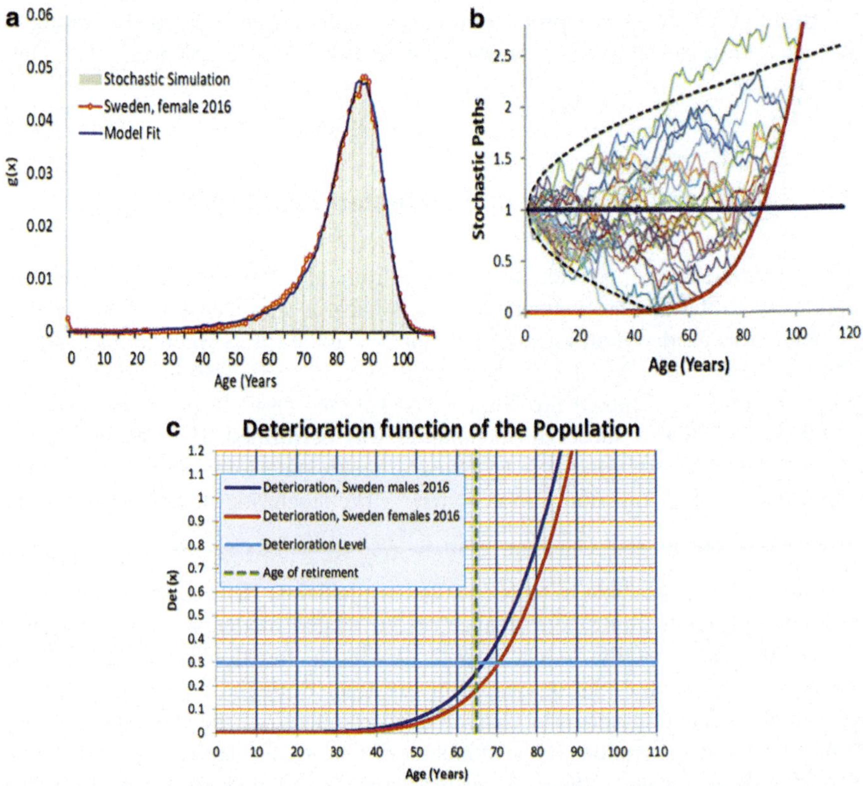

Fig. 1.2 (**a**) Sweden female data. Model fit and simulation. (**b**) Sweden female data. Stochastic simulations. (**c**) Sweden deterioration. Male and female comparisons

directly from the life tables after rescaling from 100.000 to 1. It is worth noting that the number of deaths per year and age in a country is not adequate for direct applications as we have to account for the population in the same period. Long ago, a death measure $\mu_x = Deaths_x/Population_x = D_x/P_x$, was accepted. From this, the death probability density *dx* is constructed from a Life Table or from an equation form as $g(x) = \mu_x \exp\left(-\int_0^x \mu_s ds\right)$, where $g(x) \cong d_x$. Then, the provided probability density function *dx* is a death distribution normalized for the population per age or per age group. This is smoother than the simple death distribution and it is convenient for the stochastic applications to follow. So far, and following the Gompertz legacy, numerous studies on modeling the so-called Force of Mortality μ_x, followed. However, almost all the models and equation forms based on the force of mortality cannot be used for the first exit time calculations except after laborious transformations, if any. Instead, a new approach is needed with the benefit of exploring new fields and unexplored topics, though hard work is needed to cover the gap in the research field of stochastic processes. However, the quest of further exploring the health status of a population is worth noting, and future research is needed.

1.4 Stochastic Modeling and Health Measure

The key-point of Demography in the following decades will be the acceptance of a Measure of Health along with the estimation methods. So far, the established Global Burden of Disease methodology, highly supported the health related studies and processes. The main achievement being the introduction and acceptance of *one measure* of healthy life years lost to disability provided on a periodic basis from the World Health Organization and other agencies. Though they have accepted a laborious method of estimation based on a large international system of researchers, data collectors and evaluators, the most important achievement was the acceptance of a *single index of measure* of the Healthy Life Years Lost (HLYL). The estimates started and continue to appear as the Healthy Life Expectancy or Health Adjusted Life Expectancy (HALE). However, HLYL is simpler, measuring the years lost only and not connected directly to Life Expectancy. HALE could be different between countries, but HLYL could be the same or similar. Though several HLYL values may appear, depending on the methodology used, the suggested and established single measure for the HLYL is of high importance. Firstly, it is accepted that a measure of health is fusible, at least for the healthy life years lost. Secondly, if the HLYL estimates are collected on an annual basis, a trend would appear over time, resulting in a mathematical-statistical estimating methodology. Thirdly, a systematic HLYL collection and evaluation will lead to different methods of estimation. Fourthly, the internal mechanisms of healthy aging could be explored and analytically presented. So far similar results to HLYL provided by WHO can be estimated with other methods. The Weibull shape parameter, a similar Gompertz parameter or a Direct estimation parameter, provides figures similar to the HLYL provided by WHO. A pioneer study of Matsushita et al. (1992) pointed out the connection of the Weibull shape parameter to various levels of mortality and health state.

Establishing a healthy life years lost estimation system rises the important question of a health measurement method. The traditional life expectancy and the recently healthy life expectancy proposed are based on probabilities to be alive or dead. But how about estimating a functionality level of a population at a given age or better how to estimate the health state level at a specific age, say at age of retirement? The previously presented methodology of Weitz and Fraser and expanded in the human population is of particular importance. The mean health level in the beginning of the life-span is assumed to be 1 or 100% of the total health and decreases until the end of life. Note that Janssen and Skiadas (1995) had presented an advanced stochastic model with interesting applications in subsequent publications (see Skiadas and Skiadas 2018a, b). The stochastic estimation methods provide a number for the health state at every age or an alternative measure for the deterioration of the human health at the same age. These type of measures are important to several estimates. Especially in the pension systems, where the life expectance or the healthy life expectancy do not provide direct figures, the health state estimation as provided by the first exit time methodology could be the standard methodology.

Another question is related to the gradual health deterioration during age. It was shown that a specific parameter connected to age and health level could provide valuable information for the HLYL and, most importantly, could be used to allocate the Health Expenditure per year of age or per age group of the population.

More information and special details are given in the chapters of this volume.

References

Bachelier, L. (1900). *Théorie de la spéculation*. Paris: Gauthier-Villars.

Janssen, J., & Skiadas, C. H. (1995). Dynamic modelling of life-table data. *Applied Stochastic Models and Data Analysis, 11*(1), 35–49.

Jennen, C., & Lerche, H. R. (1981). First exit densities of Brownian motion through one-sided moving boundaries. *Zeitschrift für Wahrscheinlichkeitstheorie und verwandte Gebiete, 55*, 133–148.

Matsushita, S., Agiwara, K., Hiota, T. A., Himada, H., Kijramoto, I., & Toyokura, Y. (1992). Lifetime Data Analysis of Disease and Aging by the Weibull Probability Distribution. *Journal of Clinical Epidemiology, 45*(10), 1165–I175.

Skiadas, C., & Skiadas, C. H. (2010). Development, simulation and application of first exit time densities to life table data. *Communications in Statistics Theory and Methods, 39*, 444–451.

Skiadas, C. H., & Skiadas, C. (2018a). *Exploring the health state of a population by dynamic modeling methods* (The springer series on demographic methods and population analysis) (Vol. 45). Chum: Springer. https://doi.org/10.1007/978-3-319-65142-2.

Skiadas, C. H., & Skiadas, C. (2018b). *Demography and health issues: Population aging, mortality and data analysis* (The springer series on demographic methods and population analysis) (Vol. 46). Chum: Springer. https://doi.org/10.1007/978-3-319-76002-5.

Torrance, G. W. (1976). Health status index models: A unified mathematical view. *Management Science, 22*(9), 990–1001.

Weitz, J. S., & Fraser, H. B. (2001). Explaining mortality rate plateaus. *Proceedings of the National Academy of Sciences of the USA, 98*(26), 15383–15386.

Part I
Healthy Aging, Healthy Life Years Lost and Health Expenditure Allocation

Chapter 2
Relation of the Weibull Shape Parameter with the Healthy Life Years Lost Estimates: Analytical Derivation and Estimation from an Extended Life Table

Christos H. Skiadas and Charilaos Skiadas

2.1 The Weibull Model Revisited

In this paper, we explore a very interesting property of the Weibull model (Weibull, 1951) more recently discussed in the past few decades in papers published in books and journals (see Matsushita et al. 1992; Skiadas and Skiadas 2014, 2015, 2018a, b, c, 2019; Skiadas and Arezzo 2018; and Weon and Je 2011, 2012). Our findings relate to the similarities of the shape parameter of this model with the estimated values of the Healthy Life Years Lost (HLYL), provided by the World Health Organization (WHO). Both of these estimates by WHO and Weibull are close to each other in several countries. Table 2.1 summarizes several indicative cases from European Countries with the higher Life Expectancy for males and females appearing in 2010. The table includes Life Expectancy (LE) and Healthy Life Expectancy (HALE) values, as provided by the World Health Organization (WHO). We then calculated Healthy Life Years Lost (HLYL) as the difference between LE and HALE to provide the WHO figures and our estimates for the Weibull shape parameter b and the Direct estimates (see Skiadas and Skiadas 2018a, b, c, 2019) of the HLYL parameter b_x from the Life Tables provided by the Human Mortality Database (HMD). The last row of the table shows the average values. The Weibull and Direct estimates are similar for males and not much higher than the HLYL estimates by WHO. The standard error is 0.855 for the WHO vs. Weibull comparison and 0.667 for the WHO vs. the Direct estimation. For females, the Weibull and Direct estimates

C. H. Skiadas (✉)
ISAST, Athens, Attiki, Greece
e-mail: skiadas@cmsim.net

C. Skiadas
Department of Mathematics and Computer Science, Hanover College, Hanover, IN, USA
e-mail: skiadas@hanover.edu

C. H. Skiadas, C. Skiadas (eds.), *Demography of Population Health, Aging and Health Expenditures*, The Springer Series on Demographic Methods and Population Analysis 50, https://doi.org/10.1007/978-3-030-44695-6_2

Table 2.1 Healthy Life Expectancy estimates and comparisons

	Males in 2010					Females in 2010				
Country	LE	HALE	WHO	Weibull	Direct	LE	HALE	WHO	Weibull	Direct
Austria	77.83	69.47	8.36	8.71	8.90	83.25	73.18	10.07	11.69	10.73
Belgium	77.53	69.15	8.38	8.75	8.78	82.70	72.34	10.36	11.09	10.31
Czechia	74.48	65.62	8.86	7.52	8.20	80.72	70.52	10.21	10.60	10.02
Denmark	77.35	69.17	8.18	8.55	8.48	81.39	71.64	9.75	9.60	9.15
Finland	76.76	68.05	8.71	8.34	8.73	83.24	72.64	10.60	11.64	10.69
France	78.20	70.19	8.02	8.56	8.84	84.52	74.06	10.46	11.74	10.68
Germany	77.57	69.30	8.27	8.67	8.77	82.60	72.48	10.13	11.34	10.52
Greece	77.98	69.84	8.14	8.58	8.52	83.12	73.05	10.08	11.75	10.79
Ireland	78.71	70.17	8.54	9.29	8.96	82.95	72.67	10.28	10.60	9.84
Italy	79.50	71.23	8.27	9.63	9.21	84.42	74.19	10.23	11.80	10.76
Luxembourg	78.80	70.01	8.80	9.21	9.02	83.70	73.14	10.56	11.43	10.61
Netherlands	78.81	70.36	8.45	9.40	9.08	82.73	72.42	10.31	10.98	10.31
Norway	78.88	70.30	8.58	9.54	9.51	83.15	73.23	9.92	11.09	10.42
Portugal	76.80	68.83	7.96	8.66	8.59	83.03	72.74	10.29	11.50	10.36
Slovenia	76.20	66.54	9.66	8.08	8.31	82.76	71.71	11.05	11.30	10.32
Spain	79.10	71.05	8.06	9.00	9.00	85.10	74.76	10.33	12.19	11.00
Sweden	79.55	70.83	8.72	9.86	9.50	83.48	72.97	10.51	11.42	10.71
Switzerland	80.08	71.22	8.86	9.96	9.69	84.51	73.65	10.87	12.30	11.25
United Kingdom	78.60	70.05	8.54	8.92	8.92	82.50	72.35	10.16	10.26	9.73
AVERAGE			**8.49**	**8.91**	**8.90**			**10.32**	**11.28**	**10.43**

Life Expectancy (LE), Healthy Life Expectancy (HALE), Healthy Life Years Lost (HLYL) from the World Health Organization (WHO) and our estimates for the Weibull shape parameter b and Direct estimates of the HLYL parameter b_x from the Life Tables provided by the Human Mortality Database (HMD)

differ, since the Weibull estimates with mean 11.28 are higher than the HLYL estimates by WHO (mean = 10.32). The standard error is 1.107. The standard deviation is 0.437 for the WHO vs. the Direct estimate, while the mean is 10.32 for WHO and 10.43 for the Direct estimate for females. In what follows, we introduce a quantitative methodology and mathematical analysis to discuss this significant finding and extend further the applicability of this method to estimate the HLYL and Healthy Life Expectancy.

2.2 The HLYL Estimation Method

Our methodology, also used in Skiadas and Skiadas (2014, 2015, 2018a, b, c, 2019), is based on a geometric approach shown in the following graph of mortality spaces, where both mortality and survival are presented as distinct areas.

The usual formula to express mortality μ_x in a population at age x is by estimating the fraction Death(D_x)/Population(P_x), that is, $\mu_x = D_x/P_x$. As in the following Life Tables provided by the Human Mortality Database, we will use the term *mx* instead of μ_x. The above graph showing data on Sweden 1950 females from the HMD is formulated with μ_x as the blue exponential curve. The main forms of Life Tables start with μ_x and estimate the survival forms of the population. This methodology leads to the estimation of a probability measure named life expectancy at age x or life expectancy at birth, when considering the total life time. There are several differences between the graph with the survival space above and the survival curves methodology. First of all, the vertical axis in the Survival-Mortality Space (SMS) diagram is the probability μ_x. On the other hand, in the survival diagram, the vertical axis represents population (usually starting from 100,000 in most life tables and gradually slowing down until the end). In the SMS diagram, we have probability spaces for both survival and mortality. For an age x, the total space (ABCOA) in the SMS diagrams is (OA).(BC) $= x\ \mu_x$. The mortality space is the sum S(μ_x) and survival space is ($x\mu_x$ -S(μ_x)). Accordingly, the important measure of the Health State is simply the fraction (ABDOA)/(BCODB). It is simpler to use the fraction (ABCOA)/(BCODB) $= x\mu_x$/S(μ_x) that can be estimated from μ_x for every age x of the population.

Ruben Roman et al. (2007) propose a similar methodology stating: "In the expression of the survival function; H(x) denotes the cumulative hazard function, which is equivalent to the area under the hazard function m(x). The area under the hazard function was defined by taking the corresponding integration limits ranging from x, the current age of an individual, to x + y_x, age at death or quantity of time lived from birth to death, where X and Y_x are non-negative continuous random variables. The calculated area will give the risk of dying at a given age x up to a particular future time y_x". The cumulative hazard they propose is $S\ \mu_x$ where μ_x is equivalent to the hazard function in our notation.

In modeling the healthy life years lost to disability, some important issues should be emphasized. Mortality expressed by μ_x is important for modeling disability, but what is more important cumulative mortality $S\ \mu_x$ which, as an additive process, is more convenient for the estimation of healthy life years lived with disability and the deterioration process causing deaths. The estimates for this type of mortality are included in the term b_xS μ_x.

Our approach in previous publications (Skiadas and Skiadas 2018a, b, c, 2019) was to set a time-varying fraction b_x for Health/Mortality of the form:

$$b_x = \frac{x\mu_x}{\int_0^x \mu_s ds} \tag{2.1}$$

This formula is directly provided from the last figure by considering the fraction:

$$b_x = \frac{Total\ Space}{Mortality\ Space} = \frac{OABCO}{ODBCO} = \frac{x\mu_x}{\int_0^x \mu_s ds}$$

It should be noted that an alternative approach is given by:

$$b_x = \frac{Survival\ Space}{Mortality\ Space} = \frac{OABCO - ODBCO}{ODBCO} = \frac{x\mu_x}{\int_0^x \mu_s ds} - 1$$

In the latter case, the estimated fraction b_x is smaller by one than the previous case. It remains until the applications stage to decide which is more appropriate. So far, the Total Space approach is simpler and gave good results.

The main hypothesis is that the population involved in the deterioration process is a fraction of the total population determined by the level of mortality μ_x at age x. Accordingly, the mortality process will have two alternatives expressed by the simple equation:

$$x\mu_x = b_x \int_0^x \mu_s ds \approx b_x \sum\nolimits_0^x \mu_x \tag{2.2}$$

where $x\mu_x$ is the incoming part related to the disability of the living population and the second part is the outgoing part that is summed by the mortality for the period from 0 to age x. The parameter bx is a correspondent adding to express the rate of healthy life lost to disability. The applications verify that the maximum values for $b = b_{max}$ are compatible with the WHO estimates for several countries. Even more, our estimate expressing the values for b_x across the lifespan is of particular significance in studies related to Health Expenditure estimation.

Some important properties of the last formula are given below:

First, we can formulate the Survival Probability $S(t)$

$$S(t) = \exp\left(-\int_0^x \mu_s ds\right) = \exp\left(-\frac{x\mu_x}{b_x}\right) \approx \exp\left(-\sum\nolimits_0^x \mu_x\right) \tag{2.3}$$

$$S(t) \approx \exp\left(-\sum\nolimits_0^x \mu_x\right) = \exp\left(-\mu_0\right)\exp\left(-\mu_1\right)\exp\left(-\mu_2\right)\ldots\exp\left(-\mu_x\right)$$

Next, we differentiate (2.2) to obtain

$$(x\mu_x)' = b'_x \int_0^x \mu_s ds + b_x\mu_x \tag{2.4}$$

For a constant b, we have $b'_x = 0$ and

$$x\mu'_x + \mu_x = b\mu_x$$

it follows that

$$x\mu'_x = (b-1)\,\mu_x$$

rearranging

$$\frac{\mu'_x}{\mu_x} = \frac{b-1}{x}$$

solves the differential equation

$$\ln(\mu_x) = \ln(c) + (b-1)\ln x$$

where c is a constant of integration. Finally,

$$\mu_x = cx^{b-1}$$

by setting $c = \lambda b$, the hazard function or the generating function of the Weibull appears as

$$\mu_x = \lambda b x^{b-1} \tag{2.5}$$

and the cumulative hazard of the Weibull is

$$\Lambda(x) = \lambda x^b = \frac{x\mu_x}{b} = \int_0^x \mu_s ds \tag{2.6}$$

This is to verify the formula for the survival probability (2.3) presented earlier.

Accordingly, the hazard function for the Gompertz as provided in Appendix A is given by:

$$\mu_x = a\exp(ax - b)$$

This is a convenient method to solve the Healthy Life Years Lost problem by estimating the appropriate parameter b either from the Gompertz or the Weibull model forms (Fig. 2.1).

Matsushita et al. (1992) had suggested the Weibull model for a Lifetime Data Analysis of Disease and Aging. Among the numerous studies modeling and applying the Weibull model for more than 70 years, this work is of particular significance, because it highlights the importance of introducing the Weibull shape parameter in connection to survival rates. The authors have studied specific cases of Japanese females for the period 1891–1898 and for 1980. They have diagnosed the growing process of the shape parameter, m in their notation and b in our notation, during age and time as the maximum $m = 7.40$ for the period 1891–1898 and m = 9.19 for 1980. For the latter case, our direct estimates give $b = 9.43$. They have also presented similar changes of m for males and females for several age and time periods, thus establishing a systematic variation of m over time and age. Even more, they have done calculations for m from data for several diseases. In Table 2.2, we have done estimates for the (Life Expectancy-Healthy Life Expectancy) = (LE-HALE) = HLYL for males and females in Japan. LE, HALE, and LE-HALE are

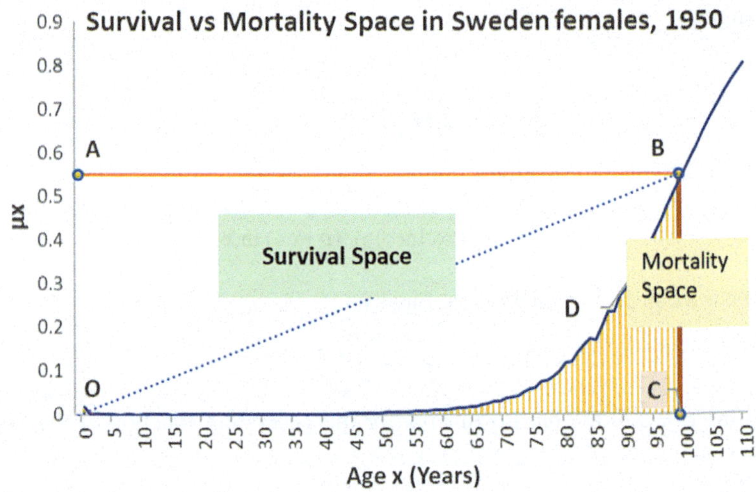

Fig. 2.1 Survival vs mortality space graph

Table 2.2 Healthy Life Expectancy and Healthy Life Years Lost in Japan

	Japan males				Japan females			
	LE	HALE	LE-HALE	HLYL	LE	HALE	LE-HALE	HLYL
1980				7.9				9.4
1985				8.0				9.8
1990	76.0	68.1	7.9	8.3	82.0	72.2	9.7	10.0
1995	76.5	68.4	8.1	8.4	82.2	72.9	9.9	10.1
2000	77.6	69.1	8.5	8.4	84.3	74.0	10.4	10.2
2005	78.7	69.9	8.8	8.6	85.5	74.8	10.7	10.5
2010	79.3	70.8	8.5	9.0	86.1	75.4	10.7	11.0
2013	80.1	71.1	8.9	9.1	86.4	75.6	10.8	11.1

provided in the paper by Tokudome et al. (2016), whereas the HLYL are estimated with our direct method. The HLYL estimates are very close across these two methodologies. The advantage of our methodology is that we can estimate the HLYL across all time periods as long as life table data exist. We have made projections for the Tokudome et al. estimates in Fig. 2.2 by fitting a line to the female data in Japan from 1990 to 2013. The estimates provided (LE-HALE) = 9.19, precisely the same as the Matsushita et al. (1992) estimates.

As already presented in previous studies (Skiadas and Skiadas 2018a, b, c, 2019), b_x can be estimated directly from the life table data. By using this method, the resulting form is illustrated in the following figures, showing almost a study growth to a high level and then a decline at very high ages. Clearly, this is not a simple equation form. However, there is a simple case to estimate b_x at the top level by taking a small linear part of the b_x curve at the top level parallel to the horizontal axis. This implies that $b_x = b =$ constant. The estimated b in this case is nothing else

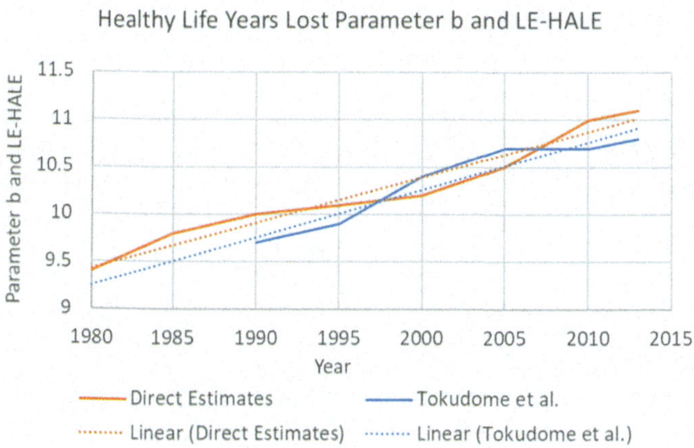

Fig. 2.2 Healthy Life Years Lost and extensions in Japan

Table 2.3 Comparing the Healthy Life Years Lost estimates

Comparisons of *b* estimates with the Healthy Life Years Lost numbers provided by WHO					
Year	2000	2005	2010	2015	2016
WHO	10.3	10.4	10.5	10.7	10.8
Direct method	10.0	10.4	10.7	10.9	10.9
Weibull	10.7	11.0	11.4	11.6	11.6

but the related parameter of the Weibull model. We can estimate this parameter by fitting the Weibull model to the death probability density function or by a direct estimate from the life tables with the method already discussed and applied in Skiadas and Skiadas (2018a, b, c). The direct estimate provided $b = 8.1$ compared to 8.7 for the estimate via the Weibull model for females in Sweden in 1950. The related figures for 2015 are $b = 10.9$ with the direct estimate for $\mathrm{b} = 11.6$ via the Weibull model. Note that the figure for the Healthy Life Years Lost provided by the World Health Organization is 10.7 years of age, very close to our direct estimates for b in 2015. Our estimates with both methods (direct and Weibull) are presented in the ~~next~~ Table 2.3 along with the WHO estimates for the HLYL. The estimates with the direct method are closer to those by WHO. The direct method estimates b_x in all lifespan periods, thus providing a flexible tool to compare the results provided by WHO at age 60 for females in Sweden. Table 2.3 summarizes the related figures. Both methods yield very similar results. The Direct method estimates b_x for the entire lifespan, and we can compare the related results with the WHO findings at 60 years of age presented in Table 2.4 for the years 2000, 2005, 2010, 2015 and 2016. These results verify that both methods yield similar results. Of course, the Direct method, based on only the life tables can used across time periods as long as life tables exist.

Table 2.4 Comparing the Healthy Life Years Lost estimates at age 60

Comparisons of b_x estimates with the Healthy Life Years Lost numbers provided by WHO at age 60					
Year	2000	2005	2010	2015	2016
WHO	5.3	5.4	5.6	5.7	5.8
Direct method	5.2	5.1	5.5	5.3	5.4

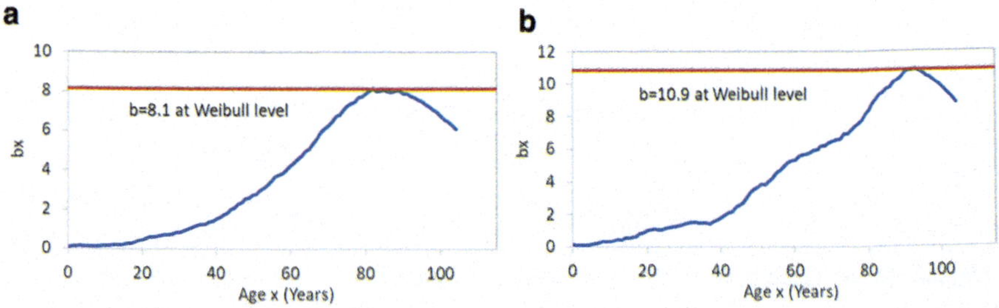

Fig. 2.3 (**a**) HLYL indicator in Sweden, females 1950. (**b**) HLYL indicator in Sweden, females 2015

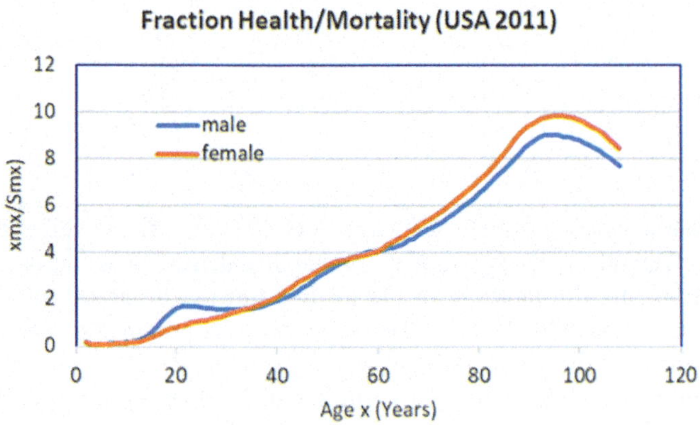

Fig. 2.4 Fraction Health/Mortality in USA, males and females in 2011

The Health/Mortality fraction, as presented in Fig. 2.3a, b, shows an increasing pattern for the main part of a lifetime until a maximum followed by the decline. In Fig. 2.4 next, the case of the USA in 2011 is studied. The fraction is similar until 60 years of age with the exception of ages 15–30 years when an excess of mortality is shown in males. After 60 years of age, females show higher values than males with a maximum of 9.85 at age 96, compared to 9.01 for males at 93 years of age. The most important point here is that the maximum points correspond to years lost to disability. We can easily observe this important feature by considering a linear form for mortality mx = ax. This is the simplest case of drawing a linear line from O to B in the graph above. The resulting fraction is 2, yielding 1 if we select the fraction (ABDOA)/(BCODB). Following up on the previous discussion, the healthy life years lost to disability (HLYL) are 1 in the last notation and 2 when considering the total space vs. the mortality space. The second higher by 1 from the simple fraction provides results similar to those estimated by the World Health Organization. Following that, the only thing one has to do is to remove this estimate from ~~the~~ life expectancy at birth to find Healthy Life Expectancy. As demonstrated in the graph, the HLYL for females is higher than for males in the case studied. It is a universal-like estimate for the majority of countries.

2.3 Excel Program for the Estimates

We have developed an Excel program for the Direct Estimates of b_x which is provided free of charge. One version can be downloaded from the Demographics 2019 Workshop website at www.asmda.es. The program uses the full life tables from the human mortality database to provide the Healthy Life Year Lost estimator b_x from the general equation form (2.1):

$$b_x = \frac{x\mu_x}{\int_0^x \mu_s ds}$$

The Cumulative Mortality Mx is given by

$$M_x = \int_0^x \mu_s ds \approx \sum_0^x \left(\frac{dx}{lx}\right) \tag{2.7}$$

where dx expresses the death population at age x in the life tables of the HMD and lx is the remaining population at age x in the same life tables. Note that the starting population at age x = 0 is set at 100000.

The average mortality Mx/x is estimated by

$$\overline{M}_x = \frac{M_x}{x} \approx \frac{\sum_0^x \left(\frac{dx}{lx}\right)}{x}$$

Then the Person Life Years Lost (PLYL) are provided by

$$PLYL = \frac{dx}{\overline{M}_x} = \frac{xdx}{M_x}$$

The final estimate for b_x is given by

$$b_x = \frac{x\mu_x}{\int_0^x \mu_s ds} \approx \frac{PLYL}{lx} = \frac{xdx}{l_x M_x} = \frac{xdx}{l_x \sum_0^x \left(\frac{dx}{lx}\right)} \tag{2.8}$$

The methodology is presented in the following figure (Fig. 2.5). The full life table from the HMD is followed by 4 more columns for the estimation of b_x. In the first, the cumulative mortality is estimated with $\mathrm{M} = \sum^x \mu_x$. The average mortality $(M/x) = \sum^x \mu_x /x$ is provided in the following column, whereas the Person Life Years Lost (PLYL) $= \mathrm{x}d_x/(\sum^x \mu_x)$ are calculated in the column next, where d_x is provided in the column indicated by dx in the life table. Towards illustrating this, an interesting graph is provided. The graph follows a growth process until a high level at 77 years of age and a decline in the remaining period of the lifespan. In the next column, the Healthy Life Year Lost estimator b_x is provided by dividing the PLYL with the lx from the life table. The results are presented in an illustrative graph showing the growing trend for b_x reaching a maximum of 9.71 and declining at higher ages. This high level can also be estimated by fitting the Weibull model.

An additional feature added in this program in Excel is the estimates of the World Health Organization from 2000–2016 for Life Expectancy at birth and at 60 years of age for all member countries, whereas information for Healthy Life Expectancy (HALE) at birth and at 60 years of age is provided for the years 2000, 2005, 2010, 2015 and 2016. We have added a small table to comparatively present the WHO estimates with our estimates using the direct method. The only action needed following copying and pasting the life table from the HMD is to select the

Healthy Life Years Lost (HLYL)	Select Contry Name:	Canada
Program: C H Skiadas and C Skiadas, 2019, April. http://www.asmda.es/demographics2019.html Download the Excel files from	Select Gender:	Female

Manually complete the full Life Table from the Human Mortality Database, HMD (recommended) or from the right side files from this Excel.
Don't forget to Select Country Name and Gender in the appropriate Box

Year	Age	mx	qx	ax	lx	dx	Lx	Tx	ex	Cumulative mortality Mx=Σ(dx/lx)	Average Mortality Mx/x	Person life years lost PLYL	bx HLYL Indicator PLYL/lx
										Mx=Σ(dx/lx)	Mx/x	xdx/Mx	(xdx)/(lxMx)
2000	0	0.0046	0.00455	0.07	100000	455	99575	8176679	81.77	0.005	0.065	7000	0.07
2000	1	0.0003	0.00033	0.5	99545	32	99529	8077104	81.14	0.005	0.003	9053	0.10
2000	2	0.0002	0.0002	0.5	99513	20	99503	7977575	80.17	0.005	0.002	9857	0.10
2000	3	0.0001	0.00013	0.5	99493	13	99486	7878072	79.18	0.005	0.001	8745	0.09
2000	4	8E-05	0.00008	0.5	99479	8	99475	7778586	78.19	0.005	0.001	6814	0.07
2000	5	0.0001	0.00014	0.5	99471	14	99464	7679111	77.2	0.005	0.001	14195	0.14
2000	6	7E-05	0.00007	0.5	99457	7	99454	7579647	76.21	0.005	0.001	8281	0.08
2000	7	9E-05	0.00009	0.5	99450	9	99446	7480194	75.22	0.006	0.001	12086	0.12
2000	8	8E-05	0.00008	0.5	99441	8	99437	7380748	74.22	0.006	0.001	12002	0.12
2000	9	0.0002	0.00015	0.5	99433	14	99426	7281311	73.23	0.006	0.001	22906	0.23
2000	10	9E-05	0.00009	0.5	99419	9	99415	7181885	72.24	0.006	0.001	16025	0.16
2000	11	0.0001	0.00011	0.5	99410	11	99405	7082470	71.24	0.006	0.001	21057	0.21
2000	12	0.0001	0.00013	0.5	99399	13	99393	6983066	70.25	0.006	0.000	26473	0.27
2000	13	0.0002	0.00018	0.5	99386	18	99377	6883673	69.26	0.006	0.000	38453	0.39
2000	14	0.0002	0.00018	0.5	99368	18	99359	6784296	68.27	0.007	0.000	40150	0.40
2000	15	0.0003	0.00027	0.5	99350	26	99337	6684937	67.29	0.007	0.000	59595	0.60
2000	16	0.0003	0.00026	0.5	99323	25	99311	6585600	66.3	0.007	0.000	58811	0.59
2000	17	0.0003	0.00034	0.5	99298	34	99281	6486289	65.32	0.007	0.000	80882	0.81
2000	18	0.0004	0.00042	0.5	99264	41	99244	6387008	64.34	0.008	0.000	97626	0.98

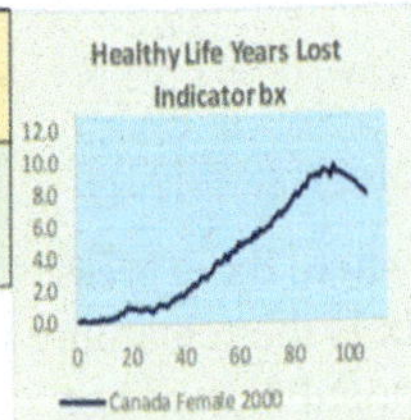

Parameters	WHO Estimates	Direct HLYL Estimates
LE	81.83	81.77
HALE	72.12	72.06
HLYL	9.71	9.71
LE at 60	24.56	24.47
HALE at 60	19.57	19.72
HLYL at 60	4.99	4.75

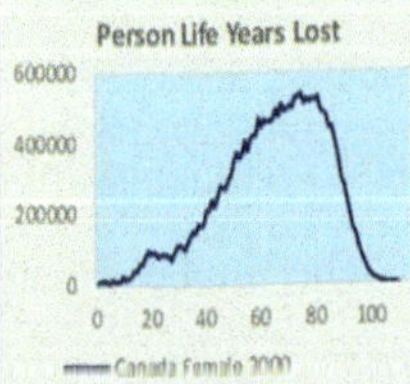

Fig. 2.5 The extended Life Table for the HLYL estimates

Healthy Life Years Lost (HLYL)

Select Contry Name: Canada

Program: C H Skiadas and C Skiadas, 2019, April. http://www.asmda.es/demographics2019.html
Download the Excel files from

Select Gender: Female

Manually complete the full Life Table from the Human Mortality Database, HMD (recommended) or from the right side files from this Excel.
Don't forget to Select Country Name and Gender in the appropriate Box

										Cumulative mortality Mx=Σ(dx/lx)	Average Mortality Mx/x	Person life years lost PLYL	bx HLYL Indicator PLYL/lx
Year	Age	mx	qx	ax	lx	dx	Lx	Tx	ex	Mx=Σ(dx/lx)	Mx/x	xdx/Mx	(xdx)/(lxMx)
2000	0	0.0046	0.00455	0.07	100000	455	99575	8176679	81.77	0.005	0.065	7000	0.07
2000	1	0.0003	0.00033	0.5	99545	32	99529	8077104	81.14	0.005	0.003	9853	0.10
2000	2	0.0002	0.0002	0.5	99513	20	99503	7977575	80.17	0.005	0.002	9857	0.10
2000	3	0.0001	0.00013	0.5	99493	13	99486	7878072	79.18	0.005	0.001	8745	0.09
2000	4	8E-05	0.00008	0.5	99479	8	99475	7778586	78.19	0.005	0.001	6814	0.07
2000	5	0.0001	0.00014	0.5	99471	14	99464	7679111	77.2	0.005	0.001	14195	0.14
2000	6	7E-05	0.00007	0.5	99457	7	99454	7579647	76.21	0.005	0.001	8281	0.08
2000	7	9E-05	0.00009	0.5	99450	9	99446	7480194	75.22	0.006	0.001	12086	0.12
2000	8	8E-05	0.00008	0.5	99441	8	99437	7380748	74.22	0.006	0.001	12002	0.12
2000	9	0.0002	0.00015	0.5	99433	14	99426	7281311	73.23	0.006	0.001	22906	0.23
2000	10	9E-05	0.00009	0.5	99419	9	99415	7181885	72.24	0.006	0.001	16025	0.16
2000	11	0.0001	0.00011	0.5	99410	11	99405	7082470	71.24	0.006	0.001	21057	0.21
2000	12	0.0001	0.00013	0.5	99399	13	99393	6983066	70.25	0.006	0.000	26473	0.27
2000	13	0.0002	0.00018	0.5	99386	18	99377	6883673	69.26	0.006	0.000	38453	0.39
2000	14	0.0002	0.00018	0.5	99368	18	99359	6784296	68.27	0.007	0.000	40150	0.40
2000	15	0.0003	0.00027	0.5	99350	26	99337	6684937	67.29	0.007	0.000	59595	0.60
2000	16	0.0003	0.00026	0.5	99323	25	99311	6585600	66.3	0.007	0.000	58811	0.59
2000	17	0.0003	0.00034	0.5	99298	34	99281	6486289	65.32	0.007	0.000	80882	0.81
2000	18	0.0004	0.00042	0.5	99264	41	99244	6387008	64.34	0.008	0.000	97626	0.98

2000	76	0.0284	0.02796	0.5	75368	2107	74315	902465	11.97	0.309	0.004	521703	6.92
2000	77	0.0312	0.03069	0.5	73261	2248	72137	828151	11.3	0.340	0.004	512947	7.00
2000	78	0.036	0.03534	0.5	71013	2510	69758	756014	10.65	0.375	0.005	525439	7.40
2000	79	0.0404	0.03962	0.5	68503	2714	67146	686255	10.02	0.415	0.005	520400	7.60
2000	80	0.047	0.0459	0.5	65789	3020	64279	619109	9.41	0.461	0.006	527910	8.02
2000	81	0.05	0.0488	0.5	62769	3063	61238	554830	8.84	0.509	0.006	490141	7.81
2000	82	0.0571	0.05548	0.5	59706	3313	58050	493592	8.27	0.565	0.007	483927	8.11
2000	83	0.0651	0.06308	0.5	56393	3557	54615	435542	7.72	0.628	0.008	473039	8.39
2000	84	0.0709	0.06849	0.5	52836	3618	51027	380928	7.21	0.696	0.008	439032	8.31
2000	85	0.0856	0.08208	0.5	49218	4040	47198	329901	6.7	0.778	0.009	443736	9.02
2000	86	0.094	0.08974	0.5	45178	4054	43151	282703	6.26	0.868	0.010	403920	8.94
2000	87	0.1046	0.09939	0.5	41124	4087	39080	239552	5.83	0.968	0.011	369605	8.99
2000	88	0.1195	0.11273	0.5	37036	4175	34949	200473	5.41	1.080	0.012	342029	9.24
2000	89	0.1279	0.12023	0.5	32861	3951	30886	165524	5.04	1.201	0.013	294553	8.96
2000	90	0.1504	0.13991	0.5	28910	4045	26888	134638	4.66	1.340	0.015	273101	9.45
2000	91	0.1659	0.15322	0.5	24865	3810	22960	107751	4.33	1.494	0.016	233397	9.39
2000	92	0.1828	0.16752	0.5	21055	3527	19292	84790	4.03	1.661	0.018	196396	9.33
2000	93	0.1945	0.17722	0.5	17528	3106	15975	65498	3.74	1.838	0.020	157972	9.01
2000	94	0.2352	0.21046	0.5	14422	3035	12904	49523	3.43	2.049	0.022	139987	9.71
2000	95	0.25	0.22222	0.5	11387	2530	10122	36619	3.22	2.271	0.024	106391	9.34
2000	96	0.2752	0.24192	0.5	8856	2143	7785	26498	2.99	2.513	0.026	82292	9.29
2000	97	0.3019	0.26234	0.5	6714	1761	5833	18713	2.79	2.775	0.028	61867	9.21
2000	98	0.3301	0.28333	0.5	4953	1403	4251	12879	2.6	3.059	0.031	45184	9.12
2000	99	0.3595	0.30474	0.5	3549	1082	3009	8628	2.43	3.363	0.034	32009	9.02
2000	100	0.39	0.32639	0.5	2468	805	2065	5620	2.28	3.690	0.037	21927	8.88
2000	101	0.4214	0.34809	0.5	1662	579	1373	3555	2.14	4.038	0.040	14554	8.76
2000	102	0.4535	0.36967	0.5	1084	401	883	2182	2.01	4.408	0.043	9325	8.60
2000	103	0.4859	0.39095	0.5	683	267	550	1299	1.9	4.799	0.046	5759	8.43
2000	104	0.5185	0.41175	0.5	416	171	330	749	1.8	5.210	0.050	3430	8.25
2000	105	0.5509	0.43193	0.5	245	106	192	419	1.71	5.643	0.053	1982	8.09
2000	106	0.5829	0.45135	0.5	139	63	108	227	1.63	6.096	0.057	1101	7.92
2000	107	0.6142	0.46989	0.5	76	36	58	119	1.56	6.569	0.061	589	7.75
2000	108	0.6446	0.48746	0.5	40	20	31	61	1.5	7.069	0.065	307	7.67
2000	109	0.6738	0.50402	0.5	21	10	16	30	1.45	7.546	0.069	145	6.91
2000	110+	0.7018	1	1.42	10	10	15	15	1.42				

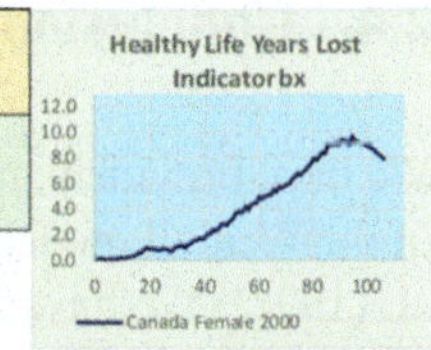

Parameters	WHO Estimates	Direct HLYL Estimates
LE	81.83	81.77
HALE	72.12	72.06
HLYL	9.71	9.71
LE at 60	24.56	24.47
HALE at 60	19.57	19.72
HLYL at 60	4.99	4.75

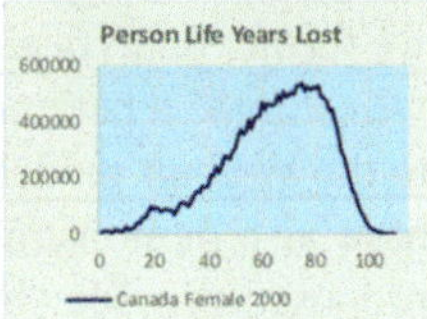

Fig. 2.6 The extended Life Table for the HLYL estimates. Full presentation

name of the country in L1 and the gender (male, female or both sexes) in L2 in the Excel chart. To avoid mistakes, we have used the list of countries as they appear in WHO with their official names (Fig. 2.6).

2.4 Conclusions and Further Study

We have provided an analytical explanation of the behavior of the shape parameter of the Weibull model verifying and expanding the arguments made by Matsushita et al. (1992). We have also presented an analytical formulation of these observations along with the development of appropriate extensions of classical life tables in order to provide a valuable tool for estimating the Healthy Life Years Lost. We have also discussed how the Weibull model properties, expressing the fatigue of materials and in particular the cumulative hazard of this model, can express the additive process of disabilities and diseases in the human population. In this paper, we further analytically derived a more general model of survival-mortality in which we estimate a parameter related to the Healthy Life Years Lost (HLYL) and leading to the Weibull model and the corresponding shape parameter as a specific case. We have also demonstrated that the results found for the general HLYL parameter we proposed provide similar results to those given by the World Health Organization for Healthy Life Expectancy (HALE) and the corresponding HLYL estimates. An analytical derivation of the mathematical formulas was presented here together with an easy to apply Excel program. This program is an extension of the classical life table that includes four more columns to estimate cumulative mortality, average mortality, a person's life years lost and, finally, the HLYL parameter b_x. The last version of this program appears on the Demographics2019 website at http://www.asmda.es/demographics2019.html and in the Demographics2020 website at http://www.smtda.net/demographics2020.html.

A.1 Appendix A

To derive the appropriate forms, the generating functions for Gompertz and Weibull functions, the starting point is Eq. (2.2)

$$x\mu_x = b_x \int_0^x \mu_s ds \tag{2.A1}$$

Next we rearrange (2.A1) in the form

$$(x\mu_x / b_x) = \int_0^x \mu_s ds \tag{2.A2}$$

and differentiate to obtain

$$\frac{b_x\left(x\mu_x' + \mu_x\right) - x\mu_x b_x'}{b_x^2} = \mu_x \tag{2.A3}$$

A.2 The Weibull Generating Function

From (2.A3) and for a constant b we have $b'_x = 0$ and

$$x\mu'_x + \mu_x = b\mu_x \tag{2.A4}$$

It follows

$$x\mu'_x = (b-1)\,\mu_x$$

And rearranging

$$\frac{\mu'_x}{\mu_x} = \frac{b-1}{x}$$

Solving the differential equation

$$\ln(\mu_x) = \ln(c) + (b-1)\ln x$$

where c is a constant of integration. Finally

$$\mu_x = cx^{b-1}$$

By setting $c = \lambda b$ the hazard function or the generating function of the Weibull appear

$$\mu_x = \lambda b x^{b-1} \tag{2.A5}$$

And the cumulative hazard of the Weibull is

$$\Lambda(x) = \lambda x^b = \frac{x\mu_x}{b} = \int_0^x \mu_s ds \tag{2.A6}$$

A.3 The Gompertz Generating Function

From (2.A3) and by assuming that $b_x = ax$ the following equation form results

$$\mu'_x = a\mu_x$$

where a is a parameter

The solution is

$$\ln(\mu_x) = \ln(a) - c + ax \tag{2.A7}$$

and

$$\mu_x = \frac{a}{\exp(c)}\exp(ax) \tag{2.A8}$$

This is the hazard function for the Gompertz. The parameter c provides the healthy life years lost. This is another form of the parameter b provided in the Weibull case. Both estimates give similar values for the HLYL. Simpler is the following form of (2.A8) where the parameter c is replaced by b in accordance to the terminology used for the HLYL estimates

$$\mu_x = a\exp(ax - b) \tag{2.A9}$$

For simplicity, in most applications, the linear form $\ln(\mu_x) = \ln(a) - b + ax$ is preferred.

We have thus reduced the Healthy Life Years Lost problem to fit a linear function to lifetable data sets.

Note that, by rearranging, the Modal age at death or Mode M = b/a is inserted (see Missov et al. 2015 for details)

$$\mu_x = a\exp\left(a\left(x - \frac{b}{a}\right)\right)$$

$$\mu_x = a\exp(a(x - M)) \tag{2.A10}$$

Accordingly, the characteristic parameter b is provided by

$$b = aM \tag{2.A11}$$

References

Matsushita, S., Agiwara, K., Hiota, T. A., Himada, H., Kijramoto, I., & Toyokura, Y. (1992). Lifetime data analysis of disease and aging by the Weibull probability distribution. *Journal of Clinical Epidemiology, 45*(10), 1165–1175.

Missov, T. I., Lenart, A., Nemeth, I., Canudas-Romo, V., & Vaupel, J. W. (2015). The Gompertz force of mortality in terms of the modal age at death. *Demographic Research, 32*(36), 1031–1048.

Román, R., Comas, M., Hoffmeister, L., & Castells, X. (2007). Determining the lifetime density function using a continuous approach. *Journal of Epidemiology and Community Health, 61*, 923–925. https://doi.org/10.1136/jech.2006.052639.

Skiadas, C. H., & Skiadas, C. (2014). The first exit time theory applied to life table data: The health state function of a population and other characteristics. *Communications in Statistics-Theory and Methods, 43*, 1985–1600.

Skiadas, C. H., & Skiadas, C. (2015). Exploring the state of a stochastic system via stochastic simulations: An interesting inversion problem and the health state function. *Methodology and Computing in Applied Probability, 17*, 973–982.

Skiadas, C. H., & Skiadas, C. (2018a). *Exploring the health state of a population by dynamic modeling methods* (The Springer series on demographic methods and population analysis) (Vol. 45). Chum: Springer. https://doi.org/10.1007/978-3-319-65142-2.

Skiadas, C. H., & Skiadas, C. (2018b). The health-mortality approach in estimating the healthy life years lost compared to the global burden of disease studies and applications in world, USA and Japan. In *Exploring the health state of a population by dynamic modeling methods*. Chum: Springer. https://doi.org/10.1007/978-3-319-65142-2_4.

Skiadas, C. H., & Skiadas, C. (2018c). *Demography and health issues: Population aging, mortality and data analysis* (The Springer series on demographic methods and population analysis) (Vol. 46). Chum: Springer. https://doi.org/10.1007/978-3-319-76002-5.

Skiadas, C. H., & Arezzo, M. F. (2018). Estimation of the healthy life expectancy in Italy through a simple model based on mortality rate. In *Demography and health issues: Population aging, mortality and data analysis*. Chum: Springer. https://doi.org/10.1007/978-3-319-76002-5_4.

Skiadas, C. H., & Skiadas, C. (2019, March). *Modeling the health expenditure in Japan, 2011. A healthy life years lost methodology*. arXiv:1903.11565.

Tokudome, S., Hashimoto, S., & Igata, A. (2016). Life expectancy and healthy life expectancy of Japan: the fastest graying society in the world. *BMC Research Notes, 9*, 482. https://doi.org/10.1186/s13104-016-228-2.

Weibull, W. (1951). A statistical distribution function of wide applicability. *Journal of Applied Mechanics, 18*(3), 293–297.

Weon, B. M., & Je, J. H. (2011). Plasticity and rectangularity in survival curves. *Scientific Reports, 1*, 104. https://doi.org/10.1038/srep00104.

Weon, B. M., & Je, J. H. (2012). Trends in scale and shape of survival curves. *Scientific Reports, 2*, 504. https://doi.org/10.1038/srep00504.

WHO Life Expectancy and Healthy Life Expectancy Data provided in Excel.http://apps.who.int/gho/athena/data/GHO/WHOSIS_000001,WHOSIS_000015,WHOSIS_000002,WHOSIS_000007?filter=COUNTRY:*&format=xml&profile=excel

Chapter 3
Direct Healthy Life Expectancy Estimates from Life Tables with a Sullivan Extension. Bridging the Gap Between HALE and Eurostat Estimates

Christos H. Skiadas and Charilaos Skiadas

3.1 The HLYL Estimation Method

Our methodology Skiadas and Skiadas (2014, 2015, 2018a, b, c, 2019a, b) was based on a geometric approach from the following graph of mortality spaces where both mortality and survival are presented as appropriate areas of this graph.

The usual form to express mortality μ_x in a population at age x is by estimating the fraction Death(D_x)/Population(P_x) that is $\mu_x = D_x/P_x$. Because in what follows we will use the Life Tables provided from the Human Mortality Database, we will use the term mx of these tables instead of μx, above. In the above graph where data from 1950 females in the Sweden HMD was used, μ_x is shown as the blue exponential curve. The main forms of Life Tables start with μ_x and then they estimate the survival forms of the population. This methodology leads to the estimation of a probability measure termed as life expectancy at age x or life expectancy at birth when considering the total life time. There are several differences between the graph with the survival space above and the survival curves methodology. First of all, the vertical axis in the Survival-Mortality Space (SMS) diagram is the probability μ_x. Instead in the survival diagram the vertical axis represent population (usually it starts from 100.000 in most life tables and gradually slow down until the end). By the SMS diagram we have probability spaces for both survival and mortality. For the age x, the total space is (ABCOA) in the SMS diagrams, that is (OA).(BC) = $x\ \mu_x$. The mortality space is the sum $S(\mu_x)$ while the

C. H. Skiadas (✉)
ISAST, Athens, Attiki, Greece
e-mail: skiadas@cmsim.net

C. Skiadas
Department of Mathematics and Computer Science, Hanover College, Hanover, IN, USA
e-mail: skiadas@hanover.edu

C. H. Skiadas, C. Skiadas (eds.), *Demography of Population Health, Aging and Health Expenditures*, The Springer Series on Demographic Methods and Population Analysis 50, https://doi.org/10.1007/978-3-030-44695-6_3

survival space is ($x\mu_x$ −S(μ_x)). Accordingly, the important measure of the Health State is simply the fraction (ABDOA)/(BCODB). It is simpler to use the fraction (ABCOA)/(BCODB) = $x\mu_x$ /S(μ_x) that can be estimated from μ_x for every age x of the population.

Ruben Roman et al. (2007) propose a similar methodology stating: "In the expression of the survival function; H(x) denotes the cumulative hazard function, which is equivalent to the area under the hazard function m(x). The area under the hazard function was defined by taking the corresponding integration limits ranging from x, current age of an individual, to x + y_x, age at death or quantity of time lived from birth to death, where X and Y_x are non-negative continuous random variables. The calculated area will give the risk of dying at a given age x up to a particular future time y_x". The cumulative hazard they propose is $S\ \mu_x$ where μ_x is equivalent to the hazard function in our notation.

In modeling the healthy life years lost to disability some important issues should be realized. Mortality expressed by μ_x is important for modeling disability but more important is the cumulative mortality $S\ \mu_x$ which, as an additive process, is more convenient for the estimation of the healthy life years lived with disability and the deterioration process causing deaths. The estimates for this type of mortality are included in the term b_xS μ_x.

Our approach in previous publications (Skiadas and Skiadas (2018a, b, c, 2019a, b, c)) was to set a time-varying fraction b_x for Health/Mortality of the form:

$$b_x = \frac{x\mu_x}{\int_0^x \mu_s ds} \tag{3.1}$$

This formula is immediately provided from Fig. 3.1 by considering the fraction:

$$b_x = \frac{Total\ Space}{Mortality\ Space} = \frac{OABCO}{ODBCO} = \frac{x\mu_x}{\int_0^x \mu_s ds}$$

It should be noted that an alternative approach is given by:

$$b_x = \frac{Survival\ Space}{Mortality\ Space} = \frac{OABCO - ODBCO}{ODBCO} = \frac{x\mu_x}{\int_0^x \mu_s ds} - 1$$

In the latter case the estimated fraction b_x is smaller by one from the previous case. It remains to be seen in the applications stage which is more appropriate. So far, the Total Space approach is simpler, giving good results.

The main hypothesis is that the population involved in the deterioration process is a fraction of the total population determined by the level of mortality μ_x at age x. Accordingly, the mortality process will have two alternatives expressed by the simple equation:

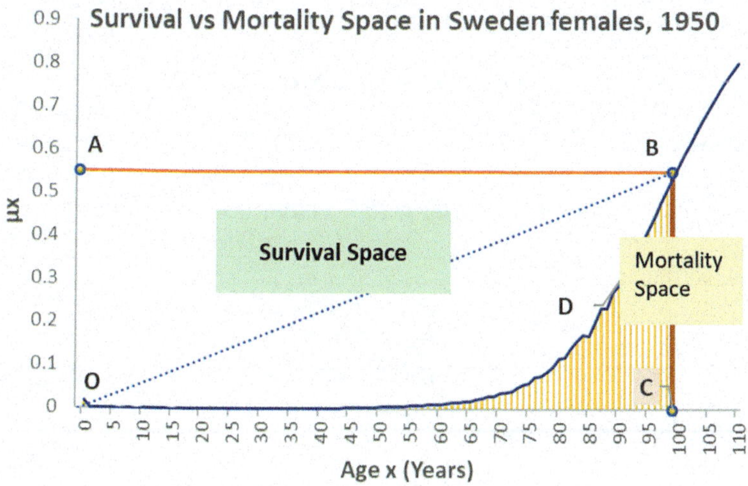

Fig. 3.1 Survival vs mortality space graph

$$x\mu_x = b_x \int_0^x \mu_s ds \approx b_x \sum\nolimits_0^x \mu_x \tag{3.2}$$

where $x\mu_x$ is the incoming part related to the disability of the living population and the second part is the outgoing part that is summed to the mortality for the period from 0 to age x. The parameter b_x expresses the rate of healthy life lost to disability. The applications verify that the maximum values for $b = b_{max}$ are compatible to the estimates of the WHO for several countries. Even more, our estimates to the values for b_x in all the life time are of particularly importance to the studies related to the Health Expenditure estimation.

Some important properties of the last formula are given below:

First we can formulate the Survival Probability $S(t)$

$$S(t) = \exp\left(-\int_0^x \mu_s ds\right) = \exp\left(-\frac{x\mu_x}{b_x}\right) \approx \exp\left(-\sum\nolimits_0^x \mu_x\right) \tag{3.3}$$

$$S(t) \approx \exp\left(-\sum\nolimits_0^x \mu_x\right) = \exp(-\mu_0)\exp(-\mu_1)\exp(-\mu_2)\ldots\exp(-\mu_x)$$

Next, we can differentiate (3.2) to obtain

$$(x\mu_x)' = b'_x \int_0^x \mu_s ds + b_x \mu_x \tag{3.4}$$

For a constant b we have $b_x' = 0$ and

$$x\mu_x' + \mu_x = b\mu_x$$

It follows

$$x\mu_x' = (b-1)\,\mu_x$$

And rearranging

$$\frac{\mu_x'}{\mu_x} = \frac{b-1}{x}$$

Solving the differential equation

$$\ln(\mu_x) = \ln(c) + (b-1)\ln x$$

where c is a constant of integration. Finally

$$\mu_x = cx^{b-1}$$

By setting $c = \lambda b$ the hazard function or the generating function of the Weibull appear

$$\mu_x = \lambda b x^{b-1} \tag{3.5}$$

And the cumulative hazard of the Weibull is

$$\Lambda(x) = \lambda x^b = \frac{x\mu_x}{b} = \int_0^x \mu_s ds \tag{3.6}$$

This is to verify the formula for the survival probability (3.3) presented earlier.

As we already have presented in previous studies (Skiadas and Skiadas (2018a, b, c, 2019a, b, c)), b_x can be estimated directly from the life table data. The estimates with the direct method are close to the WHO estimates. Applications verify that both methods yield similar results. Of course the Direct method, based on only the life tables can be used in all the time periods as long ~~far~~ as life tables exist.

3.2 Program for the Estimates

We have developed an Excel program for the Direct Estimates of b_x which is provided free of charge. One version can be downloaded from the Demographics 2019 Workshop website at www.asmda.es. The program uses the full life tables from the human mortality database to provide the Healthy Life Year Lost estimator b_x from the general equation form (3.1):

$$b_x = \frac{x\mu_x}{\int_0^x \mu_s ds}$$

The Cumulative Mortality Mx is given by

$$M_x = \int_0^x \mu_s ds \approx \sum_0^x \left(\frac{dx}{lx}\right) \tag{3.7}$$

where dx expresses the death population at age x in the life tables of the HMD and lx is the remaining population at age x in the same life tables. Note that the starting population at age x = 0 is set at 100000.

The average mortality Mx/x is estimated by

$$\overline{M}_x = \frac{M_x}{x} \approx \frac{\sum_0^x \left(\frac{dx}{lx}\right)}{x}$$

Then the Person Life Years Lost (PLYL) are provided by

$$PLYL = \frac{dx}{\overline{M}_x} = \frac{xdx}{M_x}$$

The final estimate for b_x is given by

$$b_x = \frac{x\mu_x}{\int_0^x \mu_s ds} \approx \frac{PLYL}{lx} = \frac{xdx}{l_x M_x} = \frac{xdx}{l_x \sum_0^x \left(\frac{dx}{lx}\right)} \tag{3.8}$$

The methodology is presented in Fig. 3.2 that follows. The full life table from the HMD is followed by 4 more columns for the estimation of b_x. In the first, the cumulative mortality is estimated from M=$\sum_0^x \mu_x$. The average mortality $(M/x) = \sum_0^x \mu_x / x$ is provided in the next column whereas the Person Life Years Lost (PLYL) = x$d_x / \left(\sum_0^x \mu_x\right)$ are calculated in the next column. d_x is provided by the column indicated by dx in the life table. For this very important information, an interesting graph is provided. The graph follows a growth process until a high level at 77 years of age and a decline in the remaining lifespan period. It the next column the Healthy Life Year Lost estimator b_x is provided by dividing the PLYL by the lx from the life table. The results are presented in an illustrative graph with the growing trend for b_x to reach a maximum of 9.71 with a decline at higher ages. This high level can be also estimated by fitting the Weibull model (Weibull 1951).

Another option added in this Excel is the estimates of the World Health Organization from 2000–2016 for Life Expectancy at birth and at 60 years of age for all the member countries whereas information for the Healthy Life Expectancy (HALE) at birth and at 60 years of age are provided for the years 2000, 2005, 2010,

Healthy Life Years Lost (HLYL)

Select Country Name: **Czechia**

Select Gender: **Female**

Program: C H Skiadas and C Skiadas, 2019, April. Download the Excel files from http://www.asmda.es/demographics2019.html

Manually complete the full Life Table from the Human Mortality Database, HMD (recommended) or from the right side files from this Excel.

Don't forget to Select Country Name and Gender in the appropriate Box

Year	Age	mx	qx	ax	lx	dx	Lx	Tx	ex	Cumulative mortality Mx=Σ(dx/lx)	Average Cumulative Mortality Mx/x	Years of disability influence at age x xdx/Mx	bx HLYL Indicator PLYL/lx (xdx)/(lxMx)	Person life years lost PLYL
2016	0	0.0025	0.00248	0.14	100000	248	99788	8186842	81.87	0.002	0.010	24000	0.24	1589
2016	1	0.0003	0.00028	0.5	99752	28	99738	8087055	81.07	0.003	0.002	15214	0.17	1007
2016	2	0.0002	0.00015	0.5	99724	15	99717	7987316	80.09	0.003	0.001	12882	0.16	853
2016	3	0.0001	0.00013	0.5	99710	13	99703	7887599	79.11	0.003	0.001	14960	0.13	991
2016	4	9E-05	0.00009	0.5	99697	9	99692	7787896	78.12	0.003	0.001	12932	0.12	856
2016	5	5E-05	0.00005	0.5	99687	5	99685	7688204	77.12	0.003	0.001	8643	0.14	572
2016	6	5E-05	0.00005	0.5	99682	5	99679	7588520	76.13	0.003	0.000	10055	0.14	666
2016	7	0.0001	0.0001	0.5	99677	10	99672	7488841	75.13	0.003	0.000	22506	0.15	1490
2016	8	7E-05	0.00007	0.5	99667	7	99663	7389169	74.14	0.003	0.000	17486	0.15	1158
2016	9	5E-05	0.00005	0.5	99660	5	99657	7289505	73.14	0.003	0.000	13757	0.18	911
2016	10	4E-05	0.00004	0.5	99655	4	99653	7189848	72.15	0.003	0.000	12024	0.20	796
2016	11	8E-05	0.00008	0.5	99651	8	99647	7090195	71.15	0.004	0.000	25747	0.27	1705
2016	12	9E-05	0.00009	0.5	99643	9	99638	6990549	70.16	0.004	0.000	30708	0.38	2033
2016	13	0.0002	0.00015	0.5	99634	15	99627	6890910	69.16	0.004	0.000	53092	0.43	3516
2016	14	0.0002	0.00018	0.5	99619	18	99610	6791284	68.17	0.004	0.000	65335	0.51	4326
2016	15	0.0001	0.00011	0.5	99601	11	99596	6691674	67.18	0.004	0.000	41532	0.60	2750
2016	16	0.0002	0.00016	0.5	99590	16	99582	6592078	66.19	0.004	0.000	61886	0.69	4098
2016	17	0.0002	0.0002	0.5	99574	20	99564	6492496	65.2	0.004	0.000	78357	0.71	5188
2016	18	0.0003	0.00025	0.5	99554	24	99542	6392932	64.22	0.005	0.000	94311	0.80	6245
2016	19	0.0002	0.0002	0.5	99529	20	99520	6293390	63.23	0.005	0.000	79450	0.81	5261
2016	20	0.0002	0.00021	0.5	99510	21	99499	6193871	62.24	0.005	0.000	84085	0.81	5568

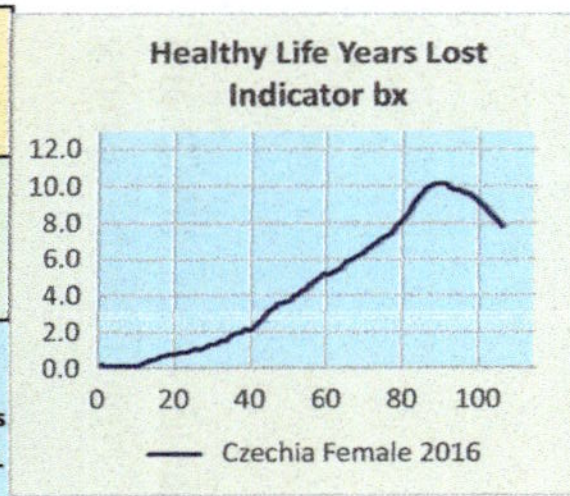

Parameters	WHO Estimates	Direct HLYL Estimates
LE	82.06	81.87
HALE	71.56	71.66
HLYL	10.50	10.21
LE at 60	24.15	24.03
HALE at 60	18.18	18.91
HLYL at 60	5.97	5.12

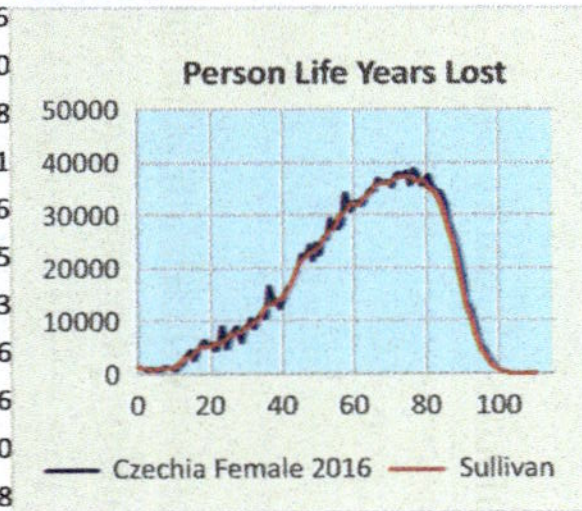

Fig. 3.2 The extended Life Table for the HLYL estimates. Download full program from http://www.smtda.net/demographics2020.html

2015 and 2016. We have added a small Table to present comparatively the WHO estimates with our estimates for the direct method. What is only needed after that is to copy and paste the life table from the HMD and to select the name of the country in L1 and the gender (male, female or both sexes) in L2 in the Excel chart. To avoid mistakes we have used the list of the WHO countries with their official names.

3.3 A Further HLE Estimate Based on the Sullivan Method

The classical Sullivan method is a standard tool to estimate the healthy life expectancy (see Sullivan 1971 and Jagger et al. 2014). The simplicity of this method and the possibility to use it as a continuation or extension of the life table gave us the opportunity to add a Sullivan extension to the above-extended life table. By this extension, we have estimated the Healthy Life Years Lost (HLYL) for all the life span along with the Healthy Life Expectancy (HLE). The main part of the estimate is based on the proportion lived with disability. This is generated from the b_x indicator from the previous columns multiplied with a discount Health Parameter. This is estimated directly from the program for our "direct" estimates and for WHO as well. For Eurostat estimates it is necessary to add the Healthy Life Expectancy at Birth in the appropriate box at the top of Fig. 3.3. Another opportunity is by selecting the "Equal" option which automatically provides an estimate higher than HALE and closer to Eurostat. Every one of these options should be selected manually in the "Select" and "Parameter" places. The "Other" selection is set by the program user if needed. After selecting the appropriate Health Parameter, the next estimates follow automatically the Sullivan method, first for estimating the person-years lived without disability, then the total years lived without disability and finally the Disability Free Life Expectancy. The column of life-years lived with disability follows. An alternative method based on the Sullivan system is presented in the columns on the right-hand side of Fig. 3.3. The estimates are based on the Person Years Lost to Disability and the direct estimates of the person-years lived without disability, the total years lived without disability and the Disability Free Life Expectancy with similar results with the previous approach. The estimated Healthy Life Years from two methods are presented in the appropriate graph of Fig. 3.4. The Health Parameters are estimated as follows:

Direct Parameter = sumproduct(Lx.bx)/(Tx-(LE-HLYL).lx)
WHO Parameter = sumproduct(Lx.bx)/(Tx-(HALE).lx)
Eurostat Parameter = sumproduct(Lx.bx)/(Tx-HLE.lx)

The inverse of the Equal Parameter corresponds to the Healthy Life Years Lost, that is:

$$\text{HLYL} = 1/\text{Equal}$$

26.4282

64 Eurostat HLE

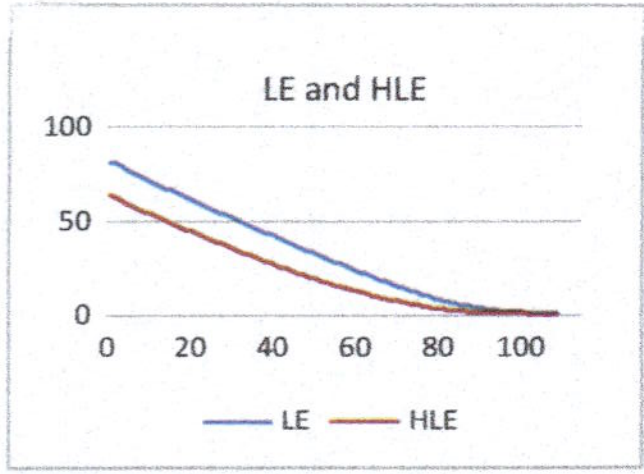

15.1021

			Select	Parameter						
			Eurostat	0.0662						
	Sullivan Method									
Proportion with disability	Person Years Lived without disability	Total Years Lived without disability	Disability free life expectancy	Healthy Life Years Lost	Index	Health Parameter	Person Years Lost to disability	Person Years Lived without disability	Total Years Lived without disability	Disability free life expectancy
0.02	98202	6400002	64.00	17.87	Direct	0.0378	1585.81	98199	6372712	63.73
0.01	98590	6301800	63.17467	17.90	Equal	0.0609	1148.45	98731	6274513	62.90
0.01	98659	6203210	62.20379	17.89	Eurostat	0.0662	1058.35	98864	6175783	61.93
0.01	98847	6104552	61.22306	17.89	Other	0.055	855.78	98712	6076919	60.95
0.01	98904	6005704	60.23957	17.88	WHO	0.0389	787.51	98836	5978206	59.96
0.01	98770	5906800	59.25346	17.87			915.01	99113	5879371	58.98
0.01	98730	5808030	58.26558	17.86			948.50	99013	5780258	57.99
0.01	98713	5709299	57.278	17.85			959.43	98182	5681245	57.00
0.01	98659	5610587	56.29333	17.85			1004.16	98505	5583063	56.02
0.01	98445	5511928	55.30733	17.83			1211.96	98746	5484558	55.03
0.01	98332	5413483	54.32224	17.83			1320.64	98857	5385812	54.04
0.02	97855	5315151	53.33765	17.81			1792.24	97942	5286955	53.05
0.02	97163	5217296	52.35988	17.80			2475.34	97605	5189013	52.08
0.03	96761	5120133	51.38942	17.77			2866.04	96111	5091408	51.10
0.03	96266	5023372	50.42584	17.74			3344.46	95284	4995297	50.14
0.04	95621	4927107	49.46845	17.71			3975.43	96846	4900013	49.20
0.05	95061	4831486	48.51377	17.68			4521.46	95484	4803167	48.23
0.05	94856	4736426	47.56689	17.63			4708.41	94376	4707683	47.28
0.05	94270	4641570	46.62364	17.60			5271.59	93297	4613307	46.34
0.05	94206	4547300	45.68819	17.54			5314.41	94259	4520010	45.41
0.05	94188	4453094	44.75022	17.49			5311.34	93931	4425751	44.48

Fig. 3.3 The extended Life Table for the HLYL estimates with the Sullivan method

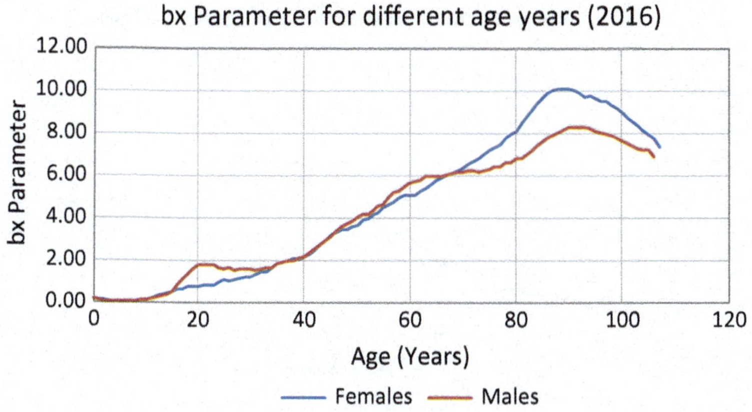

Fig. 3.4 Healthy Life Years Lost Parameter (bx) for males and females in Czechia in 2016

Table 3.1 HLE and HLYL for males and females in Czechia in 2016

Method	HLE-males	HLYL-males	HLE-females	HLYL-females
Direct	67.7	8.4	71.7	10.2
WHO	67.1	9.2	71.6	10.5
Eurostat	62.7	13.4	64.0	17.9
Equal	60.9	15.2	65.4	16.4
Related estimates from the SKI-6 program				
SKI-6 severe+light	60.8	15.3	61.5	20.4
SKI-6 severe	68.2	7.9	71.3	10.6
SKI-6 moderate	68.7	7.4	72.1	9.8

For the application for Czechia males and females in 2016 the related estimates are included in the Table 3.1.

Clearly the estimates for the HLE and the HLYL for males and females are similar for the Direct and the WHO methods. As it is also found from the estimate based on the SKI-6 program (see Fig. 3.10) presented in previous publications, both estimates (Direct and WHO) are close to Severe disability cases. Instead, the estimates provided by the Eurostat (Eurostat 2019) and Equal are close to Severe and Moderate disability cases. To this end, the Direct and WHO estimates refer to an important part of the life span with the life years lost to disability governing the development of everyday life.

Our Direct method, in addition to estimating the Healthy Life Expectancy, has the advantage of providing the Healthy Life Years Lost at every year of age via the bx parameter as is illustrated in Fig. 3.4. Males and females show similar behavior for the age period 0–70 years of age with the exception of the years from 17–30 where an excess of life years lost to disability appear with the form of a higher bx for males than for females. After 70 years of age, the bx for females becomes higher with a maximum level at 88 years of age (b = 10.21) with a decline for the higher

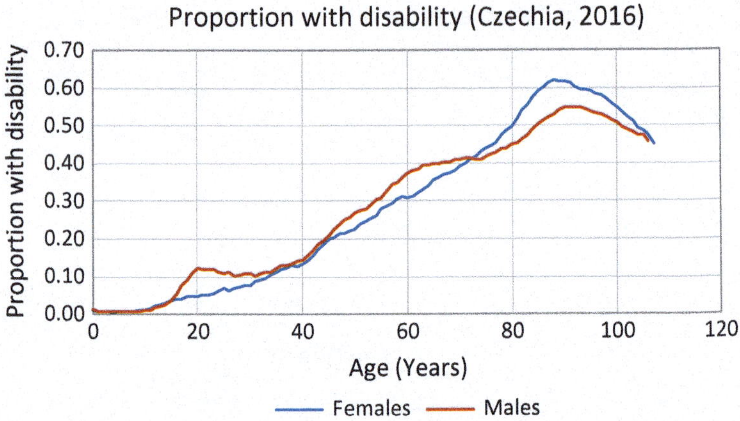

Fig. 3.5 Proportion with disability for males and females in Czechia in 2016

age years. For males, the maximum is at 92 years (b = 8.37) with a decline at higher ages.

Note that the proportion with disability for males and females in Czechia in 2016 as constructed from Fig. 3.3 data in the Excel program is illustrated in Fig. 3.5. Similarities appear with the bx estimates presented in Fig. 3.4 with a higher proportion to disability for males for the years 50–70 further to the higher values from 17–30 years. The next part for the +70 years includes higher values for females thus explaining the higher Healthy Life Years Lost for females than for males.

3.4 The Case of the United States

The above methodology may apply in many countries when good full life tables exist as in the United States. Furthermore, the U.S. censuses of the population included important information to estimate the percentage of disabilities per year of age. Stock and Beegle (2004) had summarized the material from the 1970–90 censuses. Our Direct estimates with the previous method are presented in Fig. 3.6 for three ten year periods 1970–79, 1990–99 and 2010–17. The first two curves from 1970–79 and 1990–99 correspond to Stock and Beegle calculations. Their main points refer to a disability level lower than 10% for those under age 40 and rising "dramatically" starting around age 45. In our estimates, the three curves coincide close to 45 years at 10% of disability level. The next critical points refer to a disability affecting the higher part of the population with a maximum of one-third of the population for the periods 1970–79, 1990–99 and at 35% for the period 2010–17.

Both methods, that of the census by collecting data from questionnaires and the direct method we apply by extracting information from the life tables give

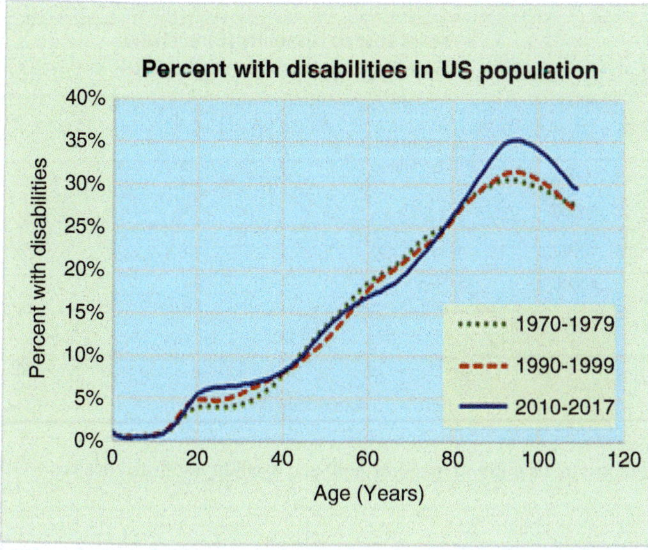

Fig. 3.6 Proportion with disabilities in USA

similar results. The life tables include information for life and death but also the probabilities to survive. That we ask people, with the questionnaires, about their health state is their assessment of the disabilities leading, sooner or later, to an end. Both methodologies can coincide depending on the systematic collection of data from questionnaires, though the direct method is simpler, less expensive and can be used officially by the Law system. For further analysis, see the paper by Stock and Beegle on: "Employment protections for older workers: Do disability discrimination laws matter?"

3.5 The Case of Norway

A very important research on the Disease Burden in Norway for 2016, provides interesting Tables and Graphs along with full documentation and further details (Tollånes et al. 2018). We apply our Direct method of estimation to calculate the Years Lost to Disability presented with blue line in Fig. 3.7. The red line in the same figure expresses the Disability-Adjusted Life Years as it appears in the Tollånes et al. paper estimated with the Global Burden of Disease (GBD) method. (The data for this red line are estimated from the paper graph with a relative precision). However, the total calculations with the GBD method (Table 4 in Tollånes et al. paper) account for 1.21 million in 2016 compared to 1.01 million with our direct estimates. Considering that Tollånes et al. estimates range from 1.03 to 1.40 million, our Direct estimates 1.01 million, are very close to their minimum estimation level. Note that

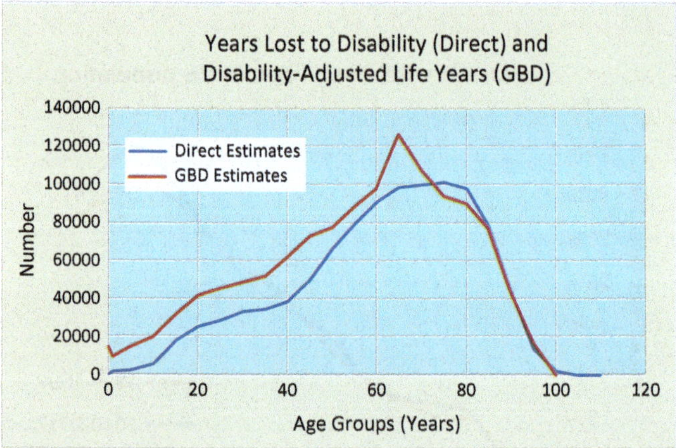

Fig. 3.7 Years lost to disability (Direct method) and disability-adjusted life years GBD)

no special disability adjustment is admitted in our methodology, explaining the almost perfect fit between both methods in a high age groups (see Fig. 3.7).

3.6 Bridging the Gap Between WHO and Eurostat Estimates for Healthy Life Years Lost

Note that the estimates of Healthy Life Years Lost for males and females provided by WHO (2019) and Eurostat (2019) differ considerably in several cases as is illustrated in Figs. 3.8 and 3.9 due to the different methodologies applied. However, in few cases provide similar results namely for Bulgaria, Malta, Norway, and Sweden for males and Bulgaria, Malta and Sweden for females (see Figs. 3.8 and 3.9).

To bridge the gap between WHO and Eurostat estimates we use the SKI-6 program (Fig. 3.10) presented in previous publications providing the figures included in Table 3.2. Healthy Life Years Lost due to severe or moderate disability causes are estimated separately using the SKI-6 Excel Program (Skiadas and Skiadas 2018a, b, c; Skiadas and Arezzo 2018). The estimate for WHO is close to Severe disability cases. Instead, the estimates provided by the Eurostat are close to the sum of Severe and Moderate disability cases as presented in the same table and illustrated in Figs. 3.11 and 3.12 for the year 2016 for European countries where the life tables are provided from the Human Mortality Database for the specific year. The results support the argument of two interrelated estimates useful for various health-related applications. However, the WHO estimates providing the HLYL from severe disability causes are most valuable when the health expenditure allocation is calculated. Severe disabilities need extensive health resources and intense treatment.

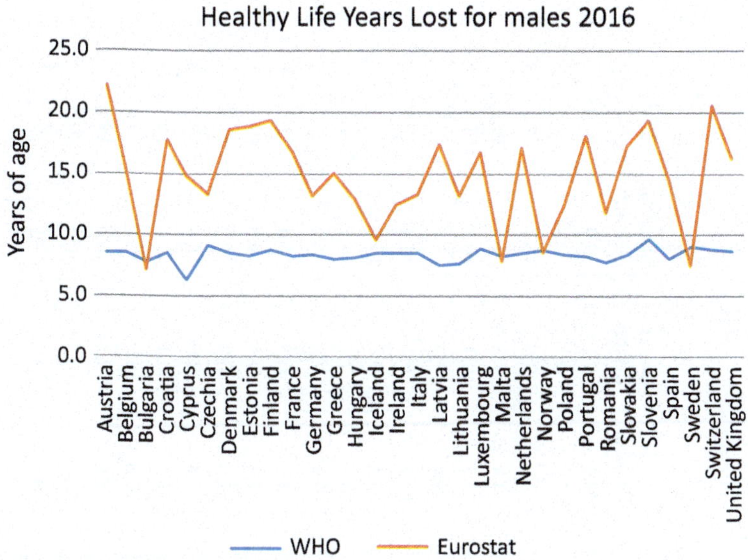

Fig. 3.8 Healthy Life Years Lost for males in 2016 estimates from WHO and Eurostat

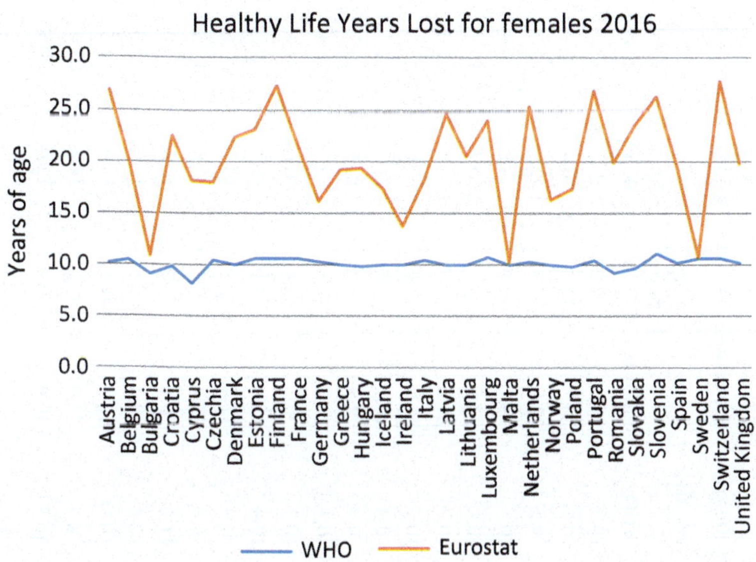

Fig. 3.9 Healthy Life Years Lost for females in 2016 estimates from WHO and Eurostat

Table 3.2 Healthy life years lost to disability for males and females in 2016

Country	Females					Males				
	WHO	Eurostat	Severe disability	Moderate disability	Total disability	WHO	Eurostat	Severe disability	Moderate disability	Total disability
Austria	10.3	27.0	12.2	10.2	22.3	8.6	22.3	9.1	9.9	19.0
Belgium	10.5	20.2	11.3	10.8	22.1	8.6	15.3	9.1	9.8	18.9
Croatia	9.9	22.6	10.3	9.7	20.0	8.6	17.9	9.0	6.7	15.7
Czech Rep.	10.5	18.1	10.6	9.8	20.4	9.2	13.4	7.8	7.5	15.2
Denmark	10.1	22.5	9.1	10.7	19.7	8.6	18.7	9.1	10.2	19.3
Estonia	10.7	23.2	10.7	10.3	21.0	8.4	18.9	7.7	5.9	13.6
France	10.7	21.6	11.6	12.0	23.6	8.3	16.9	9.1	12.4	21.5
Germany	10.3	16.2	11.0	11.3	22.3	8.4	13.3	9.3	10.0	19.3
Hungary	9.9	19.5	9.1	10.1	19.2	8.2	13.1	7.1	4.5	11.7
Latvia	10.1	24.7	10.5	9.7	20.1	7.7	17.5	7.8	3.9	11.7
Lithuania	10.2	20.7	10.6	10.6	21.1	7.7	13.3	6.8	5.4	12.3
Netherlands	10.4	25.4	11.0	11.5	22.5	8.7	17.2	9.5	9.8	19.3
Poland	9.9	17.4	9.9	10.7	20.6	8.5	12.6	6.9	6.7	13.7
Slovak Rep.	9.8	23.7	9.7	9.8	19.5	8.5	17.4	8.0	5.4	13.5
Slovenia	11.1	26.4	11.2	10.8	22.0	9.7	19.5	9.1	9.0	18.1
Spain	10.3	19.8	10.5	11.5	22.0	8.1	14.6	8.9	10.5	19.3
Sweden	10.8	10.8	10.8	10.6	21.5	9.0	7.6	9.3	10.6	19.9
Switzerland	10.7	27.9	11.2	11.4	22.6	8.8	20.7	10.3	10.3	20.6
United Kingdom	10.3	19.9	8.9	12.0	20.9	8.7	16.4	9.1	9.6	18.7

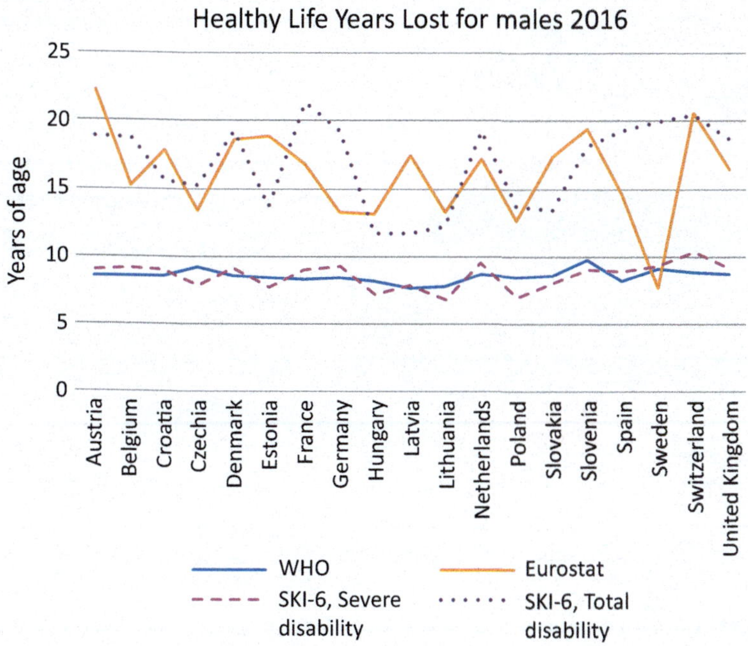

Fig. 3.10 Healthy Life Years Lost for males in 2016 (WHO and Eurostat and SKI-6 Program)

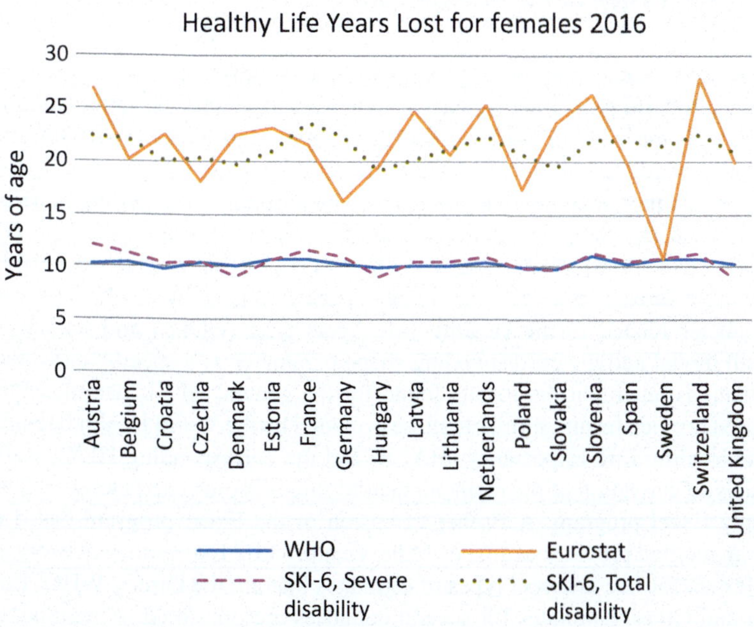

Fig. 3.11 Healthy Life Years Lost for females in 2016 (WHO and Eurostat and SKI-6 Program)

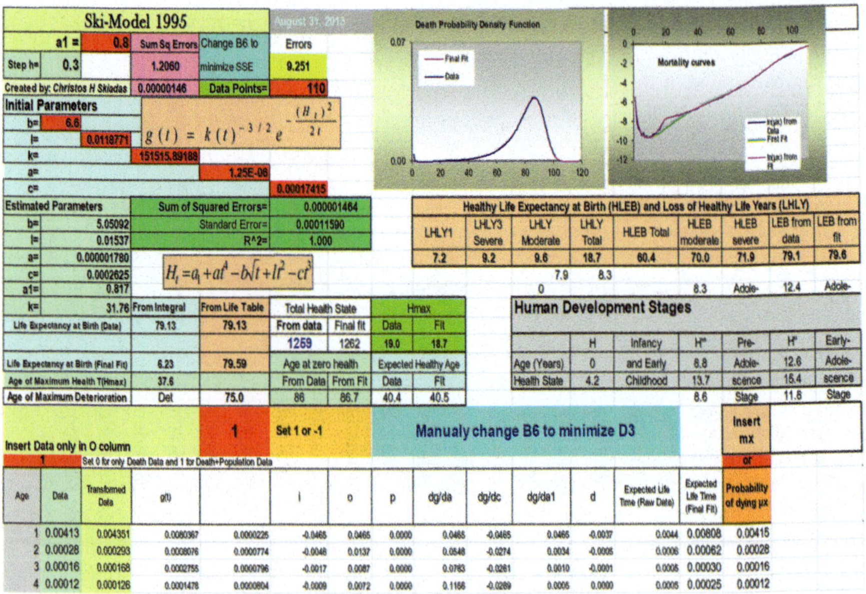

Fig. 3.12 The main view of the SKI-6 Excel Program and the Healthy Life Years Lost estimates. (Download from: http://www.smtda.net/demographics2020.html)

3.7 Conclusions and Further Study

We have provided an analytic explanation of the behavior of a parameter b_x related to the healthy life years lost. We have also presented an analytic formulation for the observations made along the development of appropriate extensions of the classical life tables in order to give a valuable tool for estimating the Healthy Life Years Lost. We have also presented how the Weibull model properties expressing the fatigue of materials and especially the cumulative hazard of this model can express the additive process of disabilities and diseases to the human population. In this paper, we further analytically derive a more general model of survival-mortality in which we estimate a parameter related to the Healthy Life Years Lost (HLYL) and leading to the Weibull model and the corresponding shape parameter as a specific case. We have also demonstrated that the results found for the general HLYL parameter that we proposed provide results similar to those provided by the World Health Organization for the Healthy Life Expectancy (HALE) and the corresponding HLYL estimates. An analytic derivation of the mathematical formulas is presented along with an easy to apply Excel program. A further extension of the Excel program based on the Sullivan method provides estimates of the Healthy Life Expectancy at every year of the lifespan for five different types of estimates that are the Direct, WHO, Eurostat, Equal and Other. Estimates for several countries are presented. A methodology is also introduced to bridge the gap between the World Health Organization (HALE) and Eurostat (HLE) healthy life expectancy estimates. The latest versions of this program appear in the Demographics2020 Workshop website at http://www.smtda.net/demographics2020.html.

References

Eurostat. (2019). *Healthy life years (from 2004 onwards)*. https://appsso.eurostat.ec.europa.eu/nui/submitViewTableAction.do

Jagger, C., Van Oyen, H., & Robine, J.-M. (2014). *Health expectancy calculation by the Sullivan method: A practical guide*. Newcastle upon Tyne: Institute for Ageing, Newcastle University.

Román, R., Comas, M., Hoffmeister, L., & Castells, X. (2007). Determining the lifetime density function using a continuous approach. *Journal of Epidemiology and Community Health, 61*, 923–925. https://doi.org/10.1136/jech.2006.052639.

Skiadas, C. H., & Arezzo, M. F. (2018). Estimation of the healthy life expectancy in Italy through a simple model based on mortality rate. In *Demography and health issues: Population aging, mortality and data analysis*. Chum: Springer. https://doi.org/10.1007/978-3-319-76002-5_4.

Skiadas, C. H., & Skiadas, C. (2014). The first exit time theory applied to life table data: The health state function of a population and other characteristics. *Communications in Statistics-Theory and Methods, 43*, 1985–1600.

Skiadas, C. H., & Skiadas, C. (2015). Exploring the state of a stochastic system via stochastic simulations: An interesting inversion problem and the health state function. *Methodology and Computing in Applied Probability, 17*, 973–982.

Skiadas, C. H., & Skiadas, C. (2018a). *Exploring the health state of a population by dynamic modeling methods* (The Springer series on demographic methods and population analysis) (Vol. 45). Chum: Springer. https://doi.org/10.1007/978-3-319-65142-2.

Skiadas, C. H., & Skiadas, C. (2018b). The health-mortality approach in estimating the healthy life years lost compared to the global burden of disease studies and applications in world, USA and Japan. In *Exploring the health state of a population by dynamic modeling methods* (The Springer series on demographic methods and population analysis) (Vol. 45, pp. 67–124). Chum: Springer. https://doi.org/10.1007/978-3-319-65142-2_4.

Skiadas, C. H., & Skiadas, C. (2018c). *Demography and health issues: Population aging, mortality and data analysis* (The springer series on demographic methods and population analysis) (Vol. 46). Chum: Springer. https://doi.org/10.1007/978-3-319-76002-5.

Skiadas, C. H., & Skiadas, C. (2019a). Relation of the Weibull shape parameter with the healthy life years lost estimates: Analytical derivation and estimation from an extended life table. *ArXiv*https://arxiv.org/ftp/arxiv/papers/1904/1904.10124.pdf

Skiadas, C. H., & Skiadas, C. (2019b, March). *Modeling the health expenditure in Japan, 2011. A healthy life years lost methodology*. arXiv:1903.11565.

Skiadas, C. H., & Skiadas, C. (2019c, April). *Relation of the Weibull shape parameter with the healthy life years lost estimates: Analytic derivation and estimation from an extended life table*. arXiv:1904.10124.

Stock, W. A., & Beegle, K. (2004, January). Employment protections for older workers: Do disability discrimination laws matter? *Contemporary Economic Policy, 22*(1), 111–126.

Sullivan, D. F. (1971). A single index of mortality and morbidity. *Health Services Mental Health Administration Health Reports, 86*, 347–354.

Tollånes, M. C., Knudsen, A. K., Vollset, S. E., Kinge, J. M., Skirbekk, V., & Øverland, S. (2018). Disease burden in Norway in 2016. *Tidsskrift for Den norske legeforening, 2019*. https://doi.org/10.4045/tidsskr.18.0274.

Weibull, W. (1951). A statistical distribution function of wide applicability. *Journal of Applied Mechanics, 18*(3), 293–297.

WHO Life Expectancy and Healthy Life Expectancy Data by Country. (2018). https://apps.who.int/gho/data/node.main.688, Last updated: 2018-04-06.

Chapter 4
Modeling the Health Expenditure in Japan, 2011. A Healthy Life Years Lost Methodology

Christos H. Skiadas and Charilaos Skiadas

4.1 Introduction

Life Tables have dominated the quantitative and qualitative demography issues for the last 4 centuries. Following the works of John Graunt (1676) and Edmond Halley (1693), life tables became the standard measure of several demographic parameters. Later on, models as the Gompertz (1825) and Makeham (1860) came to support and straighten the use of mortality models in demography and insurance while giving rise to various approaches and studies in probability and statistics.

However, further exploration of the hidden properties and functions in the classical life tables is needed. It was already demonstrated in several publications in the last decades and more recently with few publications (see Janssen and Skiadas (1995), Skiadas, C.H. and Skiadas, C. (2010, 2014, 2015, 2018a, b, c), Skiadas and Arezzo (2018), Strehler and Mildvan (1960), Sullivan (1966, 1971), Torrance (1976), Chia and Peng Loh (2018)). The main task was the use of appropriate transformations in the life tables in order to extract valuable information. To this end, the estimation of the life years lost to disabilities and health deterioration is of particular importance. Our works on estimating this important period of the life time are in parallel to the other approaches developed under the Global Burden of Disease (GBD) terminology (WHO 2014). The latter is a very complicated and heavy international methodology, important to estimate the impact of every particular kind of disease and disability on the health state of the population and

C. H. Skiadas (✉)
ISAST, Athens, Attiki, Greece
e-mail: skiadas@cmsim.net

C. Skiadas
Department of Mathematics and Computer Science, Hanover College, Hanover, IN, USA
e-mail: skiadas@hanover.edu

C. H. Skiadas, C. Skiadas (eds.), *Demography of Population Health, Aging and Health Expenditures*, The Springer Series on Demographic Methods and Population Analysis 50, https://doi.org/10.1007/978-3-030-44695-6_4

consequently on the health expenditure in a country. The results are produced as Healthy Life Years Lost (HLYL) or as the estimated Healthy Life Expectancy (HLE) that is the Life Expectancy (LE) minus the HLYL. The GBD results regarding the particular kind of diseases and disabilities are the most important part of these studies. A complicated methodology weights and interconnects the particular kinds of disabilities and diseases in a final number for the HLE provided by agencies as the World Health Organization (WHO) as HALE and health statistical bureaus under several terminologies. Following our estimates, the HLYL can be found directly from the life tables with a relatively simple methodology presented in recent publications (Skiadas and Skiadas (2018a, b, c). The method is based on developing a Survival vs Mortality diagram and does direct estimates based on mx or qx as termed in the well-developed life tables of the Human Mortality Database (HMD). To further straighten the validity and importance of this methodology we also have provided parallel HLYL estimates based on the Gompertz (1825) and Weibull (1951) models. A brief presentation of the related methodology follows here.

The main task of this paper is to explore how the estimated healthy life years lost can be used to calculate the Health Expenditure in a Country. Clearly the up to now estimates of life expectancy, healthy life expectancy and healthy life years lost are not easily applied to the health expenditure calculations. Instead, our approach of the HLYL estimates is directly used in the health expenditure calculations as we present in the following.

4.2 The HLYL Estimation Method (Fig. 4.1)

The simplest form to express mortality mx in a population at age x is by estimating the fraction Death(D)/Population(P) that is mx = D/P. Accordingly, the above graph is formulated with mx as the blue exponential curve whereas the horizontal axis expresses years of age. The main forms of Life Tables start with mx and subsequently estimate qx and the survival forms of the population. This methodology leads to the estimation of a probability measure termed as life expectancy at age x or life expectancy at birth when considering the total life-time. There are several differences between the graph with the survival space above and the survival curves methodology. First of all, the vertical axis in the Survival-Mortality Space (SMS) diagram is the probability mx. Instead in the survival diagram, the vertical axis represents the population (usually it starts from 100.000 in most life tables and gradually slows down until the end). By the SMS diagram we have probability spaces for both survival and mortality. For the age x, the total space is (ABCOA) in the SMS diagram that is (OC)x(OA) = x.mx. The mortality space (ODBCO) is the sum of S(mx) and the survival space, (x.mx-S(mx)). Accordingly, the important measure of the Health State is simply the fraction (ABDOA)/(BCODB). This is to provide the form FHM_{*} = (x.mx-S(mx))/S(mx) = x.mx/S(mx)-1. It is simpler to use

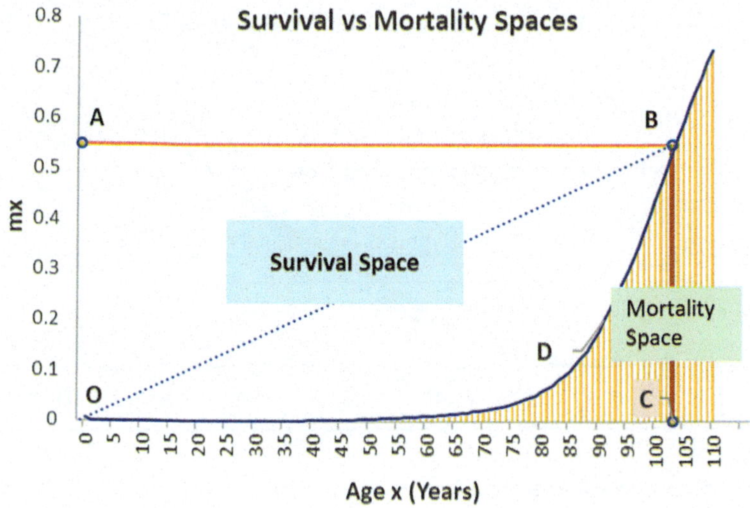

Fig. 4.1 The Survival – Mortality spaces

the fraction FHM = (ABCOA)/(BCODB) = x.mx/S(mx) that will be also estimated from mx for every age x of the population.

The FHM is calculated by:

$$FHM(mx) = \lambda \frac{xmx}{\sum_1^x mx}$$

Note that a similar approach based on qx estimates in a life table is to use the following formula:

$$FHM(qx) = \lambda \frac{xqx}{\sum_1^x qx}$$

In both cases the parameter λ has to be estimated. In the majority of applications, λ = 1 provides quite good results.

Both estimates are presented in the following Fig. 4.2 obtained from the Japan 2011 Extended Full Life Table (see also Fig. 4.3). Both methods show similar results until 70 years of age and then the mx estimates give higher values than the qx. In both cases a maximum is reached in the interval 95–100 years of age followed by a decline in the remaining years. The estimated HLYL are 10.1 years for the mx based estimate and 9.5 years for the qx based estimate. As the life expectancy at birth for Japan 2011 is 82.7 years of age, the estimates for the healthy life expectancy are 72.6 and 73.2 years of age for the estimates based on mx and qx respectively. In the related Fig. 4.2, the healthy years lost estimated with the Healthy Life Expectancy method are also presented with a cyan line. The latter is mainly a smoothing like expression of the HLYL estimates from mx and qx (Fig. 4.4).

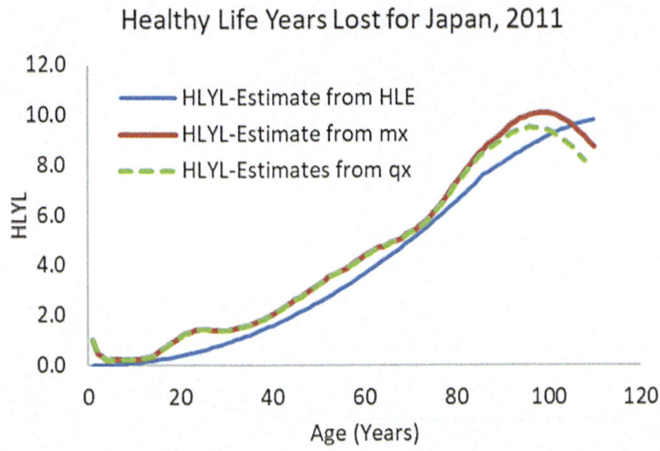

Fig. 4.2 Healthy life years lost estimated

The Health/Mortality fraction has an increasing form for the main part of the lifetime until a maximum and then a declining form. The fraction for Japan 2011 is similar until 70 years of age for both estimates from mx or qx. The very important point here is that the maximum points correspond to the maximum years of age lost to disability. We can easily observe this important future by considering a linear form for mortality mx = ax. This is the simplest case of drawing a linear line from O to B in the graph above. The resulting fraction is 2 whereas it is 1 if we select the fraction (ABDOA)/(BCODB). Following the previous discussion, the healthy life years lost to disability (HLYL) are 1 with the last notation and 2 when considering the total space vs the mortality space. The latter which is higher by 1 from the simple fraction provides results similar to those estimated by the World Health Organization. After that, we only need to remove this estimate from the life expectancy at birth to find the Healthy Life Expectancy.

Using the 2011 Life Table data-set for both (male and female) from the Japan official website we found the following Table of the estimates. Included are the estimates from the Gompertz and Weibull models following our methodology presented in our books. From the same reference are also the direct estimates based on mx and qx provided from the life table. The healthy life years lost per year of age and the healthy life expectancy are also included. The full life table with estimates is presented in the end of this paper in Fig. 4.3.

The results obtained are from the Full Life Table for Japan. The related for the Abridged Life Table for 2011 are included in the following Table.

Japan-b-2011

Estimates of HLYL and HLE (C H Skiadas and C Skiadas, 2016).
Add data in this Excel file and do not change the supporting files Gompertz and Weibull as they run automatically

Manually complete the blue Excel part only for mx or add all the Life Table from the Human Mortality Database, HMD (recommended)

											Healthy Life Years Lost (HLYL)				
											HLYL Gompertz	HLYL Weibull	HLYL from mx	HLYL from qx	HLYL Average
										Years of age	9.87	9.89	10.09	9.50	9.84
Year	Age	mx	qx	ax	lx	dx	Lx	Tx	ex	HLYL	HLE				
2011	0	0.00237	0.00237	0.21	100000	237	99813	8271181	82.71	9.8	72.87	0.00237	0.00237	0.00237	0.00237
2011	1	0.00042	0.00042	0.5	99763	42	99742	8171368	81.91	9.8	72.07	0.00279	0.00279	0.00042	0.00042
2011	2	0.00028	0.00028	0.5	99721	27	99707	8071626	80.94	9.8	71.10	0.00307	0.00307	0.00056	0.00056
2011	3	0.00023	0.00023	0.5	99694	23	99682	7971919	79.96	9.8	70.13	0.0033	0.0033	0.00069	0.00069
2011	4	0.00018	0.00018	0.5	99671	18	99662	7872236	78.98	9.8	69.16	0.00348	0.00348	0.00072	0.00072
2011	5	0.00016	0.00016	0.5	99653	16	99645	7772575	78	9.8	68.19	0.00364	0.00364	0.0008	0.0008
2011	6	0.00016	0.00016	0.5	99637	16	99629	7672930	77.01	9.8	67.21	0.0038	0.0038	0.00096	0.00096
2011	7	0.00014	0.00014	0.5	99621	14	99614	7573301	76.02	9.8	66.23	0.00394	0.00394	0.00098	0.00098
2011	8	0.00011	0.00011	0.5	99607	11	99602	7473687	75.03	9.8	65.26	0.00405	0.00405	0.00088	0.00088
2011	9	0.00012	0.00012	0.5	99596	12	99590	7374085	74.04	9.8	64.28	0.00417	0.00417	0.00108	0.00108
2011	10	0.00011	0.00011	0.5	99584	11	99579	7274494	73.05	9.7	63.31	0.00428	0.00428	0.0011	0.0011
2011	11	0.00011	0.00011	0.5	99573	11	99567	7174916	72.06	9.7	62.35	0.00439	0.00439	0.00121	0.00121
2011	12	0.0001	0.0001	0.5	99562	10	99556	7075348	71.07	9.7	61.38	0.00449	0.00449	0.0012	0.0012
2011	13	0.00013	0.00013	0.5	99551	13	99545	6975792	70.07	9.7	60.41	0.00462	0.00462	0.00169	0.00169
2011	14	0.00015	0.00015	0.5	99538	15	99530	6876247	69.08	9.6	59.45	0.00477	0.00477	0.0021	0.0021
2011	15	0.00018	0.00018	0.5	99523	18	99514	6776717	68.09	9.6	58.49	0.00495	0.00495	0.0027	0.0027
2011	16	0.00025	0.00025	0.5	99505	25	99493	6677203	67.1	9.6	57.53	0.0052	0.0052	0.004	0.004
2011	17	0.00029	0.00029	0.5	99480	29	99466	6577710	66.12	9.5	56.58	0.00549	0.00549	0.00493	0.00493
2011	18	0.00037	0.00037	0.5	99451	37	99432	6478245	65.14	9.5	55.64	0.00586	0.00586	0.00666	0.00666
2011	19	0.00036	0.00036	0.5	99414	36	99396	6378812	64.16	9.5	54.70	0.00622	0.00622	0.00684	0.00684
2011	20	0.0004	0.0004	0.5	99378	40	99358	6279416	63.19	9.4	53.77	0.00662	0.00662	0.008	0.008
2011	21	0.00045	0.00045	0.5	99339	45	99316	6180058	62.21	9.4	52.83	0.00707	0.00707	0.00945	0.00945
2011	22	0.00049	0.00049	0.5	99294	48	99270	6080741	61.24	9.3	51.91	0.00756	0.00756	0.01078	0.01078
2011	23	0.00052	0.00052	0.5	99246	51	99220	5981471	60.27	9.3	50.99	0.00808	0.00808	0.01196	0.01196
2011	24	0.00053	0.00053	0.5	99195	53	99168	5882251	59.3	9.2	50.07	0.00861	0.00861	0.01272	0.01272
2011	25	0.00051	0.00051	0.5	99141	50	99116	5783083	58.33	9.2	49.15	0.00912	0.00912	0.01275	0.01275
2011	26	0.00051	0.00051	0.5	99091	51	99066	5683967	57.36	9.1	48.23	0.00963	0.00963	0.01326	0.01326
2011	27	0.0005	0.0005	0.5	99040	50	99016	5584901	56.39	9.1	47.32	0.01013	0.01013	0.0135	0.0135
2011	28	0.00052	0.00052	0.5	98991	52	98965	5485886	55.42	9.0	46.40	0.01065	0.01065	0.01456	0.01456
2011	29	0.00056	0.00056	0.5	98939	56	98911	5386921	54.45	9.0	45.49	0.01121	0.01121	0.01624	0.01624
2011	30	0.00057	0.00057	0.5	98883	56	98855	5288010	53.48	8.9	44.59	0.01178	0.01178	0.0171	0.0171
2011	31	0.00058	0.00058	0.5	98827	57	98798	5189155	52.51	8.8	43.68	0.01236	0.01236	0.01798	0.01798
2011	32	0.00061	0.00061	0.5	98770	61	98740	5090356	51.54	8.8	42.78	0.01297	0.01297	0.01952	0.01952
2011	33	0.00061	0.00061	0.5	98709	60	98679	4991617	50.57	8.7	41.88	0.01358	0.01358	0.02013	0.02013

Fig. 4.3 The Heathy Life Years Lost and Healthy Life Expectancy estimates Extended Life Table form

Japan Life Table Data (period 5x, b)						Ch Skiadas 18 February 2019			
Year	Age	mx	Age x	mx	k.x.mx / sum(m1:mx)	Move Up	Japan Estimates	Japan Data	SSE
2011				**0.01989**	**28.657**	**68**			
2011	0--4	0.00265	2	0.01989	143.28	68	211	223	137
2011	5--9	0.00014	5	0.00014	1.40	68	69	123	2873
2011	10--14	0.00012	10	0.00012	2.05	68	70	86	254
2011	15-19	0.00029	15	0.00029	6.91	68	75	68	48
2011	20-24	0.00048	20	0.00048	14.46	68	82	71	131
2011	25-29	0.00052	25	0.00052	18.76	68	87	90	10
2011	30-34	0.00061	30	0.00061	25.36	68	93	106	160
2011	35-39	0.00083	35	0.00083	38.46	68	106	116	91
2011	40-44	0.00126	40	0.00126	62.81	68	131	132	1
2011	45-49	0.00189	45	0.00189	97.78	68	166	167	1
2011	50-54	0.00297	50	0.00297	152.59	68	221	210	112
2011	55-59	0.00445	55	0.00445	217.27	68	285	266	371
2011	60-64	0.00694	60	0.00694	305.25	68	373	352	452
2011	65-69	0.01021	65	0.01021	387.38	68	455	454	2
2011	70-74	0.01596	70	0.01596	494.71	68	563	614	2631
2011	75-79	0.02771	75	0.02771	648.58	68	717	771	2962
2011	80-84	0.05012	80	0.05012	815.65	68	884	910	694
2011	85-89	0.08928	85	0.08928	952.56	68	1021	1009	134
2011	90-94	0.15692	90	0.15692	1059.18	68	1127	1086	1696
2011	95-99	0.26304	95	0.26304	1118.64	68	1187	1167	386
2011	100+	0.40944	100	0.40944	1125.79	68	1194	1198	18
								R2=	**0.943**
								se=	**25.04**
							9117	9219	13164

Fig. 4.4 The extended abridged life table for health expenditure estimates

4.3 Modeling of the Health Expenditure in Japan, 2011

The results from the two methods are similar. The abridged life tables are more useful as the provided health expenditure data for several countries follow an abridged life table population group schedule with estimates of five-year age groups. This is presented in the following Fig. 4.5 Extended Life Table for Japan 2011 along with the health expenditure estimates and comparisons with the real-life data sets.

Health Expenditure Data from: Jennifer Friedman (2015) Health Expenditure per Capita: Application in Japan 2011, Public Lecture.

We recall that the FHM in Fig. 4.5 is calculated by:

$$FHM(mx) = \frac{xmx}{\sum_1^x mx}$$

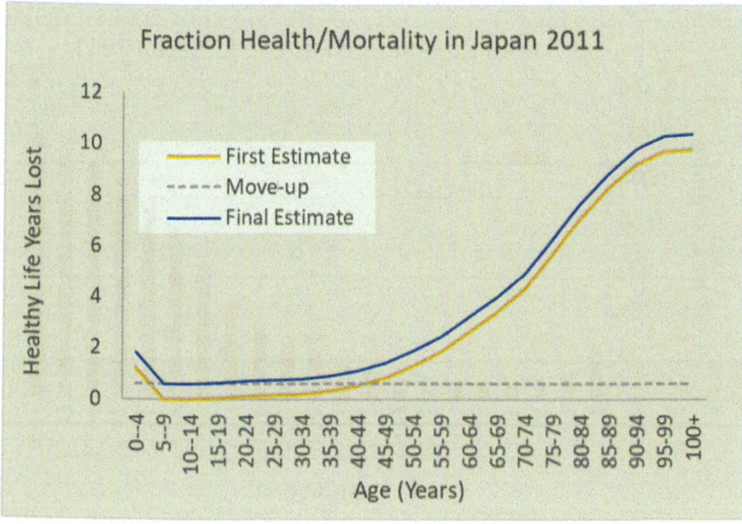

Fig. 4.5 FHM in Japan 2011 and move-up correction for the health expenditure

The estimates are based on the FHM multiplied by a parameter k (orange color) that is FHM = kx(mx)/sum(m1:mx). This parameter k = 28.723 is estimated by a regression analysis so that to minimize the sum of squared errors between the data and the estimates. The first estimate for the mx_* (green color) is selected from the data sets following the formula:

$$mx* \;=\; (D1 - \text{MIN}\,(\text{Data}))\,/\,(\text{sum}\,(\text{Data}) - 21 * \text{MIN}\,(\text{Data})) \;=\; 0.01989$$

where D1 = 223, Min(Data) = 68 is the minimum data point and n = 21 is the number of data points. The other points of the fifth column are equal to those provided for mx (column three). The sixth column includes the estimates based on FHM theory. Note that the formula for FHM takes the form:

$$FHM(mx) = k\frac{(x+2)\,mx}{\sum_1^x mx}$$

where x is provided in column 4 of Fig. 4.5.

The application to Japan 2011 showed that the FHM was a good approach. We have estimated the model with $R^2 = 0.943$. The standard error is se = 25.04 corresponding to 5.7% of the mean of the data set.

Clearly our method indicates that there is no need to try to find any special model and estimate the parameters as all the needed data and modeling are included into the life tables. The only thing that remains to be done is to search further to find the hidden forms, to extract them and apply them.

In Fig. 4.6 we present the first estimate of the FHM done in Fig. 4.5 life table followed by a move-up to the minimum level of the health expenditure data provided

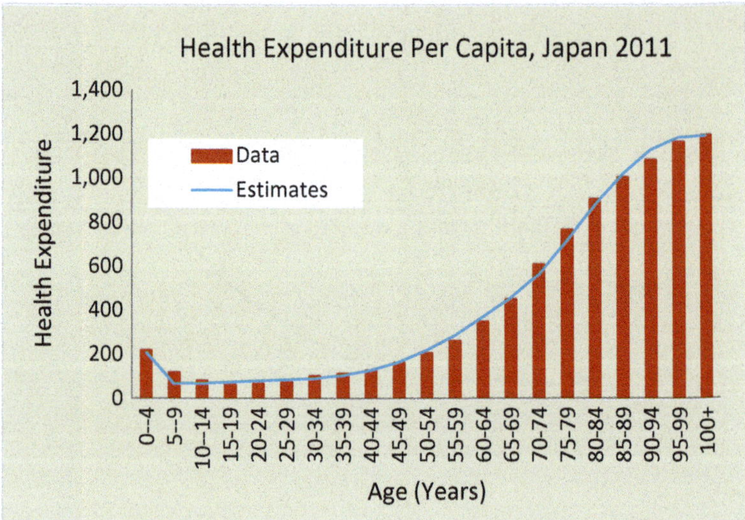

Fig. 4.6 Health Expenditure per Capita in Japan in 2011 (red bars) and Estimates (cyan curve)

for Japan 2011. The Life Expectancy is 82.71 years according to the abridged life table for Japan in 2011 and the HLYL is 9.81 years of age. The estimated HLE is 72.90 years (Note that the HLE is higher when estimating with the qx method). Following the last graph, the health expenditure should be very low in the age interval from 5 to 15 years of age. However, the Health Systems have a minimum spending to be taken into account corresponding to an upward movement of the FHM curve that is estimated in the Japan 2011 case to account for 5.68% of the maximum health expenditure per capita.

The estimates for the health expenditure in Japan in 2011 in 1000s of Yen are illustrated in Fig. 4.7. Clustered columns represent the data sets in Japan currency and the cyan curve represent our estimates. The statistics for these estimates are already referred to earlier and obtained from the extended life table presented in Fig. 4.5. The approach provided verifies the applicability and importance of our theory and the methodological and applied tools provided.

4.4 Conclusions and Further Study

We have developed an easy to apply methodology and designed expanded life tables to estimate the health expenditure in a country based on the Life Tables of this country. The application in Japan is quite promising that analogous studies will be possible for other countries thus supporting the decision-makers and Health Agencies to develop effective strategies to better allocate the existing funds and make better predictions of the future health trends. A project is in progress

Japan-b-2011

Estimates of HLYL and HLE (C H Skiadas and C Skiadas, 2016).
Add data in this Excel file and do not change the supporting files Gompertz and Weibull as they run automatically

Manually complete the blue Excel part only for mx or add all the Life Table from the Human Mortality Database, HMD (recommended)

										Years of age	Healthy Life Years Lost (HLYL)				
											HLYL Gompertz	HLYL Weibull	HLYL from mx	HLYL from qx	HLYL Average
											9.87	9.89	10.09	9.50	9.84
Year	Age	mx	qx	ax	lx	dx	Lx	Tx	ex	HLYL	HLE				
2011	0	0.00237	0.00237	0.21	100000	237	99813	8271181	82.71	9.8	72.87	0.00237	0.00237	0.00237	0.00237
2011	1	0.00042	0.00042	0.5	99763	42	99742	8171368	81.91	9.8	72.07	0.00279	0.00279	0.00042	0.00042
2011	2	0.00028	0.00028	0.5	99721	27	99707	8071626	80.94	9.8	71.10	0.00307	0.00307	0.00056	0.00056
2011	3	0.00023	0.00023	0.5	99694	23	99682	7971919	79.96	9.8	70.13	0.0033	0.0033	0.00069	0.00069
2011	4	0.00018	0.00018	0.5	99671	18	99662	7872236	78.98	9.8	69.16	0.00348	0.00348	0.00072	0.00072
2011	5	0.00016	0.00016	0.5	99653	16	99645	7772575	78	9.8	68.19	0.00364	0.00364	0.0008	0.0008
2011	6	0.00016	0.00016	0.5	99637	16	99629	7672930	77.01	9.8	67.21	0.0038	0.0038	0.00096	0.00096
2011	7	0.00014	0.00014	0.5	99621	14	99614	7573301	76.02	9.8	66.23	0.00394	0.00394	0.00098	0.00098
2011	8	0.00011	0.00011	0.5	99607	11	99602	7473687	75.03	9.8	65.26	0.00405	0.00405	0.00088	0.00088
2011	9	0.00012	0.00012	0.5	99596	12	99590	7374085	74.04	9.8	64.28	0.00417	0.00417	0.00108	0.00108
2011	10	0.00011	0.00011	0.5	99584	11	99579	7274494	73.05	9.7	63.31	0.00428	0.00428	0.0011	0.0011
2011	11	0.00011	0.00011	0.5	99573	11	99567	7174916	72.06	9.7	62.35	0.00439	0.00439	0.00121	0.00121
2011	12	0.0001	0.0001	0.5	99562	10	99556	7075348	71.07	9.7	61.38	0.00449	0.00449	0.0012	0.0012
2011	13	0.00013	0.00013	0.5	99551	13	99545	6975792	70.07	9.7	60.41	0.00462	0.00462	0.00169	0.00169
2011	14	0.00015	0.00015	0.5	99538	15	99530	6876247	69.08	9.6	59.45	0.00477	0.00477	0.0021	0.0021
2011	15	0.00018	0.00018	0.5	99523	18	99514	6776717	68.09	9.6	58.49	0.00495	0.00495	0.0027	0.0027
2011	16	0.00025	0.00025	0.5	99505	25	99493	6677203	67.1	9.6	57.53	0.0052	0.0052	0.004	0.004
2011	17	0.00029	0.00029	0.5	99480	29	99466	6577710	66.12	9.5	56.58	0.00549	0.00549	0.00493	0.00493
2011	18	0.00037	0.00037	0.5	99451	37	99432	6478245	65.14	9.5	55.64	0.00586	0.00586	0.00666	0.00666
2011	19	0.00036	0.00036	0.5	99414	36	99396	6378812	64.16	9.5	54.70	0.00622	0.00622	0.00684	0.00684
2011	20	0.0004	0.0004	0.5	99378	40	99358	6279416	63.19	9.4	53.77	0.00662	0.00662	0.008	0.008
2011	21	0.00045	0.00045	0.5	99339	45	99316	6180058	62.21	9.4	52.83	0.00707	0.00707	0.00945	0.00945
2011	22	0.00049	0.00049	0.5	99294	48	99270	6080741	61.24	9.3	51.91	0.00756	0.00756	0.01078	0.01078
2011	23	0.00052	0.00052	0.5	99246	51	99220	5981471	60.27	9.3	50.99	0.00808	0.00808	0.01196	0.01196
2011	24	0.00053	0.00053	0.5	99195	53	99168	5882251	59.3	9.2	50.07	0.00861	0.00861	0.01272	0.01272
2011	25	0.00051	0.00051	0.5	99141	50	99116	5783083	58.33	9.2	49.15	0.00912	0.00912	0.01275	0.01275
2011	26	0.00051	0.00051	0.5	99091	51	99066	5683967	57.36	9.1	48.23	0.00963	0.00963	0.01326	0.01326
2011	27	0.0005	0.0005	0.5	99040	50	99016	5584901	56.39	9.1	47.32	0.01013	0.01013	0.0135	0.0135
2011	28	0.00052	0.00052	0.5	98991	52	98965	5485886	55.42	9.0	46.40	0.01065	0.01065	0.01456	0.01456
2011	29	0.00056	0.00056	0.5	98939	56	98911	5386921	54.45	9.0	45.49	0.01121	0.01121	0.01624	0.01624
2011	30	0.00057	0.00057	0.5	98883	56	98855	5288010	53.48	8.9	44.59	0.01178	0.01178	0.0171	0.0171
2011	31	0.00058	0.00058	0.5	98827	57	98798	5189155	52.51	8.8	43.68	0.01236	0.01236	0.01798	0.01798
2011	32	0.00061	0.00061	0.5	98770	61	98740	5090356	51.54	8.8	42.78	0.01297	0.01297	0.01952	0.01952
2011	33	0.00061	0.00061	0.5	98709	60	98679	4991617	50.57	8.7	41.88	0.01358	0.01358	0.02013	0.02013
2011	34	0.0007	0.0007	0.5	98649	69	98615	4892938	49.6	8.6	40.98	0.01428	0.01428	0.0238	0.0238
2011	35	0.00072	0.00072	0.5	98580	71	98545	4794323	48.63	8.6	40.08	0.015	0.015	0.0252	0.0252
2011	36	0.00075	0.00075	0.5	98509	74	98472	4695778	47.67	8.5	39.19	0.01575	0.01575	0.027	0.027
2011	37	0.00085	0.00085	0.5	98435	84	98393	4597306	46.7	8.4	38.30	0.0166	0.0166	0.03145	0.03145
2011	38	0.00086	0.00086	0.5	98351	84	98309	4498913	45.74	8.3	37.42	0.01746	0.01746	0.03268	0.03268
2011	39	0.00096	0.00096	0.5	98267	94	98220	4400604	44.78	8.2	36.54	0.01842	0.01842	0.03744	0.03744
2011	40	0.00105	0.00105	0.5	98172	103	98121	4302384	43.82	8.2	35.66	0.01947	0.01947	0.042	0.042
2011	41	0.00113	0.00113	0.5	98069	111	98014	4204263	42.87	8.1	34.80	0.0206	0.0206	0.04633	0.04633
2011	42	0.00126	0.00126	0.5	97958	123	97896	4106250	41.92	8.0	33.94	0.02186	0.02186	0.05292	0.05292
2011	43	0.00132	0.00132	0.5	97835	129	97770	4008353	40.97	7.9	33.08	0.02318	0.02318	0.05676	0.05676
2011	44	0.00151	0.00151	0.5	97705	148	97632	3910583	40.02	7.8	32.22	0.02469	0.02469	0.06644	0.06644
2011	45	0.00157	0.00157	0.5	97558	153	97481	3812952	39.08	7.7	31.37	0.02626	0.02626	0.07065	0.07065
2011	46	0.00173	0.00173	0.5	97405	168	97321	3715470	38.14	7.6	30.53	0.02799	0.02799	0.07958	0.07958
2011	47	0.00186	0.00186	0.5	97237	181	97146	3618150	37.21	7.5	29.69	0.02985	0.02985	0.08742	0.08742
2011	48	0.00205	0.00205	0.5	97056	199	96956	3521003	36.28	7.4	28.86	0.0319	0.0319	0.0984	0.0984
2011	49	0.00226	0.00226	0.5	96857	219	96748	3424047	35.35	7.3	28.04	0.03416	0.03416	0.11074	0.11074
2011	50	0.00251	0.0025	0.5	96638	242	96517	3327299	34.43	7.2	27.22	0.03667	0.03666	0.1255	0.125
2011	51	0.00267	0.00267	0.5	96396	257	96268	3230782	33.52	7.1	26.42	0.03934	0.03933	0.13617	0.13617
2011	52	0.003	0.00299	0.5	96139	288	95995	3134515	32.6	7.0	25.61	0.04234	0.04232	0.156	0.15548
2011	53	0.00323	0.00322	0.5	95851	309	95697	3038519	31.7	6.9	24.82	0.04557	0.04554	0.17119	0.17066
2011	54	0.00345	0.00344	0.5	95543	329	95378	2942822	30.8	6.8	24.03	0.04902	0.04898	0.1863	0.18576
2011	55	0.00364	0.00363	0.5	95214	346	95041	2847444	29.91	6.7	23.25	0.05266	0.05261	0.2002	0.19965
2011	56	0.004	0.00399	0.5	94868	379	94679	2752403	29.01	6.5	22.47	0.05666	0.0566	0.224	0.22344
2011	57	0.00446	0.00445	0.5	94489	420	94279	2657724	28.13	6.4	21.71	0.06112	0.06105	0.25422	0.25365
2011	58	0.00481	0.0048	0.5	94069	452	93843	2563445	27.25	6.3	20.95	0.06593	0.06585	0.27898	0.2784
2011	59	0.00535	0.00533	0.5	93617	499	93368	2469602	26.38	6.2	20.20	0.07128	0.07118	0.31565	0.31447
2011	60	0.00587	0.00586	0.5	93118	545	92845	2376234	25.52	6.1	19.47	0.07715	0.07704	0.3522	0.3516
2011	61	0.00624	0.00622	0.5	92573	576	92285	2283389	24.67	5.9	18.75	0.08339	0.08326	0.38064	0.37942
2011	62	0.00683	0.00681	0.5	91997	626	91684	2191104	23.82	5.8	18.03	0.09022	0.09007	0.42346	0.42222
2011	63	0.0075	0.00747	0.5	91371	683	91030	2099420	22.98	5.7	17.32	0.09772	0.09754	0.4725	0.47061
2011	64	0.00832	0.00829	0.5	90688	751	90313	2008391	22.15	5.5	16.62	0.10604	0.10583	0.53248	0.53056
2011	65	0.00797	0.00794	0.5	89937	714	89580	1918078	21.33	5.4	15.94	0.11401	0.11377	0.51805	0.5161
2011	66	0.00946	0.00942	0.5	89223	840	88803	1828498	20.49	5.3	15.23	0.12347	0.12319	0.62436	0.62172
2011	67	0.01052	0.01047	0.5	88382	925	87920	1739696	19.68	5.1	14.56	0.13399	0.13366	0.70484	0.70149
2011	68	0.01094	0.01088	0.5	87457	952	86981	1651776	18.89	5.0	13.92	0.14493	0.14454	0.74392	0.73984
2011	69	0.01223	0.01215	0.5	86505	1051	85980	1564795	18.09	4.8	13.26	0.15716	0.15669	0.84387	0.83835
2011	70	0.01302	0.01294	0.5	85454	1106	84901	1478815	17.31	4.7	12.63	0.17018	0.16963	0.9114	0.9058
2011	71	0.01436	0.01426	0.5	[illegible]	1203	[illegible]	[illegible]	[illegible]	[illegible]	[illegible]	[illegible]	[illegible]	1.01966	1.01346
2011	72	0.01558	0.01546	0.5	83146	1285	82503	1310166	15.76	4.4	11.38	0.20012	0.19935	1.12176	1.11312
2011	73	0.01788	0.01772	0.5	81860	1450	81135	1227663	15	4.2	10.77	0.218	0.21707	1.30524	1.29356
2011	74	0.0192	0.01902	0.5	80410	1529	79645	1146528	14.26	4.1	10.18	0.2372	0.23609	1.4208	1.40748
2011	75	0.02194	0.02171	0.5	78881	1712	78025	1066883	13.53	3.9	9.61	0.25914	0.2578	1.6455	1.62825
2011	76	0.02431	0.02402	0.5	77169	1853	76242	988858	12.81	3.8	9.05	0.28345	0.28182	1.84756	1.82552
2011	77	0.02744	0.02707	0.5	75315	2039	74296	912616	12.12	3.6	8.52	0.31089	0.30889	2.11288	2.08439
2011	78	0.031	0.03053	0.5	73277	2237	72158	838321	11.44	3.4	8.00	0.34189	0.33942	2.418	2.38134
2011	79	0.03478	0.03418	0.5	71040	2428	69825	766162	10.79	3.3	7.52	0.37667	0.3736	2.74762	2.70022
2011	80	0.04006	0.03927	0.5	68611	2695	67264	696337	10.15	3.1	7.05	0.41673	0.41287	3.2048	3.1416
2011	81	0.04394	0.04299	0.5	65917	2834	64500	629073	9.54	2.9	6.60	0.46067	0.45586	3.55914	3.48219
2011	82	0.05	0.04878	0.5	63083	3077	61544	564573	8.95	2.8	6.19	0.51067	0.50464	4.1	3.99996
2011	83	0.05537	0.05388	0.5	60005	3233	58389	503029	8.38	2.6	5.79	0.56604	0.55852	4.59571	4.47204
2011	84	0.06424	0.06224	0.5	56772	3534	55005	444640	7.83	2.4	5.42	0.63028	0.62076	5.39616	5.22816
2011	85	0.07262	0.07007	0.5	53239	3730	51373	389635	7.32	2.2	5.08	0.7029	0.69083	6.1727	5.95595
2011	86	0.0807	0.07757	0.5	49508	3840	47588	338262	6.83	2.1	4.68	0.7836	0.7684	6.9402	6.67102
2011	87	0.08911	0.08531	0.5	45668	3896	43720	290674	6.36	2.0	4.34	0.87271	0.85371	7.75257	7.42197
2011	88	0.09949	0.09478	0.5	41772	3959	39792	246954	5.91	1.9	4.01	0.9722	0.94849	8.75512	8.34064
2011	89	0.11349	0.10739	0.5	37813	4061	35782	207161	5.48	1.8	3.71	1.08569	1.05588	10.1006	9.55771
2011	90	0.1266	0.11906	0.5	33752	4019	31743	171379	5.08	1.7	3.43	1.21229	1.17494	11.394	10.7154
2011	91	0.14687	0.13683	0.5	29733	4068	27699	139636	4.7	1.5	3.17	1.35916	1.31177	13.3652	12.4515
2011	92	0.15874	0.14706	0.5	25665	3774	23778	111937	4.36	1.4	2.94	1.5179	1.45883	14.6041	13.5295
2011	93	0.18069	0.16572	0.5	21891	3628	20077	88159	4.03	1.3	2.72	1.69859	1.62455	16.8042	15.412
2011	94	0.20033	0.18209	0.5	18263	3326	16600	68082	3.73	1.2	2.53	1.89892	1.80664	18.831	17.1165
2011	95	0.22363	0.20114	0.5	14937	3004	13435	51482	3.45	1.1	2.36	2.12255	2.00778	21.2449	19.1083
2011	96	0.24876	0.22124	0.5	11933	2640	10613	38047	3.19	1.0	2.20	2.37131	2.22902	23.881	21.239
2011	97	0.27583	0.2424	0.5	9293	2253	8167	27434	2.95	0.9	2.07	2.64714	2.47142	26.7555	23.5128
2011	98	0.3048	0.26449	0.5	7040	1862	6109	19267	2.74	0.8	1.95	2.95194	2.73591	29.8704	25.92
2011	99	0.33554	0.28734	0.5	5178	1488	4434	13158	2.54	0.7	1.85	3.28748	3.02325	33.2185	28.4467
2011	100	0.36791	0.31075	0.5	3690	1147	3117	8724	2.36	0.6	1.75	3.65539	3.334	36.791	31.075
2011	101	0.40168	0.3345	0.5	2544	851	2118	5607	2.2	0.5	1.68	4.05707	3.6685	40.5697	33.7845
2011	102	0.43658	0.35835	0.5	1693	607	1389	3488	2.06	0.4	1.62	4.49365	4.02685	44.5312	36.5517
2011	103	0.47229	0.38206	0.5	1086	415	879	2099	1.93	0.4	1.57	4.96594	4.40891	48.6459	39.3522
2011	104	0.50844	0.40538	0.5	671	272	535	1220	1.82	0.3	1.53	5.47438	4.81429	52.8778	42.1595
2011	105	0.54465	0.42808	0.5	399	171	314	685	1.72	0.2	1.49	6.01903	5.24237	57.1883	44.9484
2011	106	0.58054	0.44994	0.5	228	103	177	372	1.63	0.2	1.46	6.59957	5.69231	61.5372	47.6936
2011	107	0.61572	0.47079	0.5	126	59	96	195	1.55	0.1	1.44	7.21529	6.1631	65.882	50.3745
2011	108	0.64985	0.49048	0.5	66	33	50	99	1.49	0.1	1.43	7.86514	6.65358	70.1838	52.9718
2011	109	0.6826	0.50891	0.5	34	17	25	49	1.43	0.0	1.40	8.54774	7.16249	74.4034	55.4712
2011	110+	0.71372	1	1.4	17	17	23	23	1.4	0.0	1.40				

Fig. 4.7 Full Life Table and HLYL and HLE estimates

on estimating the Health Expenditure in a Country by using the health state methodology. This application in Japan is in the framework of this project. Further applications in 20 countries are at present in the last estimation stages. Related programs and supporting tools are available upon request.

References

Chia, N. C., & Peng Loh, S. P. (2018). Using the stochastic health state function to forecast healthcare demand and healthcare financing: Evidence from Singapore. *Review of Development Economics, 22*, 1081–1104. https://doi.org/10.1111/rode.12528.

Gompertz, B. (1825). On the nature of the function expressing of the law of human mortality. *Philosophical Transactions of the Royal Society, 36*, 513–585.

Graunt, J. (1676). *Natural and political observations made upon the bills of mortality* (1st ed., 1662; 5th ed., 1676). London.

Halley, E. (1693). An estimate of the degrees of mortality of mankind, drawn from the curious tables of the births and funerals at the City of Breslau, with an attempt to ascertain the price of annuities upon lives. *Philosophical Transactions, 17*, 596–610.

Janssen, J., & Skiadas, C. H. (1995). Dynamic modelling of life-table data. *Applied Stochastic Models and Data Analysis, 11*(1), 35–49.

Makeham, W. M. (1860). On the law of mortality and construction of annuity tables. *The Assurance Magazine and Journal of the Institute of Actuaries, 8*, 301–310.

Skiadas, C. H., & Arezzo, M. F. (2018). Estimation of the healthy life expectancy in Italy through a simple model based on mortality rate. In *Demography and health issues: Population aging, mortality and data analysis*. Chum: Springer. https://doi.org/10.1007/978-3-319-76002-5_4.

Skiadas, C., & Skiadas, C. H. (2010). Development, simulation and application of first exit time densities to life table data. *Communications in Statistics Theory and Methods, 39*, 444–451.

Skiadas, C. H., & Skiadas, C. (2014). The first exit time theory applied to life table data: The health state function of a population and other characteristics. *Communications in Statistics-Theory and Methods, 43*, 1985–1600.

Skiadas, C. H., & Skiadas, C. (2015). Exploring the state of a stochastic system via stochastic simulations: An interesting inversion problem and the health state function. *Methodology and Computing in Applied Probability, 17*, 973–982.

Skiadas, C. H., & Skiadas, C. (2018a). *Exploring the health state of a population by dynamic modeling methods* (The Springer series on demographic methods and population analysis) (Vol. 45). Chum: Springer. https://doi.org/10.1007/978-3-319-65142-2.

Skiadas, C. H., & Skiadas, C. (2018b). The health-mortality approach in estimating the healthy life years lost compared to the global burden of disease studies and applications in world, USA and Japan. In *Exploring the health state of a population by dynamic modeling methods*. Chum: Springer. https://doi.org/10.1007/978-3-319-65142-2_4.

Skiadas, C. H., & Skiadas, C. (2018c). *Demography and health issues: Population aging, mortality and data analysis* (The Springer series on demographic methods and population analysis) (Vol. 46). Chum: Springer. https://doi.org/10.1007/978-3-319-76002-5.

Strehler, B. L., & Mildvan, A. S. (1960). General theory of mortality and aging. *Science, 132*, 14–21.

Sullivan, D. F. (1966). Conceptual problems in developing an index of health, U.S. Department of HEW, Public Health Service Publication No. 1000, series 2, No. 17.

Sullivan, D. F. (1971). (National Center for Health Statistics): A single index of mortality and morbidity. *HSMHA Health Reports, 86*, 347–354.

Torrance, G. W. (1976). Health status index models: A unified mathematical view. *Management Science, 22*(9), 990–1001.

Weibull, W. (1951). A statistical distribution function of wide applicability. *Journal of Applied Mechanics, 18*(3), 293–297.

WHO, "*WHO methods for life expectancy and healthy life expectancy*" Global Health Estimates Technical Paper WHO/HIS/HSI/GHE/2014.5. March 2014. http://www.who.int/healthinfo/statistics/LT_method.pdf

Chapter 5
Healthy Ageing in Czechia

Tomáš Fiala and Jitka Langhamrová

5.1 Introduction

Ageing is such an important facet of life that interest in it in terms of searching for its reasons, causes and ways to delay this inevitable part of human life is as old as mankind. Ageing has been researched by many physicians and philosophers. For instance, in Ancient Greece and Italy, Aristotle and Seneca researched ageing and considered old age to be an incurable illness. We know about attempts to preserve "eternal youth" from literature and art.

Demographic development in many countries is characterized by population ageing. The number of senior citizens (aged over 65) and very old senior citizens (over 85), as well as long-living people (over 90) keeps growing. Mankind as a whole is ageing faster than ever before. In the last century, the number of senior citizens grew faster than that of other age groups.

Demographic ageing is thus often considered to be one of the most important phenomena of the present century (e.g. Gavrilov and Heuveline (2003)), because it concerns all economically developed countries. It is a natural result of the demographic development in the last decades. The main reasons for population ageing are not only permanent increase in life span but also decrease of fertility rates, in many countries below the replacement level. In some countries or regions, the third not so often mentioned cause of ageing is the massive emigration of young adults, still in the reproductive age. These significant changes in demographic behavior fundamentally modify the age structure of the population in developed countries and cause ageing of the population as a whole. Europe is the world's

T. Fiala (✉) · J. Langhamrová
Department of Demography, Faculty of Informatics and Statistics, University of Economics, Prague, Praha 3, Czechia
e-mail: fiala@vse.cz; langhamj@vse.cz

C. H. Skiadas, C. Skiadas (eds.), *Demography of Population Health, Aging and Health Expenditures*, The Springer Series on Demographic Methods and Population Analysis 50, https://doi.org/10.1007/978-3-030-44695-6_5

oldest continent, and the European population is the most rapidly ageing. Many official documents of the European Commission deal with this issue.

To experts and economists, the financial security of, and care for, a growing number of people at a post-productive age is a major problem, because the number of people at a productive age is decreasing. Population ageing is very often regarded as a serious threat to the sustainability of national welfare systems (mainly of pension and health care systems).

It is a very sensitive issue that developed countries must, however, tackle in order not to jeopardize the quality of life of senior citizens who are bound to represent a very big electorate. It is clear that we can expect a higher need for services and care for the elderly, not only in healthcare and social services but also in housing, recreational services and the silver economy. Entrepreneurs focusing on well-targeted services and products for a numerous senior population can make a very good profit.

Qualitative and individual aspects of ageing are not discussed so often. For populations of economically developed countries, long life has become a reality. As a result, not only number but also quality of these years naturally deserves attention, asking whether we "add not only years to life but most importantly life to years". Adding life to years is sometimes referred to as successful ageing.

For many years, much attention has been paid to aggregate mortality and health indicators, because they can quantify health which is important for more than just health statistics. Indicators of this type also carry information on mortality and morbidity. The ageing of the population will, besides other things, require changes in the allocation of resources within the social and health systems. Health expectancy indicators, based on a wide range of health aspects and people's perceptions of their quality of life, can be important indicators of the potential demand for health and long-term care services, especially among the older generations of the population (Vrabcová et al. 2017).

Life expectancy indicators based on health make possible to compare health conditions between subpopulations of a single country and between populations of different countries, and also put emphasis not only on entire life length but also on its quality, expressed as the period of life in good health.

5.2 Selected Indicators Used to Evaluate Population Ageing

Literature focusing on this topic often provides several different definitions of old age. We will focus on the most frequently used definitions of ageing from a demographic and geriatric perspective. The term *calendar old age* (also called calendar, birth register, chronological age) is easily defined based on the year of birth; however, the problem with this definition is that calendar age disregards the fact that there are big differences between people of the same calendar age. A healthy, active person aged 70 can see himself, or can be seen by society, differently than a person of the same age who is ill or not self-sufficient. The arbitrary age limit marking the onset of old age keeps changing as life expectancy gets longer and the health of the older population improves. Nowadays, retirement age usually marks the onset of old age, i.e. the age of 65, based on the definition of economic generations. From this point of view, old age per se starts at the age of 75, which seems to be a crucial ontogenetic point.

In the eighteenth century, A. Haller, the father of modern physiology, broke human life down into nine phases and referred to the age group of 25–62 as manly and to the age of 63 and over as old. Back then, senior citizens represented less than 5% of the population. In the middle of the twentieth century, E. B. Hurlock referred to the age group of 40–59 as middle age and to the age of 60 and over as old age (Hurlock 1959).

This age limit was adopted in the 1960s by the World Health Organization and the UN. This categorization corresponds to the division of human life into stages of 15 years. In terms of ageing the ages are classified in the following way:

- the ages of 30–44 as adulthood (adultium),
- the ages of 45–59 as middle-aged (interevium),
- the ages of 60–74 as ageing, early old age (senescence),
- the ages of 75–89 as actual old age, senility (senium),
- the age of 90 and over as longevity (patriarchium).

In the context of demographic development and improvement of people's functional state in old age, Neugarten (1966) suggested the terms "young seniors" for the age group of 55–74 and "old seniors" for the age group of 75 and over.

This former concept is used in part to specify and classify old age as follows:

- the ages of 65–74: young seniors (it deals with retirement, leisure time, activities, self-realization),
- the ages of 75–84: old seniors (it deals with adaptation to a new situation, coping with physical burdens, specific illnesses, loneliness),
- the ages of 85 and over: very old seniors (it deals with self-sufficiency and care for senior citizens).

The World Health Organization (WHO) provides the following age categorization:

- the ages of 45–59 represent middle age,
- the ages of 60–74 represent early old age,
- the ages of 75–89 represent late old age,
- the age of 90 and over represent longevity.

The ageing of population can be measured e.g. by the share of of people aged 65 and over in the whole population

$$s_{65+} = \frac{S_{65+}}{S},$$

where S_{65+} means the number of people aged 65 and over and S means the total population size.

The shares are very often expressed in percentages (per 100 people), the share of people aged 100+ is usually expressed in per cent mile (per 100,000 people). Other possible indicators of population ageing are various indexes.

The longevity index is the share of people aged 90 and over among all senior citizens aged 60 and over:

$$ILong = \frac{S_{90+}}{S_{60+}},$$

and the old age index is another indicator to define population ageing:

$$IOA = \frac{S_{65+}}{S_{0-14}}.$$

5.3 Population Ageing in Czechia and Some European Countries

This chapter investigates the development of selected indicators of population ageing in Czechia and the European Union between 1950 and 2017 (Eurostat 2020a) and the expected values based on the latest Eurostat population projection until 2080 (Eurostat 2020b). The values for European countries reaching minimal and maximal values in 2017 and 2080, respectively, are also included in the graphs.

The trend of population ageing is apparent in European countries (including Czechia) throughout the whole period investigated and projected. The proportion of senior citizens is dependent not only on mortality rates but, to some extent, it is a result of unevenness in the age structure of the population caused by the uneven development of births and, as a result, the uneven age structure of the population.

The proportion of the elderly (aged 65 years and older) in Czechia grew relatively slowly between 1960 and 2005. While in 1960, the value of this proportion was about 10%, it did not reach 15% until 2005. The reason for such a low increase was the stagnation of mortality in the 1960s, 1970s and 1980s. The value of life expectancy at birth for males in 1990 was almost the same as in 1960 (about 68 years), and the value for females grew by about 2 years only in the period mentioned. Since 2005, the proportion of senior citizens has considerably gone up currently, because the populous generation born after WW II has reached old age and fertility rates were relatively low that time. In the present, the proportion of persons 65+ reaches almost 20% and is expected to culminate in the late fifties when about 30% of the Czech population should be aged 65 and over (Fig. 5.1). At that time, numerous generations born in the "baby-boom" 1970's, will gradually reach senior age.

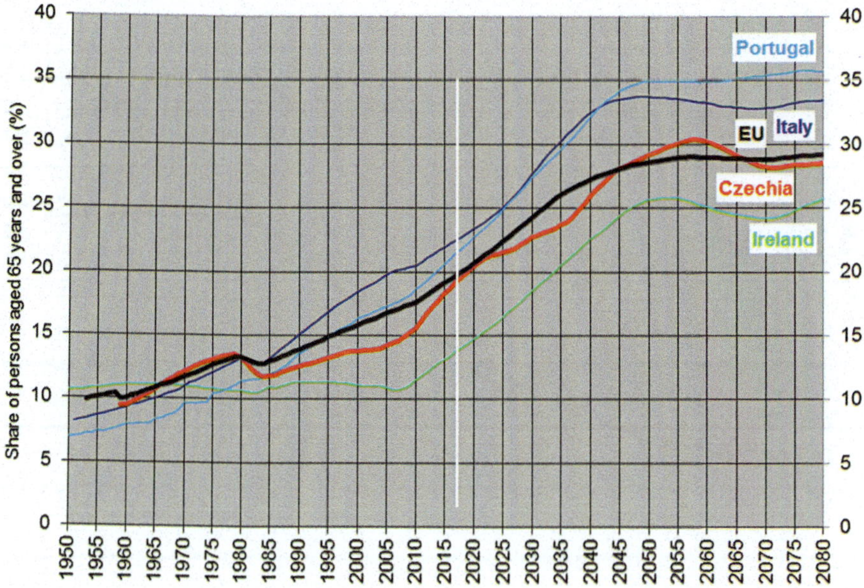

Fig. 5.1 Share of persons aged 65 years and over Source: Eurostat database, own computations and graph

The values for Czechia are not so very different from the share of seniors in the European Union. The lowest proportion of seniors in 2017 was in Ireland (less than 14%). In 2080, it is expected to be about 25% which should remain the lowest value in EU countries. The highest share of persons 65 years old and over in 2017 was in Italy (22.5%), and by 2080 the highest value should be reached in Portugal (almost 36%.)

The trends of development of the old age index (proportion of seniors 65+ to children 0–14) are quite similar. The value in Czechia grew from about 0.4 in 1960 to 1.2 in 2017, so it tripled during that period. It is expected to continue to grow to 2.0 until 2060, and it should drop a little bit in the following two decades (Fig. 5.2). The values and trends for the whole European Union are similar. The lowest value of this index in 2017 was in Ireland again (slightly over 0.6). In the last decades of the projection, this index is expected to reach 1.4. The highest number of seniors per child in 2017 was in Italy (almost 1.7), and by 2080 the highest value (close to 3) will be reached in Cyprus (Fig. 5.2).

5.4 Healthy Life Expectancy and Its Decomposition

The idea to measure "health expectancy" arose in a report published by the US Department of Health, Education, and Welfare in 1969. Not only a long life but also

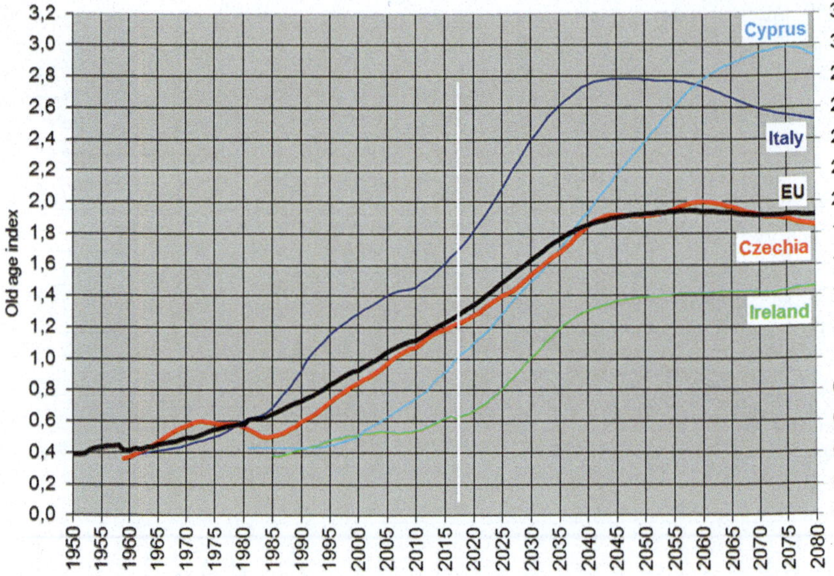

Fig. 5.2 Old age index (ratio of 65+ to 0–14) Source: Eurostat database, own computations and graph

good health are now considered to be one of the main goals of an ageing society. The increase in life expectancy brought worries and speculations about the health of the increasing number of older people. Therefore, it was necessary to find a measure not only of quantity but also of quality of life.

Health expectancies were first developed to discover whether longer life is accompanied by an increase in the time lived in good health (the compression of morbidity scenario) or in bad health (the expansion of morbidity). So, health expectancies divide life expectancy into a life spent in different states of health, from e.g., good to bad health. They add a qualitative dimension to the quantity of life lived. Health expectancy provides an effective tool for interaction between good health, bad health, and death. It can be also frequently used to monitor the health of the population with respect to increasing longevity (EHEMU 2007).

The so-called Health State Expectancies is an indicator showing the average number of years that an individual at a certain age is expected to live in a certain state of health (Czech Demographic Society 2010). It is based on the idea that life expectancy (LE) can be divided into individual time periods based on certain states of health. The sum of lengths of these periods equals life expectancy.

There are several characteristics to classify the state of health (Vrabcová et al. 2017). We can classify health status, for example, according to perceived health, the presence or absence of chronic morbidity, or according to disability (presence or absence of some limitation in various life activities). Because of this fact, there are also several possibilities for a healthy lifespan.

A known indicator of a healthy life is the healthy life expectancy (HLE). This indicator is constructed on the basis of respondents' answers regarding self-evaluation of their own health. It can be evaluated in several ways that subjectively reflect the actual health of the individual. One of the most commonly used indicators of an individual's health assessment is the General Activity Limitation Indicator (GALI), while the length of life indicator based on this health feature is called Healthy Life Years (HLY). After the inclusion of HLY among the set of structural indicators of the European Union, GALI was introduced into the regular Survey on Income and Living Conditions (SILC). This survey is being conducted by individual Member States in compliance with the EU's uniform instructions and methodology since 2005 (Vrabcová et al. 2017).

The percentage of population with bad health can be measured e.g. by the prevalence of disability, that is, the percentage of people with limited abilities in their everyday activities due to health reasons. Because our population is ageing and older people are more likely to have a disability, the overall occurrence of disability in the total population can go up even in the case that individuals in separate age groups are at lower risk of disability than before. Healthy life years take into account both changes in the length of life lived with a disability and the drop in the mortality rate which results from higher life expectancy. This is why an improvement in the health of the ageing population at any age increases life expectancy without disability, even though the overall disability prevalence is higher due to an ever-increasing number of people at risk. The value of healthy life years is independent of the age structure of the population, and thus this indicator provides an effective tool for identifying the interaction between health, disease, and mortality (EHEMU 2007). This indicator can be used for comparing differences in healthy life in various countries or for comparing changes in time. The Sullivan Method is largely used to calculate healthy life years. Healthy Life Years (Disability Free Life Expectancy) at the exact age x can be computed by the formula

$$DFLE_x = \frac{\sum_{u=x}^{\omega-1} L_u \cdot \pi_u}{l_x}$$

where L_u is the number of years lived by a population at the completed age u (between exact ages u and $u+1$),
π_u is the prevalence of good health condition (no activity limitation) at the age of u,
l_x is the number of survivors up to exact age x,
ω is the age when no one in the population survives.

For $x = 0$ we have the most often used value of Disability Free Life Expectancy at birth.

The calculations used here are described in detail in a guide prepared by Jagger et al. (2014). This method can be used universally regardless of the method of health definition.

The advantages of the Sullivan method are the relatively good availability of data (coming mostly from population surveys on health status) and its low computational complexity. The disadvantage of this method is the use of prevalence rates coming from transversal data sources, which reflect past conditions influencing the health of the population, rather than current health risks.

Aggregate measures can be computed from demographic tables. Each table aggregates a vector or a matrix of elementary rates of demographic events into one number. When analyzing the changes of its value in time or its variations across countries, it is useful to be able to decompose observed changes or differences, usually in terms of age or other demographic dimensions such as birth order, cause of death, or population group. Decomposition aims at estimating contributions of differences between elementary rates of demographic events to the overall difference between the two values of the aggregate measure.

Healthy life expectancy is an indicator which aggregates the values of a vector of mortality rates and a vector of prevalence rates. Correspondingly, decomposition of the difference between two health expectancies should include additional splitting of each age component into effects of mortality and effects of health (Andreev et al. 2002).

The effect of mortality at an age x is computed by the formula

$$\lambda_x = \frac{1}{4 \cdot l_0}\left(l_x^1 + l_x^2\right)\left(P_x^2 - P_x^1\right)\left(\pi_x^1 + \pi_x^2\right) + \frac{1}{2 \cdot l_0}\left(h_{x+1}^1 l_x^2 + h_{x+1}^2 l_x^1\right)\left(q_x^1 - q_x^2\right),$$

and the formula for the effect of health is

$$\gamma_x = \frac{1}{4 \cdot l_0}\left(l_x^1 + l_x^2\right)\left(P_x^1 + P_x^2\right)\left(\pi_x^2 - \pi_x^1\right),$$

where

$$P_x^i = \frac{L_x^i}{l_x^i},$$

h_x denotes the value of healthy life years at the exact age of x,
q_x is the probability of death at the exact age of x,
the upper indexes 1, 2 respectively denoting the two populations compared.

5.5 Life Expectancy and Disability Free Life Expectancy in Czechia in 2008–2016

The political, economic and social changes in 1989 in former Czechoslovakia brought about, besides other things, gradual improvement of the health status of the population and a relatively rapid drop of mortality in Czechia. While in previous decades (in the 60s, 70s and early 80s) there were almost no changes in life expectancy for males and only a very slow increase for females, since the late

80s, and especially in the 90s, there begun a period of a relatively rapid increase in life expectancy for both males and females. This positive trend continues until the present. While in 1989, the value of life expectancy was only about 68 years for males and 75 years for females, by 2017 this value gradually grew to 76 years for males and almost 82 years for females. This trend is, of course, a confirmation of the overall health state improving in Czechia. More precise information can be obtained based on the development of healthy life expectancy.

Data on health expectancies in European countries (including Czechia) can be found in e.g. the EurOhex website (www.eurohex.eu). It includes a database on health indicators comprising life expectancies and Healthy Life Years (HLY) for 28 European countries since 2005. But only data since 2008 are considered comparable. As a result, in this paper we shall only look at data in the period 2008–2016. To eliminate random fluctuations, the computations were performed for three-year intervals when available data were available: 2008–2010, 2011–2013 and 2014–2016. The value of life expectancy at birth is based on life tables for Czechia published by the Czech Statistical Office (ČSÚ 2020), while the gender- and age prevalence rates of disability came from the EHLESIS database. No activity limitation has been used as the definition of healthy life, and data came from the SILC survey (EHLESIS 2018). In such case, the value of Healthy Life Years is equal to Disability Free Life Expectancy.

The values of life expectancy as well as the values of Disability Free Life Expectancy show a tendency for increase in the period investigated. The growth of life expectancies is higher than for healthy life years. While the life expectancy in 2014–16 was higher than in 2008–10 by 1.59 years for males and by 1.26 years for females, the corresponding differences in disability free life expectancies reached only 1.28 years for males and 0.69 years for females (see Fig. 5.3).

In the last period (2014–16), the prevalence of disability is higher for males in the age groups 30–34 and 60–64 years and with negligible differences also in the age groups 20–24 and 65–69. In all other age groups, the prevalence of disability is higher for females (Fig. 5.4).

The decomposition shows that mortality components are positive in all age groups. The highest contribution to this total difference is reached in the age groups 55–69 years and 30–34 years to the high positive mortality component and positive or near to zero health component. On the other hand, the contributions in the age groups 15–19, 35–39, 50–54, 80–84 and 85+ are negative, because the absolute value of (negative) health component prevails the (positive) value of the mortality component (Fig. 5.5).

5.6 Conclusions

The population ageing in Czechia has accelerated in this millennium. While in the period 1960–2005 the proportion of persons aged 65 years and over has risen from 10% to 14% only, it is expected that by 2060 this proportion should be about 30%.

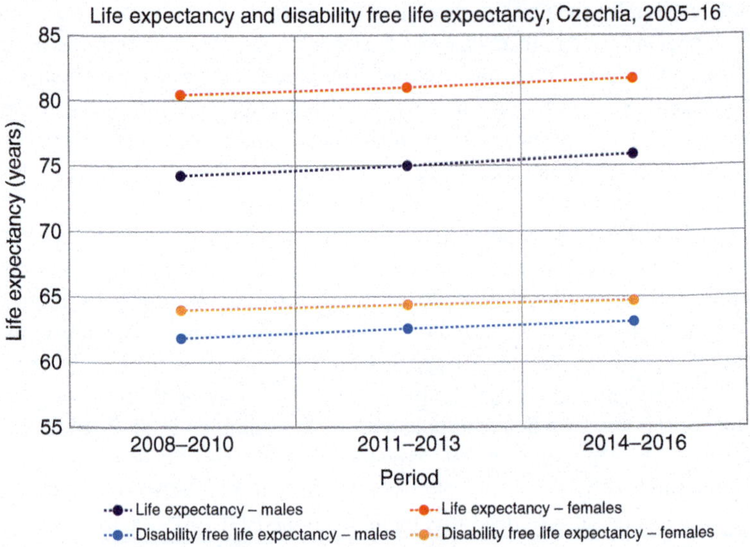

Fig. 5.3 Life expectancy and disability free life expectancy at births, Czechia, 2008–2016 Source: Eurostat and EurOhex database, own computations and graph

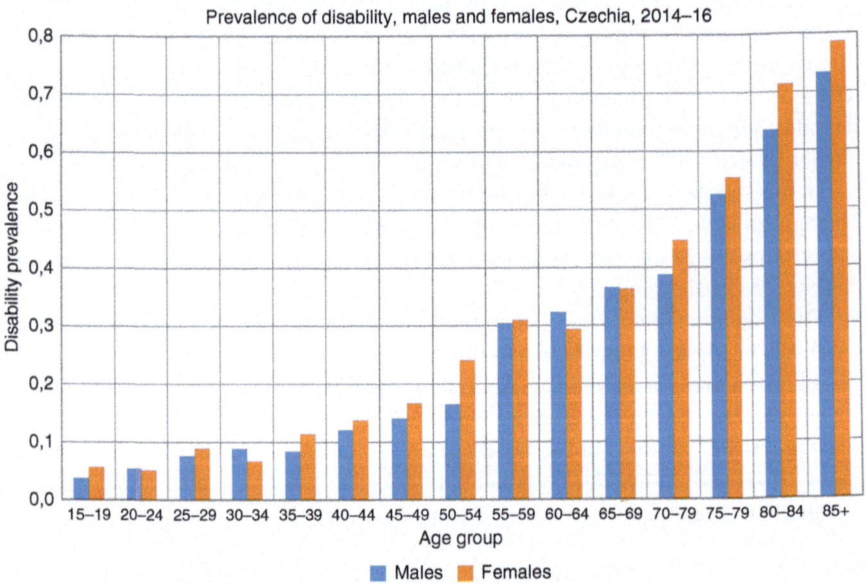

Fig. 5.4 Prevalence of disability, males and females, Czechia, 2014–2016 Source: EurOhex database, own computations and graph

The value of the Old age index (proportion of persons at the age of 65+ to 0–14) grew from 0.4 to 1.0 between 1960 and 2005, while in the second half of this century

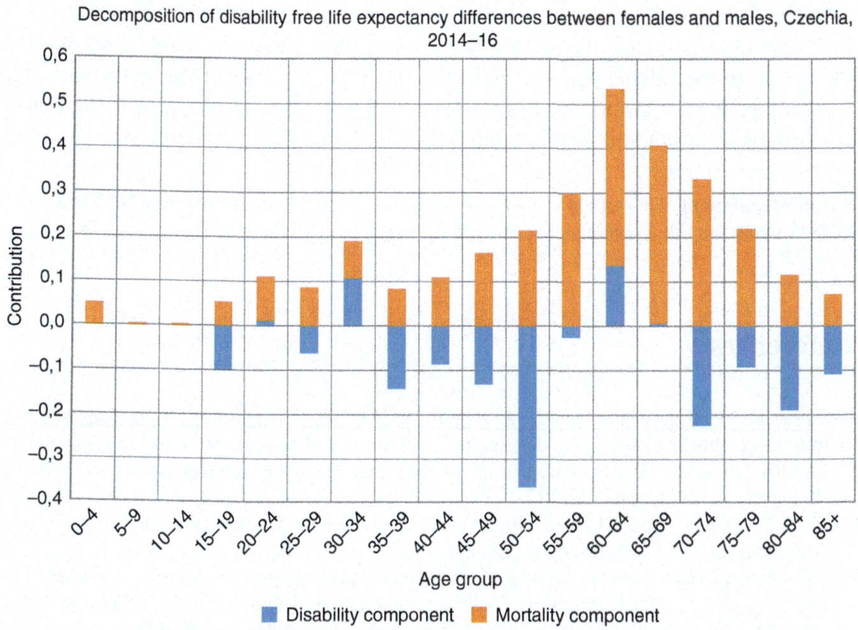

Fig. 5.5 Decomposition of differences in Disability Free Life Expectancy between females and males. Czechia, 2014–2016 Source: own computations and graph

its value should be about 2. The trends and values of these indicators do not differ too much from those of the European Union.

In Czechia between 2008 and 2016, the values of both life expectancy and the Disability Free Life Expectancy have had a growing tendency for males as well as females. Nevertheless, the growth of the Disability Free Life Expectancy was slower and, therefore, the value of percentage of life spent without disability declined somewhat during the period investigated.

While the mortality rates showed a decreasing tendency in time across all age groups, and their values for females were each time lower than for males, trends in prevalence rates of disability were not so unambiguous by far. In most age groups, the prevalence rates for females were higher than for males between 2014 and 2016.

This fact is apparent in the results of the decomposition of the total differences by age groups. While all mortality components have positive values (confirming the fact of lower mortality rates of females with respect to males), the value of relatively many health components is negative which bears witness to the fact that in many age groups the prevalence rates of disability for females is higher than for males. Although the Disability Free Life Expectancy at birth for females was higher than for males, the main reason for that was the lower mortality rates and resulting higher life expectancy of females. In some age groups, the absolute value of the (negative) health component was higher than the value of the (positive) mortality component, and so the total contribution in these age groups was negative. This is in accordance with the fact that the difference between females and males in the Disability Free Life Expectancy at birth was across all periods investigated a little bit lower than the difference in the Life Expectancy at birth. The highest values of the contributions to the total difference in the Disability Free Life Expectancy can be seen in the higher age groups of around 60 years of age.

Acknowledgement This article was supported by the Czech Science Foundation No. GA ČR 19-03984S under the title *Economy of Successful Ageing*.

References

Andreev, E. M., Shkolnikov, V. M., & Begun, A. Z. (2002). Algorithm for decomposition of differences between aggregate demographic measures and its application to life expectancies, healthy life expectancies, parity-progression ratios and total fertility rates. *Demographic Research, 7*, Art. 14, pp. 499–522.

ČSÚ. (2020). *Český statistický úřad*. Úmrtnostní tabulky za ČR od roku 1920. Available at https://www.czso.cz/csu/czso/umrtnostni_tabulky

Czech Demographic Society. (2010). *English and Czech terms for summary indicators of population health: Basic overview*. Česká demografická společnost. [online]. Available at http://kdem.vse.cz/wp-content/uploads/2010/11/ukazatele_zdravotniho_sta¬vu.pdf

EHEMU. (2007). *Final report on European population health*. Available at http://www.eurohex.eu/pdf/Report%20 EHEMU%202007.pdf

EHLESIS. (2018). *European Health and Life Expectancy Information System. Health Data*. SILC – Statistics on Income and Living Conditions. Activity Limitation. Available at http://www.eurohex.eu/IS/web/app.php/Ehleis/HealthGeographic/SILC/SILCAL

Eurostat (2020a). *Statistics database*. Database by themes. Population and social conditions. Population (demo_pop). Population on 1 January by age and sex (demo_pjan). http://appsso.eurostat.ec.europa.eu/nui/show.do?dataset=demo_pjan&lang=en ieurostat

Eurostat. (2020b). *Statistics database*. Database by themes. Population and social conditions. Population projections (proj). Population projections at national level (2019–2100) (proj_19n). Population on 1st January by age, sex and type of projection (proj_19np). https://appsso.eurostat.ec.europa.eu/nui/show.do?dataset=proj_19np&lang=en

Gavrilov, L. A., & Heuveline, P. (2003). Ageing of population. In P. Demeny & G. McNicoll (Eds.), *The encyclopedia of population*. New York: Macmillan Reference USA.

Hurlock, E. B. (1959). *Developmental psychology* (2nd Ed.). New York: McGraW-Hill, 645 p.

Jagger, C., Van Oyen, H., & Robine, J. M. (2014). Health expectancy calculation by the Sullivan method: A practical guide (4th Ed.). [online]. Available at: http://www.eurohex.eu/ehleis/pdf/Sullivan_guide_pre%20final_oct%202014.pdf

Neugarten, B. L. (1966). Adult personality. *Human Development, 9*, 61–73.

Vrabcová, J., Daňková, Š., & Faltysová, K. (2017). Healthy life years in Czechia: Different data sources, different figures. *Demografie* [online], *59*(4), pp. 315–331. eISSN 1805-2991. ISSN 0011-8265. Available at https://www.czso.cz/documents/10180/46203814/vrabcova.pdf/00096807-8007-4e3f-9461-46c1623ce331?version=1.1

Chapter 6
Evolution of Systems with Power-Law Memory: Do We Have to Die? (Dedicated to the Memory of Valentin Afraimovich)

Mark Edelman

6.1 Introduction

This paper is dedicated to the memory of Valentin Afraimovich, an outstanding mathematician and a wonderful man. It was a pleasure to be acquainted with Valentin for more than 20 years, to have conversations about life and science. The spectrum of his scientific interests was very wide. Although he didn't work in the field of fractional dynamics, he attended many talks. He told me once that he would work in discrete fractional calculus if fractional systems were considered as infinite-dimensional systems. Unfortunately, this approach to fractional dynamics was not implemented during Valentin's life.

I developed an interest in fractional calculus while working with George Zaslavsky at NYU on transport in Hamiltonian systems and billiards. Work on fractional maps, introduced in Tarasov and Zaslavsky (2008) to investigate general properties of fractional (with power-law memory) systems, was proposed by George and discrete fractional dynamics has been the main topic of my research during the last 10 years.

Fractional calculus has many applications, which include (the number of publications on applications of fractional calculus is overwhelming and the references here are just examples) transport (anomalous diffusion) in Hamiltonian systems and billiards (Zaslavsky 2005), systems with long-range interactions (Laskin and Zaslavsky 2006; Zaslavsky et al. 2007), electromagnetic fields in dielectric media

M. Edelman (✉)
Department of Physics, Stern College at Yeshiva University, New York, NY, USA

Courant Institute of Mathematical Sciences, New York University, New York, NY, USA

Department of Mathematics, BCC, CUNY, Bronx, NY, USA
e-mail: edelman@cims.nyu.edu

C. H. Skiadas, C. Skiadas (eds.), *Demography of Population Health, Aging and Health Expenditures*, The Springer Series on Demographic Methods and Population Analysis 50, https://doi.org/10.1007/978-3-030-44695-6_6

(Tarasov 2008, 2009, 2011), viscoelastic materials and rheology (Bagley and Torvik 1983; Caputo and Mainardi 1971; Mainardi 2010), fractional control (Caponetto et al. 2010), and many others.

6.2 Fractional Dynamical Systems

Fractional differences/derivatives are convolutions with power-law functions. Therefore, fractional space derivatives appear in equations describing distributed in space systems, but the real fractional dynamics is the dynamics of systems with power-law memory. In such systems, called fractional dynamical systems, new values of the system's variables depend on the whole history of the system's evolution. In what follows, we will use the fractional Caputo derivative defined as (see e.g. Kilbas et al. 2006)

$$ {}_0^C D_t^{\alpha} x(t) = {}_0 I_t^{n-\alpha}\, D_t^n x(t) = \frac{1}{\Gamma(n-\alpha)} \int_0^t \frac{D_\tau^n x(\tau) d\tau}{(t-\tau)^{\alpha-n+1}} \quad (n-1<\alpha\le n), \tag{6.1} $$

where $D_t^n x(t)$ is a regular derivative of the nth order. We will also use examples based on the Caputo fractional/fractional difference standard and logistic maps. They are particular forms of the fractional/fractional difference universal map (see Edelman 2014b, 2015a, 2018a,c). The m-dimensional Caputo fractional universal map can be written as

$$ x_{n+1}^{(s)} = \sum_{k=0}^{m-s-1} \frac{x_0^{(k+s)}}{k!} h^k (n+1)^k - \frac{h^{\alpha-s}}{\Gamma(\alpha-s)} \sum_{k=0}^{n} G_K(x_k)(n-k+1)^{\alpha-s-1}, \tag{6.2} $$

where $\alpha \in \mathbb{R}$, $\alpha \ge 0$, $m = \lceil \alpha \rceil$, $n \in \mathbb{Z}$, $n \ge 0$, $s = 0, 1, \ldots, m-1$. In this paper, we assume that $x^{(1)}$ is a momentum p and in all maps (including fractional difference maps) consider the case $h = 1$.

The m-dimensional Caputo fractional difference universal map can be written as

$$ x_{n+1} = \sum_{k=0}^{m-1} \frac{\Delta^k x(0)}{k!} (n+1)^{(k)} - \frac{1}{\Gamma(\alpha)} \sum_{s=0}^{n+1-m} (n-s-m+\alpha)^{(\alpha-1)} G_K(x_{s+m-1}), \tag{6.3} $$

where $\alpha \in \mathbb{R}$, $\alpha \ge 0$, $m = \lceil \alpha \rceil$, $n \in \mathbb{Z}$, $n \ge 0$, $s = 0, 1, \ldots, m-1$,

$$ \Delta^k x(0) = c_k, \quad k = 0, 1, \ldots, m-1, \quad m = \lceil \alpha \rceil \tag{6.4} $$

are the initial conditions, and the falling factorial $t^{(\alpha)}$ is defined as

$$t^{(\alpha)} = \frac{\Gamma(t+1)}{\Gamma(t+1-\alpha)}, \quad t \neq -1, -2, -3 \ldots . \tag{6.5}$$

In the case $\alpha = 1$ maps Eqs. (6.2) and (6.3) converge to the regular logistic map

$$x_{n+1} = K x_n (1 - x_n) \tag{6.6}$$

if (see Edelman 2013a,b, 2014a, 2018a,b,c for fractional maps and Edelman 2015a, 2018a,b,c for fractional difference maps)

$$G_K(x) = x - Kx(1-x). \tag{6.7}$$

These maps are called Caputo fractional (in the case of Eq. (6.2)) or fractional difference (in the case of Eq. (6.3)) logistic α-families of maps (or simply logistic Caputo fractional/fractional difference maps).

In the case $\alpha = 2$ maps Eqs. (6.2) and (6.3) converge to the regular standard map (see Chirikov 1979)

$$p_{n+1} = p_n - K \sin(x_n) \pmod{2\pi}, \tag{6.8}$$

$$x_{n+1} = x_n + p_{n+1} \pmod{2\pi} \tag{6.9}$$

if (see Edelman 2013a,b, 2014a, 2018a,b,c; Tarasov 2009a,b, 2011 for fractional maps and Edelman 2014b, 2015a, 2018a,b,c for fractional difference maps)

$$G_K(x) = K \sin(x). \tag{6.10}$$

These maps are called Caputo fractional (in the case of Eq. (6.2)) or fractional difference (in the case of Eq. (6.3)) standard α-families of maps (or simply standard Caputo fractional/fractional difference maps).

We'll use the fractional Riemann-Liouville standard map for $1 < \alpha \le 2$ in the form (see Edelman 2018a,b,c; Edelman and Tarasov 2009; Tarasov 2011; Tarasov and Zaslavsky 2008)

$$p_{n+1} = p_n - K \sin x_n, \tag{6.11}$$

$$x_{n+1} = \frac{1}{\Gamma(\alpha)} \sum_{i=0}^{n} p_{i+1} V_\alpha(n-i+1), \quad \pmod{2\pi},$$

where

$$V_\alpha(m) = m^{\alpha-1} - (m-1)^{\alpha-1}. \tag{6.12}$$

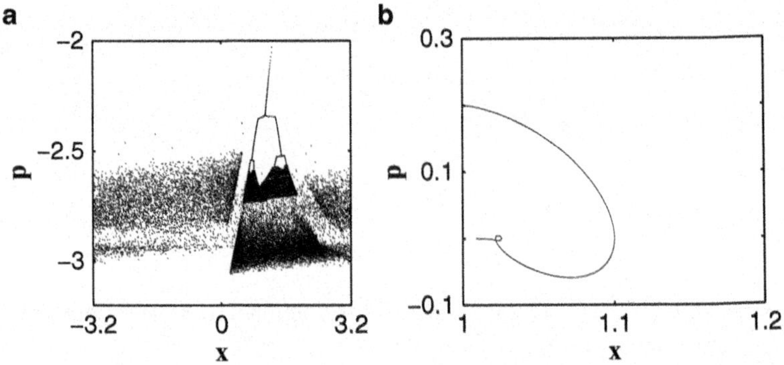

Fig. 6.1 (**a**) 20,000 iterations on each of two overlapping attractors of the Caputo fractional standard map's phase space with $K = 4.5$, $\alpha = 1.02$, and $x_0 = 0$; the cascade of bifurcations type trajectory has initial momentum $p_0 = -1.8855$ and the chaotic attractor $p_0 = -2.5135$. (**b**) A self-intersecting phase space trajectory of the fractional Caputo Duffing equation ${}_0^C D_t^{1.5} x(t) = x(1 - x^2)$, $t \in [0, 40]$ with the initial conditions $x(0) = 1$ and $dx/dt(0) = 0.199$. Fractional Caputo derivative is defined by Eq. (6.1)

The following are some differences between the properties of fractional and regular dynamical systems:

- Chaotic attractors in fractional dynamical systems may overlap and trajectories may intersect in continuous fractional systems of low orders Fig. 6.1. This leads to the inapplicability of the Poincaré-Bendixson theorem to fractional systems. Therefore, in continuous fractional systems of low orders non-existence of chaos is only a conjecture (see Deshpande and Daftardar-Gejji 2017; Edelman 2015b).
- In fractional systems, periodic sinks, except fixed points, may exist only in an asymptotic sense. Asymptotically attracting points may not belong to their own basins of attraction (see Edelman 2011, 2013a,b). A trajectory starting from an asymptotically attracting point may end attracted to a different asymptotically attracting point. The rate at which a trajectory approaches an attracting point depends on initial conditions. Trajectories originating from a basin of attraction may converge faster (as $x_n \approx n^{-1-\alpha}$ for the fractional Riemann-Liouville standard map, see Fig. 6.1 from Edelman 2013a) than trajectories originating from the chaotic sea (as $x_n \approx n^{-\alpha}$). Fractional differential/difference equations do not have periodic solutions except fixed points; they may have asymptotically periodic solutions instead (see, e.g., Area et al. 2014; Jagan Mohan 2016, 2017; Kaslik and Sivasundaram 2012; Tavazoei and Haeri 2009; Wang et al. 2013; Yazdani and Salarieh 2011).
- A new type of attractors, nonexistent in regular systems, cascade of bifurcations type trajectories (CBTT), Figs. 6.2 and 6.3, is a general feature of fractional systems. A CBTT may start converging to a period 2^n sink, but then bifurcate and start converging to a period 2^{n+1} sink, and so on. CBTT may end their evolution

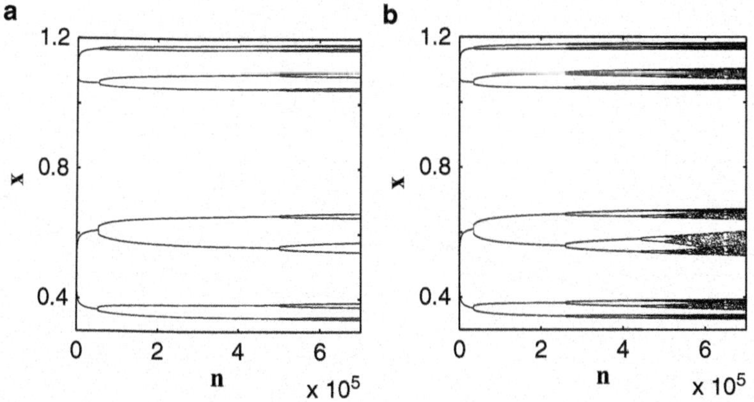

Fig. 6.2 Cascade of bifurcations type trajectories (CBTT) in the Caputo logistic fractional maps, Eq. (6.2) with $h = 1$ and $G_K(x) = x - Kx(1 - x)$. In both cases $\alpha = 0.1$ and $x_0 = 0.001$: (**a**) in the case $K = 22.37$, the last bifurcation from the period $T = 8$ to the period $T = 16$ occurs after approximately 5×10^5 iterations; (**b**) in the case $K = 22.423$, after approximately 5×10^5 iterations the trajectory becomes chaotic

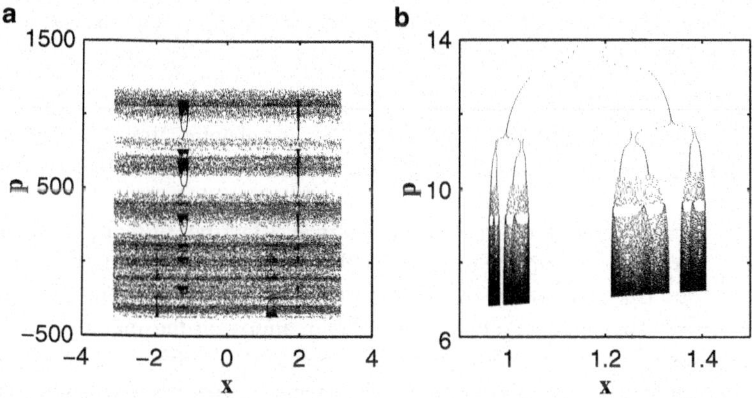

Fig. 6.3 CBTT in the Riemann-Liouville fractional standard map with $1 < \alpha < 2$, Eq. (6.11): (**a**) 120,000 iterations on a single trajectory with $K = 4.5$, $\alpha = 1.65$, and $p_0 = 0.3$. The trajectory occasionally sticks to one of CBTTs (intermittent CBTT) but then always recovers into the chaotic sea. (**b**) 100,000 iterations on a trajectory with $K = 3.5$, $\alpha = 1.1$, and $p_0 = 20$. The trajectory turns very fast into a CBTT which slowly converges to a fractal type set

as a period 2^{n+m} sink or in chaos (Edelman 2011, 2018d; Edelman and Taieb 2013).

- Integer members of fractional α-families of maps are volume-preserving (Edelman 2013b, 2014b). Fractional members are not volume-preserving – they behave like a member of the same family of the higher integer order with dissipation (Zaslavsky et al. 2006). Types of attractors which may exist in frac-

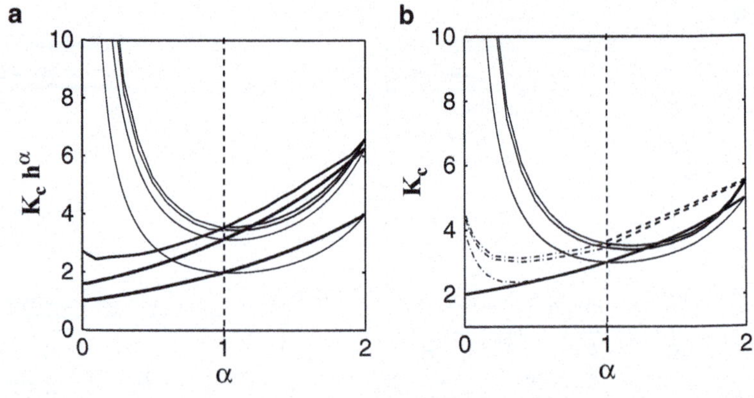

Fig. 6.4 2D bifurcation diagrams for the fractional (solid thing lines) and fractional difference (bold and dashed lines) Caputo standard (**a**) and $h = 1$ logistic (**b**) maps. The first bifurcation, transition from a stable fixed point to a stable period two ($T = 2$) sink, occurs on the bottom curves. A $T = 2$ sink (in the case of the standard α-families of maps an antisymmetric T=2 sink with $x_{n+1} = -x_n$) is stable between the bottom and the middle curves. Transition to chaos occurs on the top curves. For the standard fractional map transition from a $T = 2$ to $T = 4$ sink occurs on the line below the top line (the third from the bottom line). Period doubling bifurcations leading to chaos occur in the narrow band between the two top curves. (This figure is reprinted from Edelman (2018b), with the permission of AIP Publishing)

tional systems include sinks, limiting cycles, and chaotic attractors (Cermak and Nechvatal 2017; Edelman 2013a, 2014a; Edelman and Taieb 2013; Stanislavsky 2006).

- Dependence of fractional/fractional difference α-families of maps on the nonlinearity parameter K and the memory parameter (order of a fractional system) α can be described by two-dimensional bifurcation diagrams Fig. 6.4. This, in turn, leads to the dependence of the map's bifurcations on the memory parameter. Examples of x-α bifurcation diagrams are presented in Fig. 6.5. This allows an additional way of controlling fractional systems by manipulating their orders α.

6.3 Power-Law Memory (Fractionality) in Biological Systems

The following review of studies of memory in biological systems emphasizes the role of nonlinear fractional (with power-law memory) dynamics described by the fractional differential/difference equations of the orders $0 < \alpha < 2$, and especially, α close to zero.

Memory, as a significant property of humans, is a subject of extensive biophysical and psychological research. The power-law forgetting, the decay of the accuracy on memory tasks as $\sim t^{-\beta}$, with $0 < \beta < 1$, has been demonstrated in experiments described in Kahana (2012), Rubin and Wenzel (1996), Wixted (1990), Wixted and

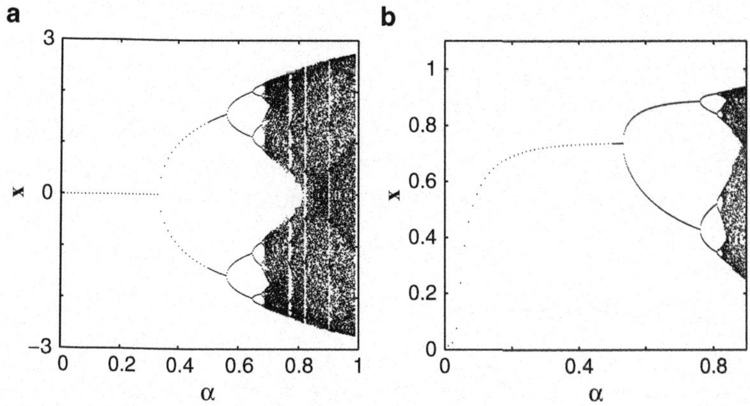

Fig. 6.5 The memory α-x bifurcation diagrams for fractional Caputo standard (**a**) and logistic (**b**) maps obtained after 5000 iterations. $K = 4.2$ in (**a**) and $K = 3.8$ in (**b**)

Ebbesen (1991, 1997). Let us note here that fractional maps of the orders $0 < \alpha < 1$ are maps with power-law, $-\beta = \alpha - 1$, decaying memory, where $0 < \beta < 1$ (Edelman 2013a). Human learning is also characterized by power-law memory: the reduction in reaction times that comes with practice is a power function of the number of training trials (Anderson 1995). Dynamics of biological systems at levels ranging from single ion channels up to human psychophysics in Fairhall et al. (2001), Leopold et al. (2003), Toib et al. (1998), Ulanovsky et al. (2004), Wixted and Ebbesen (1997), Zilany et al. (2009) is described by the application of power-law adaptation.

The underlying reasons for human's power-law memory can be related to the power-law memory of its building blocks, from individual neurons and proteins to tissues of individual organs. Processing of the external stimuli by individual neurons, as has been shown in Lundstrom et al. (2010, 2008), Pozzorini et al. (2013), can be described by fractional differentiation with the orders of derivatives $\alpha \in [0, 1]$. For example, in the case of neocortical pyramidal neurons, this order is $\alpha \approx 0.15$. The power-law memory kernel with the exponent -0.51 ± 0.07 is demonstrated in fluctuations within single protein molecules (see Min et al. 2005).

As mentioned at the end of Sect. 6.1, viscoelasticity is one of the most important applications of fractional calculus. Viscoelastic materials act as substances with power-law memory and their behavior can be described by the fractional differential equations. It was demonstrated in many publications that human tissues are viscoelastic, which is related to the viscoelasticity of the components of tissues: structural proteins, cells, extracellular matrices, and so on. Earlier references (prior to 2014) related to various organs may be found in Edelman (2014a). For more recent results on viscoelasticity of the brain tissue see Park et al. (2018), cardiovascular tissues (Wang et al. 2016), human tracheal tissues (Safshekan et al.

2016), human skin (Jeong et al. 2018), human bladder tumours (Barnes et al. 2016), human vocal fold tissues (Chan 2018), and many other publications.

Additional areas of application of fractional calculus in biology include fractional wave-propagation in biological tissues (Duck 1990; Holm and Sinkus 2010; Nasholm and Holm 2013; Szabo and Wu 2000), bioengineering (bioelecrodes, biomechanics, and bioimaging) Magin (2010), population biology and epidemiology (Brauer and Castillo-Chavez 2001; Hoppensteadt 1975).

6.4 Inevitability of Death

Humans live and die. There is no way to stop aging. According to the Guinness World Records and the Gerontology Research Group, the longest living person whose dates of birth and death were verified (although not without controversies), Jeanne Calment (1875–1997), lived to 122. Recent deaths of people with whom I had close relationships, George Zaslavsky, Valentin Afraimovich, and related to them Vladimir Arnold, stimulated me to write this note. All of them lived for approximately 73 years and died when they had active scientific plans and close family relationships – tragic, untimely, unjustifiable deaths. And this is the reality: however smart, intelligent, and athletic we are, however healthy our diet is, we all die. There are some religious, philosophical, and scientific thoughts about this inevitability of death.

The religious explanation depends on the religion. In Christianity, Adam and Eve, the first humans, lost their lives, because they sinned against God (Genesis 3:17–19). The Bible says: "Through one man sin entered into the world, and death through sin, and thus death spread to all men, because all sinned." (Romans 5:12). In Buddhism and Hinduism, death is considered to be a natural part of the cycle of life followed by rebirth. In Judaism death is not a tragedy – it is a natural process. Deaths, like lives, have meaning and are all part of G-d's plan. Islam believes in the soul's continuous existence with transformed physical existence and a Day of Judgment deciding human beings' eternal destination to Paradise or Hell.

Life and death are topics of many philosophical discussions. Evolutionary aspects can be found, for example, in the views of Heraclitus, Nietzsche, and Hegel. Life and death are two opposites which are inseparable, and Heraclitus apparently believed in the cyclic recurrence of all things, including our lives. In Hegel's philosophy, the recurrence is not cyclic but spiral.

6.4.1 Evolution and Lifespan

In 1889 August Weismann, in his interpretation and formulation of the mechanism of Darwinian evolution, theorized that aging is a part of life's program because the old need to be removed to make room for the next generation to sustain the evolution (Weismann 1889).

To list the main modern approaches in evolutionary biology, we'll follow (very briefly) Joshua Mitteldorf's review (Mitteldorf 2010). According to Mitteldorf, the four main theoretical explanations of aging are:

- *Aging as the accumulation of damage*. This approach is supported by the similarities between mortality curves and graphs of the failure rates in the non-living world (see, e.g., examples related to the Gompertz law on page 56 of the Gavrilovs' book Gavrilov and Gavrilova 1991). Corresponding mathematical models can be found in Chap. 6 of the book (see also Gavrilov and Gavrilova 2004). This approach lacks the natural selection argumentation and the only way to slow aging here is to prevent or repair the accumulated random damage. Arguments against this approach include the difference in the interaction with the environment between living and non-living matter (possibility of the energy accumulation from the environment and dumping an excess of entropy into the surroundings), difference in the repair mechanisms (internal somatic repairs are quite different from the external repair of machines), and here we would add the presence of memory as a significant feature of living matter.
- *Aging as irrelevant to evolution*. The original theoretical idea of Medawar (1952), supported by Edney and Gill (1968), is based on the fact that in the case of the low selection pressure mutational load alone could explain the evolutionary emergence of aging. The corresponding theory is called Mutation Accumulation (MA). Demographic surveys (Ricklefs 1998) and the existence of a conserved genetic basis for aging (Guarente and Kenyon 2000; Kenyon 2001) contradict the MA theory. An additional argument against the MA theory is low genetic variance, which is the proper measure of selection pressure, and the long-time existence of aging controlling families of genes (Fisher 1930; Promislow et al. 1996; Tatar 1996).

  At the cellular level, apoptosis and telomeric aging (Clark 1999; Gordeeva et al. 2004) are two well-known mechanisms of programmed cellular death. They have always been assumed to be beneficial for individual adaptation.
- *Aging as the result of evolutionary tradeoffs*. As Mittledorf mentioned, the basis for the evolutionary tradeoff theories is an "inescapable" (and this is the key word here) "tradeoff between preservation of the soma and other tasks essential to fitness, such as metabolism and reproduction". There is a significant amount of empirical data against the inescapability of the tradeoff.

  The two main branches of the pleiotropic theories are the Antagonistic Pleiotropy (AP) theory (genetic tradeoff, originated from Williams' work (Williams 1957)) and the Disposable Soma theory (tradeoff in resource allocation, originated from the work of Kirkwood (1977)).
- *Aging as a genetic program*. This approach implies that group selection to benefit the whole population is strong compared to individual adaptation based on the concept of individual competition. It contradicts modern genetic theories but has a strong support in empirical observations. For more details see Mitteldorf (2010). If the genetic program theory works, then one may expect that manipulations of this program may result in a significant extension of the life span.

Detailed arguments in support of and against each approach can be found in Mitteldorf's paper (Mitteldorf 2010) while Mitteldorf himself is a strong proponent of the last (demographic) of the four listed approaches to the problem of the evolution of aging.

6.4.2 Gompertz-Makeham Distribution

Any mathematical model of aging should satisfy the observed distribution of the lifespans. This observed distribution is the Gompertz-Makeham law

$$\mu(t) = A + R\exp(\gamma t), \tag{6.13}$$

where A, R, and γ are constants and μ is the force of mortality defined by the formula

$$\mu(t) = -\frac{d\ln l(t)}{dt}, \tag{6.14}$$

where t is the age and $l(t)$ is the number of survivors from the initial population at age t. The rate of increase of the force of mortality decreases at extremely old ages (Gavrilov and Gavrilova 1991).

Various models described in Chap. 6 of Gavrilivs' book (Gavrilov and Gavrilova 1991) are based on the statistics of extremes. A heterogeneous population consists of organisms considered to be combinations of (a very large number of) subsystems with many redundant elements under stochastic perturbations. Failure of any subsystem may (in the case of a series connection) lead to the death. Therefore, the authors of the models are looking for the limiting distributions of the least values of the lifespans of a large number of subsystems. In many cases, the authors of models recover the Gompertz-Makeham law. The distribution of the organism's lifespans depends on the distribution of lifespans of its constituent elements. The Gompertz law for the subsystems' lifespans generates the Gompertz law for the lifespans of the organisms.

Models of this type with the upper limit on the organism's lifespan result in increasing with age rate of increase of the force of mortality and contradict the observational data. The authors conclude that there can be no upper limit on the organism's lifespan.

6.4.3 Fractional Dynamics Approach

As we already mentioned in the previous section, no considered models take into account the fact that all living species possess memory. As we mentioned in

Sect. 6.3, memory is an important property of the organ tissues (subsystems, which are viscoelastic), of all animals (and humans).

Antagonistic pleiotropy is one of the theories explaining aging. In short, it states: the better we fit, the sooner we die. Species compete to make more and better children and individuals exhaust themselves in this competition. Considering humans, the competition also includes the competition in intellects, in making discoveries (which, who knows, could be the purpose the human existence in the Universe). If the fittest one is not the one who succeeds in discoveries, then the evolutionary outcome could be the extinction of human curiosity. But this is not what is observed. Making discoveries and thinking requires a lot of concentration and the use of the organism's resources. In addition to remembering and retrieving from the brain all the facts that we've learned, we keep in mind all previous steps in our reasoning. But our memory decays (or strengthens) as a power law. This could be an additional argument to consider when modeling a human being (including the evolution of aging).

Therefore, the author believes that the use of the fractional (with power-law memory) models is a natural and appropriate approach to model a human individual. It is not trivial but possible to model a living species by a single variable and, in any case, this model will be a rough approximation of reality. One of the ways is to consider a variable related (but not equivalent) to fitness understood, for example, as the total (potential) number of descendants produced by a certain age. This variable is less than one up to the age at which an organism is first capable of sexual reproduction of offspring (puberty), and then it gradually increases approaching some constant value until the organism dies (which could be long after the end of the ability to reproduce). Solutions of simple nonlinear discrete fractional equations bear a remarkable similarity to the evolution of this variable presented above. This similarity supports the author's proposition to use fractional models to describe the evolution of aging.

Regular (without memory) models (systems of differential/difference equations) may have stable and unstable solutions. It is not trivial for a model of a regular system to have a solution that converges to a stable state and then, after a while, breaks. On the other hand, as we mentioned before, cascade of bifurcations type solutions (CBTT) are essential features of discrete nonlinear fractional order (with power-law memory) systems.

In CBTT the system first converges to a low-periodicity (or fixed) point but then this solution becomes unstable, the system breaks and jumps into a higher periodicity solution. The simplest discrete nonlinear fractional/fractional difference model is the logistic map, which has power-law (or asymptotically power-law) memory and quadratic nonlinearity. In a wide range of parameters, when a fractional system asymptotically converges to a period $T > 1$ or chaotic trajectory, for a fixed set of parameters the evolution depends only on initial conditions. Figures 6.6 and 6.7 show this dependence for the fractional and fractional difference Caputo logistic maps. As one can see, for small values of initial conditions the system first slowly, in a regular way, starts converging to the unstable fixed point but then suddenly

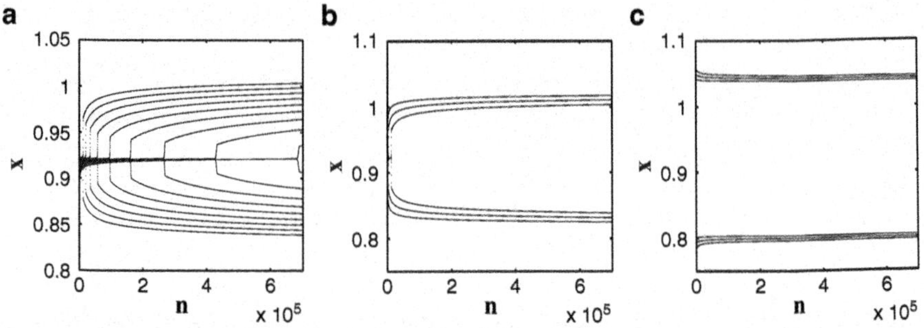

Fig. 6.6 Asymptotically period two trajectories for the Caputo logistic α-family of maps Eqs. (6.2) and (6.7) with $\alpha = 0.1$ and $K = 15.5$: (**a**) nine trajectories with the initial conditions $x_0 = 0.29 + 0.04i$, $i = 0, 1, \ldots, 8$ ($i = 0$ corresponds to the rightmost bifurcation); (**b**) $x_0 = 0.61 + 0.06i$, $i = 1, 2, 3$; (**c**) $x_0 = 0.95 + 0.04i$, $i = 1, 2, 3$. As $n \to \infty$ all trajectories converge to the limiting values $x_{lim1} = 0.80629$ and $x_{lim2} = 1.036030$. The unstable fixed point is $x_{lim0} = (K-1)/K = 0.93548$. (This figure is reprinted from Edelman (2018b), with the permission of AIP Publishing)

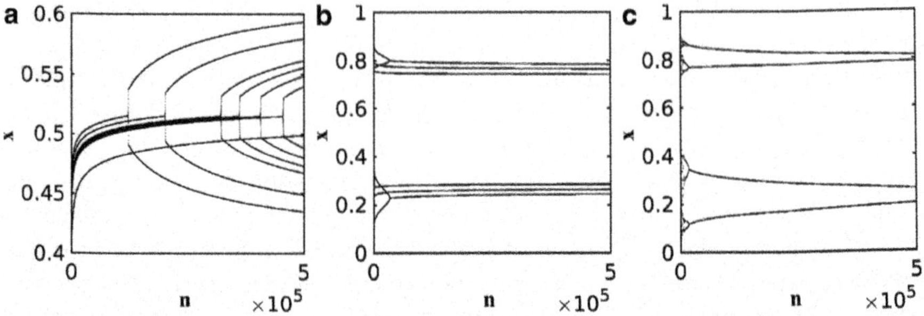

Fig. 6.7 Asymptotically period two trajectories for the fractional difference Caputo logistic α-family of maps Eqs. (6.3) and (6.7) with $\alpha = 0.1$ and $K = 2.5$: (**a**) seven trajectories with the initial conditions $x_0 = 0.0001$, $x_0 = 0.085 + 0.005i$, $i = 0, 1, 2, 3$, $x_0 = 0.12$, and $x_0 = 0.14$ (the leftmost bifurcation); (**b**) $x_0 = 0.6 + 0.1i$, $i = 1, 2, 3$; (**c**) $x_0 = 0.9$. As $n \to \infty$ all trajectories converge to the limiting values $x_{lim1} = 0.3045$ and $x_{lim2} = 0.7242$. The unstable fixed point is $x_{lim0} = (K-1)/K = 0.6$

becomes unstable and bifurcates. The time of the regular convergence to the stable fixed point decreases with the increase in initial conditions. As it can be seen from Fig. 6.8, convergence to the stable fixed point follows the power law with the power α from Eqs. (6.2) and (6.3). In the case of fractional difference maps, convergence to the power law is very clear (see Fig. 6.8c). In the case of fractional maps, for small values of α even convergence to the power law is slow (Fig. 6.8a), but Fig. 6.8b (α=0.4) confirms that eventually the convergence follows the power law (for the power-law convergence of trajectories see also Edelman 2011, 2013b, 2014a,b; Edelman and Taieb 2013).

The lifespan in this model, the times of the regular convergence to a stable fixed point until the bifurcation on a single trajectory (call it death, the breaking of the

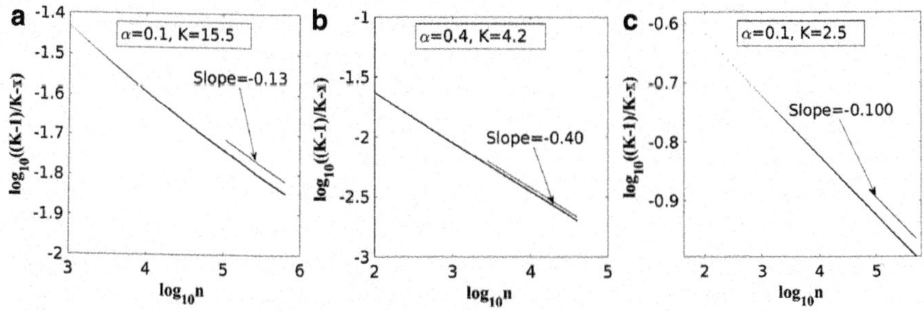

Fig. 6.8 Power-law convergence of trajectories to the fixed point (stable (**b**) or unstable (**a**) and (**c**)) for the fractional (**a**) and (**b**) and fractional difference (**c**) Caputo logistic α-family of maps. Maps' parameters are indicated on the figures

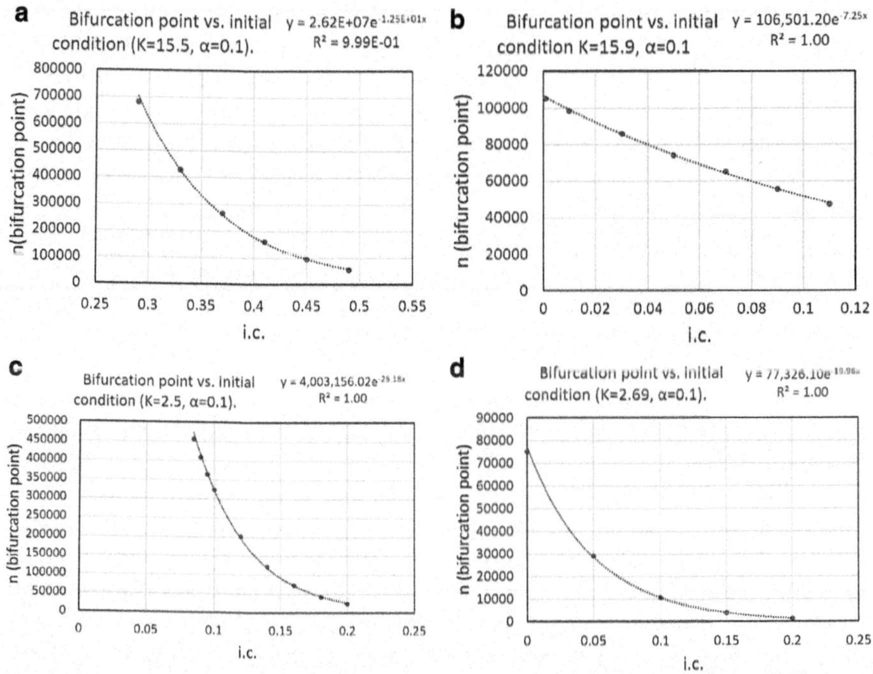

Fig. 6.9 The time of the regular convergence to a stable fixed point until the bifurcation on a single trajectory for the fractional (**a**) and (**b**), and fractional difference (**c**) and (**d**) Caputo logistic α-family of maps as a function of the initial condition. Maps' parameters are indicated on the figures

stable fixed-point evolution, etc.), as can be seen from Fig. 6.9, is the exponential function of the initial condition

$$t = t_m e^{-\gamma x}, \tag{6.15}$$

where t is the lifespan, x is the initial condition and constants t_m, which is the maximal possible age, and γ depend on the map's parameters.

If we assume the uniform distribution of the initial conditions, $dl(t) = dx(t)$, then calculating the force of mortality Eq. (6.14), we'll obtain $\mu = 1/(t * (\ln t_m - \ln t))$. This assumption seems unrealistic and the corresponding formula for the force of mortality is very different from the Gompertz-Makeham law.

It is more realistic to assume that the value of an initial condition (number of descendants produced at very early ages) is near zero and consider the evolution under random perturbations (mutations, interactions with the environment, etc.). There are various ways to introduce perturbations: they may be either uniformly distributed or normally; they may be added at each step of iterations starting either from the beginning of the evolution or at a certain age (this would correspond to a perfect cellular repair mechanism up to that age); they may be of a constant magnitude or increasing; and so on. In this publication we present only two cases: In Fig. 6.10 we present four different trajectories for the Caputo fractional logistic map with $\alpha = 0.1$, $K = 15.9$ and $x_0 = 0.0001$. In this case, the maximal lifespan (without perturbations, this case is added to each subfigure) is approximately 10,6500 iterations (then the system bifurcates very fast). The perturbations σ are uniformly distributed within the interval $(-0.0005, 0.0005)$ and are added at each step starting from $n = 20001$.

Figure 6.11 presents four different trajectories for the fractional difference logistic map with $\alpha = 0.1$, $K = 2.69$ and $x_0 = 0.0001$. In this case, the maximal lifespan is approximately 77,000 iterations. The perturbations σ are normally distributed with the zero mean value and the standard deviation 0.0005; perturbations are added at each step starting from $n = 16$.

These figures demonstrate the relative robustness of the solution with respect to perturbations. Still, in each case the bifurcation occurs much earlier compared to the unperturbed case.

To obtain the distribution of the times of the stable evolution prior to the bifurcation (lifespans) in the fractional/fractional difference under perturbation models, the author followed the standards described in Chaps. 2 and 3 of the Gavrilovs' book Gavrilov and Gavrilova (1991). In different cases (various maps, parameters, and perturbations), from tens of thousands to hundreds of thousands runs with various sets of random perturbations were performed. In each set of runs the magnitude of perturbations was much (hundreds or thousands times) smaller than the difference between two limiting values x_{lim} of the asymptotically period two solution of the unperturbed equation. To distinguish the difference $|x_{n+1} - x_n|$ due to a bifurcation from the difference due to random perturbations, the value of the *level* (mentioned in Fig. 6.12), which is much larger than any reasonably probable value of a perturbation (e.g., 40 standard deviations in Fig. 6.12b), was selected. Then, it was assumed that the steady power-law convergence to the fixed point terminates (death of an organism) when the difference $|x_{n+1} - x_n|$ for the first time exceeds the *level*.

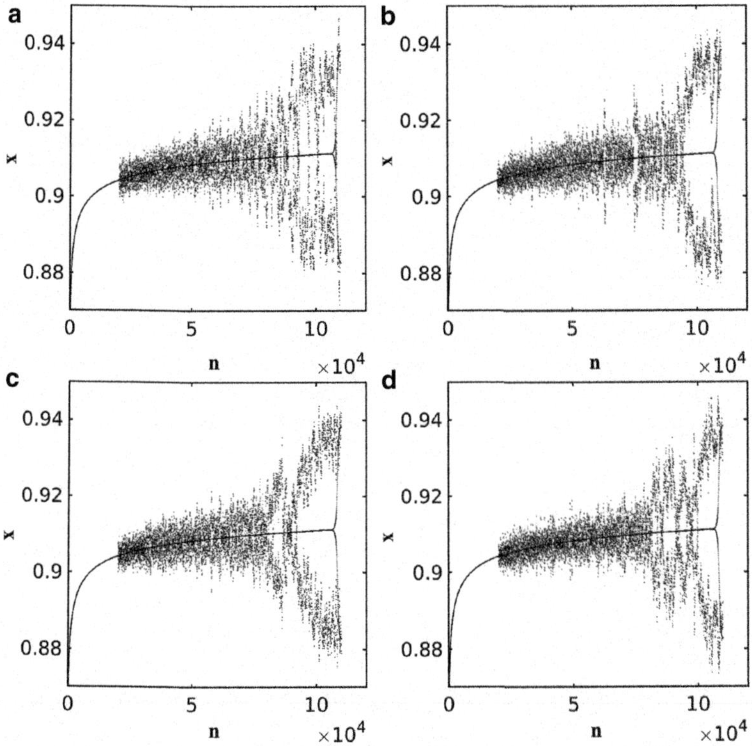

Fig. 6.10 Four examples of the influence of perturbations in Caputo fractional logistic map with $\alpha = 0.1$, $K = 15.9$ and $x_0 = 0.0001$. Perturbations, uniformly distributed within the interval $(-0.0005, 0.0005)$, are added at each step of iterations starting from $n = 20001$. The steady line, which breaks at the end, in the middle of each graph shows the unperturbed solution

Each run was terminated at the iteration n_{lev} at which the magnitude of the difference between two consecutive values of x for the first time exceeded the *level* and the value of l_t, where $t = [n_{lev}/1000]$ was increased by one. Then, the force of mortality μ_t, Eq. (6.14), was calculated according to the Eq. (9) from the Chap. 2 of the book Gavrilov and Gavrilova (1991) as

$$\mu_t = \frac{1}{2} \ln \left(\frac{l_{t-1}}{l_{t+1}} \right). \tag{6.16}$$

As in Chap. 3 of Gavrilov and Gavrilova (1991), in this paper we draw the graphs of $\log(\mu_t \times 10^5)$ as a function of the age t and see if it fits a straight line (which implies the Gompertz-Makeham law).

Here (in Fig. 6.12) we present only results of the two sets of runs, but the Gompertz-Makeham law (with the various relative widths in time of nonzero changes in the number of survivors from the initial population l_t) was obtained in

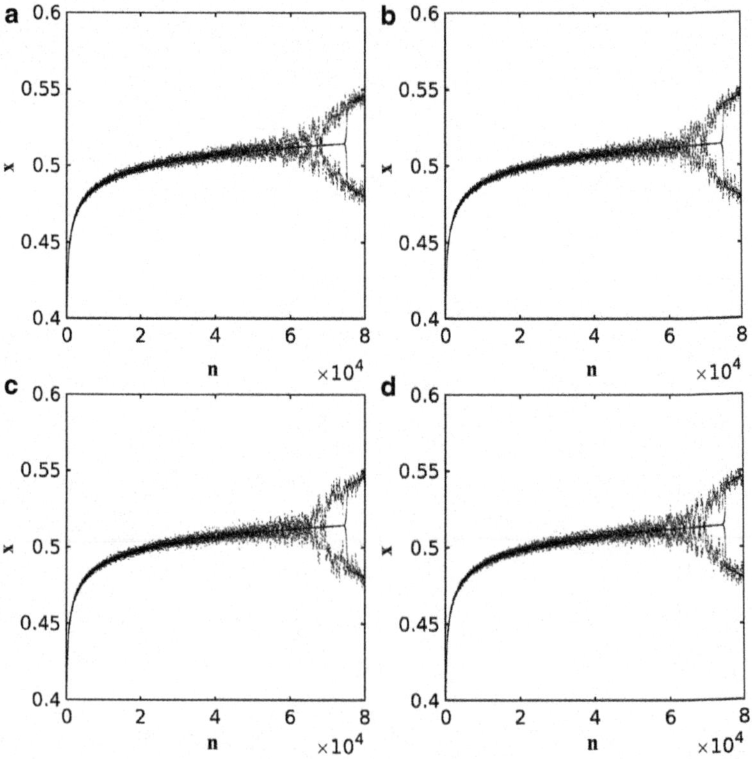

Fig. 6.11 Four examples of the influence of perturbations in Caputo fractional difference logistic map with $\alpha = 0.1$, $K = 2.69$ and $x_0 = 0.0001$. Perturbations are normally distributed with the zero mean value and the standard deviation 0.0005 and are added at each step of iterations starting from $n = 16$. The steady line, which breaks at the end, in the middle of each graph shows the unperturbed solution

all cases. As one may see, the fit is approximately linear. It is not perfect, but it is not perfect in the graphs obtained from the real tables of mortality either.

6.5 Conclusion

We are conceived and born programmed to develop in a certain way at a certain pace. If this program includes power-law memory, then we may be programmed to die. In some scenarios this death is inevitable, and the maximal possible lifespan may be calculated. The observed power-law in human memory and the viscoelastic nature of our organ tissues present reasonable arguments for modeling human individuals as fractional (with power-law memory) systems. Whether we are programmed to die or not, the power-law memory in the evolution of individual humans and separate

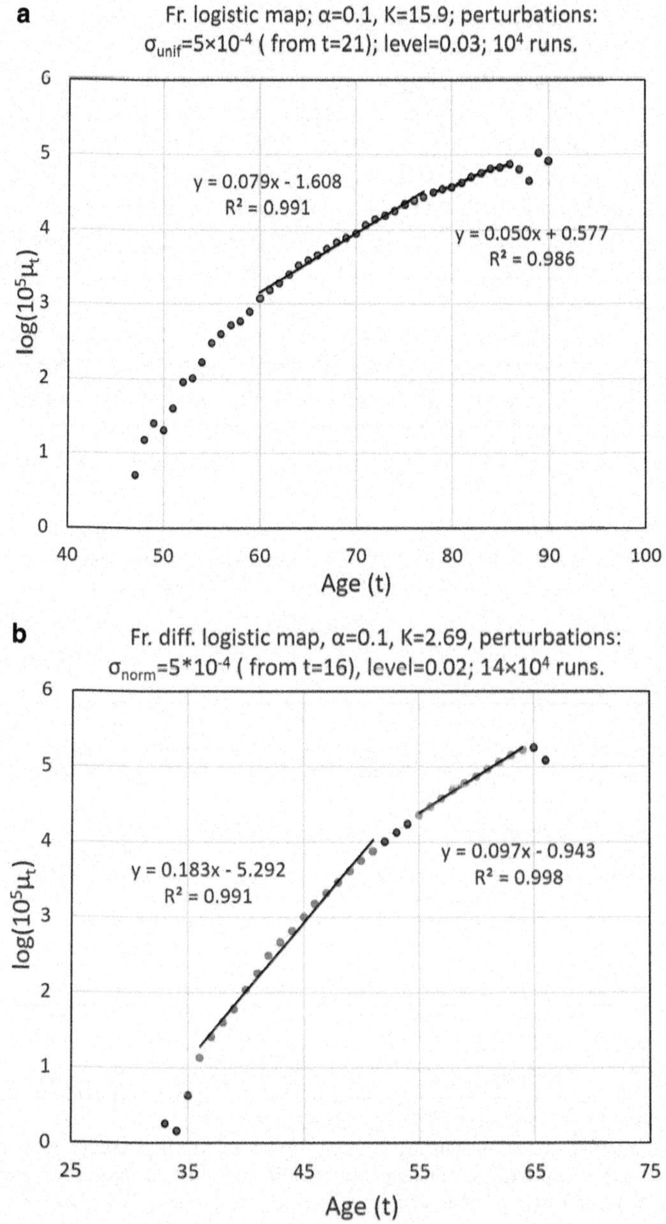

Fig. 6.12 The logarithm of the force of mortality as a function of age. All relevant parameters are indicted in the subfigures' titles. The maximal possible age (when there are no perturbations) in case (**a**) is 106 and in case (**b**) is 77

organs may lead to the limit of the human lifespan. The same may be true for any living species. As one can see from the results of numerical simulations, the real lifespans may be significantly shorter than the maximal possible lifespan. In the table created for the example presented in Fig. 6.12b based on 140,000 runs, when the maximal lifespan was 77, none survived at the age of 68; for Fig. 6.12a the numbers are 10,000 runs, maximal lifespan is 106, and none survived at the age of 94. Another significant fact is that in models with power-law memory the decrease at very large ages of the rate of increase of the force of mortality does not contradict the limited lifespan, which is impossible in memoryless models (Gavrilov and Gavrilova 1991).

If the assumption that living species develop as systems with power-law memory is correct, then the reasonable questions are where and how the power-law memory is recorded in our DNA and whether we may manage and correct this recording. The author understands that his proposition raises more questions than it produces answers and welcomes any discussions on the topic of the power law in human evolution.

With regards to the theory of stability of fractional (with power-law memory) systems, the simulations presented in this paper demonstrate the robustness of the factional systems with respect to various types of perturbations. As can be seen from Figs. 6.10 and 6.11, during a significant interval of time in the evolution of a fractional system, the memory may prevail in competition with perturbations and preserve the unperturbed evolution.

Acknowledgments The author acknowledges support from the Joseph Alexander Foundation, Yeshiva University. The author expresses his gratitude to Prof. Sylvain E. Cappell for the support at the Courant Institute where the computer simulations were performed. The author also expresses his gratitude to Virginia Donnelly for technical help and valuable discussions.

References

Anderson, J. R. (1995). *Learning and memory: An integrated approach*. New York: Wiley.

Area, I., Losada, J., & Nieto, J. J. (2014). On fractional derivatives and primitives of periodic functions. *Abstract and Applied Analysis, 2014*, 392598.

Bagley, R. L., & Torvik, P. J. (1983). A theoretical basis for the application of fractional calculus to viscoelasticity. *Journal of Rheology, 27*, 201–210.

Barnes, S. C., Lawless, B. M., Shepherd, D. E. T., Espino, D. M., Bicknell, G. R., & Bryan, R. T. (2016). Viscoelastic properties of human bladder tumours. *Journal of the Mechanical Behavior of Biomedical Materials, 61*, 250–257.

Brauer, F., & Castillo-Chavez, C. (2001). *Mathematical models in population biology and epidemiology*. New York: Springer.

Caponetto, R., Dongola, G., Fortuna, L., & Petras, I. (2010). *Fractional order systems: Modeling and control applications*. Singapore: World Scientific.

Caputo, M., & Mainardi, F. (1971). Linear models of dissipation in anelastic solids. *Rivista del Nuovo Cimento, 1*, 161–198.

Cermak, J., & Nechvatal, L. (2017). The Routh-Hurwitz conditions of fractional type in stability analysis of the Lorenz dynamical system. *Nonlinear Dynamics, 87*, 939–954.

Chan, R. W. (2018). Nonlinear viscoelastic characterization of human vocal fold tissues under large-amplitude oscillatory shear (LAOS). *Journal of Rheology, 62*, 695–712.

Chirikov, B. V. (1979). A universal instability of many dimensional oscillator systems. *Physics Reports, 52*, 263–379.

Clark, W. R. (1999). *A means to an end: The biological basis of aging and death*. New York/Oxford: Oxford University Press.

Deshpande, A. S., & Daftardar-Gejji, V. (2017). On disappearance of chaos in fractional systems. *Chaos, Solitons and Fractals, 102*, 119.

Duck, F. A. (1990). *Physical properties of tissue: A comprehensive reference book*. San Diego: Academic Press.

Edelman, M. (2011). Fractional standard map: Riemann-Liouville vs. Caputo. *Communications in Nonlinear Science and Numerical Simulation, 16*, 4573–4580.

Edelman, M. (2013a). Fractional maps and fractional attractors. Part I: α-families of maps. *Discontinuity, Nonlinearity, and Complexity, 1*, 305–324.

Edelman, M. (2013b). Universal fractional map and cascade of bifurcations type attractors. *Chaos, 23*, 033127.

Edelman, M. (2014a). Fractional maps as maps with power-law memory. In A. Afraimovich, A. C. J. Luo, & X. Fu (Eds.), *Nonlinear dynamics and complexity* (Nonlinear systems and complexity, pp. 79–120). New York: Springer.

Edelman, M. (2014b). Caputo standard α-family of maps: Fractional difference vs. fractional. *Chaos, 24*, 023137.

Edelman, M. (2015a). Fractional maps and fractional attractors. Part II: Fractional difference α-families of maps. *Discontinuity, Nonlinearity, and Complexity, 4*, 391–402.

Edelman, M. (2015b). On the fractional Eulerian numbers and equivalence of maps with long term power-law memory (integral Volterra equations of the second kind) to Grünvald-Letnikov fractional difference (differential) equations. *Chaos, 25*, 073103.

Edelman, M. (2018a). Universality in systems with power-law memory and fractional dynamics. In: M. Edelman, E. Macau, & M. A. F. Sanjuan (Eds.), *Chaotic, fractional, and complex dynamics: New insights and perspectives* (Understanding Complex Systems, pp. 147–171). Cham: Springer. eBook.

Edelman, M. (2018b). On stability of fixed points and chaos in fractional systems. *Chaos, 28*, 023112.

Edelman, M. (2018c). Maps with power-law memory: direct introduction and Eulerian numbers, fractional maps, and fractional difference maps. In: A. Kochubei & Yu. Luchko (Eds.), *Handbook of fractional calculus with applications* (Theory, Vol. 2). Berlin: De Gruyter.

Edelman, M. (2018d). Dynamics of nonlinear systems with power-law memory. In: V. E. Tarasov (Ed.), *Handbook of fractional calculus with applications* (Applications in physics, Vol. 4). Berlin: De Gruyter.

Edelman, M., & Taieb, L. A. (2013). New types of solutions of non-linear fractional differential equations. In: A. Almeida, L. Castro, & F.-O. Speck (Eds.), *Advances in harmonic analysis and operator theory* (Operator theory: Advances and applications, Vol. 229, pp. 139–155). Basel: Springer.

Edelman, M., & Tarasov, V. E. (2009). Fractional standard map. *Physics Letters A, 374*, 279–285.

Edney, E. B., & Gill, R. W. (1968). Evolution of senescence and specific longevity. *Nature, 220*, 281–282.

Fairhall, A. L., Lewen, G. D., Bialek, W., & de Ruyter van Steveninck, R. R. (2001). Efficiency and ambiguity in an adaptive neural code. *Nature, 412*, 787–792.

Fisher, R. A. (1930). *The genetical theory of natural selection*. Oxford: The Clarendon Press.

Gavrilov, L. A., & Gavrilova, N. S. (1991). *The biology of life span: A quantitative approach*. Chur: Harwood Academic Publishers.

Gavrilov, L. A., & Gavrilova, N. S. (2004). The reliability-engineering approach to the problem of biological aging. *Annals of the New York Academy of Sciences, 1019*, 509–512.

Gordeeva, A. V., Labas, Y. A., et al. (2004). Apoptosis in unicellular organisms: Mechanisms and evolution. *Biochem-Moscow, 69*, 1055–1066.

Guarente, L., & Kenyon, C. (2000). Genetic pathways that regulate ageing in model organisms. *Nature, 408*, 255–262.

Holm, S., & Sinkus, R. (2010). A unifying fractional wave equation for compressional and shear waves. *Journal of the Acoustical Society of America, 127*, 542–548.

Hoppensteadt, F. (1975). *Mathematical theories of populations: Demographics, genetics, and epidemics*. Philadelphia: SIAM.

Jagan Mohan, J. (2016). Periodic solutions of fractional Nabla difference equations. *Communications in Applied Analysis, 20*, 585–609.

Jagan Mohan, J. (2017). Quasi-periodic solutions of fractional Nabla difference systems. *Fractional Differential Calculus, 7*, 339–355.

Jeong, K. M., Bo-seung, K., Sejin, H., & Hee, Y. H. (2018). Experimentally derived viscoelastic properties of human skin and muscle *in vitro*. *Medical Engineering and Physics, 61*, 25–31.

Kahana, M. J. (2012). *Foundations of human memory*. New York: Oxford University Press.

Kaslik, E., & Sivasundaram, S. (2012). Nonexistence of periodic solutions in fractional order dynamical systems and a remarkable difference between integer and fractional order derivatives of periodic functions. *Nonlinear Analysis. Real World Applications, 13*, 1489–1497.

Kenyon, C. (2001). A conserved regulatory system for aging. *Cell, 105*, 165–168.

Kilbas, A. A., Srivastava, H. M., & Trujillo, J. J. (2006). *Theory and Application of Fractional Differential Equations*. Amsterdam: Elsevier.

Kirkwood, T. (1977). Evolution of aging. *Nature, 270*, 301–304.

Laskin, N., & Zaslavsky, G. M. (2006). Nonlinear fractional dynamics on a lattice with long-range interactions. *Physica A, 368*, 38–54.

Leopold, D. A., Murayama, Y., & Logothetis, N. K. (2003). Very slow activity fluctuations in monkey visual cortex: implications for functional brain imaging. *Cerebr Cortex, 413*, 422–433.

Lundstrom, B. N., Higgs, M. H., Spain, W. J., & Fairhall, A. L. (2008). Fractional differentiation by neocortical pyramidal neurons. *Nature Neuroscience, 11*, 1335–1342.

Lundstrom, B. N., Fairhall, A. L., & Maravall, M. (2010). Multiple time scale encoding of slowly varying whisker stimulus envelope incortical and thalamic neurons in vivo. *The Journal of Neuroscience, 30*, 5071–5077.

Magin, R. L. (2010). Fractional calculus models of complex dynamics in biological tissues. *Journal of Computational and Applied Mathematics, 59*, 1586–1593.

Mainardi, F. (2010). *Fractional calculus and waves in linear viscoelasticity: An Introduction to mathematical models*. London: Imperial College Press.

Medawar, P. B. (1952). *An unsolved problem of biology*. London: H. K. Lewis.

Min, W., Luo, G., Cherayil, B. J., Kou, S. C., & Xie, X. S. (2005). Observation of a power-law memory kernel for fluctuations within a single protein molecule. *Physical Review Letters, 94*, 198302.

Mitteldorf, J. (2010). Evolutionary origins of aging. In G. M. Fahy, M. D. West, L. S. Coles, & S. B. Harris (Eds.), *The future of aging* (pp. 87–126). Dordrecht: Springer.

Nasholm, S. P., & Holm, S. (2013). On a fractional zener elastic wave equation. *Fractional Calculus and Applied Analysis, 16*, 26–50.

Park, K., Lonsberry, G. E., Gearing, M., Levey, A. I., & Desai, P. J. (2018). Viscoelastic Properties of Human Autopsy brain tissues as biomarkers for Alzheimer's diseases. *IEEE Transactions on Biomedical Engineering*. https://doi.org/10.1109/TBME.2018.2878555. (Early Access).

Pozzorini, C., Naud, R., Mensi, S., & Gerstner, W. (2013). Temporal whitening by power-law adaptation in neocortical neurons. *Nature Neuroscience, 16*, 942–948.

Promislow, D. E., Tatar, M., Khazaeli, A. A., & Curtsinger, J. W. (1996). Age-specific patterns of genetic variance in Drosophila melanogaster. I. Mortality. *Genetics, 143*, 839–848.

Ricklefs, R. (1998). Evolutionary theories of aging: confirmation of a fundamental prediction, with implications for the genetic basis and evolution of life span. *The American Naturalist, 152*, 24–44.

Rubin, D. C., & Wenzel, A. E. (1996). One hundred years of forgetting: A quantitative description of retention. *Psychological Review, 103*, 743–760.

Safshekan, F., Tafazzoli-Shadpour, M., Abdouss, M., & Shadmehr, M. B. (2016). Viscoelastic properties of human tracheal tissues. *Journal of Biomechanical Engineering, 139*(1), 011007.

Stanislavsky, A. A. (2006). Hamiltonian formalism of fractional systems. *European Physical Journal B, 49*, 93–101.

Szabo, T. L., & Wu, J. (2000). A model for longitudinal and shear wave propagation in viscoelastic media. *Journal of the Acoustical Society of America, 107*, 2437–2446.

Tarasov, V. E. (2008). Fractional equations of Curie-von Schweidler and Gauss laws. *Journal of Physics: Condensed Matter*, 20, 145212.

Tarasov, V. E. (2009). Fractional integro-differential equations for electromagnetic waves in dielectric media. *Theoretical and Mathematical Physics, 158*, 355–359.

Tarasov, V. E. (2009a). Differential equations with fractional derivative and universal map with memory. *Journal of Physics A, 42*, 465102.

Tarasov, V. E. (2009b). Discrete map with memory from fractional differential equation of arbitrary positive order. *Journal of Mathematical Physics, 50*, 122703.

Tarasov, V. E. (2011). *Fractional dynamics: Application of fractional calculus to dynamics of particles, fields and media* (HEP). Berlin/Heidelberg: Springer.

Tarasov, V. E., & Zaslavsky, G. M. (2008). Fractional equations of kicked systems and discrete maps. *Journal of Physics A, 41*, 435101.

Tatar, M., Promislow, D. E., Khazaeli, A. A., & Curtsinger, J. W. (1996). Age-specific patterns of genetic variance in Drosophila melanogaster. II. Fecundity and its genetic covariance with age-specific mortality. *Genetics, 143*, 849–858.

Tavazoei, M. S., & Haeri, M. (2009). A proof for nonexistence of periodic solutions in time invariant fractional order systems. *Automatica, 45*, 1886–1890.

Toib, A., Lyakhov, V., & Marom, S. (1998). Interaction between duration of activity and recovery from slow inactivation in mammalian brain Na+ channels. *The Journal of Neuroscience, 18*, 1893–1903.

Ulanovsky, N., Las, L., Farkas, D., & Nelken, I. (2004). Multiple time scales of adaptation in auditory cortex neurons. *The Journal of Neuroscience, 24*, 10440–10453.

Wang, J., Feckan, M., & Zhou, Y. (2013). Nonexistence of periodic solutions and asymptotically periodic solutions for fractional differential equations. *Communications in Nonlinear Science and Numerical Simulation, 18*, 246–256.

Wang, Z., Golob, M. J., & Chesler, N. C. (2016). Viscoelastic properties of cardiovascular tissues. In M. El-Amin (Ed.), *Viscoelastic and Viscoplastic Materials* (pp. 141–163). IntechOpen. Online. https://www.intechopen.com/books/viscoelastic-and-viscoplastic-materials

Weismann, A. (1889). *Essays upon heredity and kindred biological problems*. Oxford: Clarendon Press.

Williams, G. (1957). Pleiotropy, natural selection, and the evolution of senescence. *Evolution, 11*, 398–411.

Wixted, J. T. (1990). Analyzing the empirical course of forgetting. *Journal of Experimental Psychology. Learning, Memory, and Cognition, 16*, 927–935.

Wixted, J. T., & Ebbesen, E. (1991). On the form of forgetting. *Psychological Science, 2*, 409–415.

Wixted, J. T., & Ebbesen, E. (1997). Genuine power curves in forgetting. *Memory & Cognition, 25*, 731–739.

Yazdani, M., & Salarieh, H. (2011). On the existence of periodic solutions in time-invariant fractional order systems. *Automatica, 47*, 1834–1837.

Zaslavsky, G. M. (2005). *Hamiltonian chaos and fractional dynamics*. Oxford: Oxford University Press.

Zaslavsky, G. M., Stanislavsky, A. A., & Edelman, M. (2006). Chaotic and pseudochaotic attractors of perturbed fractional oscillator. *Chaos, 16*, 013102.

Zaslavsky, G. M., Edelman, M., & Tarasov, V. E. (2007). Dynamics of the chain of forced oscillators with long-range interaction: From synchronization to chaos. *Chaos, 17*, 043124.

Zilany, M. S., Bruce, I. C., Nelson, P. C., & Carney, L. H. (2009). A phenomenological model of the synapse between the inner hair cell and auditory nerve: Long-term adaptation with power-law dynamics. *The Journal of the Acoustical Society of America, 126*, 2390–2412.

Part II
Mortality Modeling and Applications

Chapter 7
Application of Structural Equation Modeling to Infant Mortality Rate in Egypt

Fatma Abdelkhalek and Marianna Bolla

7.1 Introduction

Graphical models are applicable to multivariate measurement (manifest) variables. These either satisfy a log-linear model (in the discrete case) or a covariance selection model (in the Gaussian case). A graph is built, where the vertices are the variables. If causal relations between the variables are known (including the longitudinal data), we have a directed acyclic graph (DAG), and can solve so-called recursive linear equations (Wermuth 1980). In particular, when the variables are divided into exogenous and endogenous ones (the formers have only outgoing arrows), the method described in Kiiveri et al. (1984) can be used. For more general constructs, like regression graphs (see Bolla et al. 2019; Wermuth et al. 2012).

In another context, SEM finds causal relationships between measurement and latent concepts in psychology, sociology, or economics. Here, the recursive equation model is applied to latent variables, which in turn are related to measurement variables. If the latent variables are divided into exogenous and endogenous ones, then the corresponding measurement variables are divided accordingly. Between the latent and measurement variables, there are linear (factor model-like) equations with loading matrices, which are also to be estimated.

In the early 1900s, the geneticist S. Wright first used the term path analysis. He developed it to impose causal dependencies represented by structural coefficients on the observed relationships (Wright 1934). Thus, SEM finds causal relationships between either the observed variables or the latent variables or the two of them together. The use of latent variables enables researchers to measure the unobserved variables through one or more observed measurements. Accordingly, the latent

F. Abdelkhalek (✉) · M. Bolla
Institute of Mathematics, Budapest University of Technology and Economics, Budapest, Hungary
e-mail: fatma@math.bme.hu; marib@math.bme.hu

C. H. Skiadas, C. Skiadas (eds.), *Demography of Population Health, Aging and Health Expenditures*, The Springer Series on Demographic Methods and Population Analysis 50, https://doi.org/10.1007/978-3-030-44695-6_7

variables are hidden variables that cannot be obtained directly from the data; these include for example, educational achievement in school, socioeconomic status, and intelligence. The principle reason to use SEM is to estimate models containing causal paths and latent variables. In many studies, the assumptions required for regression analysis are not fulfilled. In multiple regression, the analyst must assume perfect measurement (i.e. no measurement error). In SEM, on the other hand, the measurement error can be specified, and its size can be estimated. In addition, constraints can be introduced based on theoretical expectations. Therefore, the advantages of SEM are as follows:

- Determines exactly which indicators will load on which latent variables;
- Specifies the internal relationships between both the exogenous and endogenous variables;
- Estimates the amount of variance in the indicators that is not explained by the latent construct (due to error in either measurement or model specification); and
- Examines the correlations between the latent variables and among the errors that are associated with the indicators (Raykov and Marcoulides 2006).

Note that the work of Havelmo (1943) also contributed to defining this class of models with exogenous and endogenous variables distinguished. He was honored with a Nobel prize in 1989.

In 2000, Egypt, along with most other countries, accepted the Millennium Development Goals (MDG), including a commitment to reduce the under-5 mortality rate (U5MR) by two-thirds (Goal 4.1), and the infant mortality rate (IMR, Goal 4.2). The IMR is one of the indicators used to assess progress toward Goal 4.1 (United Nations 2019b). In 2015, the Sustainable Development Goals (SDG) replaced the MDG and set specific goals for neonatal mortality (the number of infant deaths during the first 28 days of life, per 1000 live births – goal 12/1000) (United Nations 2019a), and for the under-5 morality rate (U5MR – 25/1000). In general, the IMR accounts for about two-thirds of the U5MR. In Egypt, the IMR declined from about 50/1000 in 1995 to about 20/1000 in 2015. This decline reflects excellent progress, but falls just short of the two-third reduction implied by the MDG. However, given that the IMR accounts for about 2/3 of the U5MR, an IMR of 20/1000 in 2015 suggests that even at the beginning of the Sustainable Development period, Egypt was close to achieving the childhood/infant mortality goals. Numerous factors affect infant deaths. Among the most important ones are family socioeconomic background characteristics including wealth, parental education, place of residence, and family size, along with mother- and pregnancy-related factors such as maternal health, antenatal and neonatal care, birth order, the conditions of birth delivery,. . . etc (see Amouzou and Hill 2003; Kembo and van Ginneken 2009). In addition, diarrheal disease, respiratory infection, and vaccine-preventable diseases are important determinants of the IMR. However, the overall economic indicators can be seen as the root cause of the mortality issue.

Table 7.1 Indicators definitions and abbreviations

Indicator	Definition	Abbreviation
Infant Mortality Rate	"The number of infants dying before reaching one year of age, per 1,000 live births in a given year."	IMR
GDP Per Capita	"Gross domestic value added by midyear population. It is the sum of gross value added by all resident producers in the economy plus any product taxes and minus any subsidies not included in the value of the products. It is calculated without making deductions for depreciation of fabricated assets or for depletion and degradation of natural resources. It is measured in the current U.S. dollars."	GDP
Current Health Expenditure (as % of GDP)	"An estimate of current health expenditures, including health care goods and services consumed during each year. It does not include capital health expenditures such as buildings, machinery, IT and stocks of vaccines for emergency or outbreaks."	HExp
Out-of-Pocket Health Expenditure (as % of HExp)	"The payments that are spent on health directly by households."	OPExp
Maternal Mortality Ratio	"Number of women who die from pregnancy-related causes while pregnant or within 42 days of pregnancy termination per 100,000 live births."	MMR
Hepatitis B Immunization	"The percentage of children who received hepatitis B vaccinations before 12 months. A child is considered adequately immunized after three doses."	HepB

Source. World Bank: https://data.worldbank.org/

In this paper, we focus on the decline in the IMR in Egypt between 1995 and 2015. We use SEM to better understand both the direct and indirect effect of the following indicators on the IMR: (1) gross domestic product (GDP) per capita; (2) current health expenditure as a percentage of GDP (HExp); (3) out-of-pocket health expenditure (OPExp); (4) maternal mortality ratio (MMR); and (5) hepatitis B immunization (HepB).

Consider Table 7.1 for more explanation and the abbreviations of the indicators. A World Bank time series data[1] for 1995–2015 is used to examine the hypothesized model.

[1] https://data.worldbank.org/indicator

7.2 Research Methodology

7.2.1 *Research Model*

Based on the literature, we propose the hypothesized model to examine the causal relationships between the economic indicators and the IMR (Fig. 7.1). The model's hypotheses can be stated as follows:

1. There is a direct effect of the GDP on HExp, HepB, and IMR.
2. There is a direct effect of the HExp on HepB, MMR, and IMR.
3. There is a direct effect of the OPExp on both HepB and MMR.
4. Both HExp and HepB mediate the influence of GDP on MMR and IMR.
5. Also HepB and the MMR mediate the influence of OPExp on IMR.

Considering Fig. 7.1, all indicators are represented by rectangles that denote that there are no latent variables in the model. Hence, we used the SEM path analysis to evaluate the proposed hypotheses. Moreover, exogenous and endogenous variables can be distinguished through the arrows that come out of or go into each rectangle. For each endogenous variable, we could create a regression equation, incorporating all variables that influence that endogenous variable. Accordingly, the system of SEM linear equations for exogenous (**X**) and endogenous (**Y**) of the hypothesized model (Fig. 7.1) can be written as:

$$\boldsymbol{A}\mathbf{Y} + \boldsymbol{B}\mathbf{X} = \boldsymbol{\varepsilon}, \tag{7.1}$$

where the matrix $\boldsymbol{A}$ represents the paths from endogenous to endogenous variables, the matrix $\boldsymbol{B}$ represents the paths from exogenous to endogenous variables, and $\boldsymbol{\varepsilon}$ is the random error vector (see Kiiveri et al. 1984; Wermuth 1980). Preciously, both the GDP and OPExp are exogenous variables, and the other four indicators (HExp, HepB, MMR, and IMR) are the endogenous ones. If $\boldsymbol{A}$ is a lower triangular matrix,

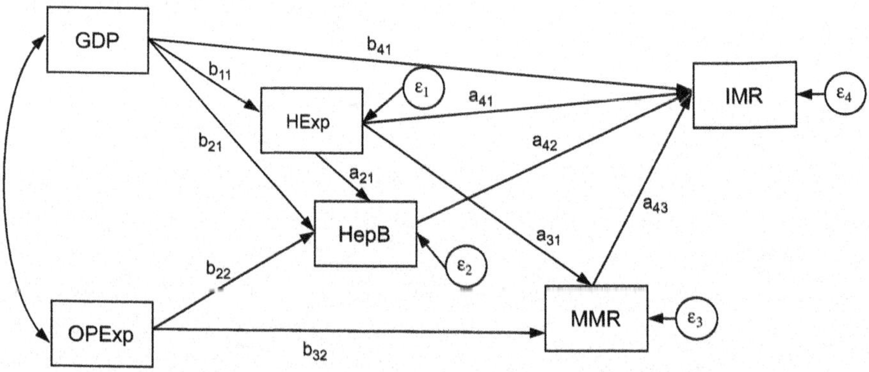

Fig. 7.1 The hypothesized model

we have a directed acyclic graph (DAG) and can get a system of recursive linear equations (Wermuth 1980), where **X** is the exogenous variable. For the estimation, a variant of Cholesky decomposition of the inverse covariance matrix of (**X**, **Y**) is applicable (Kiiveri et al. 1984). Even if $\boldsymbol{A}$ is not lower or upper triangular, we could specify some entries of $\boldsymbol{A}$ and $\boldsymbol{B}$ to be zero (according to some arrows on the directed graph), then the method of covariance-based SEM can be used. This means that we maximized the Gaussian likelihood on the model assumptions (Jöreskog 1977). In our case, the hypothesized graphical model equation (7.1) is:

$$\begin{bmatrix} 1 & 0 & 0 & 0 \\ a_{21} & 1 & 0 & 0 \\ a_{31} & 0 & 1 & 0 \\ a_{41} & a_{42} & a_{43} & 1 \end{bmatrix} \begin{bmatrix} HExp \\ HepB \\ MMR \\ IMR \end{bmatrix} + \begin{bmatrix} b_{11} & 0 \\ b_{21} & b_{22} \\ 0 & b_{32} \\ b_{41} & 0 \end{bmatrix} \begin{bmatrix} GDP \\ OPExp \end{bmatrix} = \begin{bmatrix} \epsilon_1 \\ \epsilon_2 \\ \epsilon_3 \\ \epsilon_4 \end{bmatrix}.$$

The goal is to reproduce the covariance matrix of the observed variables (**X**, **Y**) that minimizes the discrepancy function between the sample variance covariance matrix ($\boldsymbol{S}$) and the model-implied covariance matrix ($\boldsymbol{\Sigma}$) as much as possible under the model assumptions which are as follows:

- $\mathbb{E}(\mathbf{X}) = \mathbb{E}(\mathbf{Y}) = \mathbb{E}(\boldsymbol{\varepsilon}) = 0$;
- $\boldsymbol{\varepsilon}$ and **X** are independent Gaussian random vectors;
- $\boldsymbol{I} - \boldsymbol{B}$ is a non-singular matrix; and
- The free parameter matrices to be estimated are $\boldsymbol{A}$ and $\boldsymbol{B}$, akin to the diagonal covariance matrix of $\boldsymbol{\varepsilon}$.

7.2.2 Model Estimation

The analysis was conducted using the STATA software package. To summarize the data features and elucidate relations among the observed indicators, we used both descriptive statistics and the pairwise correlation test. Since the data follows Gaussian distribution, the Maximum Likelihood (ML) estimation method was used. In SEM, the model fit function of the likelihood ratio test follows χ^2 distribution under the model assumptions. The significance is determined by comparing the p-value with the significance level α. If the p-value is greater than α, the assumed model fits the data well (usually $\alpha = .05$). However, the χ^2 value is substantially increased in the presence of larger sample sizes. Therefore, alternative goodness-of-fit measures are often used to test the model fit function. The two classes of alternative fit indices were identified by Hu and Bentler (1999) are:

- Incremental Fit Indices (IFI) measure the increase of relative fit to the baseline model (in which all observed variables are uncorrelated). Unlike χ^2, the IFI is relatively insensitive to sample size. It includes the Tucker-Lewis Index (TLI), and the Comparative Fit Index (CFI), which is not too sensitive to sample size.

However, the CFI is not effective if the correlation coefficients between the variables approach zero. A value close to one is ideal. Practically speaking, a test value of at least 0.9 is considered as an acceptable fit.

- Absolute Fit Indices (AFI) measure the extent to which a specified model reproduces a sample covariance matrix. It includes the Goodness-of-Fit Index (GFI), the Adjusted Goodness-of-Fit Index ($AGFI$), the Standardized Root Mean Square Residual ($SRMR$), and the Root Mean Square Error ($RMSEA$). The AFI ranges between zero and one, with a value close to zero being ideal. Practically speaking, a test value of at most 0.08 is viewed as an acceptable fit.

Note that many of these fit indices are derived from χ^2-values. Conceptually, it is better to use indices from different classes to overcome the limitations of depending on only one index. Therefore, a good model fit is indicated when there is a significant χ^2, a higher value of IFI, and a lower value of AFI.

7.3 Results

7.3.1 Descriptive Statistics

As shown in Table 7.2, a sample of 21 years over the 1995–2015 period was considered. Over time, the infant mortality rate and maternal mortality ratio have declined by about half. However, they did not achieve the extent of decline set by the Millennium Development Goals, even though GDP per capita, immunization, current, and out-of-pocket health expenditures were steadily increasing.

7.3.2 Correlations

The pairwise correlation test presents not only the strength and direction, but also the significance of the relationships between each pair of variables. As shown in Table 7.3, most of the relationships are strongly correlated and statistically significant.

Table 7.2 Descriptive statistics

	N	Mean	SD	Min	Max
GDP	21	1858.82	882.09	944.20	3547.71
OPExp	21	62.14	1.63	59.04	65.41
HExp	21	4.99	.49	4.09	5.79
HepB	21	93.52	7.29	74	99
MMR	21	53.19	15.04	33	83
IMR	21	31.43	8.87	20.1	49.5

Table 7.3 Pearson correlation coefficients

	1	2	3	4	5	6
1.GDP	1					
2.OPExp	−.43*	1				
3.HExp	−.82***	.58*	1			
4.HepB	.25	.23	−.19	1		
5.MMR	−.85***	.21	.75**	−.65**	1	
6.IMR	−.81***	.18	.73**	−.69**	.99**	1

$^{*}p < .05$, $^{**}p < .01$, and $^{***}p < .001$

Table 7.4 Goodness-of-fit measures of the fitted model

Overall Model-Fit	Criteria	Model Value
χ^2	$p > .05$	.099
Absolute Fit Indices		
$RMSEA$	$\leq .08$	.087
$SRMR$	$\leq .08$	.054
Incremental Fit Indices		
TLI	$\geq .90$	.914
CFI	$\geq .93$	.947
Coefficient of Determination		
R^2	Close to 1	.79

This shows the extent to which these variables interact with and affect each other. For example, the correlation coefficient of GDP and the MMR is ($\rho = -.85$, p-value $< .001$). This reflects a strong inverse relation between them, i.e. as GDP per capita increases, the MMR decreases. Furthermore, the correlation coefficient between the IMR and MMR is ($\rho = .99$, p-value $< .01$). Here there is a very strong direct relationship, i.e. as the MMR decreases, the IMR decreases as well.

7.3.3 *Goodness-of-Fit*

To evaluate the overall hypothesized model, it must be noted that a good model that fits the data well is the one with a higher value of IFI and a lower value of the AFI plus a significant χ^2 with p-value greater than $\alpha = .05$. Accordingly, as shown in Table 7.4, the values of all fit indices follow the criteria, where $TLI = .914$, $CFI = .947$, $RMSEA = .077$, and $SRMR = .054$. Moreover the significance of χ^2 holds, and the overall Coefficient of Determination ($R^2 = .79$) implies that the hypothesized model explains about 79% of the infant mortality rate variation. Therefore, the overall model is statistically significant and fits the data well.

Table 7.5 reports each of the fitted, predicted and residual variances of each of the observed endogenous variables. It also shows the correlations between them and their predictors. For example, the correlation between HExp and its predictors is 0.82. This implies that the fraction of the variance of HExp explained by its

Table 7.5 Equation-level goodness-of-fit

	Variance					
Endogenous Variables	Fitted	Predicted	Residual	R^2	mc	mc_2
Observed						
HExp	.23	.15	.07	.67	.82	.67
HepB	53.64	14.16	39.48	.26	.51	.26
MMR	247.03	170.59	76.44	.69	.83	.69
IMR	83.41	83.16	.25	.99	.99	.99

mc: is the correlation between endogenous variables and its prediction
mc_2: is the Bentler-Raykov squared multiple correlation coefficient

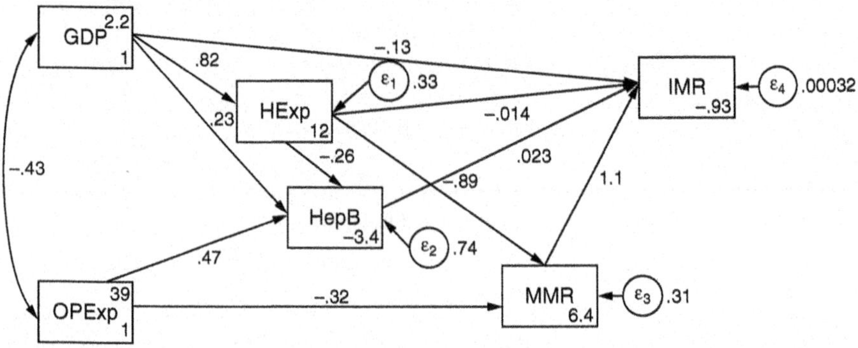

Fig. 7.2 The estimated model

predictors, which is calculated from $(.82)^2$, is equal to .671, while the variance left unexplained is calculated as $1 - .671 = .33$. Moreover it provides the value of R^2 for the simultaneous system of equations. For instance, the model equation of the endogenous variable MMR has explained 69% of the variance in the MMR.

7.3.4 *Fitted Model*

Table 7.6 shows: (1) the direct effect that measures the straight influence of one variable on another without any mediators, which is also shown in Fig. 7.2; (2) the indirect effect that measures the influence of one variable on another through one or more mediators; and (3) the total effect, which is the sum of both direct and indirect effects. These effects are illustrated by the standardized path coefficients. Each path presents the response of the endogenous variable to a standard unit change in the exogenous variable, while all other variables in the model are held constant (Brown and Moore 2012). Thus the results can be interpreted as follows:

1. The model partially supports the first hypothesis. There is a statistically significant direct effect of GDP on HExp, and the IMR (.82 and −.13 respectively),

Table 7.6 Estimated standardized path coefficients

			Direct	Indirect	Total
GDP	→	HExp	.82***	–	.82***
GDP	→	HepB	.23	.21	.45*
GDP	→	IMR	−.13***	−.64***	−.0073***
GDP	→	MMR	–	−.73***	−.73***
OPExp	→	HepB	.47*	–	.47*
OPExp	→	MMR	−.32*	–	−.32*
OPExp	→	IMR	–	−.33*	−2.06*
HExp	→	HepB	−.26	–	−.26
HExp	→	MMR	−.89***	–	−.89***
HExp	→	IMR	−.014*	−.96***	−.95***
HepB	→	IMR	.023	–	.023
MMR	→	IMR	1.1***	–	1.09***

$^{*}p < .05$, and $^{***}p < .001$

i.e. as the GDP increases by one unit, the HExp increases by .82 unit, and the IMR decreases by .13 unit, while all other variables are held constant. The direct effect of GDP on HepB is not statistically significant.

2. The model partially supports the second hypothesis. There is a statistically significant direct effect of HExp on the MMR, and IMR (−.89 and −.014 respectively). That is, as HExp increases by one unit, both the MMR and IMR decrease by .89 and .014 units respectively, with all other variables are held constant. The direct effect of HExp on the HepB is not statistically significant.
3. The model supports the third hypothesis. OPExp has a statistically significant positive effect on HepB and negative effect on the MMR (.47 and −.32 respectively).
4. The model supports the fourth hypothesis. There are a statistically significant indirect effects of the GDP on both the MMR and IMR (−.73 and −.64 respectively), operating through current health expenditures.
5. The model supports the fifth hypothesis. There is a significant indirect effect of OPExp on the IMR (−.33) through the MMR.

SEM is a set of simultaneous regression equations. Eventually this enabled us to examine the different connections between the variables more broadly, and recognize both direct, indirect and total effects of the economic indicators on the IMR. The data shows that hepatitis B immunization increased from 74% in 1995 to 93% in 2015, having reached 99% in 2001. Simultaneously, the model shows that both the direct effect of HepB and the indirect effect of GDP, through HepB, on the IMR are not statistically significant. This suggests that the immunization rate is already reaching complete coverage, and is unlikely to contribute to further reduction in infant mortality.[2] This illustrates the importance of considering the

[2]Further decline in immunizations coverage might have an adverse effect on the IMR.

indirect and internal relationships between these indicators to better understand and be able to reduce the infant mortality rate.

7.4 Conclusion and Recommendations

This study outlines a framework for the application of SEM to examine a specific issue. We studied the causal connections between some economic indicators and the IMR. The results show the extent to which the gross domestic product per capita affects the infant mortality rate directly and indirectly through some other mediators. Moreover, our findings highlight the crucial effect of maternal mortality. However, we did not fully underline the factors that affect maternal mortality in this study. These factors can be related to the general care of mothers, including nutrition, care, immunization during pregnancy, and the conditions during the delivery. For further research work on this area, it is important to recognize these factors and other child vaccinations to see how that would affect the model. In sum, policy makers (1) should consider the different relationships between the causes of the IMR; (2) innovate strategies to improve the health systems; and (3) increase the share of GDP spent on current health costs. Together, these policies in turn will contribute to reducing the infant mortality rate.

Acknowledgments The research reported in this paper was supported by the Higher Education Excellence Program of the Ministry of Human Capacities in the frame of Artificial Intelligence research area of the Budapest University of Technology (BME FIKP-MI/SC). It was also supported by the National Research, Development and Innovation Fund (TUDFO/51757/2019-ITM, Thematic Excellence Program) and by the EFOP-3.6.2-16-2017-00015-HU-MATHS-IN Hungarian Industrial and Innovation project to develop its mathematical support activity.

References

Amouzou, A., & Hill, K. (2003). Child mortality and socioeconomic status in Sub-Saharan Africa. *African Population Studies, 19*(1), 1–11.

Bolla, M., Abdelkhalek, F., & Baranyi, M. (2019). Graphical models, regression graphs, and recursive linear regression in a unified way. *Acta Scientiarum Mathematicarum, 85*, 9–57.

Brown, T. A., & Moore, M. T. (2012). *Confirmatory factor analysis handbook of structural equation modeling*. London: Guilford Press.

Haavelmo, T. (1943). The statistical implications of a system of simultaneous equations. *Econometrica, 11*, 1–11.

Hu, L.-T., & Bentler, P. (1999). Cut off criteria for fit indexes in covariance structure analysis: Conventional criteria versus new alternatives. *SEM-Multidisciplinary, 6*(1), 1–55.

Jöreskog, K. G. (1977). Structural equation models in the social sciences. Specification, estimation and testing. In P.R. Krishnaiah (Ed.), *Applications of statistics* (pp. 265–287). Amsterdam: North-Holland Publishing Co.

Kembo, J., & van Ginneken, J. K. (2009). Determinants of infant and child mortality in Zimbabwe: Results of multivariate hazard analysis. *Demographic Research, 21*(13), 367–834.

Kiiveri, H., Speed, T. P., & Carlin, J. B. (1984) Recursive casual models. *Journal of the Australian Mathematical Society, 36*, 30–52.
Raykov, T., & Marcoulides, G. A. (2006). *A first course in structural equation modeling*. London: Lawrence Erlbaum Associates.
Wermuth, N. (1980). Recursive equations, covariance selection, and path analysis. *Journal of the American Statistical Association, 75*, 963–972.
Wermuth, N., & Sadeghi, K. (2012) Sequences of regressions and their independences. *TEST, 21*, 215–279.
Wright, S. (1934) The method of path coefficients. *The Annals of Mathematical Statistics, 5*(3), 161–215.
United Nations. Sustainable developments goals. (https://www.un.org/sustainabledevelopment/health/). Accessed March 2019.
United Nations. Millennium Development Goals targets and indicators 2015, statistical tables. (http://mdgs.un.org/unsd/mdg/Resources/Static/Products/Progress2015/Statannex.pdf). Accessed March 2019.

Chapter 8
Modeling of Mortality in Elderly by Lung Cancer in the Northeast of Brazil

João Batista Carvalho and Neir Antunes Paes

8.1 Introduction

Cancer occupies the second position in the mortality of the human being at the present time. Considering its typology, lung cancer is among the highest in both incidence and mortality worldwide (Ferlay et al. 2013). In Brazil, 18,740 new cases of the disease in men and 12,530 in women were estimated for the year 2019, corresponding to an estimated risk of 18.2 new cases per 100,000 men and 11.8 per 100,000 women. Thus, without considering non-melanoma skin tumors, lung cancer is the second most frequent tumor among men and the fourth among Brazilian women National Institute of Cancer José Alencar Gomes da Silva (INCA) 2017).

Regarding the mortality of men by cancer, lung cancer is the second largest villain for Brazilians, being the largest burden of the disease in the age group of 60 years and over. In 2016, it was responsible for 12,612 deaths of the elderly, which corresponded to 15.3% of all deaths of elderly men due to neoplasias in the country. For the Northeast Region, which is the third in number of deaths of elderly people due to lung cancer in Brazil, this percentage reached 12.8% (Brazil. Department of Informatics of SUS (DATASUS) 2018).

With a population of 56 million inhabitants in 2015 and an HDI of 0.69 classified as low, the Northeast presents the country's lowest development indicators. It represents 18% of the Brazilian territory and contributes about 14% of the national

J. B. Carvalho (✉)
Department of Statistics, Federal University of Campina Grande, Campina Grande, Paraíba, Brazil

N. A. Paes
Health and Decision Modeling Postgraduate Course, Federal University of Paraíba, João Pessoa, Paraíba, Brazil
e-mail: antunes@de.ufpb.br

C. H. Skiadas, C. Skiadas (eds.), *Demography of Population Health, Aging and Health Expenditures*, The Springer Series on Demographic Methods and Population Analysis 50, https://doi.org/10.1007/978-3-030-44695-6_8

GDP. The public health sector – SUS – provides universal access to all Brazilians, which covers about 75% of the population. Three quarters of the population live in urban areas along the Atlantic coast (Atlas of Human Development in Brazil (AHD) 2013). The Northeast Region is one of the least developed in Latin America. It goes through a late and unequal process of demographic and epidemiological transition and faces the consequences of population aging without having reached reasonable levels of development Santos and Paes 2014).

Smoking is the leading cause of lung cancer. Thus, mortality due to this neoplasm, in general, reflects the consumption of cigarettes in the region, which varies according to socioeconomic and environmental factors (American Cancer Society (ACS) 2005). While on the one hand, the wealthiest nations have already shown a declining trend in their cancer mortality rates, on the other hand, in underdeveloped countries and those in development, rates are continually rising (Barbosa et al. 2015).

In Brazil, however, the scarcity of research on the relationship between cancer and factors that express the living conditions of the population makes it difficult to understand this issue. In addition, the poorest regions of the country still suffer from deficiency in the quality of cancer mortality data (Carvalho and Paes 2018; Malta et al. 2016). The objective of this study was, then, to evaluate the relationship between lung cancer mortality rates and the indicators of living conditions of the elderly in the northeast of Brazil from 2010 to 2015.

8.2 Methods

An ecological study was adopted, with transversal cuts in the years of 2010 and 2015, having as units of observational analysis the 187 micro-regions that make up the nine states of the Northeast Region of Brazil. The database was formed by the standardized rates of lung cancer mortality and living conditions indicators, referring to the elderly population of the northeastern micro-regions, disaggregated by sex and quinquennial age groups (60 to 64, 65 to 69, 70 to 74,75 to 79, 80 and more).

For the calculation of the rates, the information on the deaths was extracted from the Mortality Information System of the Ministry of Health (SIM/MS), available at www.datasus.gov.br. To reduce the effect of random fluctuations in death data, the means of deaths calculated for the triennia from 2009 to 2011 and 2014 to 2016 were used, as references for 2010 and 2015, respectively. The census population of 2010 and projected population for 2015 of each micro-region, by sex and quinquennial age groups of the elderly, necessary for the calculation of the rates, were obtained from the Brazilian Institute of Geography and Statistics (IBGE), available at www.ibge.gov.br.

The living conditions indicators were obtained from the System of Health Indicators and Accompaniment of Policies of the Elderly (SISAP-Idoso) of the Oswaldo Cruz Foundation (Fiocruz), available at https://sisapidoso.icict.fiocruz.br. This system is formed by several indicators grouped into dimensions, including the

Table 8.1 Indicators by dimension of condition of life

Dimension	Indicator
Environmental condition	Elderly people living in households with piped water
	Elderly people living in households with sewage system
	Elderly people living in households with waste disposal service
	Elderly people living in inadequate households
Socioeconomic condition	Illiterate elderly people
	Elderly persons with a nominal income of up to one minimum wage
	Economically active elderly
	Elderly people living in poverty
	Elderly people living alone
	Elderly people who are responsible for the Household
Demographic condition	Dependency ratio
	Population aging index
	Elderly people residents in urban areas

Source: SISAP-Idoso 2010

environmental, socioeconomic and demographic dimensions, which provide a series of useful indicators to measure the living conditions of the elderly. After a scan in the database, all the indicators were selected, within these dimensions, with information available by gender and at the municipal level, for subsequent aggregation in micro-regions. The selected indicators are presented in Table 8.1. The indicators of the environmental and socioeconomic dimensions are expressed as percentages and the demographic indicators are of the ratio type, except for the indicator – elderly people living in urban areas – which is expressed as a percentage.

Two steps were taken to develop this research. The first one consisted in producing mortality rates after analyzing the quality of death records. The second one used the Confirmatory Factor Analysis to assess the adjustment adequacy of a proposed model relating lung cancer mortality with life expectancy.

Step 1: Quality of data.

(i). Ill-defined causes – 9.2% of the total deaths of the elderly in the Northeast were classified as ill-defined causes. In this way, the method proposed by Ledermann 1955 was used to redistribute between neoplasms and the other Chapters of the International Classification of Diseases (ICD-10) the deaths of elderly men from each micro-region classified in the group of ill-defined (Chapter XVIII of ICD-10). This method uses a simple linear regression. In its application, the proportion of deaths of elderly men due to neoplasms was considered as a response variable and the proportion of deaths of elderly men due to ill-defined causes was a covariate. Therefore, the absolute value of the coefficient of the estimated line provided the proportion of deaths due to ill- defined causes to be redistributed to the neoplasms in each micro-region.

(ii). Correction factors – To reduce the impact of death coverage deficits, correction factors were estimated using the quotient between the deaths of elderly men due to neoplasms corrected according to Active Search Research (a strategy of the Ministry of Health to capture events that were not informed as deaths to the SIM/MS) (Szwarcwald et al. 2011) and the ones registered (observed) at SIM. Thus, the product between the registered deaths and the corresponding correction factors provided the total of deaths due to neoplasms corrected for under reporting in each micro-region. Proportional redistribution of deaths due to neoplasms corrected by under-registration among types of neoplasms was carried out.

(iii). Garbage codes – The deaths related to the garbage codes were redistributed: malignant neoplasms from other poorly defined locations (C76) and malignant neoplasms with no location specification (C80) (Naghavi et al. 2010).

(iv). Redistribution of deaths by age and sex – Thus, the deaths of elderly people due to poorly defined causes of lung cancer, underreporting deaths and garbage codes were obtained which, proportionally, were redistributed among the five-year age groups of the elderly aged 60 and over.

(v). The standardized rates of mortality – The standardized rates of mortality of the elderly by lung cancer were calculated for each micro-region using as standard population the census population of Brazil in 2010, referring to the elderly of both sexes. Rates were expressed per 100,000 inhabitants (60 and over).

Step 2: Application of Confirmatory Factorial Analysis (CFA) technique.

The model of CFA establishes the form of how latent variables (ξ) are measured by observable variables (X). This is,

$$X = \Lambda\xi + \varepsilon$$

Where Λ is the matrix of regression coefficients and ε, the vector of random errors (Hair Jr. et al. 2010).

The proposed theoretical model involves the relation between the exogenous variable (conditions of life), measured indirectly by the indicators obtained from the *SISAP-Idoso*, and the endogenous variable standardized rate of lung cancer mortality. In the CFA application, the software AMOS – Analysis of Moment Structures, version 18 was used. The estimation of the model was done by the maximum likelihood method. The analysis provided the standardized estimates of the factorial weights and the adjustment goodness indexes (Marôco 2014), which are represented in Table 8.4.

8.3 Results

For the Northeast (Table 8.2), with the correction, a total of 440 deaths were added to the deaths of the elderly men in 2010 and 408 in 2015, representing an increase of 27.7% and 20.1%, respectively. Of these, the majority were ill- defined deaths, 176

Table 8.2 Contribution of each stage of correction of lung cancer deaths among the elderly of the Northeast of Brazil, 2010 and 2015

			Stages of correction of deaths			
Sex	Registered deaths	Corrected deaths	Under-reporting	Ill-defined	Garbage codes	Total
2010						
Male	1586	2027	169	176	95	440
Female	1066	1408	151	123	67	342
Total	2652	3435	320	299	162	782
2015						
Male	2090	2498	151	172	85	408
Female	1626	1992	119	159	88	366
Total	3716	4490	270	331	173	774

Source: Data Base was extracted from the Mortality Information System of the Ministry of Health

(40.0%) in 2010, and 172 (42.2%) in 2015. The number of underreported deaths was 38.4% in 2010 and 37.0% in 2015 and garbage codes corresponded to 21.6% and 20.8% in 2010 and 2015 respectively.

For the elderly women, in 2010, in the Northeast, 342 deaths were recovered with the correction, corresponding to an increase of 32.1%. Most of these, 151 (44.2%), refer to under-recorded deaths. For 2015, the total number of recovered deaths was 366, which is equivalent to an increase of 22.5%. In that year, a higher impact of ill-defined deaths was observed, accounting for 43.4% of the total deaths recovered.

The mean standardized lung cancer mortality rates are presented per 100,000 elderly citizens in Table 8.3. The rates refer to the average of the microregion rates for each state. The highest averages were observed for male lung cancer in all states and in both years, changing the average from 66.1 to 87.8 per 100 thousand elderly people in the period for the entire Northeast. Also, for the Northeast, the variation in the period for women was from 38.6 to 53.2 per 100 thousand elderly people. Lung cancer in males presented higher average rates in Ceará, Sergipe and Piauí in both years. In women, this type of cancer had higher average rates in Ceará, Rio Grande do Norte and Pernambuco. The lowest averages were observed in Alagoas. In all states, significant increases were observed in the comparison of the years, especially Paraíba, which showed an increase in the period of 64.0% and 62.9% in male and female lung cancers, respectively.

The results of the estimates for the final model are presented in Figure 8.1, for men. Among the thirteen initial indicators used in the analysis, four of them provided statistical significance to the adjustment of the re-specified model. The results of the standardized solution for the final model were significant for the following indicators: percentage of elderly people living in households with running water (0.58), percentage of illiterate elderly people (−0.90), percentage of elderly with nominal income up to a minimum wage (−0.63) and dependency ratio (−0.94), taking into account the convergent validity criteria for the measurement model of

Table 8.3 Mean standardized mortality rates (per 100 thousand elderly) for lung cancer for the microregions of the Brazilian Northeast states

	Male lung			Female lung		
	Mortality rate			Mortality rate		
State (no of microregions)	2010	2015	Variation (%)	2010	2015	Variation (%)
Maranhão (21)	59.4	82.7	39.2	31.6	49.7	57.3
Piauí (15)	73.3	94.3	28.6	35.4	48.3	36.4
Ceará (33)	80.9	100.2	23.9	51.6	72.9	41.3
R. G. Norte (19)	70.9	94.3	33.0	45.9	55.9	21.8
Paraíba (23)	54.2	88.9	64.0	35.6	58.0	62.9
Pernambuco (18)	67.4	90.3	34.0	41.5	55.3	33.3
Alagoas (13)	51.3	65.8	28.3	26.6	41.7	56.8
Sergipe (13)	75.8	98.4	29.8	37.6	42.4	12.8
Bahia (32)	58.9	73.8	25.3	32.5	40.3	24.0
Northeast (187)	66.1	87.8	32.8	38.6	53.2	37.8

Source: Data Base was extracted from the Mortality Information System of the Ministry of Health

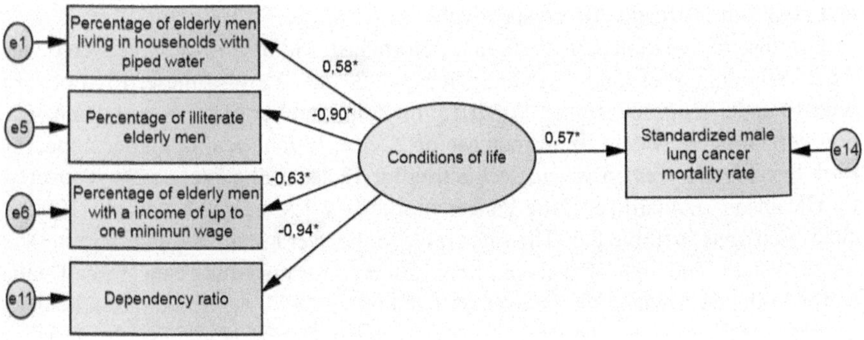

Fig. 8.1 Standardized estimates of the CFA for the final model (male lung). (*significant correlations, p-value <0.01. Random errors: e1, e5, e6, e11, e14)

the exogenous construct called "living conditions". High values of these last three indicators reflect the low living conditions of the elderly.

With a standardized coefficient of 0.57, the final model for elderly men had a direct effect on the relationship between standardized rates of lung cancer mortality and the percentage of elderly in households with running water, and an inverse effect with indicators that represent poor living conditions for the elderly. That is, in general, rates were higher in microregions where the magnitudes of the first indicator were high and of the latter, low.

In relation to women, three indicators provided statistical significance to the adjustment of the re-specified model (Fig. 8.2): percentage of elderly women living in households with running water (0.61), percentage of illiterate elderly women and dependency ratio, with corresponding values of the standardized solution of 0.61, −0.79 and −0.75, respectively, considering the criteria of convergent validity for the

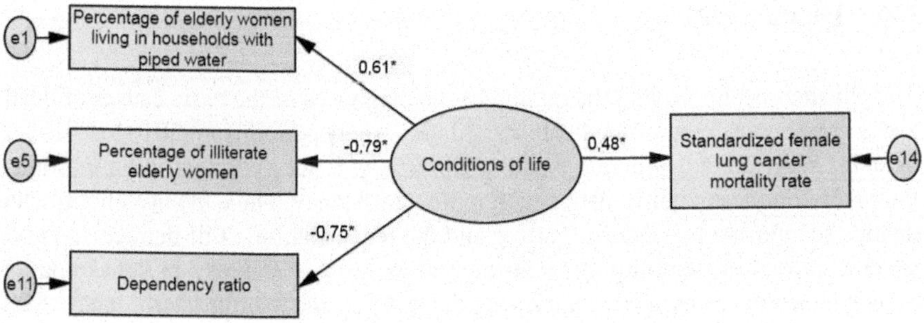

Fig. 8.2 Standardized estimates of the CFA for the final model (female lung). (*significant correlations, p-value <0.01. Random errors: e1, e5, e11, e14)

Table 8.4 Adjustment indicators of the CFA for modeling the mortality of the elderly for lung cancer in the micro-regions of the Brazilian Northeast

Adjustment indicator	Criterion for a good fit	Final model (male lung)	Final model (female lung)
Chi-square (p-value)	p-value > 0.05[a]	6.11 (0.30)	2.10 (0.35)
Normed chi-square	Between 1 and 2	1.221	1.052
Goodness of Fit Index (GFI)	> 0.90	0.987	0.994
Adjustment GFI	> 0.90	0.962	0.971
Normed Fit Index (NFI)	> 0.90	0.987	0.988
Comparative Fit Index(CFI)	> 0.90	0.998	0.998
Tucker-Lewis Index (TLI)	> 0.90	0.995	0.998
Root mean square error approximation (RMSEA)	< 0.08	0.034	0.017

[a]p-value >0.05 indicates overall adjustment of the model

model of measurement of the exogenous variable "living conditions". High values of the latter two indicators reflect the low living conditions of the elderly women.

With a standardized coefficient of 0.48, the model had a direct effect on the relationship between the standardized rates of lung cancer mortality and the percentage of elderly women in households with running water, and an inverse effect with the indicators that represent poor living conditions of the elderly women. This means that, in general, rates were higher in microregions where the magnitudes of the first indicator were high and of the last two, low.

In addition, the criteria for a good fit of the model were all satisfied for both final models, according to the normative criteria for the adjustment indices shown in Table 8.4.

8.4 Discussion

The advances in improving the quality of the diagnoses of the basic causes of death and a greater coverage of the death records in a short period from 2010 to 2015 are evident. With the adjustments made in this study, it was possible to approximate the reality on these vital statistics with more consistency. Thus, the use of corrected deaths, accounting for under-reporting and the redistribution of ill-defined and non-specific causes of mortality allowed to obtain a greater accuracy in the estimation of the mortality rates at the micro-region level of the Northeastern states. Still, the results indicated that the quality of deaths from lung cancer in the Northeast is still worrisome, given the considerable increases in their values observed after correction. That is, they indicate that it is necessary to maintain good monitoring and quality in the collection of data from the information systems.

This fact draws attention to the need for managers to better perceive the true levels of mortality of the disease when dealing with planning actions in public and epidemiological health. Thus, because these levels are higher than those registered, there is an increased need for measures to be taken in the sense of reinforcing the training of physicians for the correct filling of death certificates, exercising more vigilance regarding the registries and improving the system of collection and monitoring of data, among other measures. This way, there will be a greater accuracy of mortality statistics and, consequently, of the indicators derived from them, as well as improvement in the specific actions for the elderly and the general population.

It was observed that regional inequalities in lung cancer mortality are positively associated with living conditions reflected in education, income and household conditions. In general, rates for these causes were higher in micro-regions with lower percentages of illiterate elderly, lower percentages of elderly people with nominal income up to a minimum wage, lower dependency ratio and higher percentage of elderly people living in households with running water. These findings corroborate some studies that addressed this problem (Ribeiro and Nardocci 2013; Sharpe 2014; Medeiros 2015).

According to recent research, cancer is already the leading cause of death in about 10% of Brazilian municipalities, most of them located in more developed regions of the country, precisely where life expectancy and the HDI are highest (Cepas 2018). In the case of lung cancer, studies indicate its relation to social and economic differentials. Rafiemanesh 2016 conducted a study with the objective of investigating the incidence/mortality of lung cancer in the world and its relation with HDI, based on data from the Global Cancer Project of 2012. This study showed the existence of a significant relationship between mortality rates with life expectancy at birth and average years of schooling. Hagedoorn 2016 examined the contribution of individual characteristics of people with age over

65 years to lung cancer mortality rates in Belgium, a highly industrialized country with one of the highest rates of lung cancer mortality among men in Europe. According to the authors, the urban characteristics favored mortality from lung cancer.

As age is a factor that contributes to the increase in cancer mortality, it is important to raise awareness of the consequences that the levels of these rates will reach in the future due to the fact that the states of the Brazilian Northeast are in a process of demographic transition not yet consolidated, with the elderly population rapidly increasing. In 2030, projections indicate that Brazil will be the fifth country with the largest elderly population in the world (Atlas of Human Development in Brazil (AHD) 2013). In this context, special attention should be given to the states of Alagoas and Maranhão with the lowest rates of aging and the poorest in the country in 2010 among all states, where the male population is, therefore, more vulnerable.

8.5 Conclusions

It was evidenced that the mortality of elderly people due to lung cancer in the Northeast of Brazil was associated with the socioeconomic level. Therefore, the increase in the averages of the standardized mortality rates in 2015 compared to the averages of the 2010 rates may have as one of its possible causes the improvement in the living conditions, generally linked to the population aging and changes in the epidemiological pattern, due to a higher incidence of the disease in men and women.

In this context, it is speculated that these levels should increase further in the near future, aggravating an already worrying scenario. Thus, health care and awareness campaigns, especially the fight against smoking, promoted by governmental institutions are useful, but they are not enough, considering the urgent challenges to be faced by the health system of the Brazilian Northeast. Among the strategies for conducting government policies and actions of social and health care, this research draws attention to the relevance of the described relationship. There is evidence that the improvement of the socioeconomic level, coupled with an ongoing aging process, may increase the incidence of lung cancer and, if there is no improvement in overall survival through global therapeutic measures, the levels of mortality due to this type of pathology will increase in the Northeast of Brazil.

References

American Cancer Society (ACS). (2005). *Cancer facts & figures 2015*. Atlanta, 2015 [cited 2019 Jan 16]. Available from: http://oralcancerfoundation.org/wp-content/uploads/2016/03/Us_Cancer_Facts.pdf.

Atlas of Human Development in Brazil (AHD). (2013). [cited 2018 dec 08]. Available from: http://atlasbrasil.org.br/2013/.

Barbosa, I. R., Souza, D. L. B., Costa, I. C., & Pérez, M. B. (2015). Cancer mortality in Brazil: temporal trends and predictions for the year 2030. *Medicine (Baltimore), 94*(16), e746.

Brazil. Department of Informatics of SUS (DATASUS). (2018). *Mortality information system*. Brasília: Ministry of Health Brazil. [cited 2018 Dec 05]. Available from: http://www.datasus.gov.br.

Carvalho, J. B., & Paes, N. A. (2018). *Corrected cancer mortality rates for the elderly in the states of the Brazilian Northeast*. Cien Saude Colet [serial on the Internet], 2018 [cited 2019 Feb 5]. Available from: http://www.cienciaesaudecoletiva.com.br/artigos.

Cepas, T. (2018). Cancer as the leading cause of death in Brazilian municipalities. *Oncology Observatory*. [cited 2019 mar 05]. Available from: https://observatoriodeoncologia.com.br.

Ferlay, J., Soerjomataram, I., Ervik, M., Dikshit, R., Eser, S., Mathers, C. et al. (2013). *GLOBOCAN 2012 v1.0*. Cancer incidence and mortality worldwide: IARC CancerBase No. 11 [serial on the Internet] 2013 [cited 2018 nov 15]. Available from: http://www.globocan.iarc.fr.

Hagedoorn, P. (2016). Regional inequalities in lung cancer mortality in Belgium at the beginning of the 21st century: The contribution of individual and area-level socioeconomic status and industrial exposure. *PLoS ONE, 11*(1), 1–18.

Hair, J. F., Jr., Black, W. C., Babin, B. J., & Anderson, R. E. (2010). *Multivariate data analysis* (7th ed.). New York: Prentice Hall.

Ledermann, S. (1955). La repartition des decés de causes indeterminées. *Revue d' International de Statistique, 23*(I/3).

Malta, D. C., Abreu, D. M., Moura, L. D., Lana, G. C., Azevedo, G., & França, E. (2016). Trends in corrected lung cancer mortality rates in Brazil and regions. *Rev Saúde Pública, 50*, 33.

Marôco, J. (2014). *Análise de equações estruturais: fundamentos teóricos, software & aplicações*. Pêro Pinheiro: Report Number.

Medeiros, M. J.(2015). *Relationship between colorectal cancer and socioeconomic indicators in São Paulo: use of spatial regression models*. PhD. thesis in epidemiology, Faculdade de Saúde Pública, Universidade de São Paulo, São Paulo.

Naghavi, M., Makela, S., Foreman, K., O'Brien, J., Pourmalek, F., & Lozano, R. (2010). Algorithms for enhancing public health utility of national causes of death data. *Population health metrics, 8*(1), 1.

National Institute of Cancer José Alencar Gomes da Silva (INCA). (2017). *Estimate 2018: incidence of cancer in Brazil*. Rio de Janeiro, 2017 [cited 2018 Nov 18]. Available from: http://www.inca.gov.br/estimativa/2018/index.asp.

Rafiemanesh, H. (2016. [cited 2019 mar 12]. Available from: http://jtd.amegroups.com/article/view/7522). Epidemiology, incidence and mortality of lung cancer and their relationship with the development index in the world. *Journal of Thoracic Disease, North America, 8*, 1094–1102.

Ribeiro, A. A., & Nardocci, A. C. (2013). Socioeconomic inequalities in cancer incidence and mortality: review of ecological studies, 1998–2008. *Saúde Soc, 22*(3), 878–891.

Santos, J. P., & Paes, N. A. (2014). Association between life conditions and vulnerability with mortality from cardiovascular diseases in elderly men of Northeast Brazil. *Rev Bras Epidemiol*, (2), 17, 407–420.

Sharpe, K. H. (2014). Association between socioeconomic factors and cancer risk. A population cohort study in Scotland (1991–2006). *PLoS One, 9*, e89513.

Szwarcwald, C. L., Morais, O. L., Frias, P. G., Souza, P. R., Jr., Cortez-Escalante, J. J., et al. (2011). *Active search for deaths and births in the Northeast and in the Legal Amazon: estimation of SIM and SINASC coverage in Brazilian municipalities*. Brasília: Ministry of Health. Department of Health Surveillance. Health Brazil 2010.

Chapter 9
Demographics of the Russian Pension Reform

Dalkhat M. Ediev

9.1 Introduction

Governmental proposal for the extension of the legal retirement age (from 60 to 65 for men, and from 55 to 60 for women by 2028) has generated a far-reaching discussion – as well as wide-spread opposition – in the Russian society. Demographic and economic matters normally confined to scholarly publications have an effect into the public domain and social networks. Such matters include population ageing, improving survival and demographic dependency ratios, life expectations at birth and upon retirement, the natural rent and its place in the social welfare, taxation, the size of the labor force, productivity and wages, the replacement rate, the informal sector, income inequality, etc.

While arguments on demographics form the core of both the governmental proposal and the wider discussion, the discussion is complicated due to lack of relevant demographic statistics. The Human Mortality Database (University of California and the Max Planck Institute for Demographic Research (Rostock) 2019) contains Russian data for only since 1959, while the available regional trends are even shorter (Center for Demographic Research (Moscow/Russia) 2018). This does not allow computing a single cohort life table for Russia. In the meantime, the debate relies on guesstimates of individual survival prospects, remaining lifetimes and pension contributions/benefits. Due to the lack of objective statistics, discussants

D. M. Ediev (✉)
North-Caucasian State Academy, Cherkessk, Russia

International Institute for Applied Systems Analysis, Wittgenstein Centre for Demography and Global Human Capital (IIASA, VID/ÖAW, WU), Laxenburg, Austria

Department of Demography (HSMSS), Lomonosov Moscow State University, Moscow, Russia
e-mail: ediev@iiasa.ac.at

C. H. Skiadas, C. Skiadas (eds.), *Demography of Population Health, Aging and Health Expenditures*, The Springer Series on Demographic Methods and Population Analysis 50, https://doi.org/10.1007/978-3-030-44695-6_9

turn to period indicators as proxies to cohort ones, and even go as far as to suggest the difference between the period life expectancy at birth and the age at retirement as a proxy for the (cohort) life expectation at retirement.

Here, we fill this gap by constructing (and extending) life tables for cohorts entering their 20s after 1959 and by studying various implications of the former and newly proposed pensionable ages for these cohorts. Our aim is also to go beyond the basic demographics of pensions in Russia and discuss normative balances of payments to/from the pension system for the Russian cohorts, to estimate rates of return of those payments in order to study inter-generational actuarial fairness, as well as comparative (dis)advantages of the funded and the pay-as-you-go versions of the system.

9.2 Inputs and Data Preparation

For most of the data, we rely on the Human Mortality Database, HMD (University of California and the Max Planck Institute for Demographic Research (Rostock) 2019) that contains data, by single years of age, until age 110+, on population and death rates in Russia in the period 1959–2014. Data for the subsequent years (until 2017), in single years 0 to 100+, came from the Russian Fertility and Mortality Database, RFMDB (Center for Demographic Research (Moscow/Russia) 2018).

Cohort estimates as well as long-term pension analysis must rely on a longer time series, spanning back before 1959 and projected into the future. For the former, we interpolated, into single years of age, estimates of the population and of death rates of the Russian men and women discussed by Andreev et al. (1998) in five-year-long age groups from 0 to 85. The interpolation of the death rates was done log-linearly; the death rates were additionally smoothed by a moving averaging procedure in logs with a five-year-long age frame. To interpolate population age structures, we used our own spline package. We have also calculated population at the beginning of each year by averaging the two respective mid-year estimates from the original source. For the war period (1941–1945), we apply interpolated values from the periods before and after the war (linearly for the population and log-linearly for the death rates), so as to produce estimates excluding excessive mortality due to the war that should not be institutionalized in the design of the pension system but, rather, should be considered a *force majeure* disruption of the system.

To extend the set of rates and population estimates into the future, we have studied two sources of population projection for Russia: that by Rosstat and others (RANEPA, Rosstat and IIASA 2016) until 2035, single years of age-year, ages 0 to 100+, in good agreement with the recent acceleration of mortality decline in Russia and by the UN World Population Program (UN DESA/Population Division 2017b). The latter projection extends, conveniently for the pension analysis, until 2100, age groups 0 to 100+, but offers somewhat more pessimistic demographic trends for Russia, as compared to our first source. The most recent available data indicate that the former projection is somewhat more optimistic than the UN projection for

Russia. However, both projections taken separately yielded similar results in our study (comparison not shown here). We, therefore, present analysis based on the UN projection that offers longer time series — suitable for the needs of this paper. After concluding our analysis, UN has rolled out a new set of projections that contains somewhat new results for Russia. The differences between the latest UN projection round and the previous one are, however, minor as far our study is concerned (by 2100, the total population in the new round is 1.7% higher than in the previous one).

A closer examination shows that the datasets we have at hand (HMD, RFMDB, ADKh estimates and the projection) are not fully consistent in population numbers at joint years. To avoid discontinuities and unnatural irregularities in the studied indices, we have harmonized the datasets using the HMD estimates as the primary source — that the other sources were harmonized with. The harmonization was based on adjusting the population numbers but keeping the cohort-wise survival proportions unaltered in each of the sources.bv.

As a result of pooling together the available data and projections and applying necessary interpolations, smoothing and adjustments, we ended up with two matrices, per gender, that form the basis of our subsequent study: a matrix of death rates by age (age 0 to either 85+, 100+, or 110+ depending on the data source) and year (1926 to 2100), and a similar matrix of population sizes. To harmonize our dataset and better prepare it for studying population ageing, we have extended both death rates and population estimates to until age 110+. To do so, we applied the constrained mortality extrapolation method (Ediev 2017) based on the Kannisto's logistic mortality model (Thatcher et al. 1998) and the *combined method* (Eq. 9.7), Ediev 2018) utilizing both the conventional and Mitra's (1984) estimates of the remaining life expectancy in an open age interval. To extrapolate the population numbers to old age, we used the extrapolated death rates and stable population model (Keyfitz and Caswell 2005) with the growth rate over rolling periods of 10 years prior to the year for which the extrapolations are done (same growth rate as was used in the Mitra model above).

To study the cohort dimension, we transformed the period death rates into cohort ones by taking the average of the two corresponding period rates:

$$M_{coh}\left(c,x\right)=0.5\left(M_{per}\left(t=c+x,x\right)+M_{per}\left(t=c+x+1,x\right)\right),x\neq 0, \tag{9.1}$$

where $M_{coh}(c,x)$ is the death rate for cohort c at age x, and $M_{per}(t,x)$ is the death rate at the same age for the calendar period t. At age 0, in view of the higher concentration of infant deaths in the beginning of the year, we used a weighted average with larger weight at the lower Lexis triangle:

$$M_{coh}\left(c,0\right)=\left(3*M_{per}\left(t=c,0\right)+M_{per}\left(t=c+1,0\right)\right)/4. \tag{9.2}$$

After obtaining the cohort death rates by age and gender, we calculated the life table for each cohort and produced necessary indicators of survival and lifespan.

Other than population inputs, we also formed scenarios for the parameters of the labor force and pension system. For the pension system, we used two scenarios for legal retirement ages: one with the current parameters of the system (age at retirement: 55 for women, 60 for men) and one reflecting the governmental reform (age at retirement growing from 55/60 in 2019 to 65 and 60, respectively, by year 2028). The average pension replacement rate (ratio of the average pension to the average wage) was set at 0.39 for men and 0.51 for women in accordance with the current figures (Rosstat 2018); the contribution rate (18.0% of gross wage paid to the system) was also set at its current level. For our simulations, we have translated the average replacement rates into replacement rates to the maximum age-specific wage in the year of interest, so as to stabilize the outcomes of simulations and better approximate the replacement rate to the last wage of a person achieved (maximum in many countries, but not in Russia). Those replacement ratios to the maximum wage constituted 0.35 and 0.49 for men and women, respectively.

The pattern of labor force participation rates (LFPR) is more complex. The rates vary substantially across age and between genders (OECD 2018). We interpolate the OECD estimates available in large age groups into single-year ages using our own monotone cubic spline interpolation technique. Women begin careers at smaller proportions and leave the labor force earlier than men. For either gender, the LFPR falls to about 60% by the legal retirement age (55% and 68% for men and women in 2017). It is hard to say, how the LFPRs may react to changes in the legal retirement ages when the government implements the reform. Due to lack of better alternatives, we link the LFPR schedules to the legal retirement ages in a flexible way representing both the legal push to retire later and the resistance to this process because of increasing frailty with age. To this end, we first form the baseline curves of the LFPRs by averaging and smoothing, separately for each sex, the respective LFPRs over the entire period 2000–2017. Next, for any (new) level of the legal retirement age, we use a step-wise linear transformation of the age scale so that LFPRs do not change at ages 50 and below for women, and 55 and below for men (those are roughly the ages with highest LFPRs), that stay time-invariant at a (possibly changing) legal age at retirement, and that again turn to original values (zero) by age 75 years:

$$f(x) = \begin{cases} f_0(x), x < x_1, \\ f_0\left(x_1 + \frac{x-x_1}{R-x_1}(R_0 - x_1)\right), x_1 \le x \le R, \\ f_0\left(R_0 + \frac{x-R}{x_2-R}(x_2 - R_0)\right), R \le x \le x_2, \\ f_0(x), x > x_2, \end{cases} \quad (9.3)$$

here, $f(x)$ is the transformed age schedule of LFPRs corresponding to the new value R of the legal retirement age, $f_0(x)$ and R_0 are the baseline LFPRs and retirement age, x_1 (set $R_0 - 5$) is the age until which the LFPRs do not change, and x_2 (set 75) is the age after which LFPRs turn zero.

Men and women also differ markedly in their wages (Rosstat 2018). We take account of that effect in our model by assuming the current proportions that imply women's earnings are 30% less compared to men's earnings. Similarly, we apply current age gaps in earnings in our calculations. For scenarios with varying ages at retirement, we assume that the wages are adjusting to new retirement ages and follow transformations like the LFPR's (3).

For benchmarking purposes, we also analyzed the Swedish case, for which the population data come from the Swedish Statistical Agency (Statistics Sweden 2018); the normal retirement age was set at its current level of 65 years for both genders (OECD 2018) for the entire period prior to 2018. For later years, similarly to the Russian case, we consider two scenarios: no change in the retirement age and gradual increase by 68 years, for both genders, by 2026, as announced by the Swedish government (Government Offices of Sweden 2018). LFPRs, wages and replacement rates for Sweden came from the OECD database (the contribution rate in Sweden was set to 17.2% of wages, and the replacement ratios of the pension benefit to the maximum wage were about 45% for both genders). Because the Swedish data extend until age 100+ only, they were extrapolated – using the same models of old-age mortality as in the Russian case – to age 110 + .

9.3 Results in the Period Dimension: The Welfare State View

In this section, we look at demographic developments in calendar years. That view resonates the objective of the welfare state to keep the system in balance at each calendar year or, at least, on average over the planned future. Although availability of natural rent and other sources external to the pension system may allow governments handling negative balance sheets of the pension system, a balanced system remains a desirable target, while strong imbalances put the system at risk of insolvency.

In Fig. 9.1, we present the time trend of total population size, historical and projected, of Russia. The projections indicate the strong negative population momentum (Keyfitz 1971) accumulated by the age composition of the Russian population after the collapse of the Soviet Union when the death rates surged and the fertility rates plummeted with profound long-term demographic consequences, in spite of immigration (Ediev 2001). The recent recovery of fertility rates, migration playing a substantial role in population replacement (Ediev et al. 2013), and declining death rates have enabled Russia to escape a declining population trend, for the time being. Nonetheless the population of Russia is expected to decline to ca. 125 million by the end of the century.[1]

[1]Note that we have harmonized the source datasets to the main collection that came from the Human Mortality Database and excludes the population of the Crimean peninsula (ca. 2 millions).

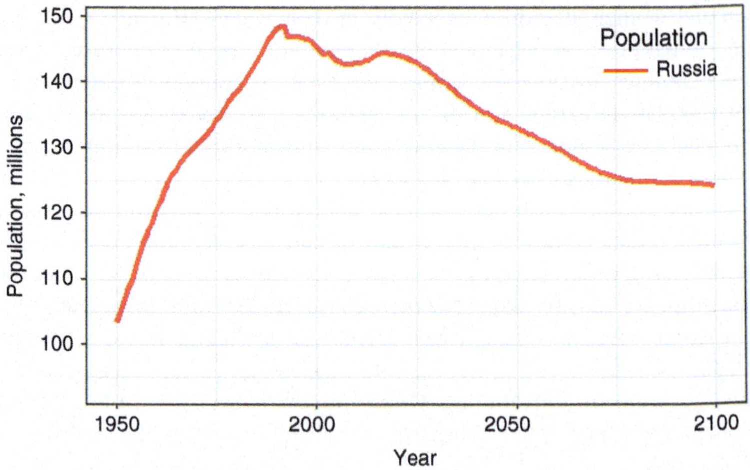

Fig. 9.1 Russian population in 1950–2100, millions: estimates prior to 2017 inclusive and projections afterwards. Projections: UN WPP (UN DESA/Population Division 2017a)

Even though not catastrophic by historical standards, the anticipated population decline is substantial and will be accompanied by important changes in the population composition and in the absolute and relative sizes of various population subgroups. Under the current retirement scheme, the working-age population is rapidly falling in the coming decades (Fig. 9.2a) by ca. 20 million (i.e. about a quarter of its current level). In other words, almost all of the population decline this century may concentrate in the working-age population. The new retirement ages proposed by the Russian government delay the fall (with short-term increase) of the working-age population by the 2030s and reduce the magnitude of the fall by almost half (ca. ten million). Similar trends characterize the labor force (estimated as a product of the LFPR and population numbers), although the difficulty involved in increasing the LFPRs at an advanced old age may make the proposed reforms less efficient to prevent the fall of the labor force (Fig. 9.2b).

One may notice that – apart from the trend change – the proposed reforms in Russia seem to lead to somewhat strengthened instability of the size of the working-age population. This effect becomes more evident, if one looks at the change of the working-age population at its age limits, i.e. due to entries to the workforce of young people and exits of retirees (Fig. 9.3; hereafter, compositional characteristics of the Russian population are shown in comparison to the Swedish case). That change has recently turned from positive to negative in Russia, and from positive to near zero for Sweden. Reforms proposed in both nations, after a period of a 'tempo-effect' (Ryder 1951) when the working-age population is boosted through inclusion of more and more cohorts retiring later, shift the temporal pattern of change in the working-age population by several years, because the new retirement schemes imply different (older) cohorts retiring at any given time when compared to the old retirement schemes. The Russian reform also implies destabilization of labor, marked by the

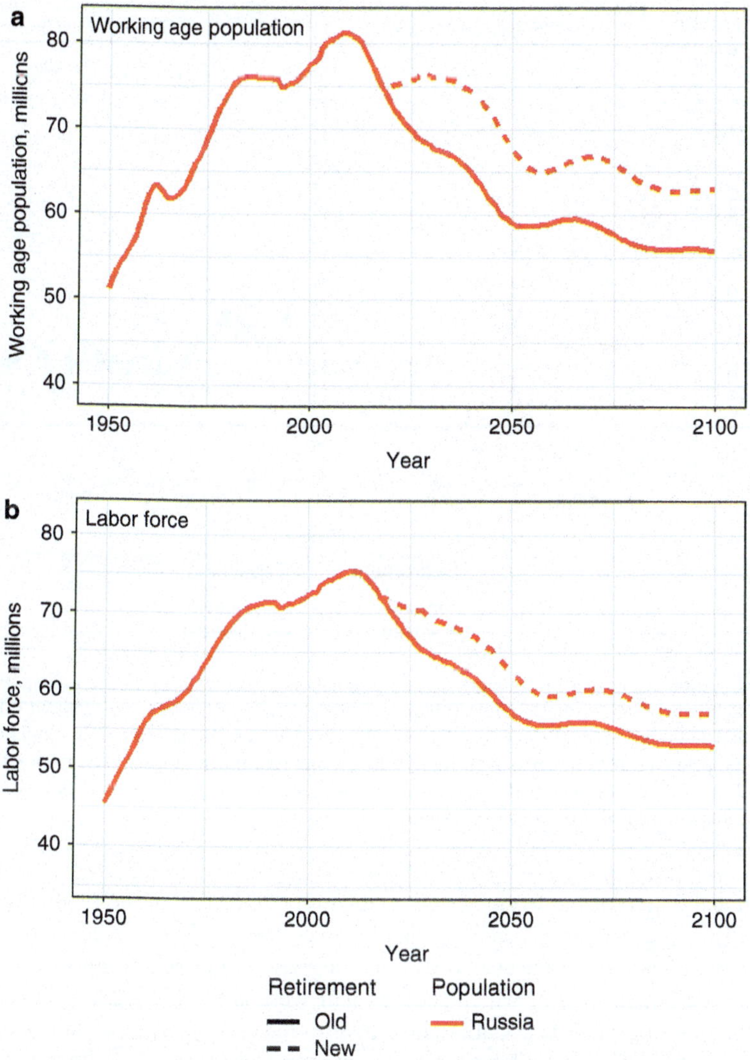

Fig. 9.2 Russian working-age population (**a**) and labor force (**b**) in 1950–2100, millions, under old and newly proposed retirement ages: solid lines – the current retirement ages, broken lines – the new retirement scheme proposed by the government; estimates prior to 2017 inclusive (the bifurcation point in the graph) and projections afterwards. Projections: UN WPP (UN DESA/Population Division 2017a)

strengthening of the swings of the working-age population change, because the new retirement ages introduce resonance between demographic waves at the entrance of the working age and at exit from it (ages at retirement and entry to the work force differ in the new scheme by about 45 years, i.e. roughly one and a half of the demographic generation). Other than volatility, it is notable that the magnitude of

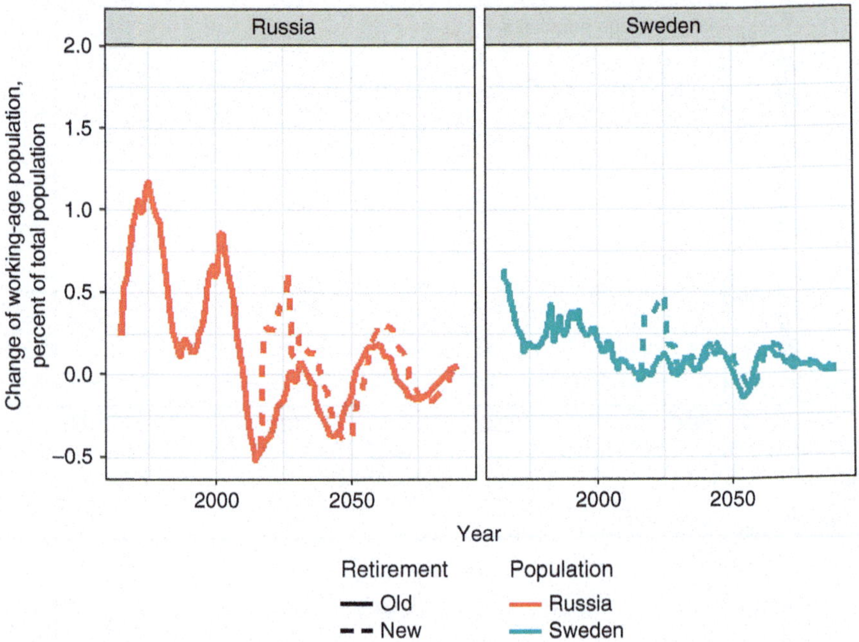

Fig. 9.3 Annual changes to the working-age population in percentages to the total population size due to entries to the workforce of young people and exits of retirees, i.e., without effects of migration and mortality, in Russia (panel to the left) and Sweden (panel to the right) under old and newly proposed retirement ages: solid lines – the current retirement ages, broken lines – the new retirement scheme proposed by the government. Estimates prior to 2017 inclusive (the bifurcation point in the graph) and projections afterwards

the annual deficit of the working-age population (ca. 0.25% of the total population annually in the coming decades) is smaller, even though comparable to the annual net migration at working ages (in 2017, net migration to Russia comprised ca. 1.5% of the total population, of which roughly half were at working ages). That means that immigration may be a viable alternative to the pension reform and will likely make up for the short-term labor force deficits and surpluses induced by the reform.

The Russian old-age population at current retirement ages (Fig. 9.4) has increased rapidly in the past and is expected to continue increasing at a slowing pace and with sizable swings until it stabilizes at above 40 million. The governmental reform may put this figure down to above 30 million.

The old-age dependency (OADR, ratio of population at and above age at retirement to the working-age population) in Russia already exceeds that in Sweden, despite longer lives and more elderly people in the latter (Fig. 9.5). That is because Sweden, its people showing longer lifespan, has been able to secure later retirement and, henceforth a larger share of the working-age population. This is an interesting empirical illustration of the counterintuitive finding by Sanderson and Scherbov (2015; Ediev et al. 2018) that faster increases in life expectancy lead to slower

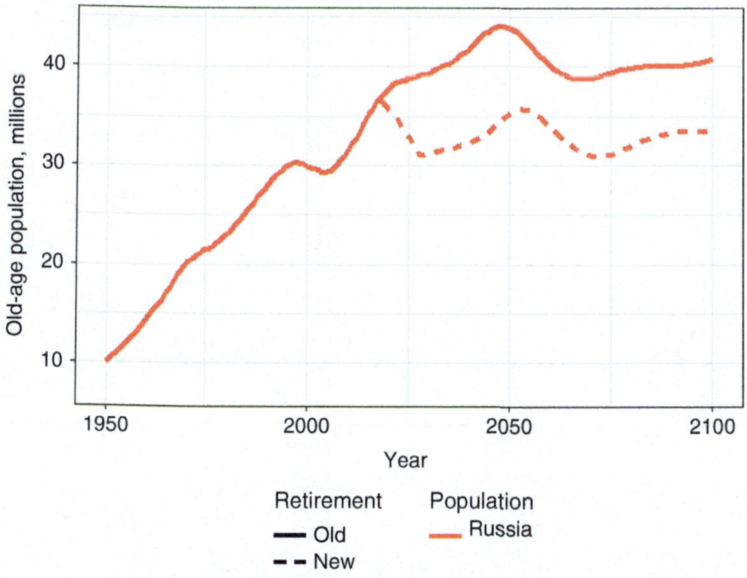

Fig. 9.4 Russian retirement-age population in 1950–2100, millions, under old and newly proposed retirement ages: solid lines – the current retirement ages, broken lines – the new retirement scheme proposed by the government; estimates prior to 2017 inclusive (the bifurcation point in the graph) and projections afterwards. Projections: UN WPP (UN DESA/Population Division 2017a)

population ageing in terms of the prospective age. With no reform of the retirement system, Russian OADR was going to increase further to 60–70%, a historical high and economically hard-to-manage level. Reforms may successfully push the Russian OADRs down to near its current levels, although with substantial swings. Even without reforms, Sweden would only reach Russian OADRs by the end of the century, and the reform could push the ratio down to 40% (still, about 5 percentage points higher than the current level in Sweden).

One of the arguments of opponents of the pension reform in Russia has been decline or stability of the total demographic dependency that combines old- and young-age dependency. Our study shows, however, that the total dependency (Fig. 9.6), that followed a declining trend before 2010s, is already on a rising path, very similar to that of the OADR. This is an outcome one may have expected, because the main driver of OADR (and of total demographic dependency) in the coming decades is declining working-age population that pushed up both the old- and young- age dependency.

In view of the anticipated instability of OADR to accompany the pension reform in Russia, it is advisable to compare the reforms laid down in the two countries with the hypothetical 'defined-dependency' retirement ages (Ediev 2014) that assure time-invariance of OADR in each country. To compute the defined-dependency retirement ages, we first assume a step-wise linear model linking the retirement age of women to that of men, so that the model reproduces the current values of

Fig. 9.5 Old-age dependency ratio in Russia (panel to the left) and Sweden (panel to the right) under old and newly proposed retirement ages: solid lines – the current retirement ages, broken lines – the new retirement scheme proposed by the government. Estimates prior to 2017 inclusive (the bifurcation point in the graph) and projections afterwards

the retirement ages as well as their target values, as suggested by the governmental reform:

$$R^f = \mathrm{MIN}\left(R^m, R^m - C - kR^m\right) \tag{9.4}$$

where R^f and R^m are retirement ages for women and men, respectively and C, k are parameters set to fit the model to the base year and the target levels of the ages at retirement. After setting the model of the gender gap in retirement age, for each projection year, we find such a combination of the ages at retirement that the projected OADR would be equal to the base year (2018) level of the OADR in either country. The results are presented in Fig. 9.7. Interestingly, both countries could keep the OADRs constant at their current levels by increasing the retirement ages at about half the pace proposed by the governments. In Sweden, that gradual increase should continue to the end of the century, while the increase may be halted after the 2050s in Russia. We can also see that the Russian case is more volatile because of stronger swings of births and deaths in Russia as compared to Sweden in the turbulent twentieth century.

Other than population ageing resulting from improved survival and population momentum, migration and fertility also play an important role in shaping the

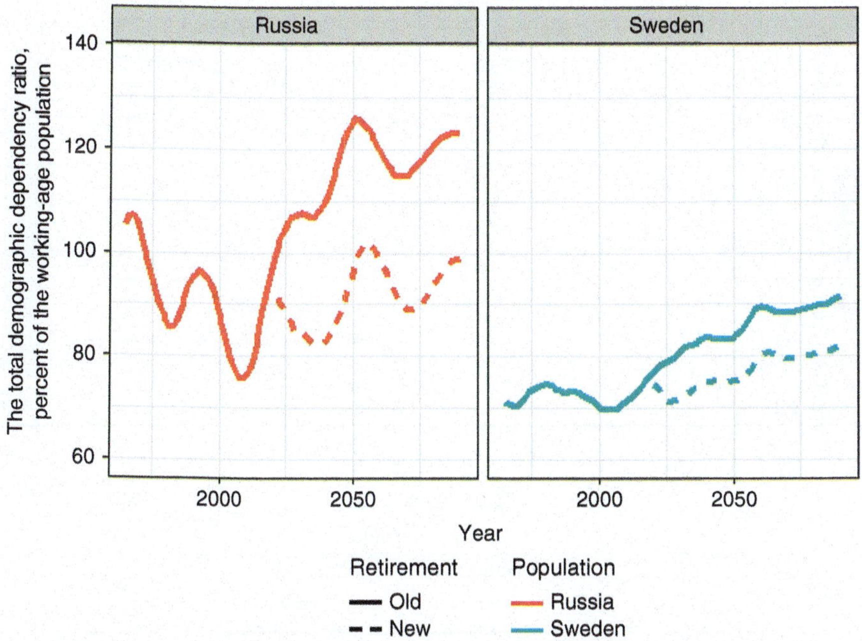

Fig. 9.6 The total demographic dependency ratio (young- and old-age population as percent of working population) in Russia (panel to the left) and Sweden (panel to the right) under old and newly proposed retirement ages: solid lines – the current retirement ages, broken lines – the new retirement scheme proposed by the government. Estimates prior to 2017 inclusive (the bifurcation point in the graph) and projections afterwards

population composition and OADRs. In the absence of migration and at time-constant births, the population composition is determined by cohort survival, and its age structure is formed by parts of the cross-sectional average length of life, CAL (Brouard 1986; Guillot 2003; Ediev 2014). To assess the contribution of migration and fertility deviations from the replacement level, we have calculated CAL-based OADRs and their difference to the observed OADRs (Fig. 9.8). These results shed light on the role of dramatic disturbances of the past decades in Russia (decline of births during the world war and recent socio-economic transformation being the major ones). The fall of Russian fertility far below the replacement level, and an immigration insufficient to compensate for the shortage of births accelerated population ageing and will contribute about ten percentage points to the OADR, albeit with large swings. In Sweden, on the contrary, the net contribution of (moderately high) fertility and immigration is towards lowering the OADR by about five percentage points.

We conclude the analysis of period demographic dimension by looking into how population ageing and growing OADRs may contribute to the (in)solvency of the pension system. To this end, we compute annual 'normative' balances of the two pension systems under assumed LFPRs, wages and replacement ratios (Fig. 9.9). In

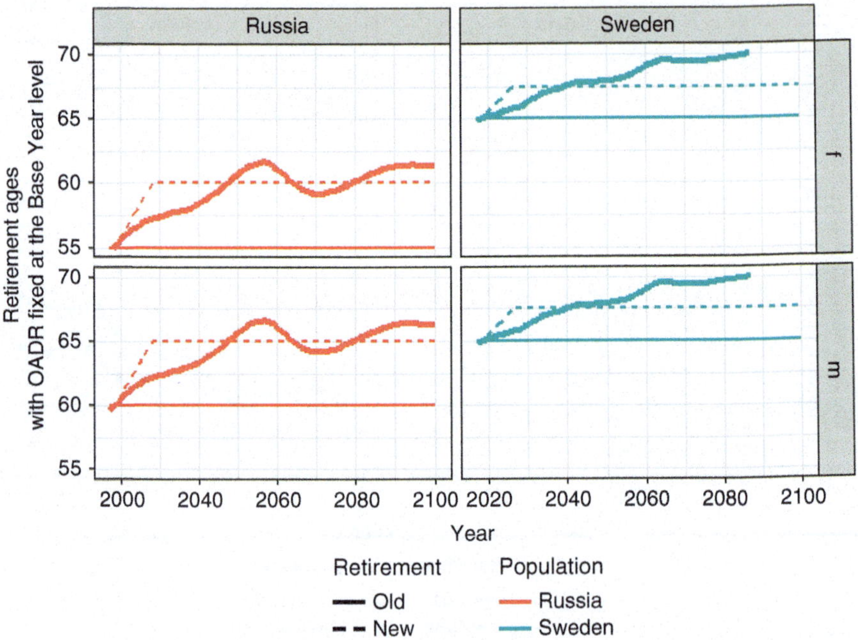

Fig. 9.7 Retirement ages in Russia (panel to the left) and Sweden (panel to the right) yielding time-constant OADRs set at their base year (2018) levels: 49% for Russia and 35% for Sweden, men (m) and women (f). Thinner lines represent the current levels of the retirement ages as well as governments' proposed reforms

both cases, the pension system is currently in a negative normative balance of below 5% of gross wages, and the negative balance is expected to strengthen if no reforms are undertaken. In the Russian case, the speed of growth of imbalance between the contributions and liabilities is stronger, partly because of the lagged effects of low fertility and net immigration in the past. The pension reform combined with fading negative effects of low fertility and immigration can prevent rapid worsening of the balance of the PAYG system in Russia. In Sweden, despite later retirement than in Russia, the higher replacement ratio for men who dominate the labor force, the lower contribution rate, and more advanced population ageing has led to stronger imbalances that may keep strengthening even after reform completion.

9.4 Results in the Cohort Dimension: Individuals' View

While period analysis is important in studying if the welfare system may make the ends meet, it is the cohort dimension that describes efficiency of the system for individuals as well as the intergenerational fairness of it.

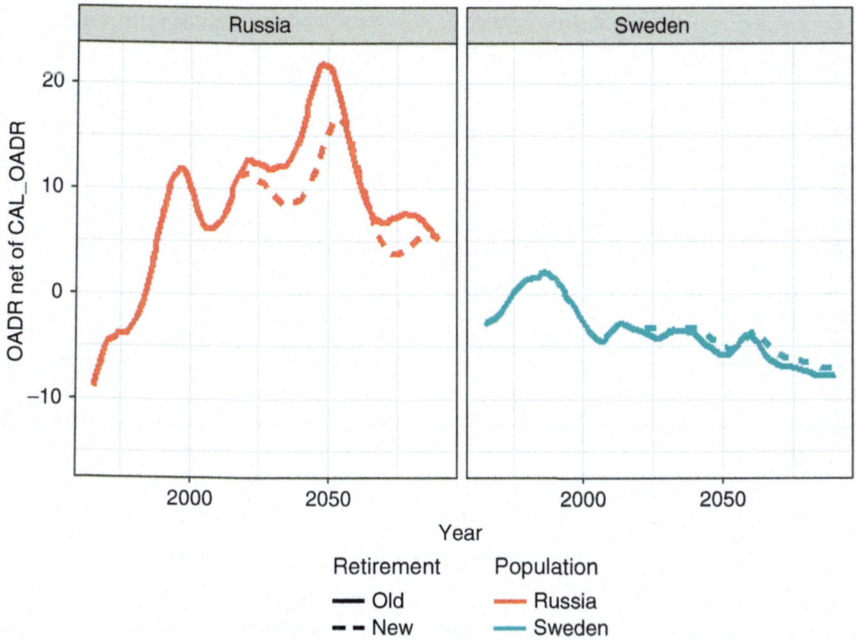

Fig. 9.8 Net effect of migration and fertility beyond replacement level for OADR in Russia (panel to the left) and Sweden (panel to the right) under old and newly proposed retirement ages: solid lines – the current retirement ages, broken lines – the new retirement scheme proposed by the government. Estimates prior to 2017 inclusive (the bifurcation point in the graph) and projections afterwards

Thanks to our long-time series of cohort life tables, we were able to produce (for the first time) estimates of time series of cohort survival to the retirement ages (Fig. 9.10). The figure fits in the widespread opposition to the reform in Russia because of the low survival ratio of men to the retirement age. Only about 60% of men in the currently retiring cohorts (aged 60 years in 2018, i.e. born in 1958) have made it to retirement. Although the proportion is projected to increase during the century, it is expected to reach the current Swedish level, ca. 85% (despite later retirement in Sweden), only by about 2075. The pension reform could push this moment further towards the end of the century. Even in the case of Russian women, who live longer but retire earlier than men, surviving till retirement (ca. 87%) lags behind the Swedish level and may decline considerably under the proposed reform. The low survival ratio to the retirement in the past was partly because of infant and child mortality that used to be relatively high retrospectively but has little to do with the economics of the pension system. To clear the picture from the effects of mortality at ages below the working age, we present the survival to the retirement from age 20 years, see Fig. 9.11. Corrections accounting for young-age mortality change the survival picture for earlier cohorts, but not much for those retiring currently or in the future. The survival of Russian women from age 20 to retirement was even

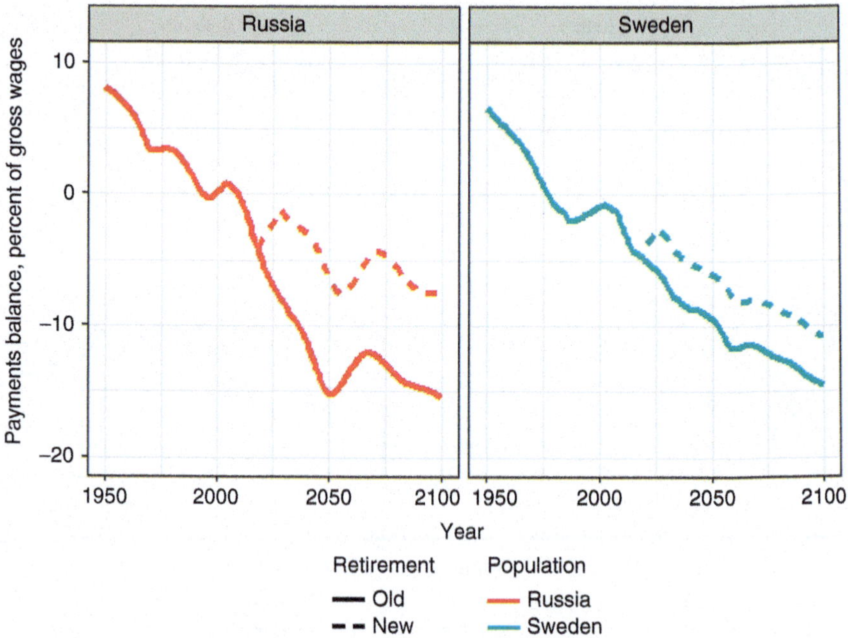

Fig. 9.9 Annual balances of the pension system as percent of gross wages in Russia (panel to the left) and Sweden (panel to the right) under old and newly proposed retirement ages: solid lines – the current retirement ages, broken lines – the new retirement scheme proposed by the government. Estimates prior to 2017 inclusive (the bifurcation point in the graph) and projections afterwards

higher than in Sweden, thanks to the earlier retirement age, before the collapse of the USSR, but it dropped below the Swedish level afterwards. Should the pension reform go on, the gap between Russian and Swedish women entering the workforce in survival to retirement will widen substantially. The survival of Russian working-age men to retirement was affected even harder during the transition period. The gap to Swedes has already peaked at about 30 percentage points in the 1990s. Should the reforms go on, the gap may again widen and reach 30 percentage points. Swedish reforms, thanks to world-record lifespan and survival estimations, will have much smaller effect upon survival to retirement.

The expected remaining lifetimes after retirement (Fig. 9.12) are lower by about 3 years compared to the Swedish levels in the case of men, and surpass them by about 5 years in the case of women. The reforms may bring levels closer for women, but they may widen the gap by about 2.5 years for men. Russian men, when compared to the Swedes, are disadvantaged in both survival to the retirement and remaining lifetime afterwards, despite retiring earlier by 5 years. Russian women, on the other hand, retire early enough (10 years earlier) to catch up with the Swedes in terms of survival and to outdo them in life expectancy after retiring.

Comparing the funded pension scheme to the currently run PAYG scheme, it is useful to see how total life years at working age compare to those at retirement for

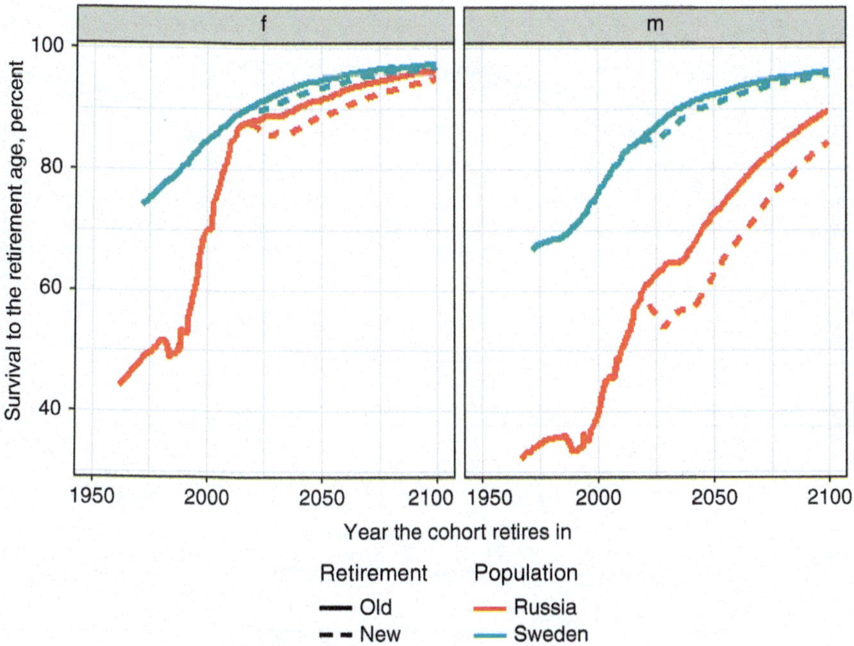

Fig. 9.10 Cohort survival from birth to the retirement in Russia (the lower curves) and Sweden (the upper curves), percent, to the retirement ages: men (m) and women (f)

the birth cohorts. To this end, we compute cohort OADRs as the ratio of cohort person years at retirement to cohort person years at working age and compare it to the period OADR (Fig. 9.13). In agreement with the theory of dynamic survival (Settergren and Mikula 2005; Ediev 2014), the Swedish cohort OADRs are substantially higher than their period counterparts, which indicates an advantage of the PAYG system over the funded pension system due to the lifespan extension. In Russia, this advantage also used to be positive and exceeded ten percentage points until recently because of the young population age composition generated by high fertility in the past. However, it is on a declining trend now and heads towards negative values (down to about minus ten percentage points by 2050) because of recent periods of low fertility and the associated acceleration of ageing of the population age composition this century. By the end of the century, when the effect of births swings originating in the 20th–early twenty-first century will largely be gone, the lifespan extension effect will dominate again and determine a positive advantage to the PAYG system. Russian cohorts retiring in the 2060s may uniquely benefit from higher returns in the Funded rather than the PAYG pension schemes. This is not the case for Sweden, where the advantage to the PAYG system from increasing longevity is continuously boosted by relatively high fertility and immigration (Ediev et al. 2013).

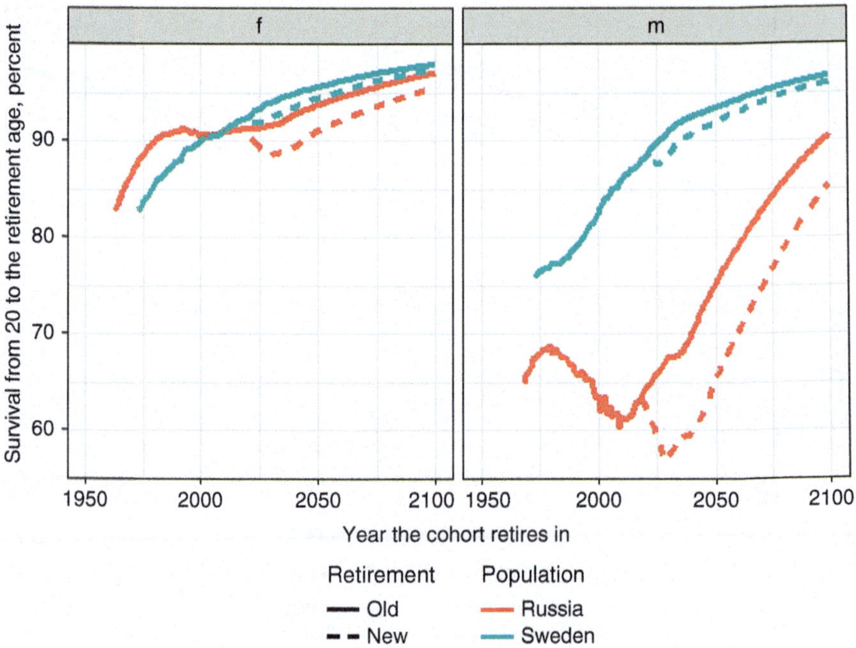

Fig. 9.11 Cohort survival from age 20 to retirement in Russia (the lower red curves) and Sweden (the upper curves), percent, to the retirement ages: men (m) and women (f)

The rates of return (IRR) to pension contributions accumulated by birth cohorts within the PAYG system are presented in Fig. 9.14. In modern populations with low and further declining mortality – even in a balanced pension system – PAYG IRRs are typically positive at ca. 1% (Ediev 2014). This was roughly the case for both nations, men and women combined, although the Swedish PAYG IRR was higher because of higher replacement ratio and not much difference in remaining life expectancy after retirement. If no reform is undertaken, the rates of return will increase at the expense of further misbalancing the PAYG systems. In that case, external funding will be needed to support the increasing IRRs. With the reforms implemented, Swedish IRR will drop back to the level above 1% per annum, while the Russian IRR may plummet to nearly zero for the most disadvantaged cohorts retiring by the completion of the reform. The picture is most gruesome for Russian men, who already have negative rate of return to their contributions in the system because of lower survival rates and shorter lifespan after retirement. The most disadvantaged male cohort in Russia (retiring in 2028) may have negative rate of return from the PAYG system of 2% per annum (in real terms). Russian women retire early and have IRRs exceeding those in Sweden. Their PAYG IRR may also drop substantially during the reform, but it will remain positive and above the IRRs of the Swedish women.

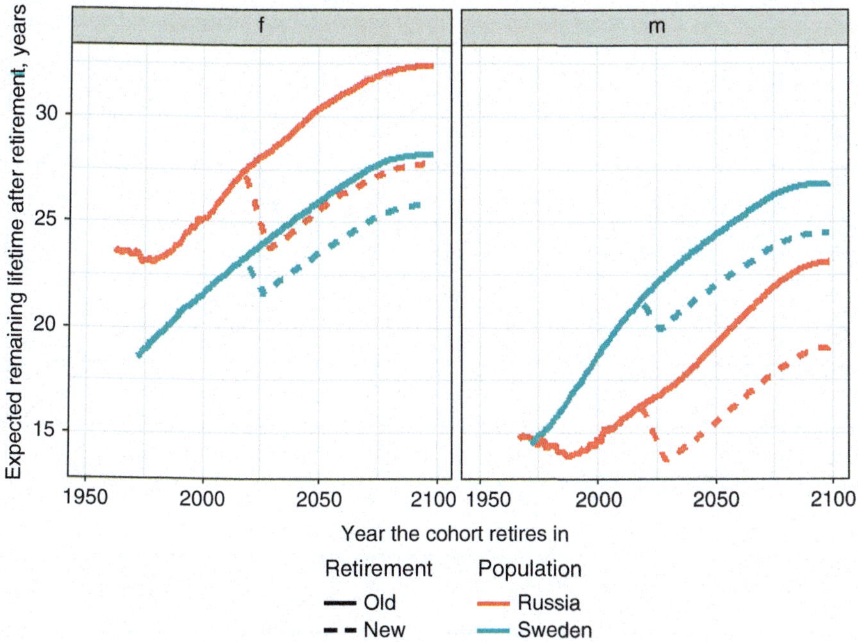

Fig. 9.12 Remaining expectation of life after retirement in Russia (the red curves) and Sweden (the turquoise curves), in years: men (m) and women (f)

9.5 Discussion

Thanks to a long time series of reconstructed and interpolated death rates and population structures with age details, we were able to study both cross-sectional and cohort dimensions of the demographic change in Russia. The two perspectives point to a stark contrast between the growing economic need to rebalance the pension system and insufficient demographic conditions to do so, especially in the case of men.

Russian men, who were already a disadvantaged group, may be hit particularly hard by the pension reform. Currently, retiring men's survival from age 20 to retirement is yet below the survival levels of cohorts retired before the collapse of the Soviet Union. Even if there were no reforms, the survival of Russian men would have caught up with the latter levels by the late 2020s only. The governmental reform, however, may push this moment further into the future with cohorts retiring only in the second half of the century. The reform may also suppress men's post-retirement lifetime down to about 15 years, substantially below the levels in low-mortality countries (exceeding 20 years, as in the case of Sweden, despite later retirement).

The low chances of Russian men to reach retirement age necessitates developing policies to overcome or, at least, mitigate consequences of the pension system for

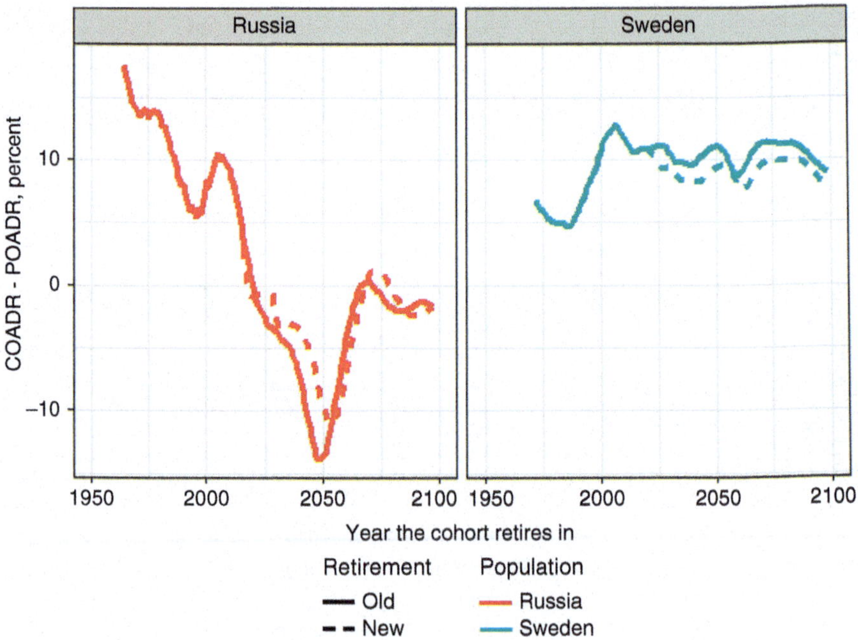

Fig. 9.13 Difference between the cohort OADR (COADR, ratio of cohort years at retirement to years at working age, men and women combined) and the period OADR (POADR) in the year the cohort retires: Russia (panel to the left) and Sweden (panel to the right), in percent

men. Improving the survival of men must be a priority when forming demographic policies. The low survival of men also indicates that many of them may have poor health prior to retirement. Hence, earlier retirement is to be extended to more men for health reasons. An easier granted early retirement might be an option with, perhaps, less of a pension benefit in cash but more medical support.

Negative and worsening rates of return to pension contributions by cohorts of Russian men suggest that a compensatory mechanism may be necessary to add to the welfare system in Russia. An option might be to introduce the possibility to bequeath full or partial pension benefits to widowed spouses or family members. Such an alternative may also be vital to mitigate the negative consequences of the pension reform for families where men with higher salaries and pensions are the major income earners.

Women, who enjoy higher survival and longer lifespan may nonetheless be at a financial loss because of more brief official employment, a lower income and pension. Policies that enable combining retirement at full pension benefits with the possibility to work and earn an additional income may be an alternative for women, an option potentially beneficial to the overall finances of the pension system.

We found an interesting temporary reversal of the demographic benefit in the PAYG system as opposed to the funded scheme for Russian cohorts retiring before

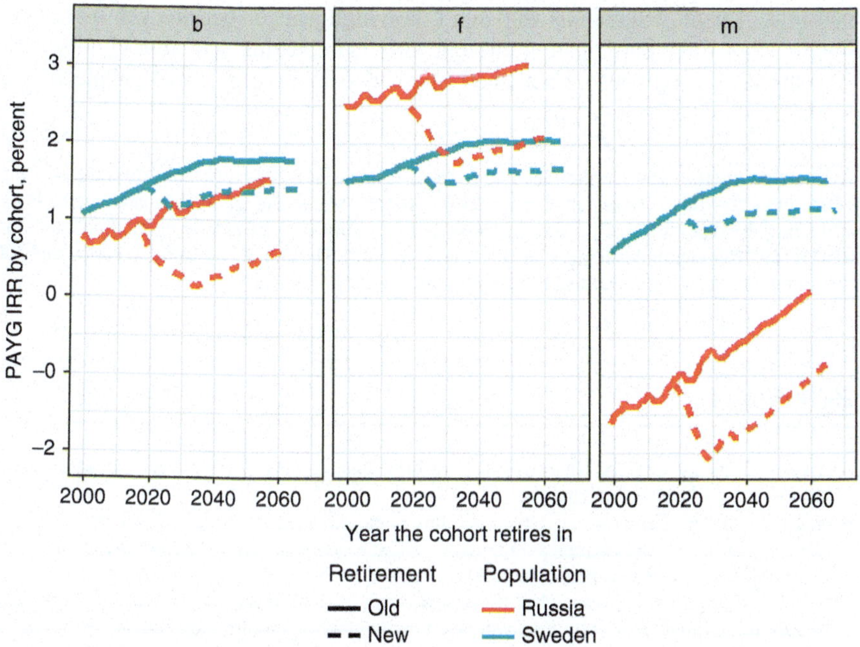

Fig. 9.14 PAYG internal rates of return for birth cohorts by the year of retirement in Russia and Sweden: both genders combined (b), women (f) and men (m) under current replacement ratios, in percent

the 2060s. Financial and actuarial implications and possible policies emanating from this observation are yet to be investigated. Notably, these are the very cohorts in which men are most disadvantaged in terms of rates of return to their contributions in the pension system (hence, less motivated to participate in the PAYG system).

Our analysis suggests several viable alternatives to the governmental proposal to increase the retirement age in Russia. In view of the Russian age structure strongly affected by demographic swings of the past, the governmental reform may cause temporal instability in the old-age dependency ratio. Twice as much gradual increase of the age at retirement may bring about the same dependency ratio as observed now with less temporal instability of the indicator. Migration compares well in magnitude to the effects of the reform and may be a better alternative, assuming that younger migrants are a more efficient workforce than preretirement workers with poor health. Proactive investments into healthcare and life extension, being necessary in mitigating the negative consequences of the reform, may also be a good way to improve the efficiency of the social welfare system, as we saw in the comparison of the Russian and Swedish cases. In the Russian context, an economic and technological spurt may also be a good alternative to later retirement, as it would create an economic-demographic dividend in the transitory period, when more productive (higher paid and contributing more) young workers can support the pensions of older generations who don't have high pension benefits yet. Increasing female labor force participation through early childbearing, and after childbearing is completed, might also be a good alternative to keeping older people in the workforce.

Acknowledgments The research leading to these results has received funding from the Russian Foundation for Basic Research Grants 18-01-00289 "Mathematical models and methods of correcting the distortions of the age structure and mortality rates of elderly population" and 18-010-01169 "Demographic changes and economic growth". The author is grateful to participants of the seminar at the Vienna Institute of Demography for useful feedback.

References

Andreev, E., Darsky, L. E., & Kharkova, T. L. (1998). *Demographic history of Russia: 1927–1959 (in russian)*. Moscow: Informatika.

Brouard, N. (1986). Structure et dynamique des populations. La pyramide des années à vivre, aspects nationaux et exemples régionaux. *Espace, Populations, Sociétés, 4*(2), 157–168. https://doi.org/10.3406/espos.1986.1120.

Center for Demographic Research (Moscow/Russia). (2018). *Russian fertility and mortality database*. Available at: http://demogr.nes.ru/index.php/ru/demogr_indicat/data. Accessed 2 Aug 2018.

Ediev, D. M. (2001). Application of the demographic potential concept to understanding the Russian population history and prospects: 1897–2100. *Demographic Research, 4*(9), 289–336. https://doi.org/10.4054/DemRes.2001.4.9.

Ediev, D. M. (2014). Why increasing longevity may favour a PAYG pension system over a funded system. *Population Studies, 68*(1), 95–110. Available at: http://www.ncbi.nlm.nih.gov/pubmed/23587003.

Ediev, D. M. (2017). Constrained mortality extrapolation to old age: An empirical assessment. *European Journal of Population, 34*, 441–457. https://doi.org/10.1007/s10680-017-9434-4.

Ediev, D. M. (2018). Expectation of life at old age: Revisiting Horiuchi-Coale and reconciling with Mitra. *Genus, 74*(1), 3. https://doi.org/10.1186/s41118-018-0029-7.

Ediev, D. M., Coleman, D., & Scherbov, S. (2013). New measures of population reproduction for an era of high migration. *Population, Space and Place*.

Ediev, D. M., Sanderson, W. C., & Scherbov, S. (2018). The inverse relationship between life expectancy-induced changes in the old-age dependency ratio and the prospective old-age dependency ratio. *Theoretical Population Biology, 125*, 1–10.

Government Offices of Sweden. (2018). *Sweden's national reform programme 2018*. Stockholm. Available at: https://www.government.se/49bfaf/contentassets/8c870068125e4941aadadf4dad740b4e/swedens-national-reform-programme-2018.pdf. Accessed 11 Aug 2018.

Guillot, M. (2003). The cross-sectional average length of life (CAL): A cross-sectional mortality measure that reflects the experience of cohorts. *Population Studies, 57*(1), 41–54.

Keyfitz, N. (1971). On the momentum of population growth. *Demography, 8*(1), 71. https://doi.org/10.2307/2060339. Springer Population Association of America.

Keyfitz, N., & Caswell, H. (2005). *Applied mathematical demography*. Berlin: Springer.

Mitra, S. (1984). Estimating the expectation of life at older ages. *Population Studies, 38*(2), 313–319. https://doi.org/10.2307/2174079.

OECD. (2018). *OECD database*. Available at: https://stats.oecd.org/. Accessed 3 Aug 2018.

RANEPA, Rosstat and IIASA. (2016). *Russian demographic data sheet 2016*. Moscow, Russia; Laxenburg, Austria.

Rosstat. (2018). *Rosstat*. Available at: http://www.gks.ru/. Accessed 14 Aug 2018.
Ryder, N. B. (1951). *Cohort Approach*. PhD thesis.
Sanderson, W. C., & Scherbov, S. (2015). Faster increases in human life expectancy could lead to slower population aging. *PLoS One, 10*(4), e0121922. https://doi.org/10.1371/journal.pone.0121922.
Settergren, O., & Mikula, B. D. (2005). The rate of return of pay-as-you-go pension systems: A more exact consumption-loan model of interest. *Journal of Pension Economics and Finance, 4*(2), 115–138. https://doi.org/10.1017/S1474747205002064.
Statistics Sweden. (2018). *Statistics Sweden*. Available at: http://www.scb.se/en/. Accessed 9 Aug 2018.
Thatcher, A. R., Kannisto, V., & Vaupel, J. W. (1998). *The force of mortality at Ages 80–120. Monographs on population aging*. Odense: Odense University Press. Available at: http://www.demogr.mpg.de/Papers/Books/Monograph5/ForMort.htm.
UN DESA/Population Division. (2017a). *World population prospects: Model life tables*. Available at: https://esa.un.org/unpd/wpp/Download/Other/MLT/. Accessed 20 Aug 2017.
UN DESA/Population Division. (2017b). *World population prospects* 2017. Available at: https://esa.un.org/unpd/wpp/. Accessed 4 July 2017.
University of California, B. and the Max Planck Institute for Demographic Research (Rostock). (2019). *Human mortality database. Online database sponsored by University of California, Berkeley (USA), and Max Planck Institute for Demographic Research (Germany)*. Available at: www.mortality.org. Accessed 15 May 2018.

Chapter 10
Using the Developing Countries Mortality Database (DCMD) to Probabilistically Evaluate the Completeness of Death Registration at Old Ages

Nan Li, Hong Mi, and Xiaotong Tang

10.1 Introduction

The deterministic completeness of deaths registered in the census or death registration (DR) is usually defined as the ratio of registered deaths to total deaths. Normally, a DR can be judged as complete when the deterministic completeness is 1. In practice, however, the deterministic completeness cannot be exactly 1 and it is impossible to judge whether a DR is complete. Consequently, even for the developed countries that are commonly believed to have reliable DRs, such as those in the Human Mortality Database, there is no quantitative evaluation to conclude that their DRs are complete.

To solve this dilemma, we propose a probabilistic completeness, of which the samples are deterministic completeness. When the difference between 1 and the mean of probabilistic completeness is statistically insignificant, the DR is probabilistically complete.

The views expressed in this paper are those of the author and do not necessarily reflect those of the United Nations. The work on this paper was supported by the Nature Science Foundation of China (NSFC) project (NO. 71490732), NSFC project (NO. 71490733), Zhejiang Social Science Planning Project Key Program (NO. 17NDJC029Z), and Nature Science Foundation of Zhejiang project (NO. LZ13G030001).

N. Li
Population Division, Department of Economic and Social Affairs, United Nations, New York, NY, USA
e-mail: li32@un.org

H. Mi (✉) · X. Tang
School of Public Affairs, Zhejiang University, Hangzhou, Zhejiang, P. R. China

C. H. Skiadas, C. Skiadas (eds.), *Demography of Population Health, Aging and Health Expenditures*, The Springer Series on Demographic Methods and Population Analysis 50, https://doi.org/10.1007/978-3-030-44695-6_10

Estimating the deterministic completeness of the DR is an important and long-standing issue. The analytical evaluations of the completeness of the DR originate from stable population models. In stationary populations, the number of deaths over a given age is the number of persons at the given age. Therefore, the number of registered deaths over a certain age could be evaluated by the number of persons at this age if the population is stationary. In a stable population, the number of deaths over a given age is the number of persons at the given age minus an additional term, which is the product of the number of persons over the given age and the growth rate of the stable population. This relationship was first utilized by Brass (1975) to evaluate the completeness of the DR. The evaluation was extended to non-stable populations, using two successive censuses and at synthetic extinct generation, seg (Bennett and Horiuchi 1981, 1984), a generalized growth balance, ggb (Hill 1987), and a combination of ggb, seg, and ggbseg (Hill et al. 2009). These methods are convincing methodologically but require unrealistic assumptions, which typically include the absence of migration and that the errors of the census population obey special relationships (Li and Gerland 2017).

At old ages such as 60 years and over, migrants are negligible compared to deaths. Furthermore, using census population could accurately estimate the levels of mortality at old ages, which provides the basic estimates of the Developing Countries Mortality Database (DCMD, www.lifetable.org, see Li 2014; Li et al. 2017, 2018). In DCMD, the basic estimates of old-age mortality (or the probability of dying between ages 60 and 75) are obtained using (1) the variable-r method with adjustment of age heaping (Bennett and Horiuchi 1981; Li and Gerland 2013), (2) the survival model (Li et al. 2018), and (3) the two-input-parameter model life table with child and adult mortality (Wilmoth et al. 2012). The DCMD estimates of old-age mortality are calculated as the averages of the above three basic estimates and extended to single years using local regressions (Li and Mi 2018).

In DCMD, the unrealistic assumptions are largely reduced. Consequently, the DCMD estimates of old-age mortality should be more reliable than that of the previous methods. Taking the DCMD estimates of old-age mortality as accurate, we applied the probabilistic evaluation to all the developing countries included in DCMD and obtained encouraging results. Because the socioeconomic conditions of some developing countries are similar to those of developed countries, the probabilistic evaluation should also work for developed countries, including those in the Human Mortality Database.

10.2 The Method

10.2.1 The Fundamental Difference

Using the numbers of the population aged 60–75 years (hereafter all age ranges for both population and deaths are 60–75 years) enumerated from two successive censuses could estimate the number of deaths occurred, and the prevailing mortality levels between the two censuses. Big estimating errors could occur in the number of deaths but not in the mortality levels. This fundamental difference may indicate the success of evaluating the completeness of the DR based on estimating the levels of mortality in the DCMD and this paper, and may also explain the failure in evaluations based on estimating the number of deaths in previous studies (Li and Gerland 2017).

10.2.1.1 Estimating the Numbers of Death

Let the number of persons in an age interval enumerated in the first census be p_1 and the number of the survivors in the next census be p_2. Furthermore, let the net undercount rates[1] be u_1 and u_2 for the first and second censuses, respectively. Neglecting intercensal migration at old ages, the number of deaths (d) and estimated deaths ($\hat{d}$) are:

$$\begin{aligned} d &= p_1 - p_2, \\ \hat{d} &= \hat{p}_1 - \hat{p}_2 = p_1(1-u_1) - p_2(1-u_2) = d - (p_1u_1 - p_2u_2)\,. \end{aligned} \tag{10.1}$$

Furthermore, let the survival ratio be

$$s = p_2/p_1. \tag{10.2}$$

Then, the relative error in estimating the number of deaths is

$$e_d = \frac{\hat{d}-d}{d} = -\frac{p_1u_1 - p_2u_2}{d} = -\frac{p_1(u_1 - s\cdot u_2)}{p_1(1-s)} = -\frac{u_1 - s\cdot u_2}{1-s}. \tag{10.3}$$

Equation (10.3) indicates that, except for two special cases ($u_1 = s\cdot u_2$ and $u_1 = u_2$), the effect of the survival ratio could lead to large errors in estimating the number of deaths, especially when the survival ratio is close to 1.

10.2.1.2 Estimating Mortality Levels

The estimated survival ratio can be written as

[1]The net undercounting rate represents the relative difference between the reported and the true numbers of population, which could be the result of misreporting of people or misreporting of age. A positive net undercounting rate indicates net under-counting or that the reported number is smaller than the true number. The net undercounting rate could also be negative to reflect net over-counting that may or may not be caused by misreporting of age.

$$\hat{s} = \hat{p}_2/\hat{p}_1 = \frac{p_2\,(1-u_2)}{p_1\,(1-u_1)} = s\frac{1-u_2}{1-u_1}. \tag{10.4}$$

Using (10.2), the relative errors in estimations s is

$$e_s = \frac{\hat{s}-s}{s} = \frac{1-u_2}{1-u_1} - 1 = \frac{u_1-u_2}{1-u_1}. \tag{10.5}$$

It is clear that the estimation errors of the survival ratio (or mortality level) are not affected by the survival ratio itself, because the true value of the survival ratio (s) cancels itself in the estimation. Eq. (10.5) differs fundamentally from Eq. (10.3), indicating that large errors are not likely to occur in the estimation of mortality levels, and provides the basis of DCMD and this paper.

10.2.2 A Deterministic Model of Completeness

The CDMD provides the estimates of single-year life tables in developing countries, which should be more accurate than the estimates of the number of deaths, as per the discussion above. In this paper, death rates $m(x,t)$ obtained from these life tables are assumed to be accurate and are used to evaluate the completeness of the DR at census years, where the numbers of deaths and of the population are available to compute the registered death[2]:

$$m_r\,(x,t) = \frac{d_r\,(x,t)}{p_r\,(x,t)}, \tag{10.6}$$

where $d_r(x,t)$ and $p_r(x,t)$ are the registered numbers of deaths and the reported numbers of the population aged x in a census, in the corresponding period and at time t. Although the evaluation can be carried out for individual ages in a single year, it is more robust and convenient to calculate the completeness of the DR for the age group 60–75 years, because the deaths could be rare at single-year age in some countries, and because $\left\{1-\exp\left[\sum_{x=60}^{75} m\,(x,t)\right]\right\}$ equals old-age mortality (the probability of dying between ages 60 and 75). To evaluate the completeness of the DR for the age group 60–75 years, however, it adopts an implicit assumption is adopted that the completeness is constant over the age group 60–75 years.

[2]The middle-point of the death period may differ from the date of the census enumeration of population. The difference could be reduced by adjustments using various assumptions. Data sources: the estimated death rates are from the DCMD(www.lifetables.org), the registered death rates are from the Demographic Yearbook (https://unstats.un.org/unsd/demographic/products/dyb/default.htm)

The completeness of the DR at time t can be defined as $c(t) = \sum_{x=60}^{75} d_r(x,t) / \sum_{x=60}^{75} d(x,t)$, where $d(x, t)$ represents the accurate number of deaths in the corresponding period and age. The accurate number of deaths, $\sum_{x=60}^{75} d(x,t)$, however, is difficult to estimate. Nonetheless, a reliable estimate of the completeness of the DR can be proposed as below

$$c_r(t) = \frac{\sum_{x=60}^{75} m_r(x,t)}{\sum_{x=60}^{75} m(x,t)} = \frac{\log[1 - q_{or}(t)]}{\log[1 - q_o(t)]}, \tag{10.7}$$

where $q_{or}(t)$ and $q_o(t)$ are the reported and accurate old-age mortality at time t. The values of reported old-age mortality are calculated using the number of deaths from the DR, and the population number from the census.

Further, Cauchy's mean value theorem indicates that

$$c_r(t) = \frac{\sum_{x=60}^{75} m_r(x,t)}{\sum_{x=60}^{75} m(x,t)} = \frac{\sum_{x=60}^{75} d_r(x,t)/p_r(x,t)}{\sum_{x=60}^{75} d(x,t)/p(x,t)} = \frac{d_r(y,t)/p_r(y,t)}{d(y,t)/p(y,t)}, \; 60<y<75. \tag{10.8}$$

Furthermore, noticing that the census enumeration may suffer undercount ($p_r(y, t) \leq p(y, t)$), and that the completeness is implicitly assumed to be constant over age, the reliable estimate of completeness is an upper limit of the true completeness:

$$c_r(t) - \frac{\sum_{x=60}^{75} m_r(x,t)}{\sum_{x=60}^{75} m(x,t)} = \frac{d_r(y,t)}{d(y,t)} \cdot \frac{p(y,t)}{p_r(y,t)} \geq \frac{d_r(y,t)}{d(y,t)} - c(t). \tag{10.9}$$

When the census undercounts are small or negligible compared to the incompleteness of the DR, the reliable estimate of completeness, $c_r(t)$, would be close to the true completeness, $c(t)$. On the other hand, when the census undercount rates are larger than that of the DR ($\frac{p(y,t)-p_r(y,t)}{p(y,t)} > \frac{d(y,t)-d_r(y,t)}{d(y,t)}$ or $\frac{p_r(y,t)}{p(y,t)} < \frac{d_r(y,t)}{d(y,t)}$), $c_r(t)$ will be bigger than 1.

Finally, the average completeness of the DR is obtained as:

$$\overline{c}_r = \frac{\sum_{t=1}^{n} c_r(t)}{n}. \tag{10.10}$$

Complete DRs can be defined as $\overline{c}_r = 1$. This definition, however, is not useful in practice, because even if the DR was indeed complete, random errors in registration and census would make $\overline{c}_r$ departure from one. To take random errors into account, a probabilistic model is proposed below.

10.2.3 A Probabilistic Model of Completeness and Hypothesis Test

Among the population enumerated in a census, each person may or may not die within the census year, and each death may or may not be registered. The surviving and registering processes of each person can be described by a random variable that takes value 1 for dead and registered and 0 otherwise. These individual random variables are assumed to be independent from each other and follow the same probability distribution. Denoting the sum of these individual variables (which represents the registered deaths, also being a random variable) by *Dr(x,t)* (hereafter all random variables are expressed in capital letters), the probabilistic completeness of the DR can be defined as

$$C_r(t) = \frac{\sum_{x=60}^{75} D_r(x,t)/p_r(x,t)}{\sum_{x=60}^{75} d(x,t)/p(x,t)} = \frac{\sum_{x=60}^{75} M_r(x,t)}{\sum_{x=60}^{75} m(x,t)}. \tag{10.11}$$

Further, when number of deaths is about 30 or larger, the central limit theorem guarantees that the probability distribution of $C_r(t)$ will be approximately normal (see Agresti and Finlay 1997: 104). Assuming that, for $t = 1,2,\ldots n$, $C_r(t)$ are independent from each other and follow the same normal distribution, it is intuitive that the average of $C_r(t)$, namely $\overline{C}_r$, is more robust than $C_r(t)$. Denoting the mean and variance of $C_r(t)$ by μ and σ^2, the unbiased statistics of the mean and variance of $C_r(t)$ are

$$\begin{aligned} \overline{C}_r &= \frac{\sum_{t=1}^{n} C_r(t)}{n}, E\left(\overline{C}_r\right) = \mu, \\ S^2 &= \frac{\sum_{t=1}^{n} \left[C_r(t) - \overline{C}_r\right]^2}{n-1}, E\left(S^2\right) = \sigma^2. \end{aligned} \tag{10.12}$$

If the data on death registration is available only at one census year, or $n = 1$, the variance of $C_r(t)$ cannot be estimated.

Since the variance of the $\overline{C}_r$ is

$$\mathrm{var}\left(\overline{C}_r\right) = \mathrm{var}\left(\frac{\sum_{t=1}^{n} C_r(t)}{n}\right) = \frac{\sum_{t=1}^{n} \mathrm{var}\left[C_r(t)\right]}{n^2} = \frac{\sum_{t=1}^{n} \sigma^2}{n^2} = \frac{\sigma^2}{n}, \tag{10.13}$$

the unbiased statistic of the variance of $\overline{C}_r$ is

$$S_C^2 = \frac{S^2}{n},\ E\left(S_C^2\right) = E\left(\frac{S^2}{n}\right) = \frac{\sigma^2}{n} = \mathrm{var}\left(\overline{C}_r\right). \tag{10.14}$$

Using $\overline{C}_r$ and S_C, a statistic for the hypothesis test is obtained:

$$T(n) = \frac{1 - \overline{C}_r}{S_C} = \frac{1 - \overline{C}_r}{S\sqrt{n}}. \tag{10.15}$$

The sample values of $C_r(t)$ and $\overline{C}_r$ are the $c_r(t)$ and $\overline{c}_r$ in (10.9) and (10.10), respectively. Using these sample values, the sample value of T can be calculated by (10.12), (10.13), (10.14) and (10.15), and the hypothesis test below can be carried out.

If the mean of $T(n)$ is 0, $T(n)$ should follow a t-distribution with $(n-1)$ degrees of freedom (see Agresti and Finlay 1997: 180). Accordingly, the hypotheses can be set as

$$Ho : mean\left(\overline{C}_r\right) = 1,\ Ha : mean\left(\overline{C}_r\right) < 1, \tag{10.16}$$

of which the null hypothesis implies that the completeness is 1 or perfect, and the alternative hypothesis represents that the completeness is smaller than 1. When the sampled value of $T(n)$ is bigger than the t-score value with the corresponding n in Table 10.1, the probability for such a sample to occur is smaller than 0.05. Subsequently, the null hypothesis should be rejected, inferring that the completeness is significantly smaller than 1.

When the sample value of $T(n)$ is smaller than the t-scores in Table 10.1 and the sample value of $\overline{C}_r$ is close to 1, the null hypothesis cannot be rejected, inferring that the difference between the mean of $\overline{C}_r$ and 1 is not statistically significant, or that the

Table 10.1 The t-score with right-trail probability equal to 0.05

n	2	3	4	5	6	7	8	9	10
t-score	6.31	2.92	2.35	2.13	2.02	1.94	1.9	1.86	1.83

Table 10.2 The t-score with two-trail probability equal to 0.05

n	2	3	4	5	6	7	8	9	10
t-score	12.71	4.31	3.18	2.78	2.57	2.45	2.37	2.31	2.26

DR is probabilistically complete. In other words, a probabilistically complete DR is defined as one that the deterministic completeness is close to 1, and the difference between 1 and the mean of probabilistic completeness is statistically insignificant. A deterministic completeness is a sample of the probabilistic completeness and contains a random error. Probabilistic completeness takes random errors into account. When the difference between 1 and the mean of probabilistic completeness is statistically insignificant, the DR is judged as probabilistically complete.

When the sample value of *T(n)* is smaller than the t-score in Table 10.1, but the sample value of C_r is not close to 1, the null hypothesis cannot be rejected. This situation could be caused by two factors. The first is that *n* is small, leading to large var(C_r). The second is that the difference between sample values of C_r (t) is large, which implies assuming that the same distribution of C_r (t) is improper, and which also leads to large var(C_r). In this context, the hypothesis test fails. When the sample value of *T(n)* is negative or the sample value of $\overline{C}_r$is bigger than 1, the alternative hypothesis should be set as $Ha\ :\ mean\left(\overline{C}_r\right) \neq 1$, which requires the t-score of two-tail probability in Table 10.2.

If the absolute sample value of *T(n)* is smaller than the t-score with corresponding *n* in Table 10.2, the null hypothesis cannot be rejected, inferring that the difference between the completeness and 1 is not statistically significant, or that the DR is probabilistically complete.

However, there could be a situation in which the sample value of *T(n)* is negative, and the absolute sample value of *T(n)* is bigger than the t-score with corresponding *n* in Table 10.2. A plausible reason for this situation to occur is that the numbers of the census population are less accurate than the number of deaths registered, which would lead to an overestimate of $m_r(x, t)$. In this situation, the DR is still probabilistically complete, but the undercount of the census population is nonnegligible.

To summarize the probabilistic model, it is necessary to indicate that the survival process is uncertain. When the population (*p(x,t)*) survives according to the unbiased death rate (*m(x,t)*), $D_r(x, t)$ is a random variable with a mean of $d(x,t) = m(x,t)p(x,t)$ (Li 2015). The number of registered deaths, $d_r(x, t)$, is a sample value of the registered deaths, $D_r(x, t)$. Thus, $d_r(x, t)$ can be either larger or smaller than the mean of $D_r(x, t)$. Also, a $\overline{c}_r$ is a sample value of $\overline{C}_r$, and could be either larger or smaller than 1, even if the mean of $\overline{C}_r$ is 1. To illustrate this phenomenon quantitatively, the hypothesis test is used as a proper tool. Setting the null hypothesis as $mean\left(\overline{C}_r\right) = 1$, meaning that the DR is probabilistically complete, $\overline{C}_r$ should follow a t-distribution. When the $\overline{c}_r$ does not imply the occurrence of a rare event

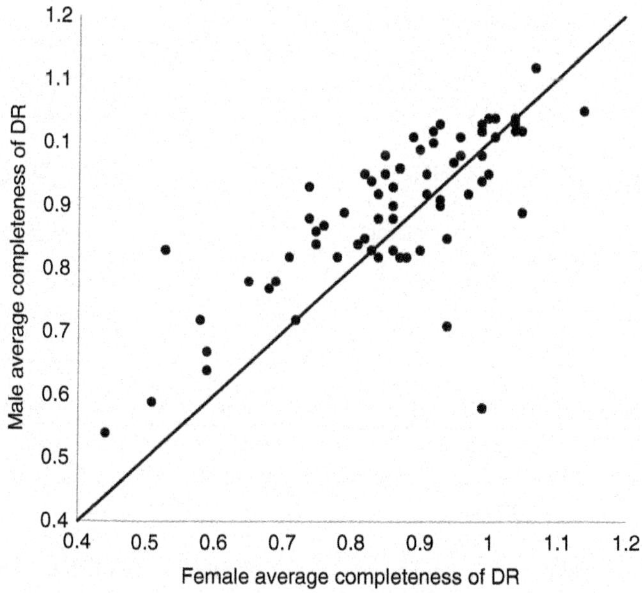

Fig. 10.1 Average completeness by sex. Note: Estimated death rates are from the DCMD (www.lifetables.org), and the census population and registered deaths are from the Demographic Yearbooks and the National Statistics Offices

under the t-distribution, the null hypothesis cannot be rejected, and a conclusion would be that the DR is probabilistically complete.

10.3 Applications

The probabilistic evaluation requires the data on deaths for more than one census year. Among the 122 countries in the DCMD, 69 have the data of the DR for two or more census years. Using (10.10), the values of the average completeness are calculated for these countries and are shown in Fig. 10.1. These values refer to the periods that are within the year group 1970–2015 and are indicated in the DCMD (www.lifetables.org). The average completeness for men is higher than that for women for 50 out of the 69 countries, indicating that the deaths of men are more likely to be registered. About 70% of the average completeness of men and women of the 69 countries is below 0.95, justifying the necessity of establishing the DCMD.

Setting 0.95 as a threshold above which the deterministic completeness could be regarded as high, and below which the deterministic completeness could be regarded as moderate or low, the results of the probabilistic evaluation can be grouped into 4 categories as are shown in Table 10.3.

Table 10.3 Results of the probabilistic evaluation

Deterministic completeness	Category	Difference between 1 and the mean of probabilistic completeness	Interpretation
Bigger than 0.95	I	Significant	Evaluation failure
Smaller than 0.95	II	Significant	Probabilistically incomplete
Smaller than 0.95	III	Insignificant	Evaluation failure
Bigger than 0.95	IV	Insignificant	Probabilistically complete

Category I includes cases of $\overline{c}_r > 0.95$; the difference between the mean of $\overline{C}_r$ and 1 is statistically significant. This is the case when the probabilistic evaluation fails, and the result of undercounted census population enlarges $\overline{c}_r$.

Category II includes cases of $\overline{c}_r \leq 0.95$; the difference between the mean of $\overline{C}_r$ and 1 is statistically significant. This is the case when the probabilistic and deterministic evaluations are consistent: the deterministic completeness is moderate or low, and the DR is probabilistically incomplete.

Category III includes cases when the difference between 1 and the mean of probabilistic completeness is insignificant, while the deterministic completeness is moderate or low. This is also the case when the probabilistic evaluation fails. The reasons could be that n (the number of data on death at census years) is small, or that $c_r(t)$ differ remarkably and are not samples of the same distribution. Among the 69 countries with $n \geq 2$, only 22 countries have data on deaths at census years ($n = 2$). For the 44 male and female populations with $n = 2$, 82% have an average completeness below of 0.95, but the difference between their average completeness and 1 is statistically insignificant. In other words, the probabilistic evaluation fails for 82% of the cases with $n = 2$. Because using only two samples to calculate variance is unreliable, this result makes sense. For this reason, we apply probabilistic evaluation to cases with data on deaths for three or more census years, which include 47 countries, or 94 cases of men and women.

Category IV includes cases of $\overline{c}_r > 0.95$; the difference between the mean of $\overline{C}_r$ and 1 is statistically insignificant. These are probabilistically complete cases.

To describe the results of the probabilistic evaluation, we introduce a significance index as

$$s_i = \frac{t(n) - t_s}{t_s}, \tag{10.17}$$

where $t(n)$ and t_s represent sample value of $T(n)$ in (10.15) and the corresponding t-score in Tables 10.1 and 10.2, respectively. Statistical significance corresponds to $s_i > 0$, and vice versa. Moreover, a bigger $|s_i|$ indicates a smaller probability

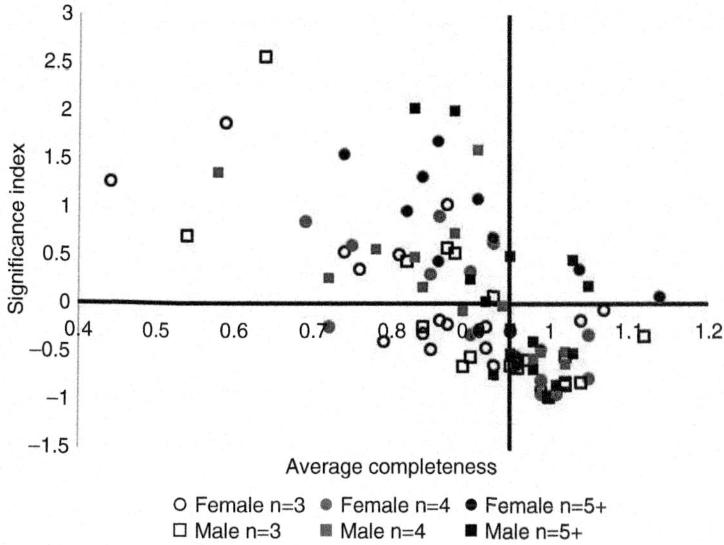

Fig. 10.2 Average completeness and statistical significance by sex

of mistakenly rejecting the null hypothesis, and therefore the more robust the conclusion of the hypothesis test. Using the significance index, the above four categories can be described by the four quadrants of a completeness-significance figure, in which the vertical axis (significance index) crosses the horizontal axis (average completeness) at 0.95, as can be seen in Fig. 10.2. In other words, in the completeness-significance figure of this paper, the origin is (0.95,0), not (0,0). In a completeness-significance figure, quadrants II and IV include successful probabilistic evaluations, while quadrants I and III contain probabilistic evaluations that do not provide reasonable conclusion and require further analysis. The probabilistic evaluation is carried out for the men and women of the 47 countries (94 cases) that have DR data on three or more census years. The results of these probabilistic evaluations are summarized in Fig. 10.2 and are shown in detail in the Table 10.4 in Annex.

10.4 Discussion

In the quadrant I of Fig. 10.2, the average completeness is close to or unreasonably bigger than 1 but the difference between the mean of $\overline{C}_r$and 1 is statistically significant (category I). Thus, the probabilistic evaluation fails to deliver reasonable conclusions. The 4 cases in quadrant I are for the men and women of the

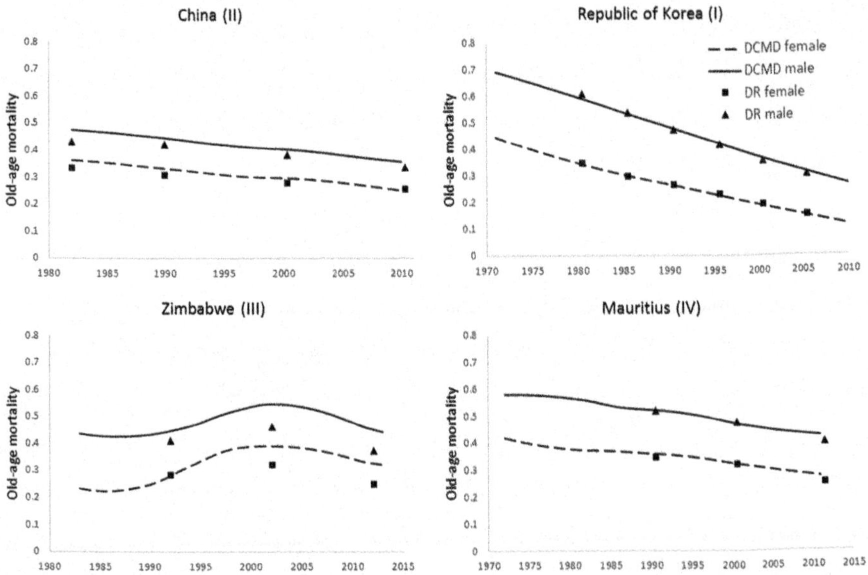

Fig. 10.3 Old-age mortality and the result of probabilistic evaluation for selected countries. (Note: Same as Fig. 10.1)

Republic of Korea (see Fig. 10.3) and Seychelles. These cases can be explained as follows: the undercounting rates of old-age population in census are larger than the incompleteness of the DR, which could be common among developed countries. Undercounting old-age population by the same rate at successive censuses does not change the DCMD estimates (the denominator of $c_r(t)$), but it will raise the registered old-age mortality (the numerator of $c_r(t)$). This analysis immediately explains the nonsensical $c_r(t) > 1$, which may still be acceptable if though not statistically significant. Noticing that an enlarged $c_r(t)$ will raise the $|T|$ in (10.15) to be bigger than its t-score, this analysis further explains why the nonsensical $c_r(t) > 1$ is statistically significant. Assuming that the censuses of the two countries undercounted old-age population by 1.5%, and correspondingly adjusting upward old-age populations by 1.5%, the results would be in quadrant IV in which the DR is probabilistically complete, because the difference between the mean of $\overline{C}_r$and 1 is statistically insignificant.

There are 37 (40% of the 94) cases in quadrant II (or category II), of which the average completeness is smaller than 0.95 (differs from 1 notably) and the difference is statistically significant. Thus, the probabilistic evaluation concludes that the DR is probabilistically incomplete. China is chosen to provide an example for Category II as is shown in Fig. 10.3. The average completeness of China is 0.93 for females and 0.91 for males.

Despite limiting the probabilistic evaluation to $n > 2$, there are still 17 (18% of the 94) cases in quadrant III (category III). For 12 of the 17 cases, the n is 3, indicating $n = 3$ is not big enough to guarantee successful probabilistic evaluation. The other 5 cases in category III are perhaps caused by large differences between sample values of completeness at various census years. In other words, assuming the same probability distribution for the completeness at different census years is not proper. The requirement of having DR data for three or more census years is valid, but it could be removed if the DR and reliable data on old-age population exist in single years between two successive censuses. Zimbabwe provides the values of old-age mortality as an example for quadrant III in Fig. 10.3. The average completeness of Zimbabwe is 0.84 for females and 0.82 for males.

There are 36 (38%) cases in quadrant IV category. These are the cases of probabilistically complete: $\overline{c}_r \geq 0.95$, where the difference between the mean of $\overline{C}_r$ and 1 is statistically insignificant. These results indicate that setting 0.95 as a threshold of deterministic completeness ($\overline{c}_r$) is meaningful. When $\overline{c}_r \geq 0.95$, all conclusions are probabilistically complete. Of course, this observation is based on limited observations (94 cases). When investigating data from countries excluded here, a threshold different from 0.95 may be more suitable. Mauritius is selected to give an example in Fig. 10.2 in category IV. The average completeness of Mauritius is 0.95 for females and 0.97 for males. Putting quadrants II and IV together, the probabilistic evaluation is successfully applied for 78% of the 94 cases. If small undercounting rates existed in the censuses of the Republic of Korea and Seychelles, successful applications would contain 82% of the 94 cases.

The concept of probabilistic completeness is useful. Because of random errors, deterministic completeness cannot be exactly 1 in practice. It is therefore impossible to conclude whether a DR is complete using deterministic completeness. Probabilistic completeness solves this dilemma: when deterministic completeness is high, and when the difference between 1 and the mean of probabilistic completeness is statistically insignificant, the DR is probabilistically complete. Furthermore, when the deterministic completeness is moderate or low, and when the difference between 1 and the mean of probabilistic completeness is statistically significant, the DR is probabilistically incomplete.

Annex

Table 10.4 Average completeness and results of probabilistic evaluation

Female					Male				
	Number of DR data at census years, n	Average completeness	Sample T	Category		Number of DR data at census years, n	Average completeness	Sample T	Category
Seychelles	7	1.14	−2.62	I	Seychelles	7	1.05	−2.93	I
Republic of Korea	8	1.04	−3.22	I	Republic of Korea	8	1.03	−3.48	I
El Salvador	3	0.81	4.43	II	Kazakhstan	3	0.93	3.16	II
Kuwait	5	0.74	5.42	II	Zimbabwe	3	0.82	4.24	II
Azerbaijan	3	0.76	3.98	II	Paraguay	4	0.72	3	II
Kyrgyzstan	3	0.74	4.5	II	Sri Lanka	4	0.83	3.51	II
Mali	3	0.44	6.64	II	Bahamas	5	0.92	2.2	II
Peru	3	0.59	8.43	II	Azerbaijan	3	0.87	4.64	II
China	4	0.93	3.87	II	Kyrgyzstan	3	0.88	4.49	II
Cuba	4	0.86	4.5	II	Mali	3	0.54	4.99	II
Dominican Republic	4	0.69	4.38	II	Peru	3	0.64	10.42	II
Ecuador	4	0.75	3.78	II	China	4	0.91	6.11	II
Panama	5	0.84	4.93	II	Cuba	4	0.88	4.08	II
Thailand	5	0.86	5.73	II	Dominican Republic	4	0.78	3.7	II
Philippines	7	0.86	2.8	II	Ecuador	4	0.84	2.77	II
Brazil	5	0.91	4.44	II	Panama	5	0.88	6.4	II
Viet Nam	3	0.87	5.97	II	Thailand	5	0.83	6.46	II
Argentina	4	0.9	3.14	II	Philippines	7	0.9	2.44	II

Egypt	4	0.85	3.08	II	Qatar	4	0.58	5.57	II
South Africa	6	0.82	3.97	II	Brazil	5	0.95	3.21	II
India	5	0.93	3.61	II	Burkina Faso	3	0.89	1.05	III
Burkina Faso	3	0.79	1.82	III	Maldives	3	0.9	1.31	III
Maldives	3	0.93	1.1	III	El Salvador	3	0.84	2.25	III
Kazakhstan	3	0.86	2.45	III	Kuwait	5	0.93	0.58	III
Zimbabwe	3	0.84	2.03	III	Colombia	4	0.94	2.31	III
Paraguay	4	0.72	1.82	III	Bahrain	4	0.89	2.18	III
Sri Lanka	4	0.9	1.61	III	Botswana	3	0.95	1.1	IV
Bahamas	5	0.91	1.53	III	Tunisia	3	1	0.13	IV
Botswana	3	0.85	1.57	III	United Republic of Tanzania	3	0.96	1.24	IV
Tunisia	3	0.92	1.59	III	Viet Nam	3	0.96	1	IV
United Republic of Tanzania	3	0.87	2.34	III	Argentina	4	0.99	1.21	IV
St. Vincent and the Grenadines	3	0.92	2.23	III	Egypt	4	0.98	1.04	IV
Colombia	4	0.99	0.5	IV	South Africa	6	0.95	1.01	IV
Qatar	4	0.99	0.18	IV	Mauritius	3	0.97	1.23	IV
Mauritius	3	0.95	2.13	IV	Singapore	5	0.95	1.61	IV
Singapore	5	1	0.08	IV	Venezuela	5	0.98	1.31	IV

(continued)

Table 10.4 (continued)

Female					Male				
	Number of DR data at census years, n	Average completeness	Sample T	Category		Number of DR data at census years, n	Average completeness	Sample T	Category
Venezuela	5	0.99	1.13	IV	Mexico	6	0.98	0.67	IV
Mexico	6	0.96	0.74	IV	St. Vincent and the Grenadines	3	1.02	−0.81	IV
Malaysia	3	0.99	0.38	IV	India	5	1.03	−1.37	IV
Chile	4	0.99	0.53	IV	Malaysia	3	1.02	−2.01	IV
Uruguay	5	0.96	0.98	IV	Chile	4	1.02	−1.6	IV
Bahrain	4	1.05	−0.74	IV	Uruguay	5	1.01	−0.47	IV
Barbados	3	1.04	−3.59	IV	Barbados	3	1.02	−0.7	IV
Mongolia	3	1.01	−0.35	IV	Mongolia	3	1.04	−0.83	IV
Saint Lucia	3	1.07	−4.08	IV	Saint Lucia	3	1.12	−2.91	IV
Costa Rica	4	1.01	−0.23	IV	Costa Rica	4	1.01	−0.29	IV
Trinidad and Tobago	4	1.05	−2.2	IV	Trinidad and Tobago	4	1.02	−1.24	IV

Note: Same as Fig. 10.1

References

Agresti, A., & Finlay, B. (1997). *Statistical methods for the social sciences*. Upper Saddle River, NJ: Prentice Hall.

Bennett, N. G., & Horiuchi, S. (1981). Estimating the completeness of death registration in a closed population. *Population Index, 47*(2), 207–221.

Bennett, N. G., & Horiuchi, S. (1984). Mortality estimation from registered deaths in less developed countries. *Demography, 1984*(21), 217–233. https://doi.org/10.2307/2061041.

Brass, W. (1975). *Methods for estimating fertility and mortality from limited and defective data*. Chapel Hill: International Program of Laboratories for Population Statistics.

Hill, K. (1987). Estimating census and death registration completeness. *Asian and Pacific Population Forum, 1*(3), 8–13. 23–24.

Hill, K., You, D., & Choi, Y. (2009). Death distribution methods for estimating adult mortality: Sensitivity analysis with simulated data error. *Demographic Research, 21*(9), 235–252. https://doi.org/10.4054/DemRes.2009.21.9.

Li, N. (2014). *Estimating life tables for developing countries*. Technical paper of the United Nations Population Devision 2014/4. Available at http://www.un.org/en/development/desa/population/publications/pdf/technical/TP2014-4.pdf.

Li, N. (2015). The probabilistic life table and its applications to Canada. *Canadian Studies in Population, 42*(1–2), 117–129.

Li, N. & Gerland, P. (2013). *Using census data to estimate old-age mortality for developing countries*. Paper prepared for session 17–05: Indirect methods of mortality and fertility estimation: New techniques for new realities in XXVII IUSSP international population conference. Busan, Korea.

Li, N. & Gerland, P. (2017). *Evaluating the completeness of death registration for developing coun tries at old ages*. Paper presented at the XXVIII IUSSP international population conference, 29 Oct-4 Nov 2017, Denver, USA.

Li, N., & Mi, H. (2018). Single-year Estimates in Developing Countries Mortality Database (DCMD). To appear in In *Proceedings of the 18th applied stochastic models and data analysis international conference with demographics workshop*.

Li, N., Mi, H & Gerland, P. (2017). Using child, adult, and old-age mortality to establish a Developing Countries Mortality Database (DCMD). In Skiadas, C. H. (Ed.) *Proceedings of the 17th applied stochastic models and data analysis international conference with demographics workshop* (ASMDA2017, pp. 645–655). Available at http://1drv.ms/b/s!ApL_0Av0YGDLijd07zI0KBZJZakF.

Li, N., Mi, H., Gerland, C. L., & Sun, L. (2018). *Establishing a Developing Countries Mortality Database (DCMD) on the basis of child, adult, and old-age mortality*. Paper presented at the 2018 annual meeting of the Population Association of America, 26–28 April 2018, Cape Town, South Africa.

Wilmoth, J., Zureick, S., Canudas-Romo, V., Inoue, V., & Sawyer, C. (2012). A flexible two-dimensional mortality model for use in indirect estimation. *Population Studies, 66*, 1 28.

Chapter 11
Mortality Developments in Greece from the Cohort Perspective

Konstantinos N. Zafeiris, Anastasia Kostaki, and Byron Kotzamanis

11.1 Introduction

Longevity in Greece has been studied in detail in a series of publications (see for example, Zafeiris and Kostaki 2019; Zafeiris 2019 etc.). According to Kotzamanis et al. (2016), the observed mortality transition in both genders between the years 1961 and 2014 resulted in a continuous and almost linear improvement of life expectancy at birth. The same general pattern emerges for life expectancy in other ages, although with varying rates of change. In absolute values, the gains are higher in the younger age groups and smaller in the older ones. However, these analyses are based on period mortality rates and ignore the fact that these only represent the mortality experience of a hypothetical, and not a real cohort of people (Preston et al. 2001; p. 42).

K. N. Zafeiris (✉)
Laboratory of Physical Anthropology, Department of History and Ethnology, Democritus University of Thrace, Komotini, Greece
e-mail: kzafiris@he.duth.gr

A. Kostaki
Laboratory of Stochastic Modelling and Applications, Department of Statistics, School of Information Studies and Technology, Athens University of Economics and Business, Athina, Greece

B. Kotzamanis
Laboratory of Demographic and social analyses (Lads), Department of Planning and Regional Development, School of Engineering, University of Thessaly, Volos, Greece

C. H. Skiadas, C. Skiadas (eds.), *Demography of Population Health, Aging and Health Expenditures*, The Springer Series on Demographic Methods and Population Analysis 50, https://doi.org/10.1007/978-3-030-44695-6_11

This is essential because cohort effects may influence mortality rates, a fact which was first observed a long time ago, naming, for example, the pioneering work of Kermack et al. (1934; 2001). In this publication it was recognized that '*the expectation of life was determined by the conditions which existed during the child's early years*' (about Kermack and a concise review of the literature see Davey Smith and Kuh 2001). The cohort effects are generally conceptualized as variations in the risk of a health outcome according to the year of birth, often coinciding with shifts in the populations exposure to risk factors over time (Keyes et al. 2010). Such factors may vary from prenatal ones (Christensen 2007) up to smoking habits (Doll et al. 2004).

Closely related to the concept of the cohort effect is the problem of natural selection, which acts on a population over time. In this sense, mortality is regulated by three determinants: age-related changes within individuals, the environment in which they live, and qualitative differences between individuals (Partridge 1997). For a cohort of people the individuals at highest risk tend to die (or exit from the population) first. This differential selection can produce patterns of mortality for the whole population, which are surprisingly different from the patterns for its subpopulations or individuals (Vaupel and Yashin 1985). In that way, in periods of significant mortality change the age specific rates are affected significantly because the elevated mortality of the past tends to "keep" the more robust people in the population and remove the frailer ones.

Additionally, observed period mortality rates may be affected by the so-called tempo effect (Bongaarts and Feeney 2002, 2003, 2005). According to Bongaarts and Feeney (2002, p.20), there is a bias in the estimated life expectancy at birth when "a rising (falling) mean age of persons at the occurrence of an event results in a temporary decline (increase) in numbers of events during the period of change". Thus, the tempo effect is positive when the mean age at death is rising and negative when the mean is declining (Bongaarts and Feeney 2003).

Overall, as Guillot and Canudas-Romo (2016) note, the observed period mortality of a population reflects not only the current mortality conditions but also "the product of past exposures and behaviours that have accumulated over the entire life span of individuals". During times of mortality transition, i.e. when mortality decreases, it is typically expected that period life expectancies will underestimate real life expectancies, which correspond to the longevity of real cohorts (Borgan and Keilman 2019); i.e. cohort analysis of mortality is essential in order to evaluate the real mortality trends observed in a population.

Therefore, this paper will first analyse the mortality transition in Greece from the cohort perspective. The method used will be presented in the data and methods section. We will also utilize cohort data from the period 1941–1960 which is provided by the former Director of ELSTAT G. Siampos to the Laboratory of Social Analyses of the University of Thessaly.

11.1.1 The Geographic Setting. A Brief Introduction to the Recent History of Greece

The modern population profile of Greece was primarily shaped by the historical events of the period 1912–1923, which correspond to the Balkan wars, the First World War, the Asia Minor Campaign, and eventually the signing of the Treaty of Lausanne. During this period, Greece initially extended to include Macedonia, part of Thrace, Epirus, the Aegean islands, and Crete, as well as a large part of Asia Minor, but ultimately it was limited to its current borders, except for the Dodecanese islands, which were annexed in 1948. After the defeat of Greece in the Greek–Turkish War of 1919–1922 a mass influx of refugees were observed entering the country (see Clogg 2002, pp. 85–113).

In 1940, after the outbreak of the Greek-Italian war, the country entered a new period of insecurity and there was a severe deterioration of living conditions along with social and economic destruction. Later, the defeat by Germany led to the triple occupation of the country by the Germans, Italians and Bulgarians (see Clogg 2002, pp. 98–141). In 1944, Greece was liberated, but in the period 1946–1949, it was afflicted by a bloody civil war.

As Zafeiris and Kostaki (2019) summarize, the period 1961–2014 is characterized by enormous developments in the economic, political, and social characteristics of Greece. As they note, a timetable of the most important events during that period may contain the following (see also Clogg 2002, pp. 166–238): 1. 1961–1967: Economic development and political instability, 2. 1967–1974: Military dictatorship (military junta), 3. 1974: Progressive restoration of democracy: political stability and social and economic growth 4, 1980: Greece re-joined NATO 5, 1981: Greece became a member of the European Union, 6. 2001: Euro was adopted as the national currency, 7. 2004: Olympic games took place in Athens, 8. After 2008: Economic crisis led to strong austerity policies. All the socioeconomic indicators of the country were burdened (decline of GDP, high unemployment rates, increase in the number of people at risk of poverty or social exclusion, cuts in the budget for social protection and health etc.).

11.2 Data and Methods

The period full lifetables were initially constructed separately for the two genders of the Greek population, based on the most recent available and revised death and mid-year population data of the Greek Statistical Authority (ELSTAT) and other sources for the years 1941–2017 [Laboratory of Demographic & Social Analyses (LDSA) data collection. Department of Planning and Regional Development University of Thessaly, School of Engineering www.ldsa.gr]. However, it must be noted that data from 1961 onwards have been revised several times, while data for the period 1941–1960 remains in its original and thus unrevised form. Therefore, a minor discrepancy exists in the calculated life table variables between the years 1960 and 1961, but this does not significantly affect the results. Of course, the 1941–1960 data could have been omitted, but in that case the necessary time depth and interpretative power of the analysis would have been limited. Additionally, by including this period in the analysis, a gap in the literature about mortality transition in Greece is covered because it is the first time that the results of mortality analysis for this period have been published.

The original calculations were carried out according to Calot and Sardon (Calot and Sardon 2004; See also Calot and Franco 2001; Calot 1999), using the original software which was developed by Calot himself.

Another problem needs to be resolved. This problem deals with the accuracy of period life expectancy calculations and the cohort analysis of mortality. A methodology to "correct" the period observed life expectancy at birth is the one developed by Bongaarts and Feeney (2002); however, this method has been extensively criticized (see for example Guillot 2006). Other approaches focus solely on cohort mortality (see for example, Guillot 2011), like the truncated cross-sectional average length of life (Canudas-Romo and Guillot 2015). However, one of the most parsimonious and effective methods is the one used by Borgan and Keilman (2019) for the comparison of the longevity of women in Italy, Japan and Scandinavia, which also has the advantage that it can be applied to both period and cohort data.

According to this method, the period life tables are used as a basis for the cohort analysis. In that way, someone born in year c will be x years old in the calendar year t = c + x. Thus, the one-year probability of death for a cohort born in year c in the calendar year c + x is:

$$q_x^{(c)} = q_{x,c+x}$$

and the probability that a woman who is born in year c will survive at least to age x is:

$$l_x^{(c)} = \prod_{i=0}^{x-1} \left(1 - q_i^{(c)}\right), x = 1, 2 \ldots$$

The problem with this approach is that it needs at least about 100 years in order to have a complete cohort life table, which is impossible based on the available data for Greece. When Borgan and Keilman (2019) faced the same problem they used, instead of the ordinary life expectancy at birth, the partial one between birth and several ages denoted with the term α, which can be calculated as:

$$e_{0/a}^{(c)} = 0.3 + 1.2 * l_1^{(c)} + \sum_{i=2}^{a-1} l_i^{(c)} + 0.5 * l_a^{(c)}$$

Then, they calculated the expected years lost before age a for a cohort c as:

$$a - e_{0/a}^{(c)}$$

This measure will be used in this paper and will serve to help compare male and female mortality transition in Greece. This idea comes from Arriaga's (1984) publication, in which life expectancy between two specific ages x and x + i is given by:

$$e_{x}^{x+i} = \frac{T_{x} - T_{x+i}}{l_{x}}$$

By using this formula, several life expectancies between specific ages will also be discussed in this paper. It is well known that the ordinary life expectancy at birth is affected by the infant, child, young, and middle-aged adult mortality (Horiuchi et al. 2013). The same happens with partial life expectancy. Thus, such an approach will give the opportunity to compare mortality experiences between specific ages which are free of the developments at younger ages.

11.3 Results

In the following paragraphs, the period and cohort calculations will be discussed in parallel. However, it must be kept in mind that in the period analysis the mortality experience of a hypothetical cohort in a calendar year is studied. On the contrary, in the cohort analysis, the mortality history of people born in the same year but who died numerous years later is studied, also taking into account the different socio-economic circumstances.

The period probabilities of death for chosen years illustrated in Fig. 11.1 have the typical shape of mortality patterns and a reasonable decrease through the years. The period before 1960 is much more diversified; mortality rates were very high for both genders in 1941, and they progressively decreased in the years after, and especially during the 1950s.

This trend is then followed, though any developments are smaller and refer mainly to younger ages and secondarily to middle life years. In all cases, the mortality rates are, as expected, higher in males, and the accident hump more severe for males than for females. Also, there is a problem with the mortality rates of the very old ages, because they are too scarce for several years. However, this problem does not significantly affect the results of cohort analysis as they are not taken into consideration in our analysis.

If the same mortality rates are considered from the cohort perspective, some interesting findings must be noted (Fig. 11.2). First, mortality decreases over time in the younger ages, but the decrease is not as sharp as the one observed in period analysis (see Fig. 11.1). Second, the developments in middle adulthood are not as

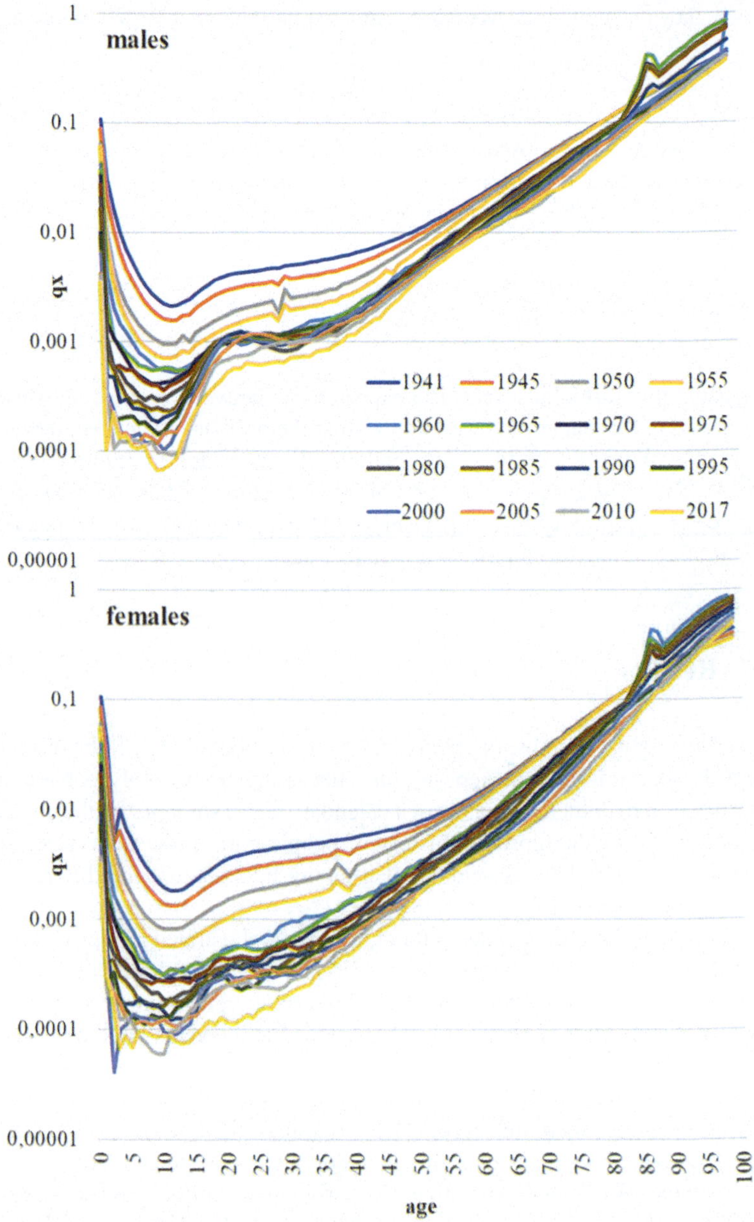

Fig. 11.1 Probabilities of death, period perspective

crucial as those observed in the period analysis. It should be noted, however, that the age of many of the younger cohorts of people is low during the time of the analysis for this work so that mortality rates cannot be recorded for their entire life span. For

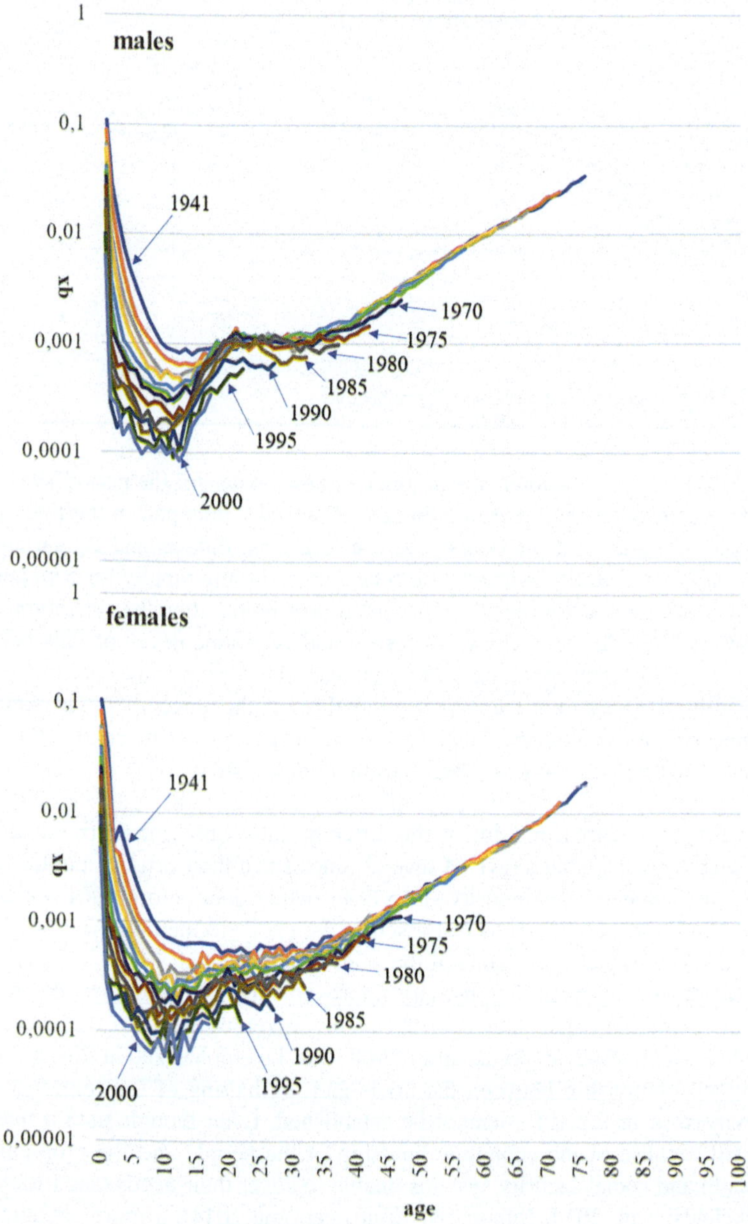

Fig. 11.2 Probabilities of death, cohort perspective

this reason, the estimation of mortality rates "stops" at different ages in Fig. 11.2. Therefore, any estimation of ordinal cohort life expectancy at birth is not possible, and the partial life expectancy at birth and the expected years lost from birth until different ages will be used subsequently.

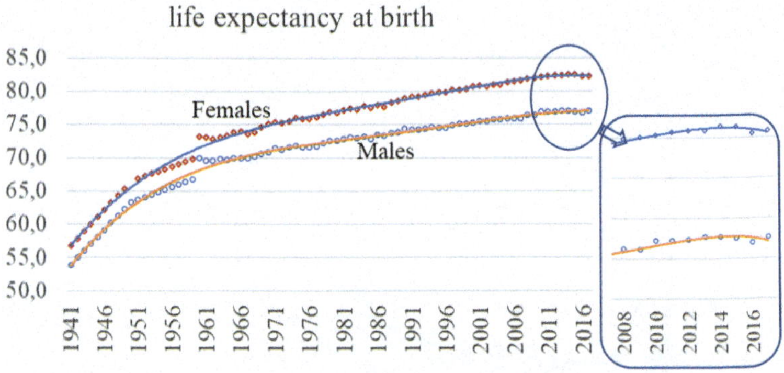

Fig. 11.3 Life expectancy at birth, period perspective

In Fig. 11.3, the developments of longevity by gender for the period 1941–2017 are presented. After the severe famine crisis of 1941, emerged in the first year of the triple occupation of the country by the Germans, Italians and Bulgarians (see Hionidou 2006), life expectancy at birth begins increasing rapidly for both genders, even if it remains at low levels. This trend still holds after the liberation of the country in 1944. A new phase of historical and economic developments begins at this time, mainly corresponding to the civil war and its preparatory political events, which afflicted Greece until at least 1949. In this era, the longevity of the population remains low but rapidly increasing. This trend weakens over time, and in the 1960s, the rate of increase of longevity becomes more moderate.

A significant gap is observed for life expectancy at birth when comparing the values for 1961 with 1960, due to the different data sources used in the analysis. Data after 1961 have been revised several times since their original publication by the National Statistical Authority (ELSTAT), while data before 1961 are in their original and unrevised form as discussed in the data and methods section. Thus, a small inconsistency of the results of the analysis is expected.

In any case, the rates of longevity improvement in Greece slow down in the years after, and life expectancy at birth steadily increases until 2013. The economic recession which afflicted Greece after 2008 must be emphasized at this point, even if a direct relationship between the crisis and the halting of the improvement in longevity seen in Fig. 11.3 cannot be established. Even though such a finding is expected because of the economic hardship of the people and the problems with the health and social security systems unable to cover their needs (see for example Kentikelenis et al. 2014; Simou and Koutsogeorgou 2014), a more extended time series is needed in order to thoroughly verify the halting as an effect of this crisis on the Greek population.

If we study mortality developments from the period perspective (Fig. 11.4), enormous mortality developments are observed. In 1941 males lose almost 25 years

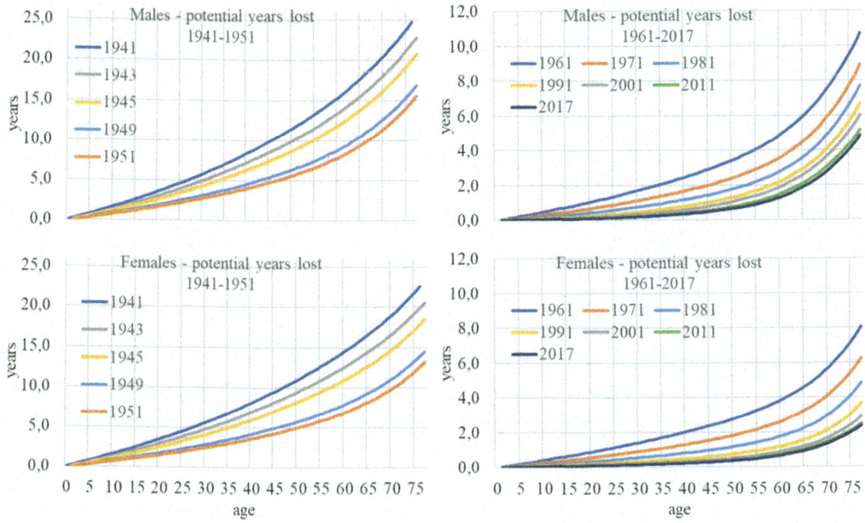

Fig. 11.4 Potential years lost, period perspective

of life between their birth and the age of 77 years, which corresponds to the highest observed age in the cohort analysis. At the same time, females lose almost 22.6 years. Ten years later, in 1951, these figures are reduced to 15.5 and 13.2 years respectively, thus in a small amount of time a very sharp decrease in mortality is observed. This mortality transition continues in the 1960s, and in the year 1961 a male loses 10.7 years of life and a female 8.1, thus the mortality transition is continuous but at a slower pace. Gradually, the losses will be confined to only 4.8 years in males and 2.4 in females in 2017; thus the mortality transition is further slowing down in the last years studied.

The same shape and trajectory of curves is observed in cohort analysis (Fig. 11.5). The potential years lost curves are "moving" in parallel towards smaller values, even though the mortality rates for some ages are not yet observed. However, a significant difference is also observed. As will be demonstrated in the following paragraphs, the number of potential years lost of the different cohorts tend to be smaller than those estimated by the period analysis.

Indeed, if the temporal trends of potential years lost are studied for several ages (Fig. 11.6), an almost parallel course is observed comparing period and cohort measures. Of course, the age at which our observation stops in the cohort approach depends on the time of the cohort formation. It should also be mentioned that for the analysis of the period perspective, many individuals belong to older cohorts than those included in the cohort analysis. For example, people aged 15 in 1941 were born in 1935–1936, and those aged 65 in the same year were born in the nineteenth century. This problem will be discussed later in more detail. At the same time, the members of the different cohorts spent a significant part of their lives in different socio-economic circumstances, mainly in a regime of social and economic

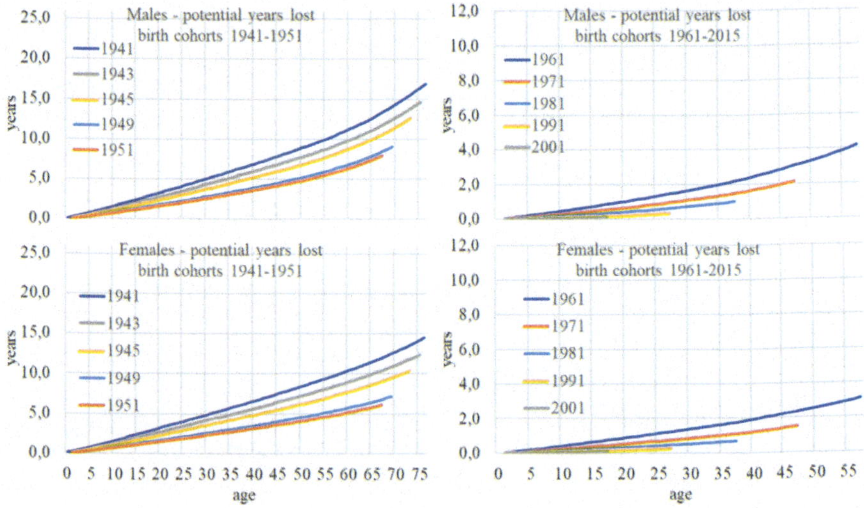

Fig. 11.5 Potential years lost, cohort perspective

development, accompanied by significant improvements in the health and social protection system of the country, especially after the 1980s.

During this process, the years lost are always higher in the period perspective than in the cohort one (Fig. 11.7). Also, they are higher in males than in females, as expected because mortality is always lower for women. In other words, the real mortality levels tend to be overestimated in the period analysis. However, besides the formal definition of a life table, which serves as a model of what would happen to a hypothetical cohort if a certain set of mortality conditions pertained throughout its life (Preston et al. 2001; p. 42), it must be stressed that the mortality trends of a real cohort are very different. As discussed in the introduction, the recorded mortality of this real cohort depends on the circumstances at the time of observation and during its lifetime, the known cohort effects, and the influence of the selection mechanism of human evolution.

Undeniably, in Fig. 11.7, we observe the period effects on mortality, but are any other factors evidenced too? In the earlier periods studied, the effects of a probable pestilence and famine on mortality, especially during the foreign occupation of the country, are without question. According to Hionidou (2006), the nutritional crisis prevailed throughout the occupation, exhibiting at the same time significant temporal and geographic fluctuations. For example, while famine hit Athens in the winter of 1941, other regions, such as Epirus, Macedonia and the Peloponnese, suffered mainly in the winter of 1942–1943 due to the influence of other factors. Also, the deaths of people who were killed in action or executed during this period must be added, as well as the Jewish genocide casualties (see Zafeiris 2015). Later, according to Christodoulakis (2016), there were more than 43,000 battle deaths during the civil war of 1946–1949, while the seriously wounded were almost 87,000.

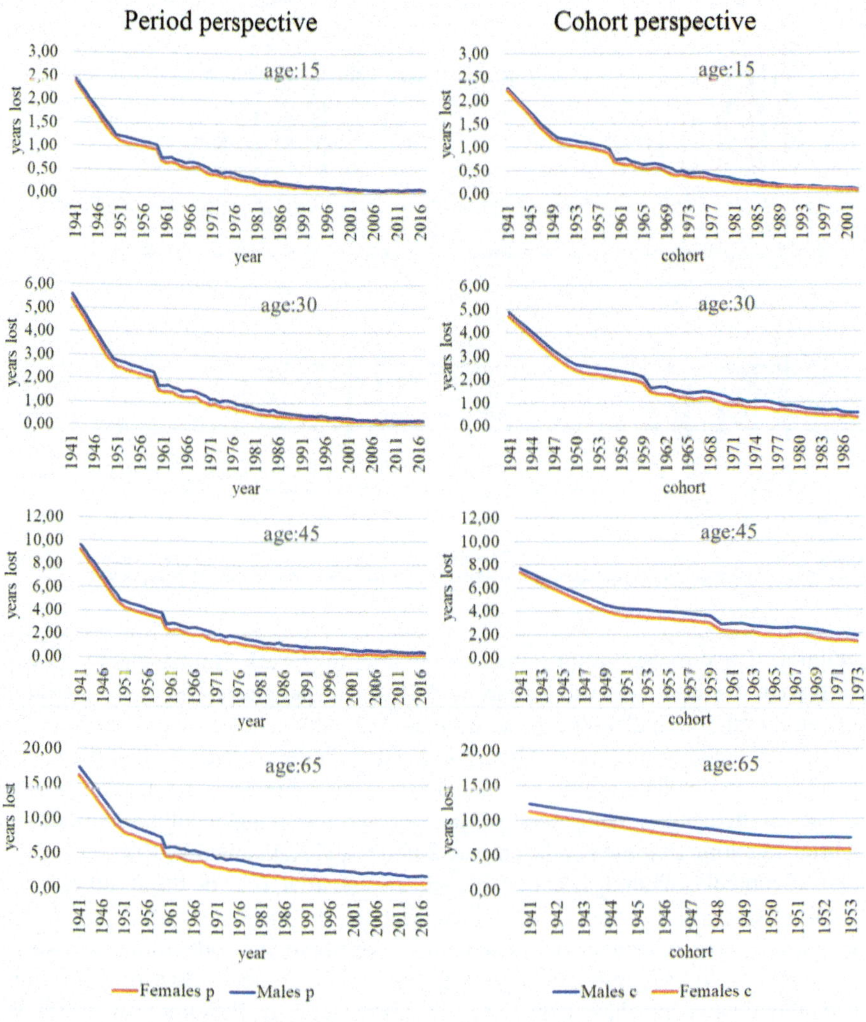

Fig. 11.6 Potential years lost at specific ages, period and cohort perspective

However, according to Averof-Tositsas (2010) the civil war casualties were almost 51,000. Therefore, the Greek population has spent a significant amount of time in health and life aggravating circumstances, a factor which undoubtedly affected the future mortality trends of the cohorts formed during that time, or the people that were alive during that era.

However, another factor must be taken into consideration. Following a very crude division, the people in this time period can be classified into two major categories. The first one includes those born within the borders of the country and the second one refers to the refugees from Asia Minor, Euxenous Pontus, Eastern Thrace, Anatolic Rumelia (southern Bulgaria) and elsewhere, along with their

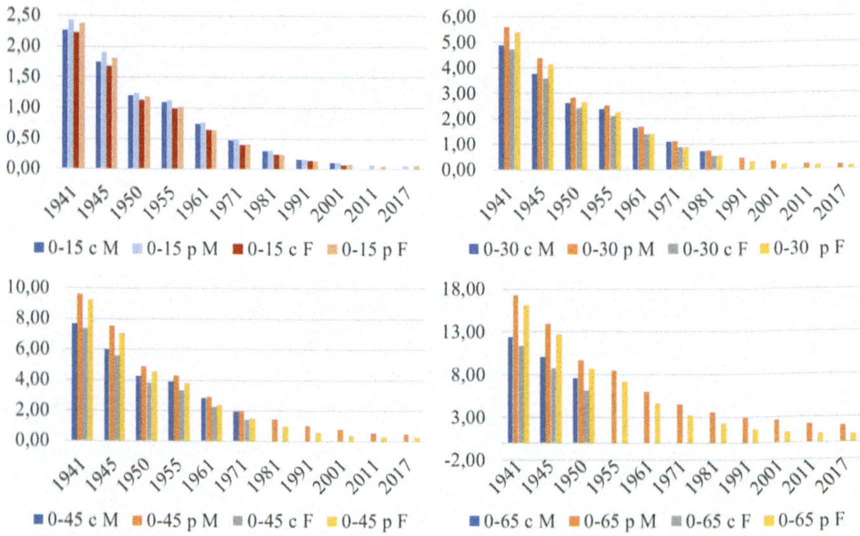

Fig. 11.7 Potential years lost at specific ages, comparing period and cohort perspective

children. The refugees group comes mainly after the Greek-Turkish and the Greek-Bulgarian disputes (see also Zafeiris 2015). These people represent the survivors of an extremely challenging process. It is not only that they survived the wars and were forced to migrate to Greece in harsh conditions, but also that they spent their lives in challenging circumstances concerning fundamental aspects of their existence: housing, nursing, food sufficiency, and social and health conditions (for a brief valuation of these conditions see Zafeiris and Kaklamani 2019).

Undoubtedly, natural selection played an essential role in these adverse circumstances by removing -among others- the 'frailer' members of this population. However, other agents may have acted too. For example, the population's low socio-economic profile suggests that significant potentiating effects, due to malnutrition and other factors, may have been observed (for the term see Pelletier et al. 1995). We also know that low levels of socio-economic development relate to elevated levels of mortality (see for example Behm 1980). Therefore, while natural selection tends to develop a no 'frail' population, the health and mortality levels of this population are aggravated significantly by several factors.

This scheme also refers to the first population, i.e. those born within the Greek borders, as Greece at that time was an impoverished country (for an appreciation of the Greek economy in the nineteenth and early twentieth century see Kostis and Petmezas 2006), a situation which was accompanied by insufficient health infrastructure and severe problems within its social protection sector. It is characteristic that the first official united social security institution (IKA) was enacted in 1932, and the relevant law was fully applied later (see https://www.ika.gr/gr/infopages/gene-ral/history.cfm). Thus, this is an area open for further scientific research within demography, epidemiology, evolution and other disciplines.

Afterwards, the country was changing and is moved towards economic growth and development, and this is imprinted in the progressive decrease of mortality levels, at least until a few years after the emergence of the economic crisis in 2008, when this trend is halted. During this process of mortality transition, longevity increased, but with a faster pace than period measurements reveal.

However, all this discussion is based on partial life expectancy between birth and some specific ages. Taking into consideration, as said in the data and methods section, that life expectancy at birth is affected by the mortality developments in the younger/middle ages, if we avoided that and studied different time fragments of the human life span we could discover the real longevity trends within those fragments.

In Fig. 11.8, the period and cohort life expectancy at birth between specific ages are seen. However, between period and cohort rates, a significant difference exists. While "cohort perspective" refers to all cohorts formed after 1941, the relevant period perspective, besides the previously mentioned cohorts, may include data from cohorts formed before 1941. For example, for calculating period partial life expectancy between 15 and 30 years the younger members of the population in the year 1956 were born in about the year 1941 while the older ones were born in about 1926. The same is the case for people aged between 30 and 45 in this year, who were born well before 1941.

In that way, the sharp increase of partial life expectancy observed when studying developments from the period perspective mainly refers to, or is partially influenced by, the mortality of the cohorts formed before World War II, or in some cases even before the Balkan Wars, World War I and the Greek-Turkish war of 1919–1922. This effect depends on the ages and the year studied. While for age group 0–15 years it disappears after 1956, for the age group 45–65 years it disappears after 2006. In any case, when moving towards the present the impacts become smaller in all the partial life expectancies studied.

If the same trends are studied from the cohort perspective, some interesting findings are observed. In this approach, it is seen that the observed mortality decrease is mainly governed by the developments in the younger age group of 0–15 years. Therefore, it is the infant, child, and young juvenile mortality, which mainly governs longevity in the post-World War II era. The developments in all the other age groups studied are minimal.

11.4 Conclusions

In this paper, the method proposed by Borgan and Keilman (2019) was used in order to study cohort mortality in Greece. In that way, the partial life expectancy between birth and specific ages was calculated and the years lost between birth and these ages were studied. The same analysis was made in the period data, in which the life expectancy at birth as a measure of longevity was additionally calculated and presented.

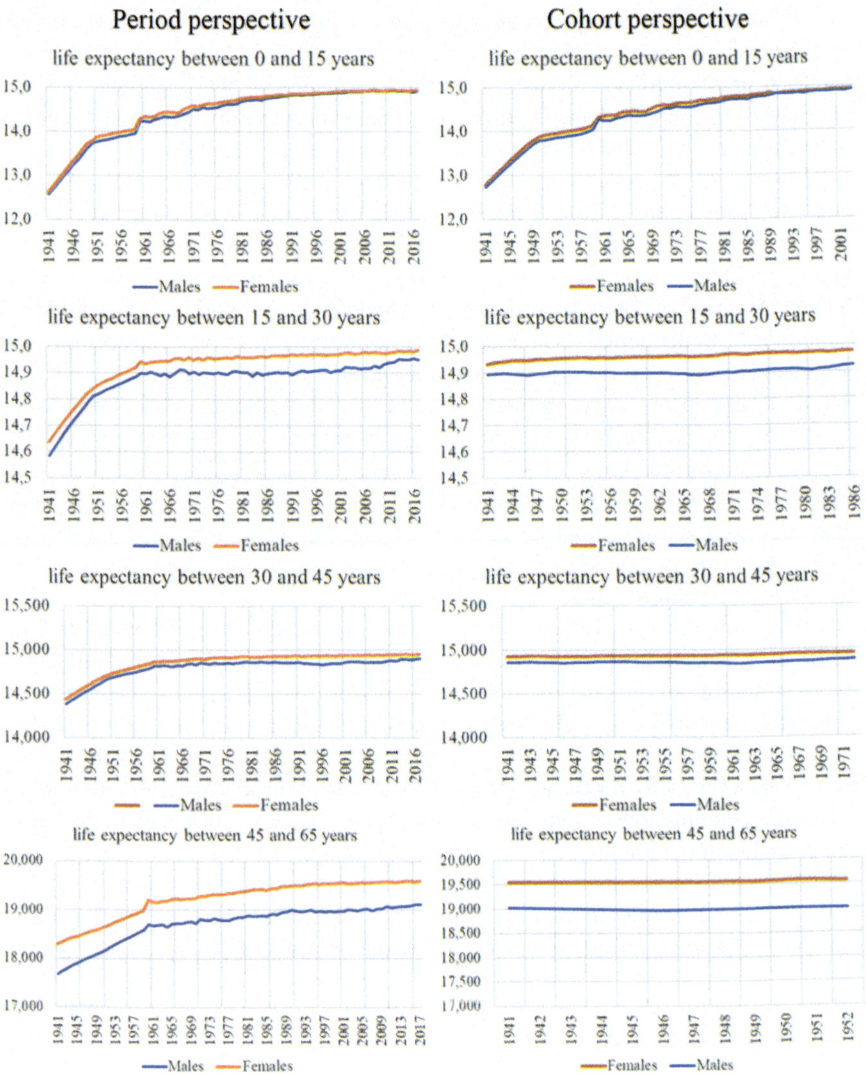

Fig. 11.8 Partial life expectancy at birth between specific ages, period and cohort perspective

Results of the period analysis indicate rapid mortality transition in the first years, which subsequently slows down and is halted in the last years of the study. In both the period and the cohort analysis, the potential years lost curves are "moving" in parallel towards smaller values, even though the mortality rates for some ages are not yet observed in the cohort perspective. The number of potential years lost of the different cohorts tends to be smaller than that estimated by the period analysis.

This finding indicates that besides the socio-economic and other circumstances which prevail in a period, significant cohort effects affect mortality rates as well as the evolutionary force of selection, which eliminates from the population its more frail members. However, in the higher mortality regime of the past, early lifetime events like malnutrition, several diseases, and the socio-economic environment may have aggravated not only period rates but also the future mortality of the cohorts.

Finally, the study of partial life expectancies between specific ages indicates two facts. The first deals with the nature of the period and cohort analysis. When analysing period partial life expectancy, older cohorts than those included in the cohort analysis are taken into consideration. The second fact is that after the 1950s the mortality transition in Greece is mainly related to the improvements in longevity in the young ages and not in the rest of ages.

References

Arriaga, E. (1984). Measuring and explaining the change in life expectancies. *Demography, 21*, 83–96.

Averof-Tositsas, E. (2010). *By fire and axe: Greece 1946–1949 and the precedents*. Athens: Estia editions. (in Greek).

Behm, H. (1980). Socio-economic determinants of mortality in Latin America. *Population Bulletin, 13*, 1–15.

Bongaarts, J., & Feeney, G. (2002). How long do we live? *Population and Development Review, 28*(1), 13–29.

Bongaarts, J., & Feeney, G. (2003). Estimating mean lifetime. *Proceedings of the National Academy of Sciences, 100*(23), 13127–13133.

Bongaarts, J., & Feeney, G. (2005). *The quantum and tempo of life-cycle events* (Policy Research Division Working Paper, p. 207). New York City: Population Council. http://www.popcouncil.org/

Borgan, Ø., & Keilman, N. (2019). Do Japanese and Italian women live longer than women in Scandinavia? *European Journal of Population, 35*, 87–99. https://doi.org/10.1007/s10680-018-9468-2.

Calot, G. (1999). L' analyse démographique conjoncturelle. In A. Kuijsten, H. de Gans, & H. de Feijter (Ed.) *The joy of demography*, édité en l'honneur de D. J. van de Kaa (pp. 295–323). La Haye/Amsterdam: Nethur DPublications, e0056. https://doi.org/10.1136/bmjopen-2014-005619.

Calot, G., & Franco, A. (2001). The construction of life tables. In G. Wunsch, M. Mouchart, & J. Duchêne (Eds.), *Life tables: Data, methods, models* (pp. 31–75). Dordrecht: Kluwer.

Calot, G., & Sardon, J. P. (2004). *Methodology for the calculation of Eurostat's demographic indicators*. Luxemburg: Office for Official Publication of the European Communities.

Canudas-Romo, V., & Guillot, M. (2015). Truncated cross-sectional average length of life: A measure for comparing the mortality history of cohorts. *Population Studies*, 1–13. https://doi.org/10.1080/00324728.2015.1019955.

Christensen, K. (2007). Early life events and later life health: Twin and famine studies. In T. Bengtsson (Ed.), *Perspectives on mortality forecasting. V. Cohort factors: How conditions in early life influence mortality in later life* (Social Insurance Studies no. 5) (pp. 31–39). Stockholm: Swedish Social Insurance Agency.

Christodoulakis, N. (2016). Conflict dynamics and costs in the Greek civil war 1946–1949. *Defence and Peace Economics, 27*(5), 688–717.

Clogg, R. (2002). *A concise history of Greece* (2nd ed.). Cambridge: Cambridge University Press.
Davey Smith, G., & Kuh, D. (2001). William Ogilvy Kermack and the childhood origins of adult health and disease. *International Journal of Epidemiology, 30*, 696–703. https://doi.org/10.1080/10242694.2014.1000010.
Doll, R., Peto, R., Boreham, J., & Sutherland, I. (2004). Mortality in relation to smoking: 50 years' observations on male British doctors. *British Medical Journal, 328*, 1519–1528.
Guillot, M., & Canudas-Romo, V. (2016). Revisiting life expectancy rankings in countries that have experienced fast mortality decline. In R. Schoen (Ed.), *Dynamic demographic analysis* (The springer series on demographic methods and population analysis) (Vol. 39). Cham: Springer.
Guillot, M. (2006). Tempo effects in mortality: An appraisal. *Demographic Research, 14*, 1–26.
Guillot, M. (2011). Period versus cohort life expectancy. In R. G. Rogers & E. M. Crimmins (Eds.), *International handbook of adult mortality*. New York: Springer.
Hionidou, V. (2006). *Famine and death in occupied Greece, 1941–1944*. Cambridge: Cambridge University Press.
Horiuchi, S., Ouellette, N., Cheung, S. L. K., & Robine, J.-M. (2013). Modal age at death: Lifespan indicator in the era of longevity extension. *Vienna Yearbook of Population Research, 11*, 37–69.
Kentikelenis, A., Karanikolos, M., Reeves, A., McKee, M., & Stuckler, D. (2014). Greece's health crisis: From austerity to denialism. *Lancet, 383*(9918), 748–753. https://doi.org/10.1016/S0140-6736(13)62291-6.
Kermack, W. O., McKendrick, A. G., & McKinlay, P. L. (1934). Death-rates in Great Britain and Sweden. Some general regularities and their significance. *Lancet, 226*, 698–773. Reprinted in International Journal of Epidemiology (2001), 30(4), 678–83.
Keyes, K. M., Utz, R. L., Robinson, W., & Li, G. (2010). What is a cohort effect? Comparison of three statistical methods for modeling cohort effects in obesity prevalence in the United States, 1971–2006. *Social Science & Medicine (1982), 70*(7), 1100–1108. https://doi.org/10.1016/j.socscimed.2009.12.018.
Kostis, K., & Petmezas, S. (2006). *The development of Greek economy in the 19thcentury*. Athens: Alexandreia. (In Greek).
Kotzamanis, B., Kostaki, A., Bergouignan, C., Zafeiris, K. N., & Baltas, P. (2016). *The population development of Greece (2015–2050)*. Athens: Dianeosis. Available at: https://www.dianeosis.org/research/demography/.
Partridge, L. (1997). National Research Council (US) Committee on Population. In K. W. Wachter & C. E. Finch (Eds.), *Between Zeus and the Salmon: The biodemography of longevity* (5: Evolutionary biology and age-related mortality). Washington, DC: National Academies Press (US). Available from: https://www.ncbi.nlm.nih.gov/books/NBK100403/.
Pelletier, D. L., Frongillo, E. A., Schroeder, D. G., & Habicht, J.-P. (1995). The effects of malnutrition on child mortality in developing countries. *Bulletin of the World Health Organization, 73*(4), 443–448.
Preston, S. H., Heuveline, P., & Guillot, M. (2001). *Demography: Measuring and modeling population processes* (1st ed.). Oxford: Blackwell Publishers.
Simou, E., & Koutsogeorgou, E. (2014). Effects of the economic crisis on health and healthcare in Greece in the literature from 2009 to 2013: A systematic review. *Health Policy, 115*(2–3), 111–119. https://doi.org/10.1016/j.healthpol.2014.02.002.
Vaupel, J. W., & Yashin, A. I. (1985). Heterogeneity's ruses: Some surprising effects of selection on population dynamics. *Journal of the American Statistical Association, 39*(3), 176–185.
Zafeiris, K. N. (2015). On the bridges between demography and biological anthropology: What a census can tell us about the composition of the modern Greek population. In K. Simitopoulou, K. Zafeiris, T. Theodorou, C. Papageorgopoulou (Eds.), *Anthropological pathways*. Festschrift for Professor N. I. Xirotiris. Mystis. Herakleion.
Zafeiris, K. N. (2019). Mortality differentials among the euro-zone countries: An analysis based on the most recent available data, communications in statistics: Case studies. *Data Analysis and Applications, 5*(1), 59–73. https://doi.org/10.1080/23737484.2019.1579682.

Zafeiris, K. N., & Kostaki, A. (2019). Recent mortality trends in Greece. *Communications in Statistics – Theory and Methods, 48*(1), 112–126. https://doi.org/10.1080/03610926.2017.1353625.
Zafeiris, K. N., & Kaklamani, S. (2019). Completed fertility during the twentieth century: An example from six settlements in northern Greece. *Journal of Biosocial Science, 51*(1), 118–137. https://doi.org/10.1017/S0021932018000019.

Chapter 12
On Demographic Approach of the BGGM Distribution Parameters on Italy and Sweden

Panagiotis Andreopoulos and Alexandra Tragaki

12.1 Introduction

Decreasing mortality at all ages has been the driving factor behind rapid population growth, over the past 100 years (Van de Kaa 1987). Mortality has been influenced by a number of different factors that need to be taken into account in the design and interpretation of variables. Age and gender are the main demographic characteristics to bc studied about their effect on mortality.

Throughout the human history, declining mortality rates have also been associated with economic growth and socio-economic prosperity. People live longer lives as fewer infants die and more persons survive to higher ages thanks to better nutrition and living conditions as well as to substantial progress in medical and public health care and technology. Increasing longevity is, beyond any doubt, one of humanity's greatest achievements. However and despite the general spectacular improvement, worth-studying differentiations still exist between genders, across regions as well as in respect to the causes of death. Environmental condition and the socio-economic context is anything but irrelevant to those differentiations. Moreover, as the age-distribution of a population changes so does the relative prevalence of age-related diseases: due to population ageing, previously uncommon conditions become frequent morbidity or mortality causes (Tragaki and Panagiotakos 2018).

The aim of this work is to model historical human mortality data using a new mortality distribution. We suggest that the proposed distribution, with its different parameters (κ, λ, θ, ξ, α, β) provides higher flexibility so as to describe more accurately the mortality curves and to capture more efficiently differentiations across ages as well as between sexes.

P. Andreopoulos (✉) · A. Tragaki
Department of Geography, Harokopio University, Athens, Greece
e-mail: pandreop@hua.gr; atragaki@hua.gr

C. H. Skiadas, C. Skiadas (eds.), *Demography of Population Health, Aging and Health Expenditures*, The Springer Series on Demographic Methods and Population Analysis 50, https://doi.org/10.1007/978-3-030-44695-6_12

12.2 The Background

A major issue in the field of bio-demography is the analysis of characteristics that influence mortality and survival rates (Benjamin and Pollard 1970), as well as, their association with the length of human lifespan (Anderson 2000). Mortality trajectories in human populations are conditioned by both endogenous (biological) attributes and exogenous (cultural and environmental) factors, as well. Studies that analyze mortality dynamics across the human lifespan and their evolution over time, are therefore of great importance for various reasons. Understanding the mechanisms underlying senescence is crucial for the implementation of health strategies and policies aiming to prevent or delay the process of ageing, reduce premature mortality, improve the quality of life and extend the duration of lifespan (Currie et al. 2004). Various mathematical techniques (Abramowitz and Stegun 2006) have been used to model processes associated with ageing and to validate mortality laws against changing demographics. Most of the mortality models are represented as parametric functions which achieve to reproduce the mortality trajectories (Forfar 2014). Mortality and survival models have been refined over the last few decades, yet the interest in elaborating new and more sophisticated models that better adapt to age-specific biological and non-biological parameters remains vivid. It seems that we still fail to fully grasp what defines the shape of human mortality curves.

Pioneer work started as early as a couple of centuries ago. Some of those first models are empirical for they describe the mortality trajectories without giving any explanations about the processes underlying the mortality dynamics such as the work by Gompertz 1825. Others are mechanistic and include terms that explain the nature of considered processes and model parameters with biological or demographic interpretations such as the work by Makeham 1860. The most notable work is the research of Benjamin Gompertz, a British actuary, who observed in 1825 that the force of mortality follows a geometric progression during most of the lifespan (Gompertz 1825).

The mortality rate $\mu(x)$, is therefore expressed by an exponential function of age x, well known as the *Gompertz law of mortality* (Table 12.1), where $\boldsymbol{\theta}$ denotes the initial level of mortality at age $x \in [0, 1)$ and $\boldsymbol{\xi}$ defines the rate of mortality increase over age. Mortality at very old ages increases at a slower pace: a phenomenon known as the "late-life decelerating mortality" (Gavrilov and Gavrilova 2006). An extension of the Gompertz's law is the well-known Makeham model (Makeham 1860), which represents the death rate as the sum of an age-dependent component which seizes mortality due to age-related diseases, and an age-independent component which captures the death risk due to external factors such as accidents (Table 12.1).

The *Gompertz–Makeham law* of mortality describes the age dynamics of human mortality quite accurately (*Makeham, 1890*). The inclusion of an additional extra parameter in the Makeham law allowed the model to be more flexible on the reproduction of empirical data and reduces the level of underestimation of mortality

Table 12.1 Probability density, cumulative density and hazard functions for Gompertz (G) and Gompertz-generalized Makeham (GGM) distributions

Model	Probability density	Cumulative density	Hazard
G	$\theta e^{\xi x} e^{-\frac{\theta}{\xi}\left(e^{\xi x}-1\right)}$	$1-e^{-\frac{\theta}{\xi}\left(e^{\xi x}-1\right)}$	$\theta e^{\xi x}$
GGM	$\left(\theta e^{\xi x}+\lambda+\kappa x\right) e^{-\frac{\kappa}{2} x^2-\lambda x-\frac{\theta}{\xi}\left(e^{\xi x}-1\right)}$	$1-e^{-\frac{\kappa}{2} x^2-\lambda x-\frac{\theta}{\xi}\left(e^{\xi x}-1\right)}$	$\theta e^{\xi x}+\lambda+\kappa x$

rates at early age. The generalized Makeham law (*Makeham, 1890*), includes a fourth parameter (parameter κ) which is multiplied by age x.

The aforementioned mortality laws are considered as the main distributions that adequately describe both the overall as well as the cause-specific probability of death as well. The probability density function, denoted by $f(x)$ gives the probability of death for an individual at age x and the cumulative density function, denoted by $F(x)$ is the complement of the survival function (Horiuchi and Coale 1990).

Fig. 12.1, presents Gompertz generalized Makeham distribution for selected values of parameters κ, λ, θ and ξ. The shape of mortality curve depends on the values of parameters θ, ξ, κ and λ. Gompertz distribution is a flexible left or right-skewed distribution. Moreover, this distribution is a generalization of the exponential distribution and is widely used in many applications, including applications in life data analysis (Balakrishnan et al. 1998). Some of the recently developed mortality models represent extensions of older ones and many of them are based on, or related to the Gompertz and Gompertz-Makeham distributions (Booth and Tickle 2008), Other distributions such as the Beta distribution can be combined (Kong et al. 2007) with the laws of mortality and provide new mathematical models for fitting and interpreting real mortality data (Andreopoulos and Bersimis 2015).

The probability density function, the cumulative density function and the hazard rate for each of these distributions here examined - the Gompertz and Generalized Gompertz Makeham laws of mortality - are summarized in Table 12.1.

Beta distribution has been widely used in a variety of scientific fields (Gupta and Nadarajah 2004) due to mainly it ability to describe a wide range of different data with bounded support. More specifically, Beta distribution with two parameters, i.e. left parameter a (shape of Beta) and right parameter b (scale of Beta), is used for modeling data that take values within the interval (0,1). Beta density is expressed as follows

$$f(x;\alpha,\beta)=\frac{1}{B(\alpha,\beta)}x^{\alpha-1}(1-x)^{\beta-1},\ \ 0<x<1,\ \ \alpha,\beta>0$$

where $B(\alpha,\beta)$ is given by the integral:

$$B(\alpha,\beta)=\int_0^1 u^{\alpha-1}(1-u)^{\beta-1}du$$

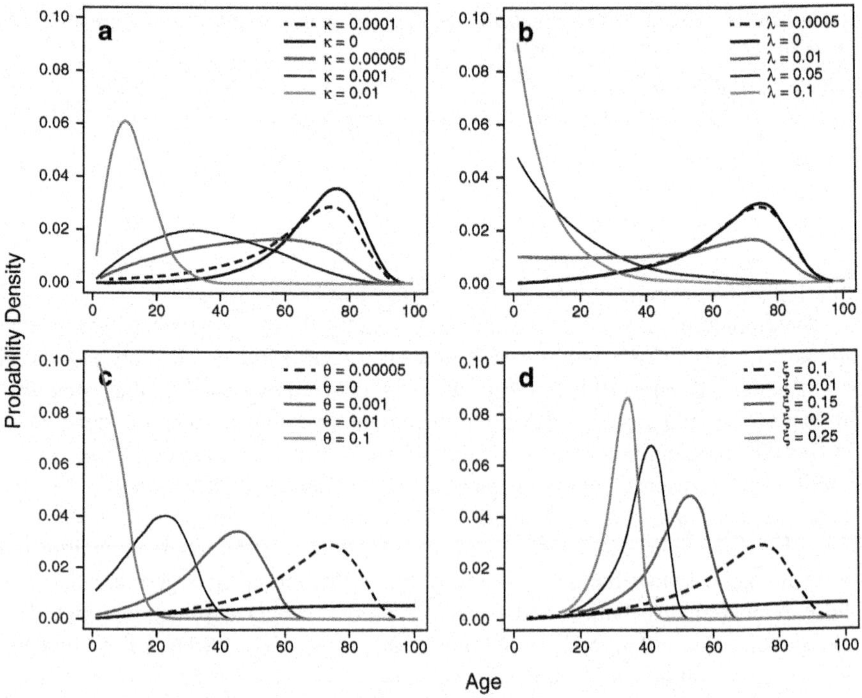

Fig. 12.1 The effect of parameters' variation on the dynamics of Gompertz generalized Makeham distribution. The black dashed line in each panel is used as a reference curve with model parameters $\kappa = 0.0001$, $\lambda = 0.0005$, $\theta = 0.00005$ and $\xi = 0.1$. In each panel three of the four parameters of the reference curve remain fixed and the fourth parameter changes. Panels (a), (b), (c) and (d) indicate the variation of κ, λ, θ and ξ respectively

Beta density function may have different shapes (Venter 1983) including left – and right- skewed or the flat shape of the uniform density, depending on the combination of its parameters.

As suggested by Zografos and Balakrishnan 2009, **Generated Beta distribution** is used in order to combine Beta distribution with the aforementioned mortality models. The corresponding probability density function is given by

$$g(x) = \frac{f(x)}{B(\alpha, \beta)}[F(x)]^{\alpha-1}[1 - F(x)]^{\beta-1}, \quad \alpha, \beta > 0$$

12.3 Method and Data

The new model used here named ***BGGM mortality model*** as has already been described (Andreopoulos et al. 2019) is based on mixing Beta and Gompertz generalized Makeham distributions. Its probability density function is given as follows:

$$f_{BGGM}(x) = \frac{\left(\theta e^{\xi x} + \lambda + \kappa x\right) e^{-\frac{\kappa}{2}x^2 - \lambda x - \frac{\theta}{\xi}\left(e^{\xi x}-1\right)}}{B(\alpha, \beta)} \cdot \left[\left(1 - e^{-\frac{\kappa}{2}x^2 - \lambda x - \frac{\theta}{\xi}\left(e^{\xi x}-1\right)}\right)\right]^{\alpha-1} \cdot \left[e^{-\frac{\kappa}{2}x^2 - \lambda x - \frac{\theta}{\xi}\left(e^{\xi x}-1\right)}\right]^{\beta-1}$$

$$\theta, \xi, \lambda, \kappa, \alpha, \beta > 0, x \geq 0$$

The purpose of the study is to apply the new model of mortality (**BGGM** distribution) on real mortality data so as to, firstly, assess its goodness of fit and secondly estimate a range of values for its parameters. Towards this end, we use data drawn from the Human Mortality Database for Italian and Swedish men and women over 114 years, from 1900 to 2013.

The data modelling and the corresponding distribution function, was conducted by using open code R (www.r-project.org) and corresponding packages of algorithms by CRAN digital library (Comprehensive R Archive).

The minpack.lm package has been used for the estimation of the 6 parameters.

12.4 Results

As already described, the model is built upon six parameters. This section presents the explanatory role and estimates the range of values for each one of those parameters.

12.4.1 Parameter θ – Initial Mortality [0, 1) Years

Parameter $\boldsymbol{\theta}$ describes mortality at initial ages [0, 1), also known as infant mortality. Human life is very fragile at the time of birth as well as during the months that follow it. This is reflected on a peak traditionally observed on the left end of the mortality curve. Neonatal mortality is sensible to economic and social conditions (ie., wars, famines) and is strictly related to the population living standards. The over-mortality of baby-boys compared to baby-girls has been well documented and is shown on the gap in the values of parameter $\boldsymbol{\theta}$ between genders. (Figs. 12.2a, 12.2b). Based on historical data, the range of values of parameter $\boldsymbol{\theta}$ is broader and lays at higher values for male compared to female populations. Moreover, the average value of $\boldsymbol{\theta}$ shows a declining, though more undulating trend for men, especially since 1980. The overtime trend is remarkably smoother for Italian and Swedish women.

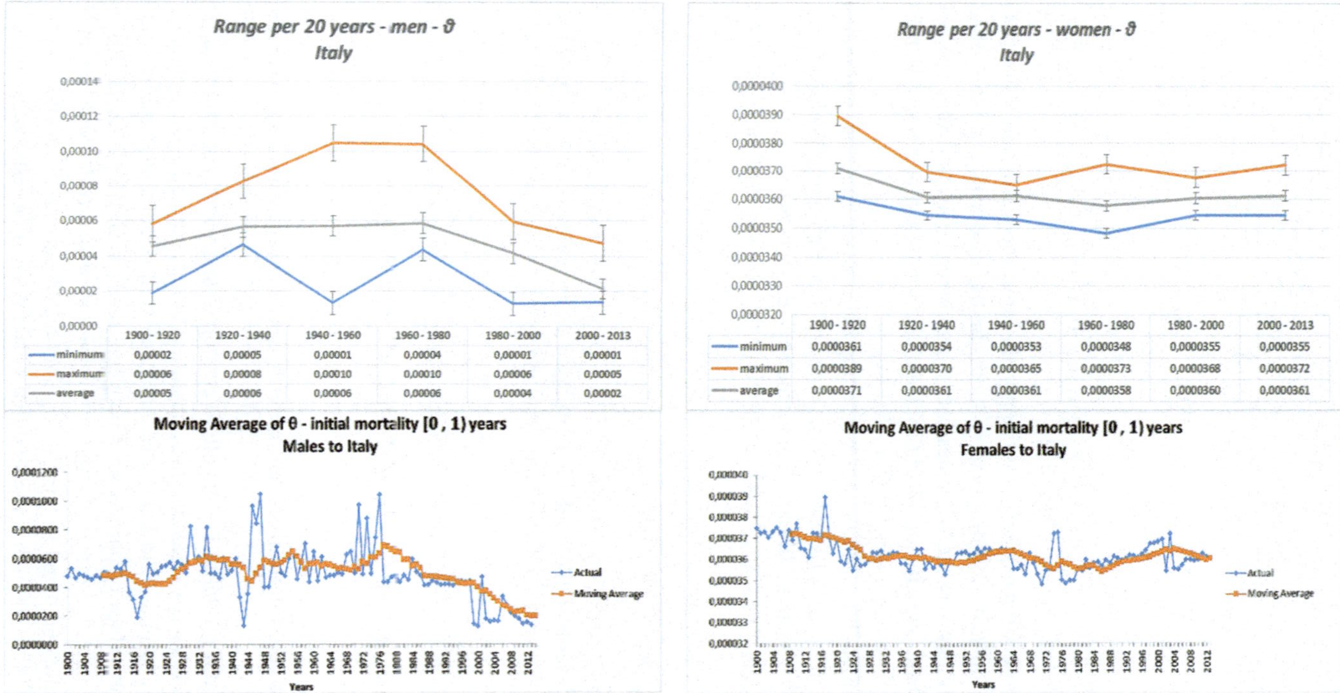

Fig. 12.2a The values of the estimated parameters θ. The graphs describe the role as well as the range of values for each one of those parameters. Mortality data, from Italy, for males and females for the years 1900 to 2013. Line in red illustrates the 10-year moving average (Analysis with *R* software)

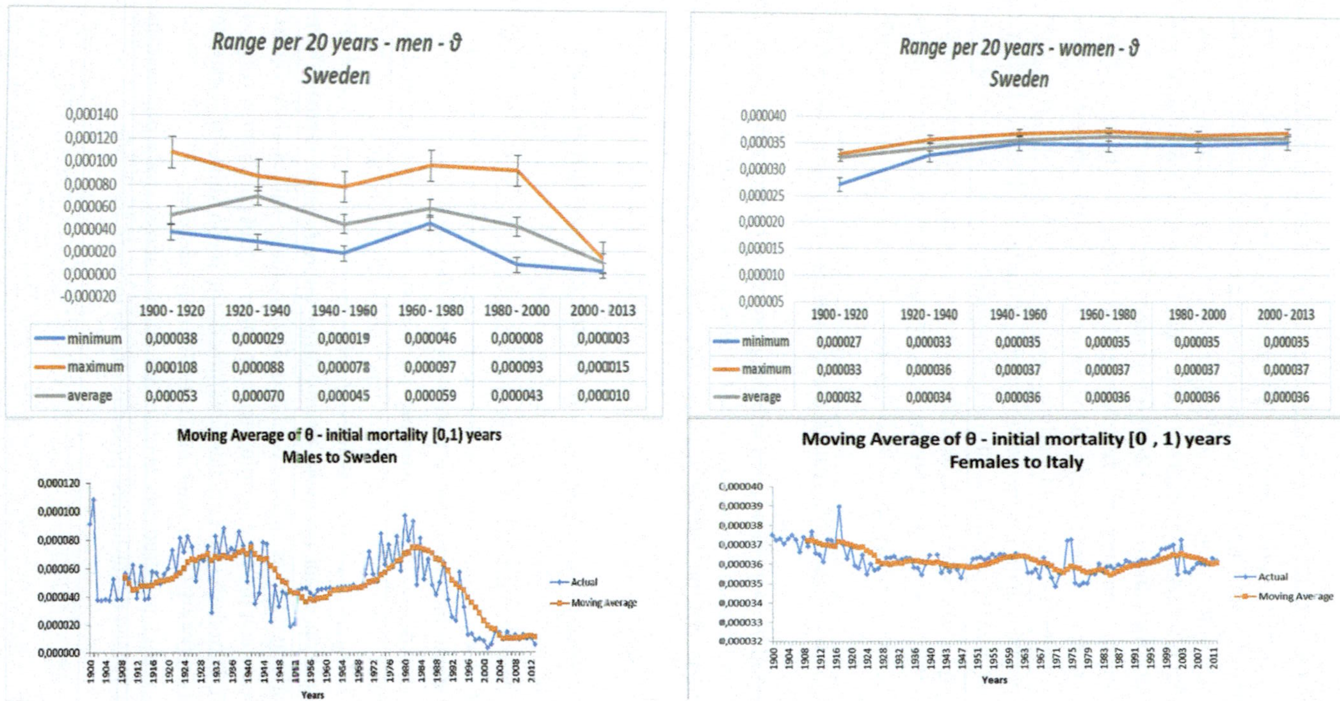

Fig. 12.2b The values of the estimated parameters $\boldsymbol{\theta}$. The graphs describe the role as well as the range of values for each one of those parameters. Mortality data, from Sweden, for males and females for the years 1900 to 2013. The red line is the 10-year moving average (Analysis was carried out with *R* software)

12.4.2 Parameter ξ – Rate of Change of Mortality with Age

The value of the parameter ξ represents the rate of change of mortality at very old ages. It is usually called "rate of ageing", mortality coefficient or Gompertz slope. The mortality rate in human populations increases exponentially with age for a significant part of the lifespan. Actually, it reflects the demographic aging rate of a population. Our results suggest that the mean value of parameter ξ fluctuates around 0.1 ($\cong$0.097) for men and 0.2 ($\cong$0.240) for women (Fig. 12.2c).

Also, the male population has remained stable during the analysis, around the mean value of the parameter ξ. Only, in the period 1940 to 1950, the male population over 80 years of age has increased, perhaps because of the war. Especially, mortality in the young population during the war increased, with the result that only the elderly survived. In the female population of Italy, demographic aging is evolving in a very interesting way. The parameter ξ describes the higher rates of demographic aging of women, in Italy. A reversal of the previous upward trend is registered during the years of the economic crisis (2003 to 2008) (Fig. 12.2d). Maybe this is the balance of human nature.

Based on Swedish data the results are similar but with less stochastic noise (graphic below).

Both men and women show a smooth progression in demographic aging. There is a smooth trend over the years. Decreased deaths and life expectancy, for Sweden, show the statistics for the coming years.

12.4.3 Parameter κ – A Random Factor at Individual Level

Parameter κ describes the risk of an unexpected death at individual level. A close-to-zero value of κ indicates that the chance of sudden death is low. When the risk of a random death increases, the κ parameter moves away from 0. For any given individual, κ is expected to be high during the first year of life, to lower in childhood; peaks in its values are observed in adolescence and early adult life as well as at ages from 50 to 75. Parameter κ reflects lifestyle choices, as well as the risk of an acute health deterioration, a car accident, or any other unpredictable fatality. Variations in κ are mostly age rather than sex-related. Parameter κ values show the smaller discrepancy between the sexes compared with other model parameters (Figs. 12.3a, 12.3b, 12.3c). Based on real data here examined, the mean value of parameter κ is approximately 0.0000137 for men and 0.0007056 for women.

Recent increases in parameter κ observed mainly across male populations may suggest an increased death hazard possibly due to modern life-style.

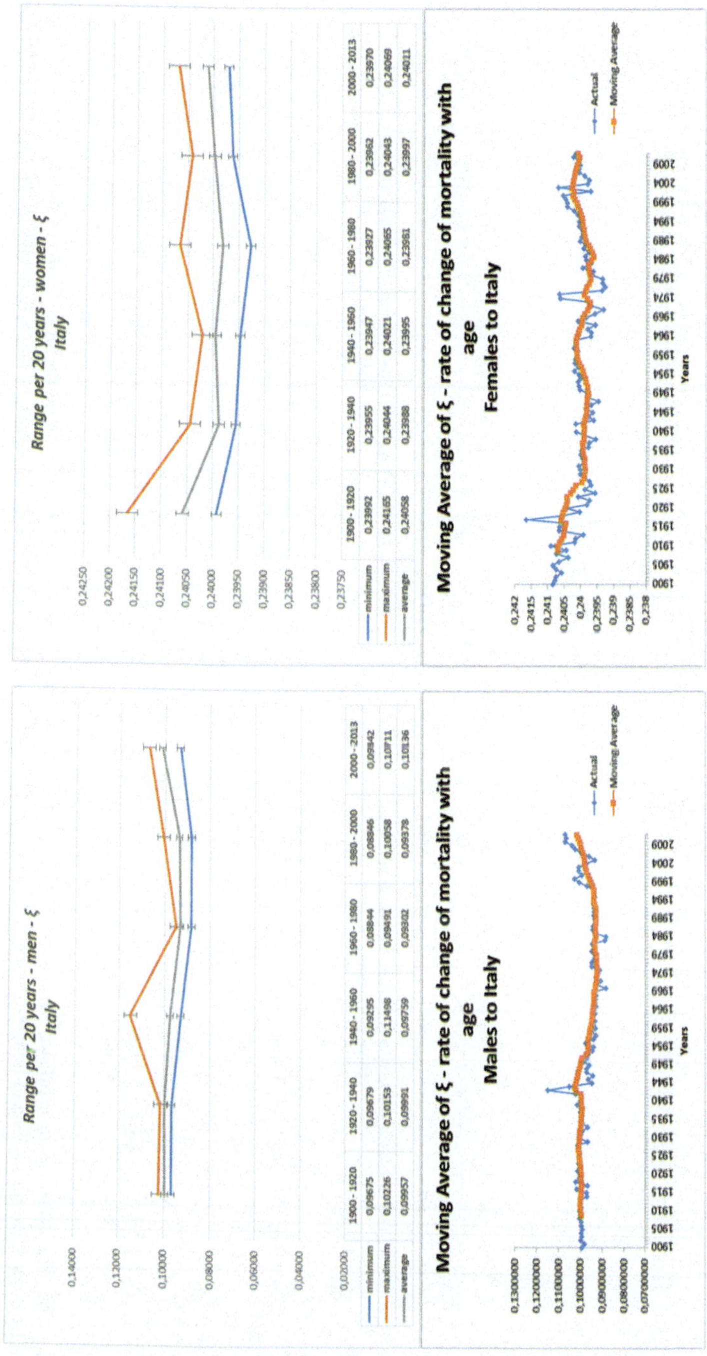

Fig. 12.2c The values of the estimated parameter ξ. Mortality data, from Sweden, for male and female for the years 1900 to 2013. The red line illustrates the 10-year moving average (Analysis with *R* software)

Fig. 12.2d The values of the estimated parameters ξ. The graphs describe the role as well as the range of values for each one of those parameters. Mortality data, from Sweden, for males and females for the years 1900 to 2013. The red font is the moving average by 10 years, a method of smoothing (Analysis with *R* software)

	1900 - 1920	1920 - 1940	1940 - 1960	1960 - 1980	1980 - 2000	2000 - 2013
minimum	0,0000000	0,0000000	0,0000000	0,0000000	0,0000000	0,0000001
maximum	0,0000816	0,0000000	0,0000416	0,0000959	0,0000924	0,0001204
average	0,0000092	0,0000000	0,0000042	0,0000163	0,0000166	0,0000486

	1900 - 1920	1920 - 1940	1940 - 1960	1960 - 1980	1980 - 2000	2000 - 2013
minimum	0,00059876	0,00069557	0,00070353	0,00069965	0,00070344	0,00065654
maximum	0,00070959	0,00074976	0,00075476	0,00075809	0,00074211	0,00070171
average	0,00066900	0,00072548	0,00071287	0,00072194	0,00072432	0,00067913

Fig. 12.3a The values of the estimated parameters κ. The graphs describe the role as well as the range of values for each one of those parameters. Mortality data, from Italy, for males and females for the years 1900 to 2013. The red line is the 10-year moving average (Analysis carried out with *R* software)

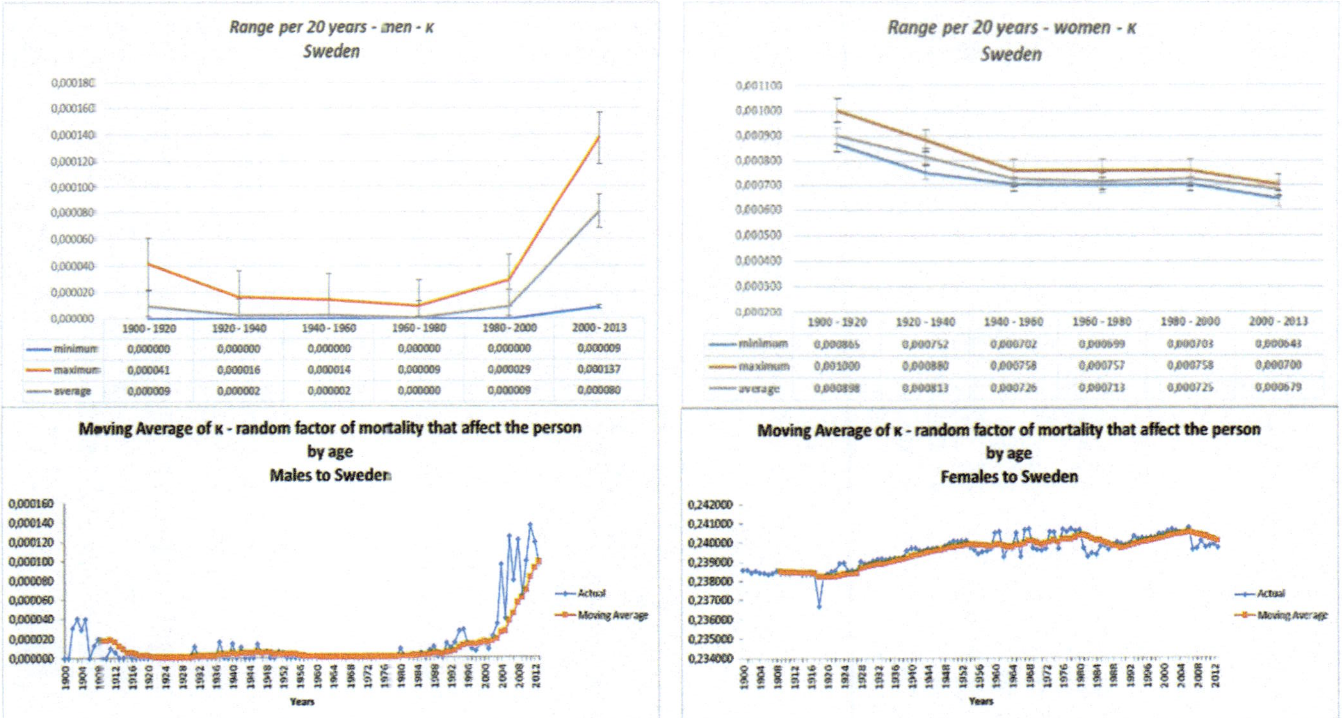

Fig. 12.3b The values of the estimated parameters κ. The graphs describe the role as well as the range of values for each one of those parameters. Mortality data, from Sweden, for males and females for the years 1900 to 2013. The red font is the moving average by 10 years, a method of smoothing (Analysis with *R* software)

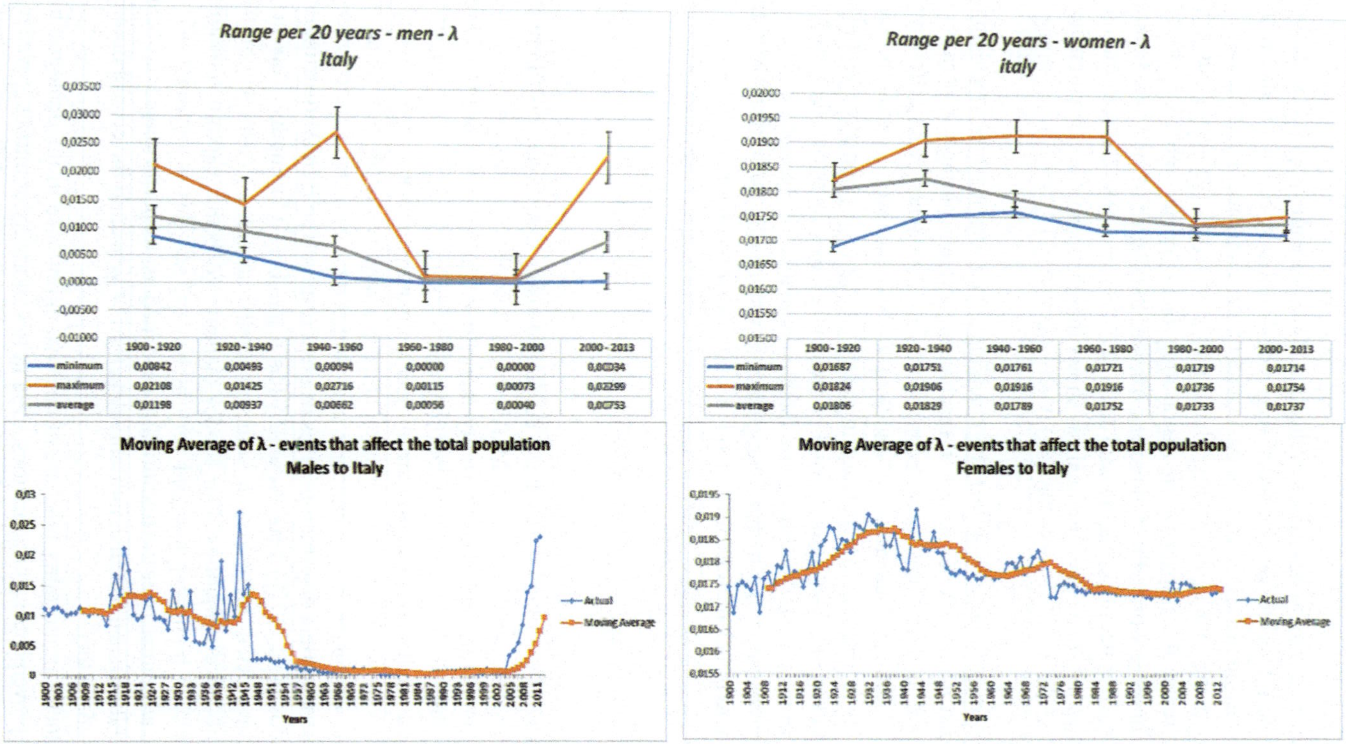

Fig. 12.3c The values of the estimated parameters λ. The graphs describes the role as well as the range of values for each one of those parameters. Mortality data, from Italy, for male and female for the years 1900 to 2013. The red font is the moving average by 10 years, a method of smoothing (Analysis with *R* software)

12.4.4 *Parameter λ – Random Factor of Mortality*

The parameter λ refers to hazardous events that affect the entire population. Those events may be related to various social (e.g. wars, illnesses, etc.) economic (e.g. low income, economic crisis, etc.) or environmental circumstances that have been proved to affect the course of mortality. The further from zero the value of parameter λ lies the greater the impact of an aggravating factor on the mortality of the population. In other words, the parameter expresses the degree of success / failure with which the population is dealing with death (λ follows the binomial distribution, λ~Bernoulli(n, p)) at any age. Our analysis shows that the mean value of parameter λ was formed for men to be 0.006 (≅0.0060) and 0.018 (≅0.0177) for women respectively.

Variations in parameter λ reflect the changing conditions a whole population is found faced against. As expected, λ reached high values during the period of WWII, as well as during the recent economic crisis. As women react differently to adverse conditions than men, the values of λ show interesting gender-related variations. There was also an increase in deaths in the total population of Italy, for males and females. During the period of financial crisis λ increases abruptly for but not for women. Calculated upon the female population, the parameter λ gives smoother results but with a higher average and less stochastic noise from 1970 to 2013; yet, there is a clear downward trend in λ values since 1980 (Fig. 12.3d).

12.5 Conclusions

A new mortality distribution has been developed using traditional distributions such as Gompertz and Gompertz - Makeham to describe the mortality rates of specific or general populations. Mortality modelling remains a top priority issue in demographic analysis and population projections, especially since survival probabilities as well as death causes are subject to changing factors.

This study shows that the new mathematical probability distribution which is based on the combination of Gompertz and Gompertz - Makeham distributions with the generated Beta distribution, gives us great results. The BGGM distribution (Andreopoulos et al. 2019) here presented has a number of strengths. This new model is constructed to overcome some generally accepted constrains of the above mentioned distributions. Our model manages to better capture mortality “irregularities” at both very low and very high ages, and this is regarded as its comparative advantage over the traditionally used models. Its performance has been tested on the Italian and Swedish mortality data for 114 years and the corresponding results are satisfactory.

Developments in mortality rates are of decisive importance in the demographic analysis. Improvements and fluctuations, even in low mortality countries (Sweden), shape the gender-composition and age-distribution and, thus, condition the popula-

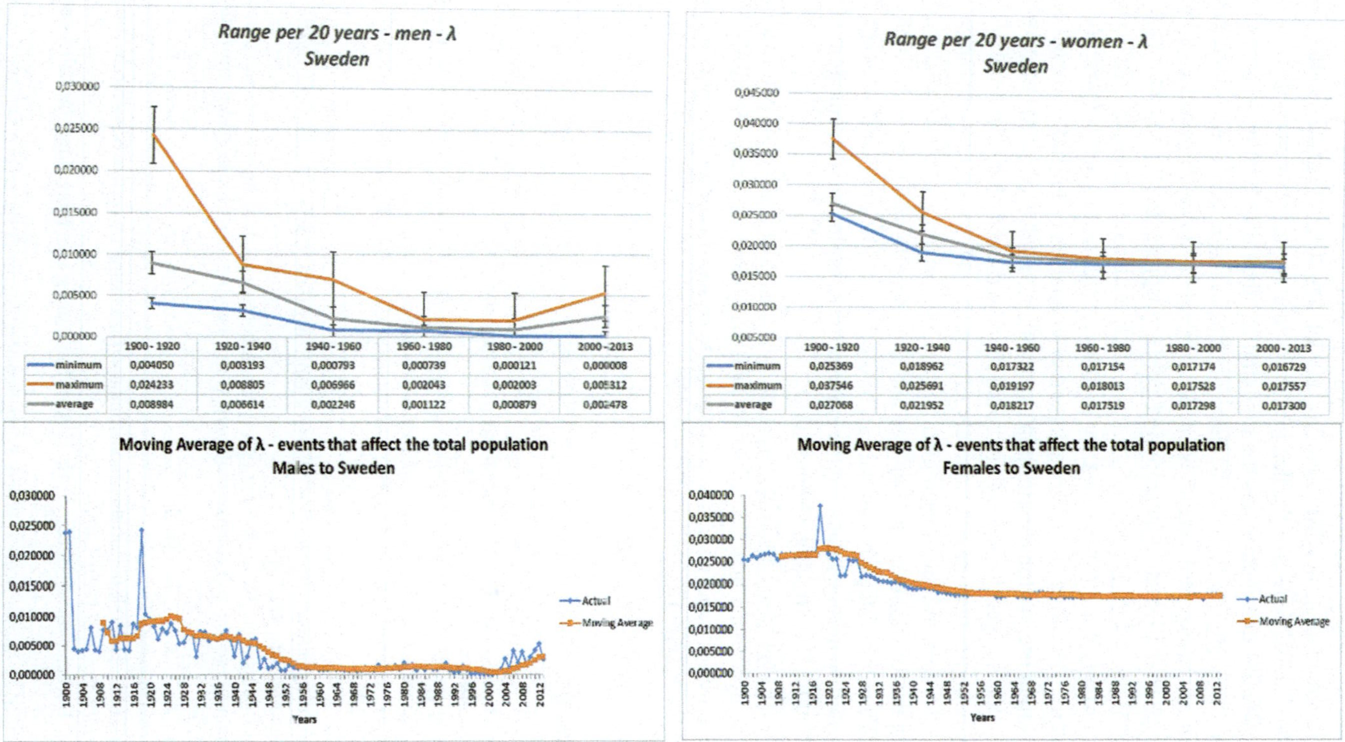

Fig. 12.3d The values of the estimated parameters λ. The graphs describe the role as well as the range of values for each one of those parameters. Mortality data, from Sweden, for males and females for the years 1900 to 2013. The red line represents the 10-year moving average (Analysis with *R* software)

tion prospects. Continuing the efforts so as to better capture and model mortality is, therefore, of crucial significance in demographic research. In conclusion, it could be assumed that BGGM mortality model has a great epidemiological - demographic value, because of the flexibility provided by its parameters' multitude, as well as, BGGM distribution has the potential to adapt data relating to common diseases or extremely rare diseases. Therefore the accurate prediction of general populations' mortality rates could be accomplished in order to generate public health strategies and better organization of public hospitals. Each parameter of the BGGM distribution, $f_{BGGM}(x; \alpha, \beta, \theta, \xi, \kappa, \lambda)$ offers a hint so as to study the "path" of human mortality, for different ages, for both sexes and for different countries (e.g. Italy, Sweden). The range given for each parameter enables us to study / predict the course of human life, within certain limits and with a confidence interval of $\pm 5\%$.

On the side of short-comings, it could be mentioned that the number of parameters is higher than in other models. However, each one of those parameters has a specific demographic interpretation; this makes modeling results easy to analyze and useful for understanding age and time specific observations related to mortality. We consider that this model can be used as an efficient mathematical tool for the analysis of mortality data and the exploration of biological or demographical processes that underlie ageing and mortality.

We intend to apply the BGGM model on mortality data from different populations so as to investigate how the values of each parameter vary geographically and how they are influenced by specific demographic or socio-economic features. This is expected to increase the ability of the model to predict future mortality trends.

References

Abramowitz, M., & Stegun, I. (2006). Handbook of mathematical functions (Washington, DC: US Government Printing Office. 1965). 9(3), 366.

Anderson, R. N. (2000). A method for constructing complete annual US life tables. Vital and health statistics. Series 2. *Data Evaluation and Methods Research*, (129), 1–28.

Andreopoulos, P., & Fragkiskos, B. G. (2015). Mortality modelling using probability distributions. *SCinTE, 1*, 189.

Andreopoulos, P., Bersimis, G. F., Tragaki, A., & Rovolis, A. (2019). Mortality modeling using probability distributions. Application in Greek mortality data. *Communications in Statistics-Theory and Methods, 48*, 1–14.

Balakrishnan, N., Johnson, N. L., & Kotz, S. (1998). A note on relationships between moments, central moments and cumulants from multivariate distributions. *Statistics & Probability Letters, 39*(1), 49–54.

Benjamin, B., & Pollard, J. H. (1970). *The analysis of mortality and other actuarial statistics*. Cambridge: University Press.

Booth, H., & Tickle, L. (2008). Mortality modelling and forecasting: A review of methods. *Annals of Actuarial Science, 3*(1–2), 3–43.

Currie, I. D., Durban, M., & Eilers, P. H. (2004). Smoothing and forecasting mortality rates. *Statistical Modelling, 4*(4), 279–298.

Forfar, D. O. (2014). *Mortality laws*. Wiley. Wiley StatsRef: Statistics Reference Online.

Gavrilov, L. A., & Gavrilova, N. S. (2006). Reliability theory of aging and longevity. *Handbook of the Biology of Aging, 6*, 3–42.
Gompertz, B. (1825). On the nature of the function expressive of the law of human mortality, and on a new mode of determining the value of life contingencies. *Philosophical Transactions of the Royal Society of London, 115*, 513–583.
Gupta, A. K. (2004). In S. Nadarajah (Ed.), *Handbook of beta distribution and its applications*. Boca Raton: CRC Press.
Horiuchi, S., & Coale, A. J. (1990). Age patterns of mortality for older women: An analysis using the age-specific rate of mortality change with age. *Mathematical Population Studies, 2*(4), 245–267.
Kong, L., Lee, C., & Sepanski, J. H. (2007). On the properties of beta-gamma distribution. *Journal of Modern Applied Statistical Methods, 6*(1), 18.
Makeham, W. M. (1860). On the law of mortality and construction of annuity tables. *Journal of the Institute of Actuaries, 8*(6), 301–310.
Tragaki, A., & Panagiotakos, D. (2018). Population ageing and cardiovascular health: The case of Greece. *The Hellenic Journal of Cardiology, 59*, 360–361.
Van de Kaa, D. J. (1987). Europe's second demographic transition. *Population Bulletin, 42*(1), 1–59.
Venter, G. (1983). Transformed beta and gamma distributions and aggregate losses. *Proceedings of the Casualty Actuarial Society, 70*(133 & 134), 289–308.
Zografos, K., & Balakrishnan, N. (2009). On families of beta- and generalized gamma-generated distributions and associated inference. *Statistical Methodology, 6*, 344–362.

Chapter 13
Alcohol Consumption in Selected European Countries

Jana Vrabcová, Kornélia Svačinová, and Markéta Pechholdová

13.1 Introduction

Alcohol represents a psychoactive substance with dependence-producing properties that has been widely used in many cultures for centuries (WHO 2014). The use of alcohol affects the social, health and economic areas of society (Dhital 2001). The harmful use of alcohol causes many diseases, as well as social and economic burdens in societies (WHO 2018). The damaging use of alcohol ranks among the top five risk factors for disease, disability and death throughout the world (WHO 2014). The health problems for which alcohol is responsible are only part of the total social damage which includes e.g. family disintegration, criminal inclinations and a loss of productivity and economic activity. Excessive alcohol drinking affect millions of people worldwide (Aryal 2008). The range of factors associated with alcohol consumption is also very wide, e.g. age group, gender, marital status, religion, ethnicity, family history etc.

Social tolerance to alcohol consumption is quite high in the Czech Republic (and also in other European countries) and so far, alcohol consumption has not been legally or practically restricted either by the Government or by any social organization. Production, sale, and consumption of alcohol has been increasing and can now be considered the number one problematic and addictive substance in the country (MHP 2011). Long-term tolerance to alcohol, both in cultural and political environments, has resulted in the unfavourable trend of increasing alcohol-related mortality for both sexes.

J. Vrabcová (✉)
Department of Statistics and Probability, University of Economics, Prague, Czech Republic

K. Svačinová · M. Pechholdová
Department of Demography, University of Economics, Prague, Czech Republic

C. H. Skiadas, C. Skiadas (eds.), *Demography of Population Health, Aging and Health Expenditures*, The Springer Series on Demographic Methods and Population Analysis 50, https://doi.org/10.1007/978-3-030-44695-6_13

Routine mortality statistics only partially capture the alcohol burden, through the so-called direct effects of alcohol. Epidemiological studies have additionally pointed at the indirect effects of alcohol on many other medical conditions (accidents, violence, hypertension, ischemic heart disease, neoplasms). Statistically, these indirect effects can be measured with the population-attributable fractions (PAF). The PAF method requires information on alcohol consumption patterns, namely the distribution of alcohol consumption across the population (the extent of indirect alcohol-related harm depends on the frequency and amount of alcohol consumption). Neither the mortality data nor the sales data provide information on alcohol consumption patterns and their distribution across population groups. Such information can only be obtained from interview sample surveys. The advantage of interview surveys is that, in addition to discovering the amount of alcohol consumed, they also inform about the distribution of alcohol consumption over time. In this respect, heavy episodic drinking (also known as binge drinking on a single occasion) is a major public health concern, since it has adverse consequences on an individual's well-being, as well as social and mental behaviour (Eurostat 2018). In this paper we will focus on the alcohol consumption levels and patterns in the Czech Republic and other selected European countries based mainly on the results of the European Health Interview Survey from 2014, which has recently been made accessible online by Eurostat. Our aims also extend to the analysis of social disparities in alcohol consumption, measured by educational attainment and income quintiles.

13.2 Data and Methods

The OECD Health Database offers the most comprehensive source of comparable statistics on health and health systems across OECD countries. It is an essential tool to carry out comparative analyses and form international comparisons of diverse health systems (OECD 2018). In the database, under the section "Non-medical determinants of health" we can find the variable "Alcohol consumption".

The World Health Organization (WHO) publishes the Global status report on alcohol and health, which is a useful source of information. Under WHO belongs The Global Information System on Alcohol and Health (GISAH) database, which is an essential tool for assessing and monitoring the health situation and trends related to alcohol consumption, alcohol-related harm, and policy responses in countries (WHO 2019). We can find a variety of variables here, e.g. levels of consumption (recorded and unrecorded alcohol consumption), drinkers only, regional alcohol per capita, etc. These online data sources do not typically provide consumption data by age, gender, education and other socio-economic characteristics of consumers. **Alcohol consumption by OECD** is defined as annual sales of pure alcohol in litres per person aged 15 years and older (OECD, not dated). **Recorded alcohol per capita (APC)** is defined as the recorded amount of alcohol consumed per capita (15+ years) over a calendar year in a country, in litres of pure alcohol.

The indicator only takes into account the consumption which is recorded from production, import, export, and sales data, often via taxation (WHO, not dated). **Total APC** is defined as the total (sum of recorded APC three-year average and unrecorded APC) amount of alcohol consumed per adult (15+ years) over a calendar year, in litres of pure alcohol. Recorded alcohol consumption refers to official statistics (production, import, export, and sales or taxation data), while the unrecorded alcohol consumption refers to alcohol which is not taxed and is outside the usual system of governmental control. In circumstances in which the number of tourists per year is at least the number of inhabitants, the tourist consumption is also taken into account and is deducted from the country's recorded APC (WHO, not dated). **Heavy episodic drinking**, or HED, is defined as drinking at least 60 grams or more of pure alcohol on at least one occasion. HED is one of the most important indicators for acute consequences of alcohol use, such as injuries (WHO 2019).

Patterns and disparities of alcohol consumption were measured based on the European Health Interview Survey 2014 (EHIS). The second wave of this survey took place between 2013 and 2015 in all EU Member States, and Iceland and Norway. The European Health Interview Survey (EHIS) aims at measuring on a harmonized basis and with a high degree of comparability among Member States (MS): the health status (including disability); health determinants (including environment); and use and limitations in access to health care services of EU citizens (EHIS, Metadata). The total sample size including Switzerland, Iceland and Norway was 210,000 respondents (6,510 in Czechia). Unlike OECD and WHO databases, the EHIS survey does not include information about alcohol consumption levels, but focuses on the frequency of alcohol consumption. Unfortunately, information on alcohol consumption in France and Netherlands is not available in EHIS data. The following characteristics were used in this article from the EHIS database:

Frequency of alcohol consumption: Distribution of the population according to the frequency of alcohol consumption using modalities 'Every day, Every week (but not daily), Every month (but not weekly), Less than once a month, Not in the last 12 months (Former drinkers), Never (Lifetime abstainers)'.

Hazardous alcohol consumption: Proportion of the population reporting to have had an average rate of consumption of more than 20 grams pure alcohol daily for women and more than 40 grams daily for men.

Frequency of heavy episodic drinking: Distribution of the population according to their frequency of heavy episodic drinking (which is ingesting more than 60 g of pure ethanol on a single occasion) using modalities 'At least once a week, Every month (but not weekly), Less than once a month, Never or not in the last 12 months'.

In this paper descriptive analysis and frequency tables were used.

13.3 Alcohol Consumption in European Countries by OECD and WHO

This section compares data about alcohol consumption according to the Organization for Economic Co-operation and Development (OECD) dataset and the World Health Organization (WHO) dataset.

Figure 13.1 shows the trends in alcohol consumption for the available European countries from 1960 to 2017 according to the OECD data. The long-term trends vary substantially, showing increases, decreases and even reversals, but we observe an overall convergence: the range of the alcohol consumption figures has narrowed from 1.1 to 23.2 liters of alcohol per capita in the early 1970s (Turkey vs. France) to 1.3 to 13.2 liters in 2016 (Turkey vs. Lithuania). The convergence is also due to the fact that alcohol consumption has particularly decreased in countries with previously extreme levels, such as France.

In 2016, according to the OECD datasets, the Czech Republic ranked as second place in terms of alcohol consumption, together with France, preceded only by Lithuania with 13.2 liters per capita (see Fig. 13.2).

The 2016 OECD consumption data was compared with the indicators of total and recorded alcohol consumption provided by the WHO. With regards to the figures on recorded alcohol consumption per person aged 15+ (three-year average in the WHO database) there is very good agreement with the OECD data for most countries, except Estonia. This is also confirmed by a correlation coefficient of 0.91 (or 0.99 if Estonia is omitted as an outlier). These results suggest that there is a

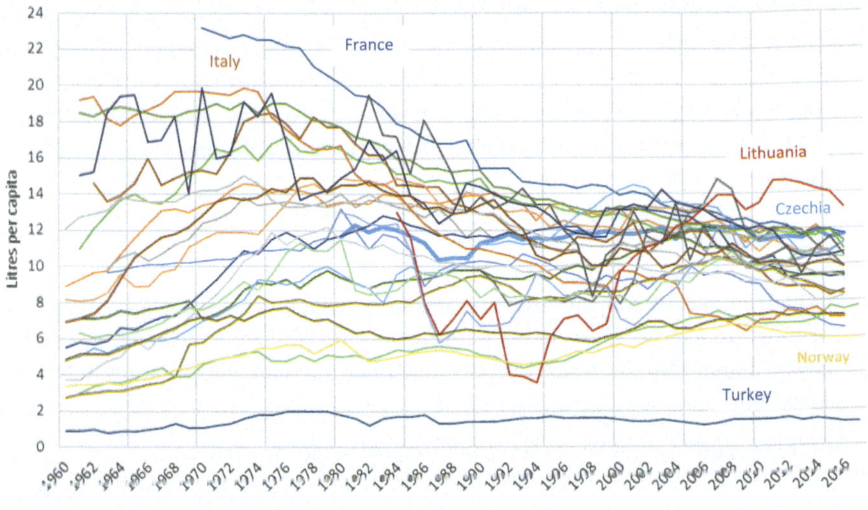

Fig. 13.1 Alcohol consumption (annual sales of pure alcohol in litres per person, 15+ years) in European countries in period 1960–2017. (Data source: OECD (2019), Alcohol consumption (indicator). doi: https://doi.org/10.1787/e6895909-en (Accessed on 21 March 2019))

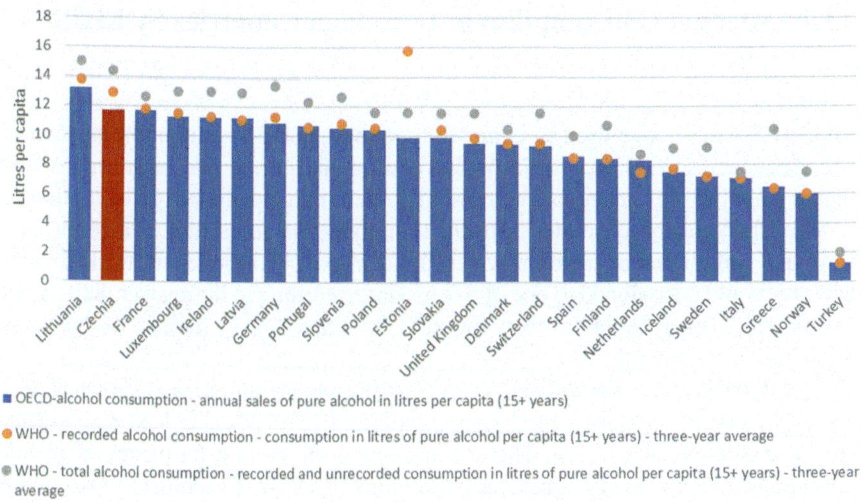

Fig. 13.2 Alcohol consumption by OECD, recorded and total consumption of alcohol by WHO of persons aged 15 and over in European countries in 2016. (Data source: WHO (2019), OECD (2019), Alcohol consumption (indicator). doi: https://doi.org/10.1787/e6895909-en (Accessed on 21 March 2019))

strong linear relationship between the alcohol consumption recorded by OECD and recorded alcohol consumption by WHO, which was confirmed at the 5% level of significance. Both data sources can thus be equally trusted regarding the recorded alcohol consumption. The Czech Republic is however among the few countries where WHO estimates differ significantly from OECD estimates.

Another characteristic, which is available in the WHO database, is the total amount of alcohol consumption per person aged 15+ (three-year average). These characteristic combines WHO expert estimates of recorded and unrecorded alcohol consumption.

As the indicator of total consumption includes estimates of unrecorded consumption, which differs greatly among countries depending on traditions and culture, we expected lower correlation with the OECD figures. The correlation coefficient is higher for the relationship between the total amount of alcohol consumption recorded by WHO and alcohol consumption published by OECD (correlation coefficient = 0.97 at 5% confidence level) than for the relationship between the total and recorded amount of alcohol consumption measured by WHO (correlation coefficient = 0.87 or 0.94 in the case of the omission of a suspicious value of recorded consumption of alcohol for Estonia at the 5% confidence level).

13.4 Alcohol Consumption in European Countries by EHIS 2014

The EHIS dataset was used to compare reported alcohol consumption prevalence and frequencies with the OECD/WHO consumption data. We expected to find higher drinking prevalence in countries with overall higher alcohol consumption.

The EHIS database provides information about the frequency of alcohol consumption and the frequency of heavy episodic drinking. Respondents aged 15 years and over could be assigned to one of five groups according to the answer they chose: "every day", "every week", "every month", "less than once a month" or "never or not in the last 12 months".

The distribution of the answers in selected European countries is shown in Fig. 13.3. The results show that the largest proportion (29.6%) of EU-28 respondents consume alcohol every week. The countries with the highest frequency of people consuming alcohol **every week**, are: Belgium (37.8%), Czechia (34.6%), Denmark (40.0%), Finland (36.6%), Germany (39.6%), Ireland (41.4%), Luxembourg (40.6%), Malta (28.8%), Slovenia (26.5%), Sweden (36.9%) and the United Kingdom (45.0%). Countries, where most respondents consume alcohol **every month** are: Austria (31.8%, but it is close to answer "every week" with 31.5%), Iceland (39.2%), Latvia (32.8%), Lithuania (37.6%), Norway (49.3%), Poland (28.5%) and Slovakia (29.6%). Estonia, with 31.6%, is the only country where people most often consume alcohol **less than once a month**. Finally, countries where most respondents answered that they **never or not in last 12 months** consumed alcohol are: Bulgaria (33.9%), Croatia (43.9), Cyprus (46.4%), Greece (32.1%), Hungary

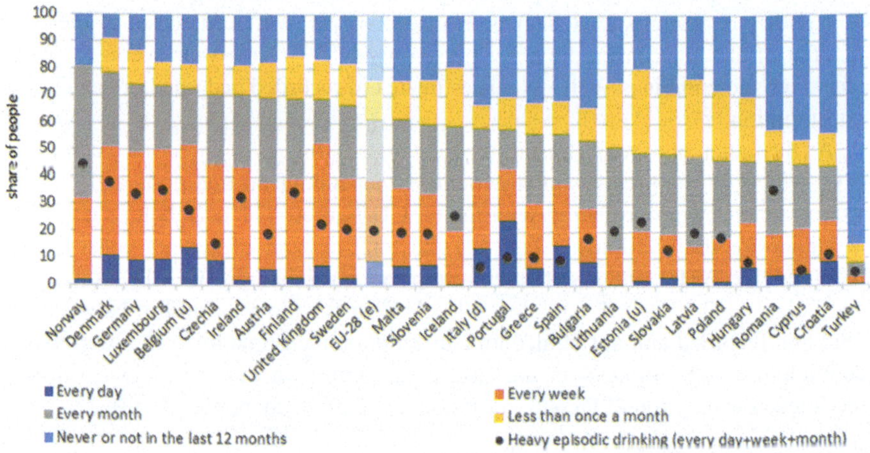

Fig. 13.3 Distribution of population aged 15 and over according to the frequency of alcohol consumption in European countries by EHIS 2014 (%). Note: *e* Estimate, *d* Definition differs, *u* Unreliable data. (Data source: Eurostat (hlth_ehis_al1e; hlth_ehis_al3e))

(29.8%), Italy (32.8%), Portugal (30.0%), Romania (42.1%), Spain (31.3%) and Turkey (85.1%).

Figure 13.3 additionally displays the prevalence of heavy episodic drinking (HED). HED is defined as consuming an excessive amount of alcohol (60 grams or more) on a single occasion at least once a month. Heavy episodic drinking is one of the most harmful ways of consuming alcohol, and can indicate a problem with alcohol drinking.

Alcohol drinking patterns vary across countries. Countries such as Norway, Denmark, Germany, Luxembourg, Ireland and Finland have a high proportion of people drinking at least every month but also a high proportion of heavy drinking episodes at least every month. Another group of countries consists of Italy, Portugal, and Spain, with a relatively higher proportion of people who drink alcohol every day, but very low levels of heavy episodic drinking. Romania stands out as a country with a high proportion of heavy episodic drinking but low alcohol consumption (frequency of consumption is typically at least every month). Looking at the EU-28, we see that every fifth person who consumes alcohol experiences at least one episode of heavy drinking every month. In Czechia, the overall prevalence of drinking is rather high (44.1% of respondents consume alcohol at least once a week), but the frequency of heavy episodic drinking is among the lowest (14.9%).

The correlation coefficients for the different combinations of responses were calculated and the following pairs of variables show some form of linear dependence: HED (every day + every week + every month) with sum of the frequencies of drinking alcohol (every day + every week + every month), gives a correlation coefficient equal to 0.66. On the other hand, the correlation between HED (every day + every week + every month) and the share of people who drink never or not in the last 12 months, recorded an indirect linear dependency (correlation coefficient = −0.59).

The data on alcohol consumption patterns according to the EHIS survey were then compared to the consumption data produced by the OECD (see Fig. 13.2). The ranking of countries for both sexes by frequency of alcohol consumption from the EHIS survey differs. For example, Norway, which according to EHIS has the number one in share of people consuming alcohol at least once a month and also in heavy episodic drinking behaviour, has one of the lowest levels of alcohol consumption per person according to the OECD and the WHO. On the contrary, Lithuania, which according to the OECD and the WHO has the highest levels of alcohol consumption per person, ranks among the countries with relatively low alcohol consumption according to the EHIS dataset. This unequal arrangement of countries is confirmed by low correlation coefficients: the correlation coefficient between EHIS alcohol consumption at least once a month and alcohol consumption by OECD is equal to 0.42 at the 5% level of significance, and the correlation coefficient between EHIS alcohol consumption and total amount of alcohol consumption by WHO is equal to 0.40 at the 5% level of significance. Other combinations across other indicators were not statistically significant.

13.5 Alcohol Consumption Disparities by Gender, Educational Attainment and Income

In this section, we assessed the size and the international variability of the socioeconomic disparities in alcohol consumption. Alcohol consumption tends to be significantly higher among males than among females. Alcohol is also linked with poverty and lower education, therefore we expected to find corresponding disparities in the distribution of alcohol consumption over the respective socioeconomic population groups.

EHIS data confirms that gender plays a major role in the amount and frequency of alcohol consumption. It can be seen from Fig. 13.4 that, for example, in Portugal (a difference of 36.4 percentage points), Romania (36.8 pp), Cyprus (35.7 pp) or Hungary (34.7 pp), there is a large gender gap in alcohol consumption. Males also experience higher HED frequencies than females in all countries. The largest HED differences between genders were recorded in Romania (a difference of 34.8 percentage points), Estonia (27.3 pp), Finland (26.1 pp) and Lithuania (25.6 pp).

Regarding educational attainment, we measured the disparities in both the drinking patterns and frequencies of heavy episodic drinking. In the EHIS dataset, heavy episodic drinking was most prevalent among respondents with upper secondary and post-secondary non-tertiary education (see Fig. 13.5). This rule applies to three quarters of European countries listed in the EHIS 2014. In the remaining quarter of countries (e. g. Norway, Spain, Greece, and Austria), HED was most frequently reported among respondents with tertiary education. There is no country, where people with less than primary, primary and lower secondary education, rank first in terms of HED prevalence.

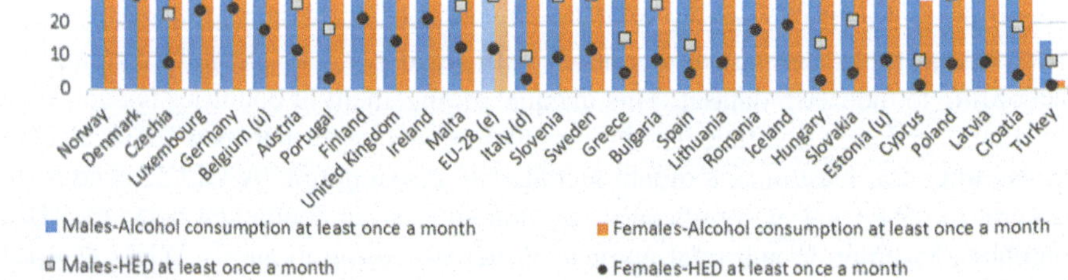

Fig. 13.4 Share of males and females aged 15 and over consuming alcohol and having HED at least once a month in European countries by EHIS 2014 (%). Note: *e* Estimate, *d* Definition differs, *u* Unreliable data. (Data source: Eurostat (hlth_ehis_al1e; hlth_ehis_al3e))

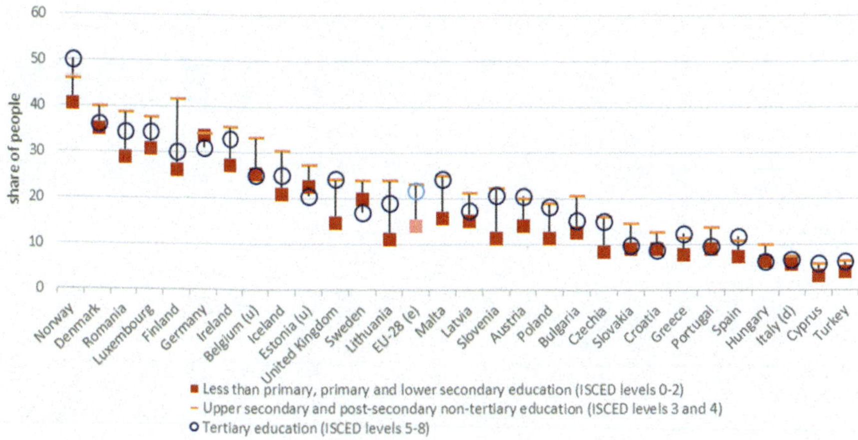

Fig. 13.5 Share of people aged 15 and over having HED at least once a month by education attainment in European countries by EHIS 2014 (%). Note: *e* Estimate, *d* Definition differs, *u* Unreliable data. (Data source: Eurostat (hlth_ehis_al3e))

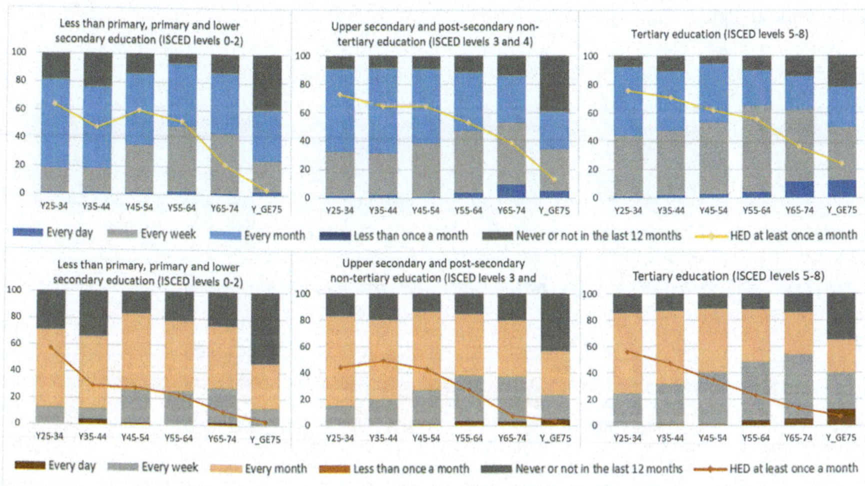

Fig. 13.6 Distribution of males (first row) and females (second row) according to the frequency of alcohol consumption by education in Norway by EHIS 2014 (%). (Data source: Eurostat (hlth_ehis_al1e; hlth_ehis_al3e))

The following charts (Figs. 13.6, 13.7, and 13.8) illustrate the patterns of alcohol consumption and HED by educational attainment in selected countries. We compare the Czech Republic to Norway and Lithuania. The selection of the countries was not random. Norway reports the highest consumption prevalence in EHIS and the lowest consumption in litres of alcohol. While Lithuania is the opposite, with the

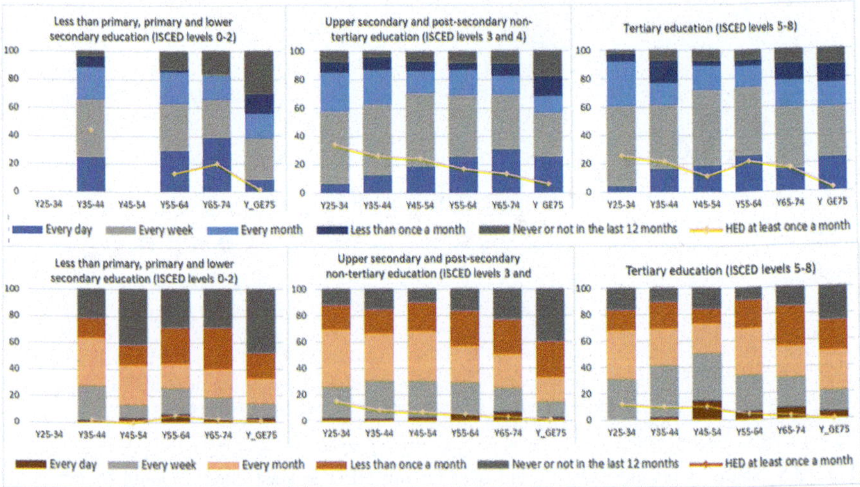

Fig. 13.7 Distribution of males (first row) and females (second row) according to the frequency of alcohol consumption by education in the Czechia by EHIS 2014 (%). (Data source: Eurostat (hlth_ehis_al1e; hlth_ehis_al3e))

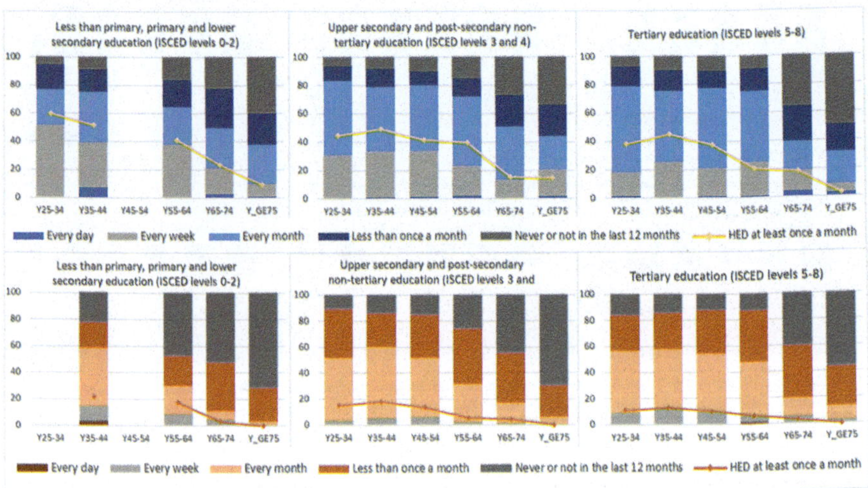

Fig. 13.8 Distribution of males (first row) and females (second row) according to the frequency of alcohol consumption by education in Lithuania by EHIS 2014 (%). (Data source: Eurostat (hlth_ehis_al1e; hlth_ehis_al3e))

highest volume of alcohol consumed and lowest drinking prevalence as reported by the frequency of consumption in EHIS.

Norway is marked by a low proportion of both men and women who replied saying that they had not consumed any alcohol in the last 12 months. The group of people with the lowest level of education has a slightly higher share of abstainers than the other two. Norway is characterized by an increasing share of people experiencing HED with increasing education. Regardless of educational attainment, the rule is that HED decreases as the respondent's age increases, and the proportion experiencing HED is lower for women than men. Norway ranks first among the countries by alcohol consumption according to EHIS 2014, but is almost the last among countries sorted by pure alcohol consumed (WHO and OECD). This would mean that Norwegians drink alcohol often, but in small amounts, which does not correspond with the high proportion of HED for both men and women. The data shows that a typical alcohol consumption pattern in Norway is irregular: daily drinking is very rare, while binge drinking episodes are prevalent across all educational categories and also in both sexes.

In Czechia, HED episodes were declared by a much smaller proportion of respondents than in Norway. Unfortunately, the data for the lowest education group is less complete, so we will focus more on the other two education groups. Concerning alcohol consumption at least once a month, Czech men are very similar to Norwegians. Unlike Norwegians, Czech men are more likely to drink alcohol daily, mainly those aged 55 years and over (16–31% depending on education and age group). Czech women, on the other hand, show lower shares of alcohol consumption at least once a month across all levels of education. The level of HED is generally lower in Czechia compared to other countries (often more than half as prevalent as in Norway). It appears that most of the alcohol consumed in the Czech Republic can be linked to a widespread culture of daily drinking among males, this is especially observed in the largest population group (i.e. the group with secondary education).

The last example is Lithuania, which came first in terms of the consumption of pure alcohol per person by the WHO and the OECD. From the point of view of alcohol consumption as reported in the EHIS, Lithuania is roughly around the average of the European countries. Interestingly, they have a lower proportion of people, both men and women, who drink at least once a month. This share falls sharply with increasing age (from age 65+) across all educational groups. Like in the Czech Republic, alcohol consumption is highly gendered in Lithuania. Like in Norway, daily drinking is rare, and heavy drinking episodes are frequent, pointing at another alcohol consumption pattern – irregular and gendered.

In the EHIS 2014 respondents are also sorted according to income quintiles.[1] By comparing the first and fifth quintiles (the 20% of respondents with the lowest income, and the 20% of respondents with the highest income), it appears that people with higher income are more likely to consume alcohol at least once a month, both men and women. This is evident in both the EU-28 as a whole, and also in the selected countries.

[1]The data (of each person) are ordered according to the value of the total equivalised disposable income. There are four cut-off point values (the so-called quintile cut-off points) of income, dividing the survey population into five groups equally represented with 20% of individuals in each quintile (Eurostat 2018).

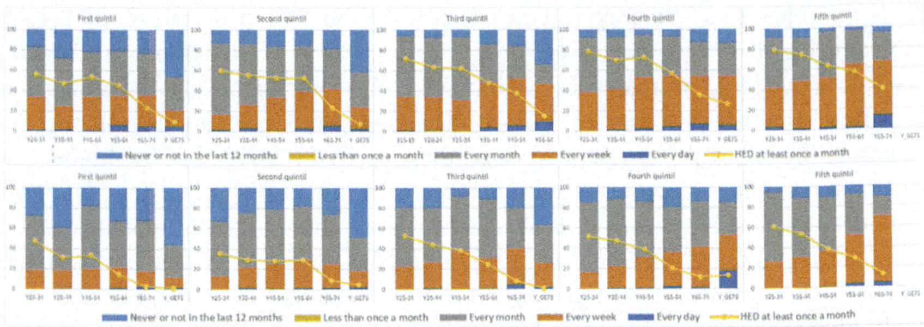

Fig. 13.9 Distribution of males (first row) and females (second row) according to the frequency of alcohol consumption by income quintiles in Norway by EHIS 2014 (%). (Data source: Eurostat (hlth_ehis_al1e; hlth_ehis_al3e))

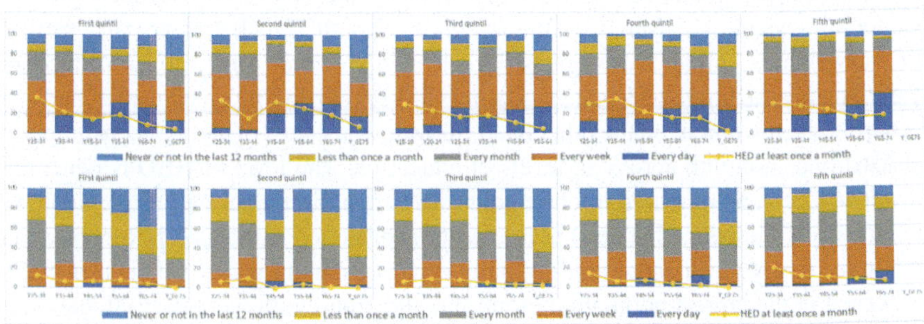

Fig. 13.10 Distribution males (first row) and females (second row) according to the frequency of alcohol consumption by income quintiles in Czechia by EHIS 2014 (%). (Data source: Eurostat (hlth_ehis_al1e; hlth_ehis_al3e))

In Norway (see Fig. 13.9), the data shows that the proportion of people who consume alcohol weekly increases with increasing income and often also with age. The HED frequency also rises with the income quintile for both genders.

In Czechia (see Fig. 13.10), the distribution of drinking frequency according to income quintiles is slightly different across genders. In women, alcohol consumption (at least once a month) decreases with increasing age from the first to fourth income quintile while it is not so obvious in the fifth quintile. The income gradient is not very obvious for Czech men either, but again there is a greater proportion of people who use alcohol at least once a month in the fifth income quintile.

Similarly, in Lithuania, there is a visible difference between the first and fifth quintiles and an increase in the proportion of alcohol-consuming respondents with increasing income, mainly in higher age groups (Fig. 13.11).

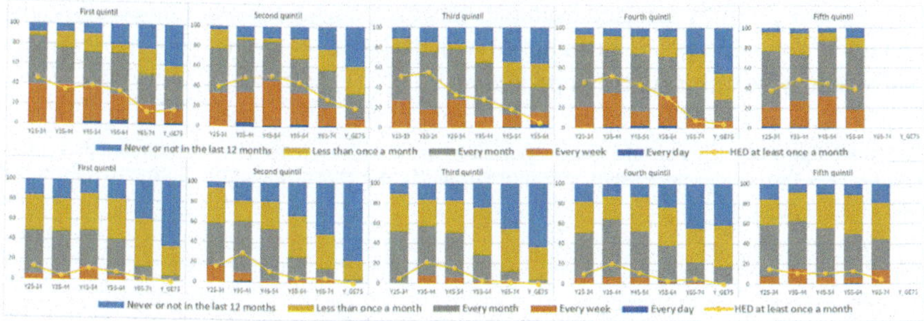

Fig. 13.11 Distribution of males (first row) and females (second row) according to the frequency of alcohol consumption by income quintiles in Lithuania by EHIS 2014 (%). (Data source: Eurostat (hlth_ehis_al1e; hlth_ehis_al3e))

13.6 Conclusions

In the study presented above, we gathered, compared and analyzed alcohol consumption data according to several data sources. Alcohol consumption as measured by official data on sales, production and excise tax was taken from the databases of the OECD and the WHO. We found good agreement between WHO and OECD data on recorded consumption, with the exception of a few countries where recorded consumption is possibly affected by large numbers of heavy drinking tourists (notably the case of Estonia but also, to a smaller extent, the Czech Republic). The WHO also produces estimates of total alcohol consumption, which includes unrecorded consumption and adjusts for drinking foreigners. This data shows more divergence from the OECD estimates, but it's still in good agreement.

Alcohol consumption in Europe is among the highest in the world. The trend data provided by the OECD since the 1960s however, points at an ongoing decrease in alcohol consumption and a homogenization of the volume of alcohol consumed. At present, (recorded) alcohol consumption in Europe ranges roughly between 6 and 12 litres of pure alcohol per capita aged 15+. However, this homogenization in consumption levels is not reflected in a homogenization of drinking patterns. As the EHIS 2014 data shows, several consumption patterns can be seen in the answers of the EHIS respondents. Thus, we observed gender-equal irregular and episodic drinking patterns in Norway, regular daily drinking gendered patterns in Czechia, and irregular, episodic and highly gendered drinking patterns in Lithuania.

Regarding socioeconomic disparities in drinking, we did not observe increased or riskier drinking behaviour among poor and less educated population groups. On the contrary, in all the countries we analysed, alcohol consumption increased with both education and income.

Data on alcohol consumption are known to have several limitations, including unrecorded or international consumption (OECD, WHO), underreporting of alcohol consumption in surveys, and under-representation of marginal population groups (least educated, poorest, richest) in surveys. The combination of the presented data sources and carefully selected examples however, suggest that in Europe currently, alcohol is not only a phenomena linked with despair and poverty, but is increasingly spread amongst well-situated populations.

Acknowledgments This article was supported by the Czech Science Foundation, Grant No. GA ČR 19-23183Y, on a project titled 'Alcohol burden in the Czech Republic: mortality, morbidity and social context'.

References

Alcohol. WHO | World Health Organization [online]. Copyright © [cit. 30.03.2019]. Available from: https://www.who.int/en/news-room/fact-sheets/detail/alcohol

Aryal, K. (2008). *To assess the prevalence of smoking and alcohol consumption among health professionals in health institution of Biratnagar*. Biratnagar: Koshi Health and Science Campus.

Dhital, R. (2001). Alcohol and young people in Nepal. *The Globe, 4*, 21–25.

European Health Interview Survey (EHIS). European Commission | Metadata [online]. Available from: https://ec.europa.eu/eurostat/cache/metadata/en/hlth_det_esms.htm

Eurostat Compact Guides. (2018). *Statistics explained: Your guide to European statistics*. Luxembourg: Publications Office, European Union. http://ec.europa.eu/eurostat/statisticsexplained/

GHO | Visualizations | Indicator Metadata Registry. [online]. Copyright © [cit. 30.03.2019]. Available from: http://apps.who.int/gho/data/node.wrapper.imr?x-id=465

GHO | Visualizations | Indicator Metadata Registry. [online]. Copyright © [cit. 30.03.2019]. Available from: http://apps.who.int/gho/data/node.wrapper.imr?x-id=462

MHP. Ministry of Health and Population. (2011). *Nepal adolescents and youth survey*. Kathmandu: Ministry of Health and Population.

The Survey of Health, Ageing and Retirement in Europe (SHARE): Why do older people hit the bottle? [online]. Copyright © [cit. 30.03.2019]. Available from: http://www.share-project.org/press-news/share-research-findings/new-scientific-findings/why-do-older-people-hit-the-bottle.html

WHO | Global Status Report on Alcohol and Health 2014. WHO | World Health Organization [online]. Copyright © [cit. 30.03.2019]. Available from: https://www.who.int/substance_abuse/publications/alcohol_2014/en/

WHO | Heavy episodic drinking among drinkers. WHO | World Health Organization [online]. Copyright © [cit. 30.03.2019]. Available from: https://www.who.int/gho/alcohol/consumption_patterns/heavy_episodic_drinkers_text/en/

WHO. 2018 | Global status report on alcohol and health 2018. WHO | World Health Organization [online]. Copyright © [cit. 30.03.2019]. Available from: https://www.who.int/substance_abuse/publications/global_alcohol_report/en/

Part III
Birth-Death Process, Self-Perceived Age and Gender Differences

Chapter 14
Modelling Monthly Births and Deaths Using Seasonal Forecasting Methods as an Input for Population Estimates

Jorge Miguel Bravo and Edviges Coelho

14.1 Introduction

Population forecasts are widely used for analytical, planning, and policy purposes (e.g., education, health, housing, security, transportation, public infrastructure, and social policy planning) at the national, regional, and local levels (Ayuso et al. 2020; Bravo 2016; Bravo et al. 2018; Smith et al. 2002). Also, National Statistical Offices use advanced estimates of residential populations to calibrate[1] many statistical survey results, among others, the Labour Force Survey (LFS) data. Forecasts of monthly births and deaths are a critical input when computing monthly estimates of a resident population (MERP) because they determine, together with international net migration, the dynamics of both the population size and its age distribution. Statistical Offices and researchers typically produce MERP using the cohort-component method, a standard demographic tool that requires credible assessments about the future behavior of age-specific fertility rates, sex, and mortality rates, as well as and international and sub-national migrations, together with detailed

[1] To make weighted sample estimates conform to known population external totals.

J. M. Bravo (✉)
Universidade Nova de Lisboa (NOVA IMS), Lisboa, Portugal

MagIC & CEFAGE-UE, Évora, Portugal

Université Paris-Dauphine PSL, Paris, France
e-mail: jbravo@novaims.unl.pt

E. Coelho
Statistics Portugal, Lisboa, Portugal & Departamento de Economia e Gestão, ECEO – ULHT, Lisboa, Portugal
e-mail: edviges.coelho@ine.pt

C. H. Skiadas, C. Skiadas (eds.), *Demography of Population Health, Aging and Health Expenditures*, The Springer Series on Demographic Methods and Population Analysis 50, https://doi.org/10.1007/978-3-030-44695-6_14

information about a base-year population. To perform this exercise for each subpopulation and gender, it is necessary to (Bravo 2007; Bravo et al. 2010): (i) obtain monthly forecasts of the total number of births and deaths; (ii) estimate age-specific mortality rates, considering period/cohort life tables derived from stochastic mortality models—eventually considering for heterogeneity in longevity (Ayuso et al. 2017a, b); (iii) estimate the level and age pattern of net international migration; and (iv) consider a number of assumptions, such as the distribution of age-specific fertility rates or the sex ratio at birth.

Birth and death forecasts can be produced using, among others: statistical time series methods (i.e., univariate or multivariate), structural models (e.g., vector autoregressive models), or machine learning methods (e.g., Artificial Neural Network and Support vector machines). To generate reliable estimates, these methods must be consistent with the annual and intra-annual observed patterns in birth and mortality data, which offer forecast accuracy and provide measures for the uncertainty in population forecasts. Empirical time series data for births and deaths exhibits strong evidence of the presence of seasonality patterns at both national and subnational (NUTS2, NUTS3) levels. These time series are typically non-stationary time series and contain trend and seasonal variations. For vital events computed for small populations on monthly time intervals, the need to uncover complex structures of temporal interdependence in time series data is critically challenged in the presence of seasonal variability. Several demographic, socio-economic, and geographic factors have been identified as having influence on the degree and direction of seasonal mortality (Rau 2007), including biomedical reactions to climate conditions (e.g., sudden temperature change or temperature extremes), which generate demographic reactions to biomedical changes (see, e.g., Zhang et al. 2018), particularly when associated with pre-existing conditions (e.g., diabetes or cardiovascular diseases) and social and biological factors (e.g., housing conditions, exposure to outdoor cold, clothing, heating, air conditioning, or healthcare). Similarly, most human populations exhibit seasonal variation in births, with a majority of European countries showing seasonal variation that usually peak in the summer and early autumn months and are the lowest during winter. Empirical evidence also shows that a child's month of birth is usually linked with later outcomes in life, such as those related to health and mortality (Reffelmann et al. 2011; Ueda et al. 2013). In the analysis of monthly birth counts, attention is frequently focused on intra-year changes in trend without knowledge of the usual seasonal change in the number of births between 2 months.

In recent decades a substantial amount of research has focused on the development and application of time series models in population forecasts, focusing either on total population growth or on individual components of growth (Abel et al. 2013; Ahlburg 1992; Alho and Spencer 1985; Bravo 2019; Bravo and El Mekkaoui de Freitas 2018; Keilman et al. 2002; Lee 1974, 1992; Lee and Tuljapurkar 1994; McNown and Rogers 1989; Pflaumer 1992; Saboia 1974; Swanson and Beck 1994; Tayman et al. 2007). The main focus of these studies is largely on the identification and measurement of uncertainty in population forecasts, with little interest on the assessment of the models forecasting accuracy or on the out-of-sample validity of the prediction intervals. Much of the research concerning the evaluation of time series models for death and birth forecasting has been focused on univariate time series ARIMA models at the national level, with little research on the predictive accuracy of these models at the sub-national level, particularly in small population areas (Land and Cantor 1983). Fewer still have explored the use of the Holt-Winters exponential smoothing and State Space time series models in small population exercises. Additionally, despite the increasing interest in short-term trends and variability in mortality and fertility patterns, accessing up-to-date statistics is sometimes difficult since detailed information on birth, and deaths counts are made available to researchers with a relative time lag. Also, researchers often need information on the present and near future, when data on birth and deaths counts could only be predicted.

In this paper, we address this gap and investigate and compare the predictive power of alternative linear and non-linear time series methods (i.e., seasonal ARIMA, Holt-Winters, and State Space models) to death and birth monthly forecasting at the sub-national level using up-to-date demographic data. Using a series of monthly death and birth data from 2000 to 2018, disaggregated by sex for the 25 Portuguese NUTS3 regions, we compare the short-term (i.e., 1 year) forecasting accuracy of Seasonal ARIMA, Seasonal Holt-Winters, and Seasonal State Space time series models. We adopt a backtesting time series cross-validation approach, that is, we consider a multi-step forecasting approach with re-estimation in which the training data or base period (i.e., the interval between the month of the earliest and the latest demographic data used to make a forecast) is extended before re-selecting and re-estimating the model at each iteration and computing forecasts. This dataset consists of 228 monthly observations from January 2000 to December 2018 for each one of the 100 different subpopulations (i.e., birth and death data for 25 NUTS disaggregated by sex), provided by Statistics Portugal, totaling 22,800 data points.

The main contributions of this paper are the following. First, we summarize and analyze the out-of-sample error performance of commonly used Seasonal ARIMA forecasting models together with alternative methods (i.e., Seasonal Holt-Winters and Seasonal State Space models), using a rich and large set of subpopulations and two different demographic events with different dynamics over time. Second, we evaluate the out-of-sample performance of the prediction intervals produced by these models. Third, we assess the consistency of the predictive performance of these methods in populations of different size and nature. Fourth, we evaluate the existence of significant differences in the model's forecasting accuracy between subpopulations of different sex. Fifth, we investigate how well the models perform in terms of predicting the uncertainty of future monthly birth and death counts. To evaluate forecast accuracy, we compare the resulting forecasts with observed data and measure forecast errors using different criteria (i.e., the Monthly Deviations (MD), the Monthly Percentage Error (MPE), the Cumulative Absolute Forecast Error (CFE), the Cumulative Absolute Percentage Forecast Error (CPFE), the Mean Square Error (MSE), the Root-mean-square error (RMSE), the Coefficient of variation of the RMSE (CVRMSE), the Mean Absolute Percent Error (MAPE), the Mean Absolute Deviation (MAD), and the Largest and the Smallest Absolute Deviation (LAD, SAD). To assess forecast uncertainty, we compute the proportion of times observed values fall outside 95% confidence intervals computed for the mean. The selection of the appropriate forecasting method depends on several factors, including the past behavior pattern of the time series, previous knowledge about the nature of the phenomenon being studied, the availability of statistical data, and the predictive capacity of the model.

Our results show that these simulations provide valuable insights regarding the forecasting performance of alternative time series models in small population forecasting exercises and on the validity of using such models as predictors of population forecast uncertainty and have, thus, significant practical implications. The remaining part of the paper is organized as follows. Section 14.2 describes the seasonal time series methods used in this paper. Section 14.3 details the research methods used to produce forecasts and assess model performance and the data features. Section 14.4 presents and discusses the results. Section 14.5 concludes the paper.

14.2 Modeling Trend and Seasonal Time Series

Modelling the trend and seasonal components of demographic time series is a challenging endeavor. Following earlier work on the decomposing a seasonal time series, Holt (Holt 2004) extended simple exponential smoothing methods to linear exponential smoothing to allow forecasting of data with time trends. The method was later extended by Winters (Winters 1960) to capture seasonality. Box and Jenkins (Box and Jenkins 1970, 1976) developed a coherent and flexible three-stage iterative cycle for time series identification, estimation, and verification (commonly known as the Box-Jenkins approach) and popularized the use of Autoregressive Integrated Moving Average (ARIMA) models and its extensions (including some to handle seasonality in time series) in many areas of science.[2] Ord et al. (1997), Hyndman et al. (2002) developed a class of State Space models that incorporate some of the exponential smoothing methods. The ability of these methods to model complex structures of temporal interdependence observed in the data has been tested, but their capability for modelling demographic seasonal time series has not been yet fully and systematically investigated. In this section, we briefly review the forecasting methods used in this study for forecasting demographic time series showing seasonality.

[2]See, for instance, De Gooijer and Hyndman (2006) for a detailed review of a quarter of century of research into time series forecasting.

14.2.1 Seasonal ARIMA Model

The seasonal ARIMA model is an extension to the classical ARIMA model that supports the direct modelling of both the trend and seasonal components of a time series, and it is widely used for forecasting. The model includes new parameters to specify the autoregression (AR), differencing (I), and moving average (MA) for the seasonal component of the series, as well as an additional parameter for the period of the seasonality (Hyndman and Athanasopoulos 2013). The model does not require any prior knowledge of the underlying structural relationships behind the data and relies solely on the assumption that past values provide sufficient information for forecasting future values. The model's mathematical and statistical properties allow us to derive not only point forecasts but also probabilistic confidence intervals (Box and Jenkins 1976).

In this paper we combine the seasonal and non-seasonal components into a multiplicative seasonal autoregressive moving average model, or SARIMA model, given by

$$\Phi_P\left(B^s\right)\phi(B)\nabla_s^D\nabla^d x_t = \delta + \Theta_Q\left(B^s\right)\theta(B)w_t \tag{14.1}$$

where w_t denotes the Gaussian white noise process. The general model can be expressed as

$$ARIMA\,(p, d, q) \times (P, D, Q)_s \tag{14.2}$$

where the ordinary autoregressive (AR) and moving average (MA) components are represented by polynomials $\phi(B)$ and $\theta(B)$ of orders p and q, respectively, and the seasonal AR and MA components are denoted by $\Phi_P(B^s)$ and $\Theta_Q(B^s)$ of orders P and Q, respectively. Respectively, the non-seasonal and seasonal difference components are represented by

$$\nabla^d = (1 - B)^d \tag{14.3}$$

$$\nabla^D = \left(1 - B^s\right)^D \tag{14.4}$$

where the seasonal period s defines the number of observations that make up a seasonal cycle (e.g., $s = 12$ for monthly observations).

The estimation process for the parameters in model (1) for each of the 100 time series follows the standard Box-Jenkins (Box and Jenkins 1976) methodology in an iterative 3-step procedure comprising the identification, estimation and evaluation, and diagnostic analysis stages. First, we analyze the stationary of the series and check whether or not a seasonal or non-seasonal difference is needed to produce a roughly stationary series. For this purpose, we analyze the autocorrelation function (ACF) and use seasonal unit root tests to determine the appropriate number of seasonal differences required, namely the Kwiatkowski et al. (1992) and Canova-Hansen (Canova and Hansen 1995) tests. Each series was tested for the white noise with Bartlett's version of the Kolmogorov-Smirnov test. When the data suggests the inexistence of seasonal unit roots in the series and the seasonality is deterministic, we can express it as a function of seasonal dummy variables (and time, eventually). In this case, an ARIMA model is fitted to the residuals of the equation:

$$Y_t = \alpha + \sum_{i=1}^{s-1} \gamma_{i,t} D_{i,t} + \beta t + \epsilon_t \tag{14.5}$$

where Y_t is the variable of interest, $D_{i,\,t}$ are seasonal dummies, t denotes time, and ϵ_t is a white-noise error term.

The SARIMA model identification is performed by assessing the patterns of the ACF and partial autocorrelation function (PACF), combined with a step-wise algorithm that selects the best order of the seasonal and non-seasonal parts of the model based on the Akaike Information Criterion (AIC). Additionally, we examine the residuals of the selected model and formally examined the null hypothesis of independence of the residuals using the Box-Pierce and Ljung-Box test (also known as "portmanteau" tests). We also tested the normality of the residuals using the Jarque-Bera Test. After examining different models, the best SARIMA model was selected, parameters were estimated using the nonlinear least squares method, and the model was used for forecasting monthly births and deaths.

14.2.2 Holt-Winters' Seasonal Method

The Holt-Winters method is a univariate automatic forecasting method that uses simple exponential smoothing (Holt 2004; Winters 1960). The forecast is obtained as a weighted average of past observed values, in which the weight function declines exponentially with time (i.e., recent observations contribute more to the forecast than earlier observations). Forecasted values are dependent on the level, slope, and seasonal components of the series being forecast. The Holt-Winters method is based on three smoothing equations: (i) one for the level, (ii) one for the trend, and (iii) one for the seasonality. The model specific formulation depends on whether seasonality is modelled in an additive or multiplicative way. The additive method is selected when the seasonal variations are approximately constant through the series, whereas the multiplicative method is preferred when the seasonal variations change proportionally to the level of the series (Chamboko and Bravo 2016). The **additive method** is specified as

$$\begin{aligned} l_t &= \alpha\,(y_t - s_{t-m}) + (1-\alpha)\,(l_{t-1} - b_{t-1}) \\ b_t &= \beta\,(l_t - l_{t-1}) + (1-\beta)\,b_{t-1} \\ s_t &= \gamma\,(y_t - l_{t-1} - b_{t-1}) + (1-\gamma)\,s_{t-m} \\ y_{t+h|t} &= l_t + hb_t + s_{t-m+h} \end{aligned} \tag{14.6}$$

where l_t, b_t and s_t denote the level, trend and seasonal components, respectively, with corresponding smoothing parameters α, β and γ; $y_{t+h\mid t}$ is the forecast for h periods ahead at time t. The Holt-Winters' **multiplicative method** is defined as

$$\begin{aligned} l_t &= \alpha \frac{y_t}{s_{t-m}} + (1-\alpha)\left(l_{t-1} - b_{t-1}\right) \\ b_t &= \beta\left(l_t - l_{t-1}\right) + (1-\beta)\, b_{t-1} \\ s_t &= \gamma \frac{y_t}{\left(l_{t-1} + b_{t-1}\right)} + (1-\gamma)\, s_{t-m} \\ y_{t+h|t} &= \left(l_t + hb_t\right) s_{t-m+h} \end{aligned} \tag{14.7}$$

After examining for each time series both the additive and multiplicative versions of the Holt-Winters' seasonal method, we finally selected the model showing lower residual sum of squares to produce forecasts of monthly births and deaths.

14.2.3 Exponential Smoothing State Space Model

We investigated the use of State Space models underlying exponential smoothing methods in monthly births and deaths forecasting. State Space models consist of a measurement equation that describes the observed data, and some state equations that describe how the unobserved components or states (i.e., level, trend, or seasonal) change over time (Hyndman and Athanasopoulos 2013). We examined both the additive and multiplicative errors versions of the model and automatically selected the best model using the procedure included in the R `forecast` package.

The general Gaussian state space model involves a measurement equation relating the observed data to an unobserved state vector $x_t = (b_t, s_t, s_{t-1}, \ldots, s_{t-(m-1)})$, an initial state distribution and a Markovian transition equation that describes the evolution of the state vector over time state. In this paper we use State Space models that underlie the exponential smoothing methods of the form (Hyndman et al. 2002):

$$Y_t = \mu_t + k\left(x_{t-1}\right)\varepsilon_t \tag{14.8}$$

$$x_t = f\left(x_{t-1}\right) + g\left(x_{t-1}\right)\varepsilon_t \tag{14.9}$$

where $\varepsilon_t \sim N(0, \sigma^2)$, $\mu_t = Y_{t-1}$ and where, for additive error models $k(x_{t-1}) = 1$, such that $Y_t = \mu_t + \varepsilon_t$, whereas for multiplicative error models $k(x_{t-1}) = \mu_t$ such that $Y_t = \mu_t(1 + \varepsilon_t)$. Model estimation involves measuring the unobservable state (prediction, filtering and smoothing) and estimating the unknown parameters using MLE methods.

14.3 Research Methodology

The objective of this research is to empirically compare the forecasting performance of alternative trend and seasonal time series models over short-term horizons. To this end, we set out a backtesting framework and use monthly demographic data for the period 2000–2018. In this section we briefly describe the research methodology used in this study.

14.3.1 Research Design

In this paper we set out a backtesting framework applicable to single-period ahead forecasts from time series methods and use it to evaluate the forecasting performance of three different univariate models applied to subnational (NUTS3) male and female monthly births and deaths data. The backtesting framework used in this paper involves the following steps[3]:

1. We begin by selecting the metric of interest, that is, the forecasted variable that is the focus of the backtest (e.g., monthly births or deaths by sex and subpopulation).
2. We define and select the historical "lookback window" to be used to estimate the parameters of each time series model for any given year. We adopt a time series cross-validation approach, that is, we consider a multi-step forecasting approach with re-estimation in which the training data or base period (i.e., the interval between the month of the earliest and the latest demographic data used to make a forecast) is extended before re-selecting and re-estimating the model at each iteration and computing forecasts. For instance, if we wish to estimate the parameters for year t, then we estimate the parameters using observations from years t_0 to $t-1$. If we wish to estimate the parameters for year $t+1$, then we estimate the parameters using observations from years t_0 to t (i.e., we adopt an expanding lookback window approach). The selection of the lookback window depends on several factors, including the past behavior pattern of the time series, previous knowledge about the nature of the phenomenon being studied, and the availability of statistical data.
3. We then select the forecasting horizon (i.e., the "lookforward window") over which we will make our forecasts, based on the estimated parameters of the model. In the present study, we focus on relatively short-term horizon forecasts since our interest is on generating 1-year ahead of monthly births and deaths forecasts (12 observations) as an input for computing monthly estimates of

[3]For a similar approach used in evaluating the forecasting performance of stochastic mortality models and interest rate and credit risk models see, e.g., Dowd et al. (2010), Bravo and Silva (2006) and Chamboko and Bravo (2016, 2019a, b).

resident population and a key input in producing the Labour Force Survey (LFS) in Portugal. The LFS is a quarterly sample survey of households living at private addresses in the Portuguese territory, with the main objective of characterizing the population in terms of the labor market. It is conducted by Statistics Portugal, in accordance with requirements under European Union (EU) regulation, and makes available quarterly and annual data. Published data are calibrated by using resident population estimates by regions NUTS3, sex and 5-year age-breakdown. The LFS quarterly results are published around 40 days after the end of the survey period. This calendar is incompatible with the current production of resident population estimates, since data on the three components—births, deaths, and migration—are not yet available. To comply with the LFS calendar, Statistics Portugal produces advanced monthly estimates of resident populations (i.e., at the beginning of each year t, monthly estimated values of resident populations are computed for year t by regions NUTS3, sex, and age). As such, monthly forecasts of live births, deaths and migration must be used to produce advanced monthly estimates of resident population.

4. We select a rolling fixed-length horizon backtesting approach, in which we consider the accuracy of forecasts over fixed-length horizons as the jump off date moves sequentially forward through time. This involves comparing the births and deaths mean forecasts and predictions intervals for some fixed-length horizon (1 year) rolling forward over time with the corresponding observed outcomes.
5. Finally, we select the evaluation criteria, which will be used to compare the forecasting performance of the different models. We computed several evaluation criteria but, due to space constraints, we report only results for the Mean Absolute Percent Error (MAPE). For a given lookback and look-forward window, the MAPE for model j is defined as

$$MAPE_j = \frac{1}{n}\sum_{t=1}^{n} \frac{\left|\hat{y}_{t,j} - y_t\right|}{y_t} \times 100 \tag{14.10}$$

where n is the number of forecasted values, $\hat{y}_t$ is the number of monthly births/deaths predicted by the model for time point t, and y_t is the corresponding value observed at time point t.

Each of the different time series models constructed, using a different lookback window and jump off year, implies a different set of prediction intervals for the forecast horizon. To better understand the performance of the models analyzed, in terms of predicting the uncertainty of future births and deaths, we computed the number of births and deaths counts falling outside the 95% prediction intervals associated with each set of forecasts. Parameter estimation and model forecasting assessment were carried out using a computer routine written in R-script (R Development Core Team 2019).

14.3.2 Data

In this paper, we use demographic data for Portugal comprising monthly data on live births and deaths broken down by sex and 25 different regions NUTS3 from January 2000 to December 2018 provided by Statistics Portugal. The demographic dataset consists of 228 monthly observations for each one of the 100 different subpopulations of different size, the smallest with 38,753 residents in December 2017 (Beira Baixa, male) and the largest with 1,505,435 residents (Lisbon Metropolitan Area, female). Of the 100 subpopulations tested, four (Lisbon and Oporto metropolitan areas male and female populations) correspond to highly populated areas with, in the case of Lisbon, more than one million residents. In contrast, the dataset tested includes several small population areas with less than 50,000 residents (e.g., Beira Baixa, Alto Tâmega, and Alentejo Litoral). This is a challenging dataset in which to assess the monthly forecasting performance of time series methods since the data exhibits significant trends, seasonal components, and high volatility in some cases, particularly in small population areas. Figures 14.1 and 14.2 represent the time series plot of monthly births and deaths of two representative (small and large) NUTS3 subpopulations.

Examination of the time plots revealed that there is a negative trend in the births series over the time period considered, although some recovery is observed in the Lisbon Metropolitan Area (LMA) in the years following the end of the Troika adjustment program; however, in the case of deaths time series we do not observe a significant trend over this period. Overall, a seasonal pattern is evident in the behavior of live births and deaths, with the highest number of births in the spring and summer months, while the highest number of deaths occurs during the winter months. Substantial changes are observed in the trend of fertility, with the number of live births showing a declining trend after 2000 in the majority of regions NUTS3. Since 2015, a relative stabilization and even a small increase are being observed. Over time, albeit the slight increase in the total number of deaths in the lasted years, time mortality patterns are relatively stable, showing a strong seasonal pattern with a higher number of deaths at winter months.

14.4 Empirical Results

The three univariate time series models are used as predictive models for making forecasts for future values of live births and deaths by sex and regions NUTS3 in Portugal. Tables 14.1 and 14.2 show, respectively, the MAPE results of 1-year ahead forecasts of monthly births and deaths by sex and regions NUTS3 for the period 2014–2018. For all models, the results are averaged over all jump off years.

The simple and weighted averages over all 25 regions and 5 launch years are shown in the Tables 14.1 and 14.2. Additionally, Tables 14.1 and 14.2 include data on the population size of each region NUTS3 on December 2017 to ascertain

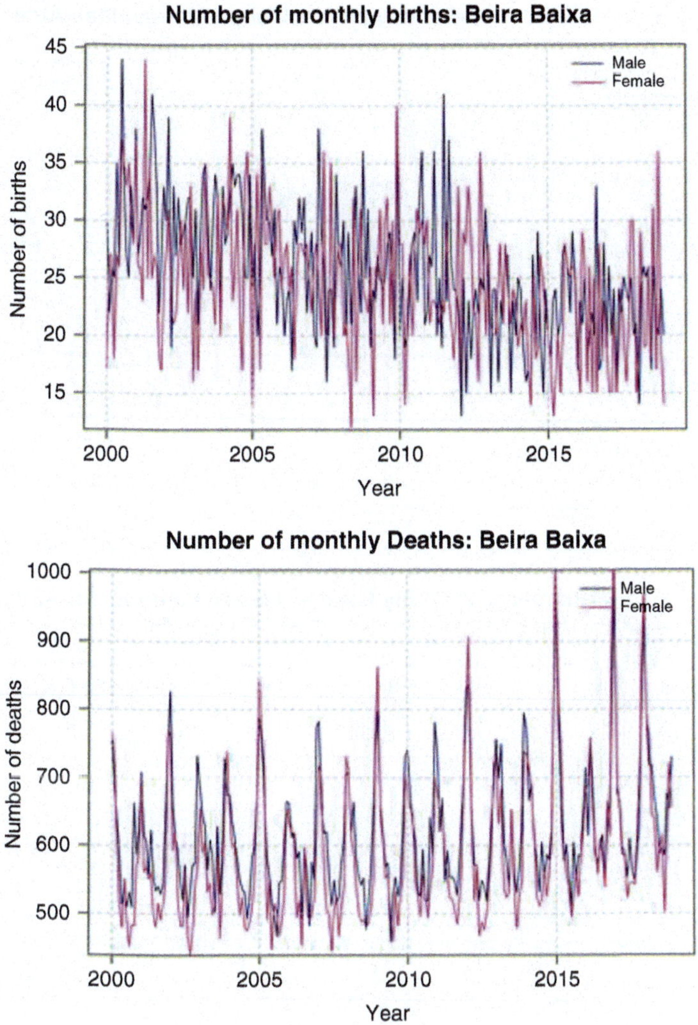

Fig. 14.1 Number of monthly births and deaths: Beira Baixa NUTS3 Region

whether the model's relative forecasting performance is a function of population size.

We first discuss the results related to monthly births forecasting. The all regions and launch years, simple and weighted averages, forecasting performance for the three models tested are similar for both male and female subpopulations, showing relatively low average MAPE results. The simple average results show that the precision of SARIMA forecasts is better than that of Holt-Winters (HW) and State Space (SS) models for the female subpopulations but, for the male counterparts, SS models show slightly lower forecasting errors. Note, however, that when considering

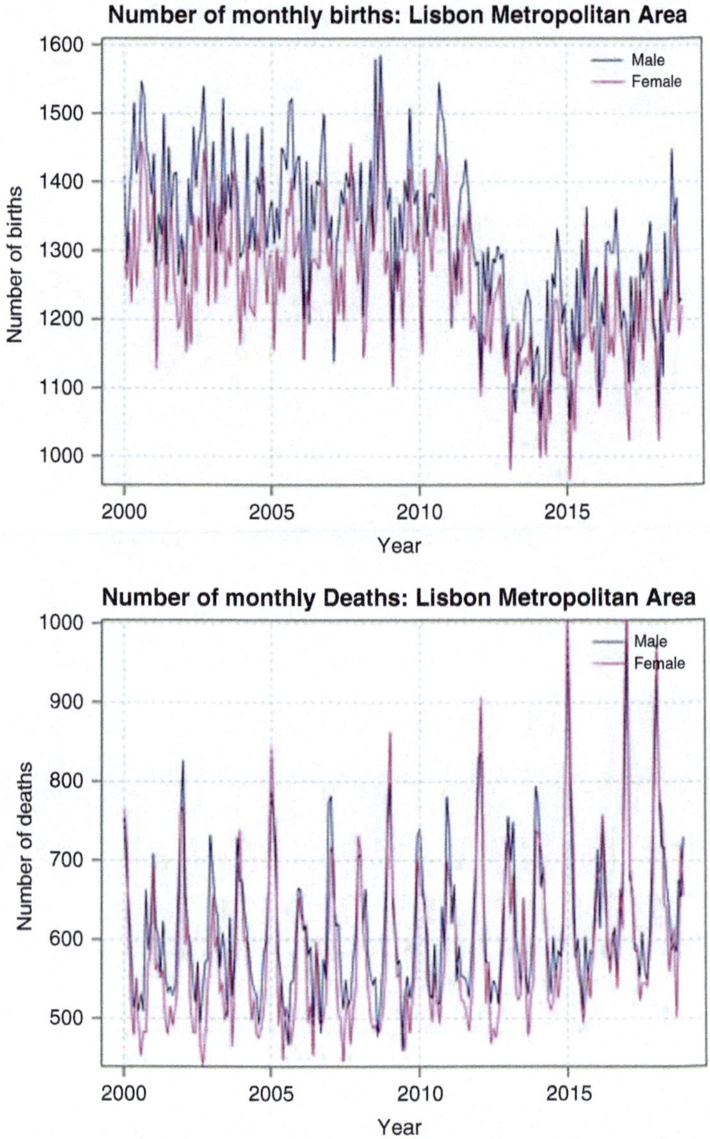

Fig. 14.2 Number of monthly births and deaths: Lisbon Metropolitan Area NUTS3 Region

the weighted average results (with weights given by the proportion of the region's subpopulation in the total resident population), SS models exhibit higher forecasting accuracy due to their superior performance in highly populated areas. Using this later metric, the SS model advantages SARIMA and HW models by 0.17 (0.18) and 0.16 (0.35) percentage points in the female (male) subpopulations, respectively.

Table 14.1 Births forecasting—average MAPE by model, sex and NUTS3

Births	Females				Males			
NUTS3	Pop. size	ARIMA	HW	SS	Pop. size	ARIMA	HW	SS
Alto Minho	124,583	13.20	13.38	14.13	107,595	12.46	13.07	12.87
Cávado	211,950	10.23	10.39	9.63	192,003	11.61	10.44	9.62
Ave	215,975	10.50	9.57	8.67	197,879	9.52	6.90	8.13
Área Metropolitana Porto	910,200	6.14	5.35	5.45	809,502	4.74	5.40	5.04
Alto Tâmega	46,044	18.49	18.86	21.03	41,113	23.66	21.11	21.26
Tâmega e Sousa	216,999	8.96	8.39	9.18	201,769	8.62	7.61	8.55
Douro	101,142	13.49	13.76	13.97	90,904	14.27	14.86	13.88
Terras de Trás-os-Montes	56,870	15.33	15.02	16.95	51,677	16.38	16.44	15.70
Oeste	186,405	9.91	10.18	9.53	171,301	7.84	8.82	8.07
Região de Aveiro	190,926	8.84	8.75	8.22	172,169	8.47	7.56	6.95
Região de Coimbra	231,654	8.15	8.21	7.84	205,294	7.86	7.70	7.34
Região de Leiria	149,784	9.20	8.85	7.84	136,525	9.86	10.79	9.75
Viseu Dão Lafões	134,679	12.15	11.62	12.21	119,952	12.64	12.53	12.12
Beira Baixa	43,061	21.60	21.31	24.19	38,753	16.40	17.92	17.23
Médio Tejo	123,699	10.53	11.54	12.01	110,956	9.99	10.63	10.21
Beiras e Serra da Estrela	114,163	12.72	12.43	12.97	102,025	11.11	10.75	12.37
Área Metropolitana Lisboa	1,505,435	3.38	3.60	3.11	1,328,244	3.59	4.18	3.29
Alentejo Litoral	47,551	16.99	17.56	17.78	46,223	17.32	18.77	19.11
Baixo Alentejo	60,669	11.59	12.45	12.57	57,199	14.33	14.17	14.59
Lezíria do Tejo	124,049	8.00	9.50	8.66	114,666	11.56	10.94	11.48
Alto Alentejo	56,092	18.80	19.01	18.54	50,965	16.32	16.33	17.15
Alentejo Central	80,677	12.81	13.87	13.01	73,859	12.01	13.61	12.80
Algarve	229,719	7.33	8.30	7.02	209,898	7.40	7.46	7.24
RA Açores	125,052	10.77	10.49	10.59	118,810	9.78	9.78	9.52
RA Madeira	135,957	12.00	12.30	14.02	118,411	10.39	11.51	11.93
All regions and launch years:								
Simple average	**216,933**	**11.64**	**11.79**	**11.96**	**194,708**	**11.53**	**11.57**	**11.45**
Weighted average		**8.00**	**7.99**	**7.83**		**7.70**	**7.87**	**7.52**
Max	**1,505,435**	**21.60**	**21.31**	**24.19**	**1,328,244**	**23.66**	**21.11**	**21.26**
Min	**43,061**	**3.38**	**3.60**	**3.11**	**38,753**	**3.59**	**4.18**	**3.29**

Source: Authors preparation; Notes: Average Mean Absolute Percent Error (MAPE) by model (ARIMA; Holt-Winters (HW); State Space (SS)) Sex and NUTS3 Region for the period 2014–2018. Weighted Average computed using the proportion of region's male or female population in the corresponding (sex) total population

Table 14.2 Deaths forecasting—MAPE by model, sex and NUTS3

Deaths	Females				Males			
NUTS3	Pop. size	ARIMA	HW	SS	Pop. size	ARIMA	HW	SS
Alto Minho	124,583	10.36	11.27	10.64	107,595	9.20	9.29	9.44
Cávado	211,950	11.45	11.21	11.07	192,003	8.88	9.14	8.99
Ave	215,975	7.73	9.27	8.55	197,879	8.89	9.31	9.05
Área Metropolitana Porto	910,200	7.24	8.07	7.85	809,502	5.76	6.33	6.12
Alto Tâmega	46,044	11.71	12.00	11.35	41,113	12.69	15.07	14.94
Tâmega e Sousa	216,999	9.02	11.21	10.33	201,769	8.94	9.61	8.99
Douro	101,142	11.25	13.74	12.21	90,904	9.88	10.27	9.73
Terras de Trás-os-Montes	56,870	11.46	12.35	11.35	51,677	10.53	11.07	10.71
Oeste	186,405	7.30	8.11	7.65	171,301	8.09	7.99	7.75
Região de Aveiro	190,926	10.29	10.47	10.04	172,169	7.94	9.20	8.20
Região de Coimbra	231,654	7.54	7.56	7.57	205,294	7.29	7.16	7.38
Região de Leiria	149,784	9.57	9.98	9.63	136,525	9.74	9.90	9.62
Viseu Dão Lafões	134,679	9.91	10.35	9.80	119,952	8.28	8.79	7.80
Beira Baixa	43,061	14.26	14.13	14.96	38,753	12.75	11.73	13.95
Médio Tejo	123,699	8.10	7.95	7.74	110,956	8.87	9.12	9.11
Beiras e Serra da Estrela	114,163	10.29	11.48	10.34	102,025	8.46	8.10	8.07
Área Metropolitana Lisboa	1,505,435	6.01	6.07	5.89	1,328,244	5.07	5.01	4.99
Alentejo Litoral	47,551	11.97	13.04	11.46	46,223	13.24	16.11	15.24
Baixo Alentejo	60,669	11.80	13.06	12.48	57,199	10.09	10.29	10.00
Lezíria do Tejo	124,049	9.48	10.33	9.97	114,666	9.07	9.85	8.87
Alto Alentejo	56,092	10.65	11.56	10.57	50,965	11.29	11.71	11.48
Alentejo Central	80,677	9.74	10.49	10.93	73,859	9.22	9.52	8.98
Algarve	229,719	9.26	9.65	8.94	209,898	7.57	7.50	7.33
RA Açores	125,052	10.90	11.72	11.33	118,810	9.67	11.31	10.52
RA Madeira	135,957	9.78	10.86	9.82	118,411	9.53	10.05	9.50
All regions and launch years:								
Simple average	**216,933**	**9.88**	**10.64**	**10.10**	**194,708**	**9.24**	**9.74**	**9.47**
Weighted average		**8.25**	**8.83**	**8.44**		**7.35**	**7.66**	**7.43**
Max	**1,505,435**	**14.26**	**14.13**	**14.96**	**1,328,244**	**13.24**	**16.11**	**15.24**
Min	**43,061**	**6.01**	**6.07**	**5.89**	**38,753**	**5.07**	**5.01**	**4.99**

Source: Authors preparation; Notes: Average Mean Absolute Percent Error (MAPE) by model (ARIMA; Holt-Winters (HW); State Space (SS)) Sex and NUTS3 Region for the period 2014–2018. Weighted Average computed using the proportion of region's male or female population in the corresponding (sex) total population

On average, for all models and for 61.3% of the subpopulations, the forecasting errors are smaller for the male subpopulations when compared to their female counterparts. As expected, the average MAPE results over the five launch years are larger the smaller the region's population size. The largest average forecasting error (24.19%) is found in the Beira Baixa female subpopulation using the SS model, whereas the highest accuracy (3.11%) is attained in the Lisbon metropolitan area (i.e., "Área Metropolitana de Lisboa")—also using the SS model. The forecasting error is less than 10% in 40% of the subpopulations considered.

Moving now to the results related to 1-year ahead monthly deaths forecasting, Table 14.2 shows, once again, that all regions and launch years—simple and weighted average forecasting performance for the three models—was relatively similar for both the male and female subpopulations, although the differences between the worst and the best performing model is higher in the male subset. Compared to births results, the average (weighted) forecasting accuracy of the alternative univariate time series methods is lower in the female subpopulations but higher in the male group. The weighted average results show that the precision of SARIMA forecasts is consistently better than that of the HW and SS models. The SARIMA model advantages the HW and SS models by 0.58 (0.31) and 0.19 (0.08) percentage points in the female (male) subpopulations, respectively. On average, for all models and for 76% of the subpopulations, the forecasting errors are notably smaller for the male subpopulations when compared to their female counterparts.

Similar to the birth results, the average MAPE results over the five launch years are smaller the more populated the region. The largest average forecasting error (16.11%) if found in the Alentejo Litoral male subpopulation, using the HW model, whereas the highest accuracy (4.99%) is attained in the Lisbon metropolitan area (i.e., "Área Metropolitana de Lisboa") male subpopulation, using the SS model. The forecasting error is less than 10% in 57% of the subpopulations considered.

Table 14.3 reports the percentage of the monthly births and deaths counts falling outside the 95% prediction interval estimated for each model, sex, and NUTS3 Region.

The goal is to measure how well the models perform when predicting the uncertainty of future monthly births and deaths counts over 1-year forecasting horizons. Each cell in Table 14.3 is based on 60 forecasts (i.e., 5 years and 12 monthly observations per year). Considering the 95% prediction intervals, a valid measure of uncertainty means they will encompass 57 of the 60 out-of-sample observed monthly births and deaths counts; thus, only 3 of the 60 observations will fall outside the 95% prediction interval boundaries. According to this criterion, the prediction intervals for SARIMA and SS models consistently provide appropriate measures of uncertainty for short-term forecasting horizons. SARIMA and SS models perform equally well in terms of predicting the uncertainty of future monthly deaths counts, with SS models slightly overperforming in births forecasting.

For the contrary, the HW model consistently fails in predicting the uncertainty of future monthly births and deaths, with up to 13.3% of observed deaths falling out of the 95% prediction interval.

Table 14.3 Percentage of monthly births and deaths counts falling outside the 95% prediction interval by model, sex, and NUTS3

	Births						Deaths					
	Females			Males			Females			Males		
NUTS3	AR	HW	SS	AR	HW	SS	AR	HW	SS	AR	HW	SS
Alto Minho	1.7	7.0	2.3	1.4	4.2	1.8	1.3	1.7	1.7	1.7	9.3	1.0
Cávado	5.0	9.0	2.0	3.6	4.2	0.8	2.0	5.0	1.7	1.3	1.0	1.7
Ave	4.0	7.0	1.0	2.8	0.8	0.8	1.0	0.3	0.7	2.3	4.0	2.0
Área Metropolitana Porto	3.3	3.7	3.0	1.0	1.0	0.2	2.0	12.7	2.0	1.0	10.3	2.3
Alto Tâmega	2.3	10.7	2.7	1.8	2.6	1.2	0.3	9.7	0.0	1.3	11.7	2.0
Tâmega e Sousa	2.3	2.3	1.3	2.0	1.4	1.4	1.7	6.0	1.0	1.0	1.7	1.3
Douro	3.7	6.3	0.3	1.8	4.8	1.4	1.0	1.3	1.3	1.7	6.3	1.0
Terras de Trás-os-Montes	3.0	8.7	2.0	1.2	5.4	0.2	1.0	1.0	1.3	1.3	1.0	0.7
Oeste	1.3	6.0	0.0	0.8	3.4	0.6	0.7	2.3	0.3	1.7	0.7	1.0
Região de Aveiro	2.0	2.0	0.7	1.4	3.2	0.6	1.3	4.7	1.3	1.0	9.0	1.0
Região de Coimbra	2.0	3.3	0.7	1.2	3.2	0.8	1.3	9.0	1.3	1.3	10.3	1.0
Região de Leiria	1.7	7.0	1.3	1.4	3.8	0.6	1.7	6.3	1.7	3.3	1.7	3.3
Viseu Dão Lafões	2.3	8.3	1.3	1.4	5.8	0.0	1.3	7.7	1.3	1.0	7.3	1.0
Beira Baixa	0.7	8.7	0.7	0.2	0.2	0.0	1.3	13.3	0.3	1.7	10.3	0.7
Médio Tejo	2.0	9.0	2.0	1.4	3.4	0.4	0.3	0.3	0.0	1.0	0.7	0.7
Beiras e Serra da Estrela	2.0	8.3	1.0	1.4	3.8	0.8	1.3	2.0	0.7	1.0	0.3	1.0
Área Metropolitana Lisboa	0.0	1.3	0.0	0.6	2.2	0.2	1.3	8.7	1.7	2.0	10.0	2.0
Alentejo Litoral	1.3	6.0	1.3	0.4	0.0	0.2	1.3	0.7	1.0	2.0	1.7	1.7
Baixo Alentejo	1.0	3.0	0.7	1.2	6.4	0.8	2.0	0.7	1.0	1.7	5.7	0.7
Lezíria do Tejo	0.3	2.7	0.0	1.6	5.6	1.8	1.3	10.3	2.3	2.3	2.0	1.3
Alto Alentejo	1.7	4.7	1.3	0.8	7.0	0.0	0.7	12.0	0.3	1.7	7.0	1.7
Alentejo Central	0.7	0.0	0.3	0.8	3.2	0.4	0.3	11.0	1.0	1.0	0.7	0.3

Algarve	0.3	3.7	0.7	0.6	0.6	0.2	1.7	9.0	2.0	2.7	12.7	2.7
RA Açores	1.7	8.7	0.7	1.2	3.0	0.6	1.3	0.3	1.3	1.3	0.7	0.7
RA Madeira	2.7	10.0	1.7	1.2	3.0	0.4	0.7	0.3	0.7	1.0	0.7	1.7
All regions and launch years:												
Simple Average	**2.0**	**5.9**	**1.2**	**1.3**	**3.3**	**0.6**	**1.2**	**5.5**	**1.1**	**1.6**	**5.1**	**1.4**
Weighted average	**1.7**	**4.3**	**1.1**	**1.2**	**2.6**	**0.5**	**1.4**	**7.2**	**1.4**	**1.6**	**7.1**	**1.7**
Max	**5.0**	**10.7**	**3.0**	**3.6**	**7.0**	**1.8**	**2.0**	**13.3**	**2.3**	**3.3**	**12.7**	**3.3**
Min	**0.0**	**0.0**	**0.0**	**0.2**	**0.0**	**0.0**	**0.3**	**0.3**	**0.0**	**1.0**	**0.3**	**0.3**

Source: Authors preparation; Notes: Average Mean Absolute Percent Error (MAPE) by model (AR = SARIMA; Holt-Winters (HW); State Space (SS)) Sex and NUTS3 Region for the period 2014–2018. Weighted Average computed using the proportion of region's male or female population in the corresponding (sex) total population

14.5 Conclusions

Monthly time series of live births and deaths show a strong evidence of seasonality patterns; therefore, appropriated forecasting methods must be considered. In this paper, we empirically evaluated the forecasting performance of seasonal ARIMA, HW, and SS models applied to death and birth monthly forecasting by sex and regions NUTS3 of Portugal. In addition, we investigated how well those models perform when predicting the uncertainty of future monthly births and deaths, using a backtesting framework and monthly data for the period 2000–2018. For all regions and launch years, simple and weighted average forecasting performance for the three models was relatively similar for both male and female subpopulations births and deaths; however, SS models showed slightly better performance for births and seasonal ARIMA for deaths. As expected, the weighted average precision is higher the more populated is the region. The prediction intervals for SARIMA and SS models consistently provide appropriate measures of uncertainty for short-term forecasting horizons. Further research we check for the robustness of these results against alternative forecasting horizons and fixed lookback windows, using rolling fixed-length horizon backtests.

References

Abel, G. J., Bijak, J., Forster, J. J., Raymer, J., Smith, P. W. F., & Wong, J. S. T. (2013). Integrating uncertainty in time series population forecasts: An illustration using a simple projection model. *Demographic Research, 29*, 1187–1226. https://doi.org/10.4054/DemRes.2013.29.43.

Ahlburg, D. A. (1992). Error measures and the choice of a forecast method. *International Journal of Forecasting, 8*(1), 99–100. https://doi.org/10.1016/0169-2070(92)90010-7.

Alho, J. M., & Spencer, B. D. (1985). Uncertain population forecasting. *Journal of the American Statistical Association, 80*(390), 306–314. https://doi.org/10.2307/2287887.

Ayuso, M., Bravo, J. M., & Holzmann, R. (2017a). Addressing longevity' heterogeneity in pension scheme design. *Journal of Finance and Economics, 6*(1), 1–21. https://doi.org/10.12735/jfe.v6n1p1.

Ayuso, M., Bravo, J. M., & Holzmann, R. (2017b). On the heterogeneity in longevity among socioeconomic groups: Scope, trends, and implications for earnings-related pension schemes. *Global Journal of Human Social Sciences: Economics, 17*(1), 31–57.

Ayuso, M., Bravo, J. M., & Holzmann, R. (2020). Getting life expectancy estimates right for pension policy: Period versus cohort approach. *Journal of Pension Economics and Finance* (in press).

Box, G. E. P., & Jenkins, G. M. (1970). *Time series analysis, control, and forecasting*. San Francisco: Holden Day.

Box, G. E. P., & Jenkins, G. M. (1976). *Time series analysis: Forecasting and control*. San Francisco: Holden Day.

Bravo, J. M. (2007). *Tábuas de Mortalidade Contemporâneas e Prospectivas: Modelos Estocásticos, Aplicações Actuariais e Cobertura do Risco de Longevidade*. Dissertação de Doutoramento em Economia pela Universidade de Évora, Maio. http://www.rdpc.uevora.pt/handle/10174/11148?locale=pt

Bravo, J. M. (2016). Taxation of pensions in Portugal: A semi-dual income tax system. *CESifo DICE Report—Journal for Institutional Comparisons, 14*(1), 14–23.

Bravo, J. M. (2019). Funding for longer lives: Retirement wallet and risk-sharing annuities. *EKONOMIAZ Basque Economic Review, 96*(II-2019), 268–291.

Bravo, J. M., & El Mekkaoui de Freitas, N. (2018). Valuation of longevity-linked life annuities. *Insurance: Mathematics & Economics, 78*, 212–229. https://doi.org/10.1016/j.insmatheco.2017.09.009.

Bravo, J. M., & Silva, C. M. P. (2006). Immunization using a stochastic process independent multifactor model: The Portuguese experience. *Journal of Banking and Finance, 30*(1), 133–156. https://doi.org/10.1016/j.jbankfin.2005.01.006.

Bravo, J. M., Coelho, E., & Magalhães, M. G. (2010). Mortality projections in Portugal. In *Work session on demographic projections* (pp. 241–252). Lisbon: Eurostat-UNECE.

Bravo, J. M., Rodrigues, T., Ribeiro, S., & Inácio, A. (2018). Portugal. Projeções de População Residente 2011–2040. In Teresa Rodrigues, T., & Painho, M. (Eds.), *Modelos Preditivos e Segurança Pública* (pp. 168–214). Fronteira do Caos Editores.

Canova, F., & Hansen, B. E. (1995). Are seasonal patterns constant over time? A test for seasonal stability. *Journal of Business & Economic Statistics, 13*(3), 237–252. https://doi.org/10.2307/1392184.

Chamboko, R., & Bravo, J. M. (2016). On the modelling of prognosis from delinquency to normal performance on retail consumer loans. *Risk Management, 18*(4), 264–287. https://doi.org/10.1057/s41283-016-0006-4.

Chamboko, R., & Bravo, J. M. (2019a). Frailty correlated default on retail consumer loans in developing market. *International Journal of Applied Decision Sciences, 12*(3), 271–287. https://doi.org/10.1504/IJADS.2019.10019811.

Chamboko, R., & Bravo, J. M. (2019b). Modelling and forecasting recurrent recovery events on consumer loans. *International Journal of Applied Decision Sciences, 12*(3), 257–270. https://doi.org/10.1504/IJADS.2019.10019807.

De Gooijer, J. G., & Hyndman, R. J. (2006). 25 years of time series forecasting. *International Journal of Forecasting, 22*(3), 443–473. https://doi.org/10.1016/j.ijforecast.2006.01.001.

Dowd, K., Cairns, A. J. G., Blake, D., Coughlan, G. D., Epstein, D., & Khalaf-Allah, M. (2010). Backtesting stochastic mortality models: An ex post evaluation of multiperiod-ahead density forecasts. *North American Actuarial Journal, 14*(3), 281–298. https://doi.org/10.1080/10920277.2010.10597592.

Holt, C. C. (2004). Forecasting seasonals and trends by exponentially weighted moving averages. *International Journal of Forecasting, 20*(1), 5–10. https://doi.org/10.1016/j.ijforecast.2003.09.015.

Hyndman, R., & Athanasopoulos, G. (2013). *Forecasting: Principles and practice*. OTexts. OTexts.org/fpp2/

Hyndman, R. J., Koehler, A. B., Snyder, R. D., & Grose, S. (2002). A state space framework for automatic forecasting using exponential smoothing methods. *International Journal of Forecasting, 18*(3), 439–454. https://doi.org/10.1016/S0169-2070(01)00110-8.

Keilman, N., Pham, D. Q., & Hetland, A. (2002). Why population forecasts should be probabilistic—Illustrated by the case of Norway. *Demographic Research, 6*, 409–454. https://doi.org/10.4054/DemRes.2002.6.15.

Kwiatkowski, D., Phillips, P. C. B., Schmidt, P., & Shin, Y. (1992). Testing null hypothesis of stationarity against the alternative of a unit root: How sure are we that economic time series have a unit root? *Journal of Econometrics, 54*, 159–178.

Land, K. C., & Cantor, D. (1983). Arima models of seasonal variation in U. S. birth and death rates. *Demography, 20*(4), 541–568. https://doi.org/10.2307/2061119.

Lee, R. D. (1974). Forecasting births in post-transition populations: Stochastic renewal with serially correlated fertility. *Journal of the American Statistical Association, 69*(347), 607–617. https://doi.org/10.1080/01621459.1974.10480177.

Lee, R. D. (1992). Stochastic demographic forecasting. *International Journal of Forecasting, 8*(3), 315–327. https://doi.org/10.1016/0169-2070(92)90050-J.

Lee, R. D., & Tuljapurkar, S. (1994). Stochastic population forecasts for the United States: Beyond high, medium, and low. *Journal of the American Statistical Association, 89*(428), 1175–1189. https://doi.org/10.1080/01621459.1994.10476857.

McNown, R., & Rogers, A. (1989). Forecasting mortality: A parameterized time series approach. *Demography, 26*(4), 645–660. https://doi.org/10.2307/2061263.

Ord, J. K., Koehler, A. B., & Snyder, R. D. (1997). Estimation and prediction for a class of dynamic nonlinear statistical models. *Journal of the American Statistical Association, 92*(440), 1621–1629. https://doi.org/10.1080/01621459.1997.10473684.

Pflaumer, P. (1992). Forecasting US population totals with the Box–Jenkins approach. *International Journal of Forecasting, 8*(3), 329–338. https://doi.org/10.1016/0169-2070(92)90051-A.

R Development Core Team. (2019). *A language and environment for statistical computing*. R Foundation for Statistical Computing. https://www.R-project.org

Rau, R. (2007). *Seasonality in human mortality: A demographic approach*. Heidelberg: Springer. https://doi.org/10.1007/978-3-540-44902-7.

Reffelmann, T., Ittermann, T., Empen, K., Dörr, M., & Felix, S. B. (2011). Is cardiovascular mortality related to the season of birth? Evidence from more than 6 million cardiovascular deaths between 1992 and 2007. *Journal of the American College of Cardiology, 57*(7), 887–888. https://doi.org/10.1016/j.jacc.2010.10.021.

Saboia, J. L. M. (1974). Modeling and forecasting populations by time series: The Swedish case. *Demography, 11*(3), 483–492. https://doi.org/10.2307/2060440.

Smith, S. K., Tayman, J., & Swanson, D. A. (2002). *State and local population projections: Methodology and analysis*. New York: Kluwer Academic Publishers.

Swanson, D. A., & Beck, D. M. (1994). A new short-term county population projection method. *Journal of Economic and Social Measurement, 20*(1), 25–50. https://doi.org/10.3233/JEM-1994-20102.

Tayman, J., Smith, S. K., & Lin, J. (2007). Precision, bias, and uncertainty for state population forecasts: An exploratory analysis of time series models. *Population Research and Policy Review, 26*(3), 347–369. https://doi.org/10.1007/s11113-007-9034-9.

Ueda, P., Bonamy, A. K. E., Granath, F., & Cnattingius, S. (2013). Month of birth and mortality in Sweden: A nation-wide population-based cohort study. *PLoS One, 8*(2), e56425. https://doi.org/10.1371/journal.pone.0056425.

Winters, P. R. (1960). Forecasting sales by exponentially weighted moving averages. *Management Science, 6*(3), 324–342. https://doi.org/10.1287/mnsc.6.3.324.

Zhang, Y., Yu, Y., Peng, M., Meng, R., Hu, K., & Yu, C. (2018). Temporal and seasonal variations of mortality burden associated with hourly temperature variability: A nationwide investigation in England and Wales. *Environment International, 115*, 325–333. https://doi.org/10.1016/j.envint.2018.03.036.

Chapter 15
Births by Order and Childlessness in the Post-socialist Countries

Filip Hon and Jitka Langhamrová

15.1 Introduction

After the collapse of the Eastern Bloc, a second demographic transition began in post-socialist countries. The aim of this article is to explore the development of fertility after the beginning of the new millennium, when these changes in society began to considerably affect the proportion of women by number of births given in the population (Van de Kaa 2002). These characteristics are important, because by following them it is possible to identify childless women and the proportion of the population they make up at the same time, which helps to understand changes in fertility (Sobotka et al., 2008). During the time of socialism, the role of a mother was one of the few socially accepted occupations for a woman to have. With new opportunities like building a career, ease of travel, and other factors, the share of women who never will become a mother in their lifetime has suddenly begun to increase (Sobotka 2006).

This article focuses mainly on developments in the Czech Republic and Slovakia, ie. in the former Czechoslovakia. Space is also devoted to other post-socialist countries, but this analysis is not as detailed as in the two named states. In addition to an analysis of recent developments, the article also includes projections of

The original version of this chapter was revised. This chapter was inadvertently published with a spelling mistake of the first author. The correction to this chapter is available at https://doi.org/10.1007/978-3-030-44695-6_31

F. Hon (✉) · J. Langhamrová
Department of Demography, Faculty of Informatics and Statistics, University of Economics, Prague, Czech Republic
e-mail: xhonf01@vse.cz; langhamj@vse.cz

C. H. Skiadas, C. Skiadas (eds.), *Demography of Population Health, Aging and Health Expenditures*, The Springer Series on Demographic Methods and Population Analysis 50, https://doi.org/10.1007/978-3-030-44695-6_15

characteristics for both target countries using the recuperation index (Sobotka et al. 2011). There is also a section devoted to the methodological description of this research.

Census data is undoubtedly effectively comparable between countries. However, for a more detailed analysis following the conclusions outlined in this article, it would, of course, be appropriate to study, for example, their methodologies to determine the exact way the data was collected. Data provided through the Human Fertility Database served as a source as well.

15.2 Census Results in Selected Countries and Cohort Childlessness

The most recent censuses from the years 2001 and 2011 were the primary censuses analyzed. Two detailed tables are devoted to the former Czechoslovakia, on which this article focuses. Below, there is an overview of several other post-socialist countries. Ireland and Greece are also included as representatives of countries without a socialist past, in order to compare the diversity of demographic developments in Europe (Table 15.1).

In the 2001 census, the changes accompanying the second demographic transition were already being reflected. This transition started in the Czech Republic after the change of the political system that followed the Velvet Revolution in 1989. The phenomenon of postponing maternity to a higher age began to be strongly promoted. Many women started reproducing as they approached the age of 30 (Fiala et al. 2018).

Table 15.1 Woman by number of children ever born in Czechia

		Number of children ever born (in %)				
Census	Age	0	1	2	3	4+
2001	15–19	98.1	1.7	0.1	0.0	0.0
	20–24	75.6	19.4	4.5	0.4	0.1
	25–29	32.1	35.7	28.7	2.9	0.6
	30–34	9.9	27.1	51.9	9.1	2.1
	35–39	5.6	19.1	57.1	14.5	3.7
	40–44	4.6	15.6	57.1	17.8	4.8
	45–49	4.4	14.5	56.5	19.5	5.2
2011	15–19	98.1	1.6	0.2	0.0	0.0
	20–24	87.0	10.2	2.3	0.3	0.1
	25–29	61.3	25.1	11.7	1.5	0.5
	30–34	26.0	31.4	36.2	5.2	1.2
	35–39	11.0	24.2	51.4	10.8	2.6
	40–44	7.2	21.3	54.3	13.5	3.8
	45–49	6.3	18.4	56.1	15.0	4.2

Source: UNSD Demographic Statistics

Table 15.2 Woman by number of children ever born in Slovakia

		Number of children ever born (in %)				
Census	Age	0	1	2	3	4+
2001	15–19	95.6	3.6	0.7	0.1	0.0
	20–24	66.2	24.4	7.5	1.3	0.6
	25–29	27.4	34.9	29.5	5.7	2.5
	30–34	11.2	23.7	45.9	13.4	5.8
	35–39	7.5	16.0	49.2	19.0	8.4
	40–44	6.3	12.9	48.1	22.4	10.2
	45–49	6.3	11.8	46.8	24.0	11.1
2011	15–19	96.4	2.7	0.6	0.1	0.3
	20–24	83.0	11.7	3.3	1.0	1.1
	25–29	59.5	24.1	11.6	2.5	2.3
	30–34	30.9	30.3	29.2	6.2	3.5
	35–39	16.4	25.3	41.3	11.4	5.6
	40–44	11.1	19.6	45.6	15.6	8.1
	45–49	10.1	14.9	47.1	18.7	9.2

Source: UNSD Demographic Statistics

Recuperation was still quite satisfactory for the first and second children in a family, and childlessness remained at a very low rate of around 4%. The share of woman with more than two children at the end of the reproduction period is roughly 25%.

Adding the most recent census shows the change in demographic behavior has greatly modified fertility. The phenomenon of postponing maternity to an older age has increased. For example, in the age range from 25 to 29, more than 60% of women remain childless. Also, second children are born to older mothers on average than before.

Recuperation stopped keeping up with postponement, and cohort childlessness started to grow. In the other words, the Czech Republic is beginning to return to more normal values after a period of unnaturally low childlessness during the socialistic era (Sobotka 2006). Of course, everything depends on what recuperation will be like at older ages. However, with these numbers it is reasonable to expect that at the end of the reproduction period, more women in the future will have just one child, and fewer will have at least three children (Table 15.2).

Slovakia is a country with very similar development in terms of demographic behavior and history in the twentieth century. It should not be surprising that the development of women according to the number of children ever born is very similar to the Czech Republic.

Similar to the Czech Republic, it is possible to observe the increase in the share of childlessness in the lower age groups, even though this increase is a few percentage points lower here. Despite this, the childlessness of women who are at the end of the reproduction period is much higher here. Among women between 45 and 49 years old in 2011, more than 10% were childless.

Table 15.3 Woman by number of children ever born in European countries

			Number of children ever born (in %)				
Age	Country	Census	0	1	2	3	4+
30–34	Belarus	1999	9.9	37.6	44.5	6.2	1.8
		2009	15.7	44.6	33.2	5.1	1.4
	Lithuania	2001	12.4	34.3	42.1	8.3	2.9
		2011	19.5	34.8	37.0	6.5	2.1
	Hungary	2001	17.9	27.7	39.1	11.5	3.8
		2011	35.5	31.5	23.4	6.9	2.7
	Greece	2001	31.7	22.2	35.9	7.8	2.4
		2011	44.3	23.1	25.6	5.5	1.5
	Ireland	2001	43.7	22.2	21.1	8.7	4.2
		2011	43.6	23.6	20.9	8.2	3.7
45–49	Belarus	1999	6.1	21.5	55.1	12.6	4.7
		2009	5.8	27.4	54.3	9.4	3.0
	Lithuania	2001	8.2	22.7	48.9	13.8	6.3
		2011	8.9	24.2	48.7	13.0	5.2
	Hungary	2001	8.0	20.6	52.2	14.3	4.9
		2011	9.0	22.4	46.0	16.1	6.5
	Greece	2001	12.0	14.4	51.5	16.5	5.6
		2011	13.2	17.4	50.6	13.7	5.1
	Ireland	2001	16.7	9.3	27.4	24.5	22.2
		2011	18.0	11.4	30.0	23.7	16.9

Source: UNSD Demographic Statistic

In Slovakia, the share of women with two children was not so dominant in the past. Unlike the Czech Republic, this number did not exceed 50%. This is also due to the fact that in 2001 women in Slovakia were more likely to have more than two children in their lives.

In 2011, it is clear that the large differences between the compared countries have diminished in women with three or more children. This is at least true with regard to the young generation, which is still in the reproduction period. Slovakia has achieved only slightly better results than the Czech Republic in these populations. Of course, it is still a question of recuperation at an older age, because higher reproduction rates for third and fourth children should be common even at an older age. At the end of the reproduction period Slovakia still reached better values. As in the comparison countries, the proportion of single-child women grew in Slovakia during the monitored period.

Table 15.3 shows the age intervals just after the age of 30 which clearly characterizes how often postponement is occurring in the population. The second segment is the age range just before reaching the age of 50, ie. just before the end of the reproduction period. In addition to post-socialist countries, Greece and Ireland are also shown for comparison purposes as countries with different political developments in recent history.

In Belarus, the phenomenon of postponing reproduction has increased by several percentage points in the period under review. However, compared with the other countries surveyed, this increase is not as significant, and most of the female population has already had at least one child at this age. Perhaps a bit surprisingly, unlike in the Czech Republic and Slovakia, the shares of women by number of children ever born at the end of the reproduction period has changed very little. It can be said that third children and beyond are born a few percentage points less often, and vice versa, more women are having just one child.

Lithuania is similar to Belarus, in that in 2011 it was possible to observe an increasing tendency to postpone maternity to a higher age. However, the recuperation has so far been sufficient to prevent the situation from changing at the end of the reproduction period, but of course whether the younger generations will show the same trend is still in question. Hungary, on the other hand, shows at the age of 30 years the tendency to significantly more frequently postpone motherhood to an older age, to the extent that this was noticeable even when compared to the Czech Republic and Slovakia. However, in addition to the rising share of women with a small number of children, and the retreat of the two-parent family model, there was also a surprising increase in the proportion of women with three or more children at age groups starting at 45. This unexpected fact might be explained by state support for families with more children (Spéder et al. 2017).

Even in countries without a socialist past, where the second demographic transition started significantly earlier (Van de Kaa 2002), changes in the situation between the two analyzed censuses can be observed. In 2001, postponing maternity was very common in Ireland, and in the interval from 30 to 35 years, the values are almost unchanged over the time period analyzed here. Conversely, at the end of the reproduction period, there is a growing tendency towards fewer women with fouth-born and higher children. There is also a high rate of childlessness. Still, Ireland is one of the countries where the personal image of children is one of the highest (Kreyenfeld and Konietzka 2017). On the other hand, in Greece, between 2001 and 2011, the tendency to postpone reproduction to an older age, to just after 30 years of age, increased. This is seen in higher childlessness and in the smaller proportion of women who have already had a second child at this age. In Greece, of course, the decision to have a baby now rather than in the future, was affected by the beginning of the economic crisis (Kotzamanis et al. 2017). From a lifetime perspective, the situation has not changed much – the percentage of women with three children has fallen by a few percentage points, and the proportion of women with only one child has increased a bit (Fig. 15.1).

There is more recent calculated data from a longitudinal view available in the Human Fertility Database. For Slovakia, unfortunately, data from the 1970 cohort is not available, but it is undoubtedly useful for the purposes of this article to include Slovakia in the comparison graph too.

In all countries, there is a gradual tendency towards an increase in childlessness. Even so, in some countries, the trend is smoother, and in others there is considerable fluctuation. What is important here is the apparent tendency for rapid growth in the last cohorts for all observed countries. This also occurs in those countries where the situation in 2011 appeared to be relatively constant.

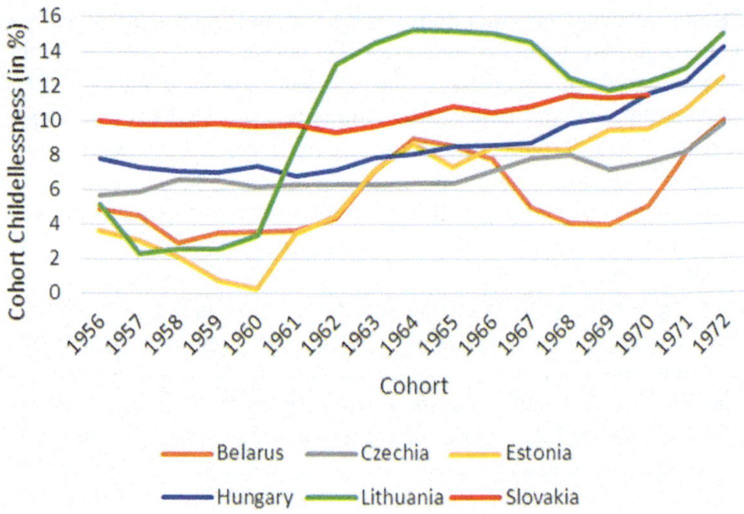

Fig. 15.1 Cohort childlessness in post-socialistic countries. (Source: HFD)

15.3 Methodology for Projection by Recuperation Index

A relatively new, but in countries at a similar level of demographic evolution as the Czech Republic and Slovakia commonly used, way of estimating future values of cohort aggregated fertility is the benchmark model based on the calculation of the recuperation index. In addition, it is possible to use a relational model, but that is even more demanding on available data, and may be less suitable for estimating fertility by number of children ever born (Sobotka et al. 2011).

The calculation is based on the comparison of the monitored cohort with the reference cohort, which is the same for the whole analysis. The reference cohort should be found from the cohorts which first began to concern the changes associated with the second demographic transition. In the Czech Republic and Slovakia, it is appropriate to choose, for example, the 1965 cohort, from which the average age of woman at the first childbirth started to increase for the next five generations (Šprocha et al. 2015).

The recuperation index is calculated using the recuperation and postponement measures. This is logical, because the final cohort statistics must depend mainly on how many women who have postponed the reproduction still have to accomplish before the end of their reproduction period. The calculation is therefore as follows:

$$P_c - F_c(m) - F_b(m),$$

where Pc is the fertility postponement measure, calculated as the difference in cumulative fertility rates at age m. This is the age of the greatest difference in cumulative fertility rates between the two compared cohorts and, of course, is different for each cohort. The designation b represents the data for the reference cohort, c is of course the cohort that is compared to the reference. In simple terms, postponing maternity is the part of a woman's reproductive period where the loss in cumulative fertility rates compared to the reference cohort is increased. When it ceases to increase, recuperation is logically triggered (Sobotka et al. 2011). It is also logical that this rate will be negative at the beginning of the second demographic transition.

The recuperation measure is then calculated as follows:

$$R_c = CTFR_c - CTFR_b - P_c,$$

where CTFR denotes the final aggregated fertility of a given cohort. An estimation of Rc future values is of course the problem because CTFRc is not yet known. The recuperation index can be derived as:

$$RI_c = (R_c / - P_c)\,.$$

This calculation makes it possible to compare cohorts for which the final cohort fertility is known, but it is also possible to easily calculate the estimate of future values using the formula:

$$pCTFR_c = CTFR_b + P_c * \left(1 - pRI_c\right),$$

where p stands for the projected values. Similarly to previous calculations, of course, this formula can be calculated for each number of children separately, which will be used in the following analysis (Sobotka et al. 2011).

The problem with this formula is that in order to obtain RIc, Rc is needed, which in turn requires CTFRc for its calculation. This is possible in the Human Fertility Database, which draws reasonable data to get the most up-to-date information for the 1965 cohort. It assumes the end of the reproduction period is at 49 years old.

However, this cohort does not adequately describe reproductive behavior during the second demographic transition because a large portion of the reproduction of this generation was made before social changes began. It is therefore possible to assume a distortion of the recuperation rate and similar shortcomings for the purposes of calculating the most realistic projection. A possible correction is to estimate fertility rates after reaching, for example, 39 years of age in another way. Which necessitates CTFR, available up to the 1975 cohort, which should be affected much more by changes in reproductive behavior. The formula just needs to be modified a bit:

$$pCTFR_c = CTFR_b(39) + P_c * \left(1 - pRI_c(39)\right) + pF_c\,(39+),$$

where pFc (39+) is an increase in overall aggregate fertility for predicted fertility at an elderly age (Sobotka et al. 2011).

Of course, the question is how to create this estimate. It is probable that the level of this fertility will get closer to Western states in the future. An assumption of a linear increase in fertility was used, with the fact that for the cohort of 1975, which reached the age of adulthood at a time that was already undergoing social changes, the average levels of the states of Europe without a socialist past will be reached.

In this paper a scenario is presented based on the latest observable recuperation index. A projection could not be made for all numbers of children until the 1987 cohort, as they could with the first-born child. Finding in which generation it is possible to quantify the greatest difference of cumulative fertility rates, in the other words, is where it is possible to determine the Pc. The age when the difference is the greatest would of course also be foreseeable, but the distortion of the other results could be bigger (Sobotka et al. 2011). That's why, in this analysis, values were only estimated for generations where the finding of this age was still possible.

It must be stressed that the projection results cannot be taken into account as certain values in the future, for example, in the case of other major changes in the level of population recuperation, the situation would undoubtedly change. Thus, if the assumption of smooth progression is not followed, then the projection will of course not be fulfilled and should rather serve as an idea of what the future would look like, for example, if the cohort recuperation of 1975 is assumed.

The 1975 recuperation index will, of course, very slightly overestimate the fertility of the older cohorts and perhaps underestimate the recuperation of the generations of younger people (Hon, 2018). But, as has been said, it is more of an idea of a likely future development that, as with any projection, does not have to be fully accurate.

15.4 Longitudinal Characteristics and Their Estimation

The last part is devoted to the estimation of fertility according to number of children ever born and childlessness in the states of former Czechoslovakia. Of course, the procedure using the recuperation index described above is used (Fig. 15.2).

In both studied countries, the final cohort childlessness can be expected to grow. In the Czech Republic, where this characteristic was very low, it is likely to grow by about ten percentage points over the 20 cohorts. Childlessness in the 1986 cohort, which are woman born just before the Velvet Revolution, could be around 15%.

In Slovakia, where childlessness has exceeded 10% in past, it is likely to grow to a higher level of around 20%. As seen above in the example of Greece and Ireland, these values are similar to those commonly achieved in countries where the second demographic transition occurred earlier. The growth of childlessness is of course related the decline in total fertility rate for first-born children in both the Czech Republic and Slovakia (Fig. 15.3).

Based on the method of calculation described above, it is not possible to design the total fertility rates for third-born and greater children without further simplification so far into the future. On the decreasing fertility of second-born

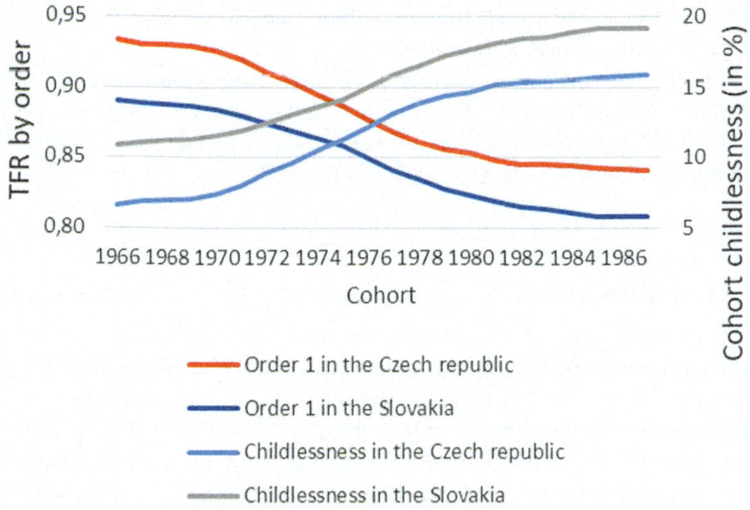

Fig. 15.2 Cohort childlessness and total fertility rate by order in Czech Republic and Slovakia. (Source: HFD)

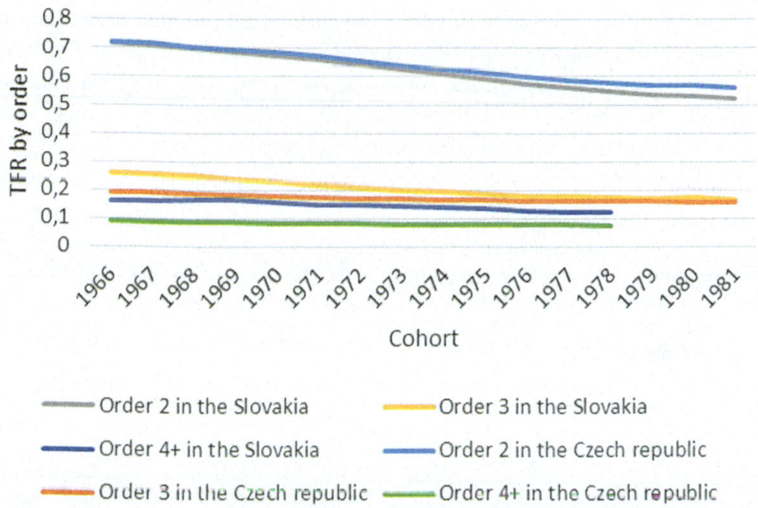

Fig. 15.3 Total fertility rate by order in Czech Republic and Slovakia. (Source: HFD)

children, a gradual decline is seen in the previously dominant family model with two children, in favor of women with fewer children.

A little more optimistic is the forecast for the Czech Republic. In Slovakia, it seems that only a little more than half of the women will have a second child if the current trends continue. The fertility rates of third-born and higher children were relatively low in this country, and continued to decline continuously throughout the reporting period. In the Czech Republic, these rates were lower than in Slovakia, and in future will continue to fall slowly.

In Slovakia, this drop is slightly faster. For the 1981 cohort, it appears that the total fertility rate of third-born children could fall to the same level as in the Czech Republic. On the contrary, the share of women with four and more children should remain higher in Slovakia than in the Czech Republic.

15.5 Conclusions

The change of the political establishment in former Czechoslovakia marked the beginning of the second demographic transition. As in other countries, it was associated with a fundamental change in the reproductive behavior of the population.

Previously, the most common family model in the population had two children. This was true primarily in the Czech Republic – in Slovakia women with three or more children also occurred frequently. Lifetime childlessness and the proportion of women with one child in both countries were low.

Now, due to insufficient recuperation following the postponement of maternity, the number of births has begun to decline even in a longitudinal view. As a result, the population can expect a significantly higher proportion of women with one or less child for their lifetimes in the future. These facts, with some unique aspects, also appeared in other post-socialist states analyzed in brief. Where this growth will stop is still in question. Based on the projection presented here, it is probable that, for example, childlessness, will soon reach similar values as in countries without a socialist past.

Acknowledgments This article was supported by the Grant Agency of the Czech Republic No. GA ČR 19-03984S under the title *Economy of Successful Ageing* and with the contribution of long term institutional support of research activities by the Faculty of Informatics and Statistics, University of Economics, Prague.

References

Fiala, T., Langhamrová, J., Pechholdová, M., Ďurček, P., & Šprocha, B. (2018). *Population development of Czechia and Slovakia after 1989*. Demografie, revue pro výzkum populačního vývoje, 3/2018.

Hon, F. (2018). *Narození podle pořadí se zaměřením na bezdětnost*. Diploma thesis. University of Economics. Faculty of Informatics and Statistics. Department of Demography.

Human Fertility Database. [online]. [quoted 2019-02-02]. Available from: http://humanfertility.org/

Kotzamanis, B., Baltas, P., & Kostaki, A. (2017). The trend of period fertility in Greece and its changes during the current economic recession. *Population Review,* Sociological Demography Press, *56*(2), 30–48.

Kreyenfeld, M., & Konietzka, D. (Eds.). (2017). *Childlessness in Europe: contexts, causes, and consequences* (Demographic research monographs). Cham: Springer.
Sobotka, T. (2006). Bezdětnost v České republice. In D. Hamplová, P. Šalamounová a G. Šamanová (Eds.), *Životní cyklus (sociologické a demografické perspektivy)*. Prague: Sociologický ústav AV ČR.
Sobotka, T., Šťastná, A., Zeman, K., Hamplová, D., & Kantorová, V. (2008). *Czech Republic: A rapid transformation of fertility and family behaviour after the collapse of state socialism* (Demographic research. Special collection 7). Rostock: Max Planck Institute.
Sobotka, T., Zeman, K., Lesthaeghe, R., & Frejka, T. (2011). *Postponement and recuperation in cohort fertility: New analytical and projection methods and their application* (European Demographic Research Papers 2). Vienna: Vienna Institute of Demography.
Spéder, Z., Murinkó, L., & Oláh, L. (2017). *Policy support and impact on third births in post-socialist Hungary* (Working Paper 07). Stockholm University.
Šprocha, B., Šídlo, L., & Šťastná, A. (2015). *Bezdetnosť žien na Slovensku a v Česku vo výsledkoch sčítaní 1991 – 2011 (a jej možný vývoj do budúcnosti)*. Zborník príspevkov z 15. Slovenskej demografickej konferencie. Bratislava: Štatistický úrad Slovenskej republiky.
The United Nations Statistics Division Demographic Statistics. [online]. [quoted 2019-03-01]. Available from: http://data.un.org/Data.aspx?d=POP&f=tableCode:40
Van de Kaa, D. (2002). *The idea of a second demographic transition in industrialized countries*. Paper presented at the Sixth welfare policy seminar of the National Institute of Population and Social Security.

Chapter 16
On the Evaluation of 'Self-perceived Age' for Europeans and Americans

Apostolos Papachristos and Georgia Verropoulou

16.1 Introduction

Chronological Age and life tables by sex are used as the main parameters to estimate an individual's future life expectancy. Subjective health measures, such as self-rated health, and socioeconomic indicators, such as education, have strong association with mortality and differentiate future life expectancy (Idler and Benyamini 1997; Verropoulou 2014; Feinglass et al. 2007). Therefore, a definition of 'age' which would incorporate the effect of such factors could refine the estimation of future life expectancy. 'Subjective survival probabilities' are quantities which vary by sociodemographic factors and health status (Rappange et al. 2016) and could form the basis for estimating a more personalized 'age'.

16.1.1 Subjective Survival Probabilities

'Subjective survival probabilities' reflect the views of individuals regarding their future survival. Past research noted that experiences, history and environmental factors are taken into account when forming subjective survival expectations (Griffin et al. 2013). Physical health status, physical activities, functional limitations and

This work has been partly supported by the University of Piraeus Research Center.

A. Papachristos (✉) · G. Verropoulou
Department of Statistics and Insurance Science, University of Piraeus, Piraeus, Greece
e-mail: apapachristos@webmail.unipi.gr; gverrop@unipi.gr

C. H. Skiadas, C. Skiadas (eds.), *Demography of Population Health, Aging and Health Expenditures*, The Springer Series on Demographic Methods and Population Analysis 50, https://doi.org/10.1007/978-3-030-44695-6_16

cognitive function are factors affecting subjective survival probabilities (Rappange et al. 2016; van Solinge and Henkens 2018; Liu et al. 2007; Hurd and McGarry 1995). Further, it has been shown that subjective survival probabilities predict mortality (van Solinge and Henkens 2018; Elder 2007; Kutlu-Koc and Kalwij 2013; Hurd and McGarry 2002; d'Uva et al. 2017).

16.1.2 Self-perceived Age

'Self-perceived age' is a quantity designed to capture future survival, combining information from the general population life tables as well as the self-reported subjective survival probabilities. Subjective survival probabilities quantify future subjective survival information for survey respondents. However, a probability, expressed as a percentage, is a rather technical concept to compare and communicate. In contrast, an individual's age, expressed in years is more legible. Therefore our objective is to express a subjective survival probability in terms of an individual's age, using population life tables.

'Self-perceived age' is the age which corresponds to similar objective and subjective survival probabilities. 'Self-perceived age' is implied by subjective survival probabilities, and it incorporates general population mortality. The main advantage of 'Self-perceived age' is that it is expressed in years, thus facilitating comprehension. Second, it is linked to a life table, thus incorporating the effect of general population mortality and of the ageing process. Third, it varies with common sociodemographic factors.

16.1.3 Objectives of the Study

The first objective is to develop a method for the calculation of 'Self-perceived age', by reference to life tables by country and sex, for American and European longitudinal survey respondents. The second objective is to provide a validation of 'Self-perceived age'. We validate 'Self-perceived age' by comparing the variations of 'Self-perceived age' with sociodemographic factors, cognitive function as well as lifestyle and behavioral risk factors to actual mortality patterns. For example, we would expect individuals with better self-rated health and socioeconomic status to have younger 'Self-perceived age' and lower actual mortality.

16.2 Methods

16.2.1 Data

Data obtained from the 6th Wave of the Survey of Health, Ageing and Retirement in Europe (SHARE) and from the 12th Wave of the Health and Retirement Study have been used in the analysis. The HRS (Health and Retirement Study) is sponsored by the National Institute on Aging (grant number NIA U01AG009740) and is conducted by the University of Michigan (Health and Retirement Study n.d.). It is an age-cohort-based longitudinal panel survey of persons aged 50 and older in the USA. SHARE (Börsch-Supan et al. 2013) is a cross-national database with information on health, socio-economic status, and social networks. The data collection of the 6th Wave was completed in November 2015 (Börsch-Supan 2017) and the sampling was carried out in 18 countries. The harmonised version of the longitudinal studies, RAND HRS and RAND SHARE, provided by the Gateway to Global Aging Data (Gateway to Global Aging Data n.d.), was used in order to produce a consistent combined dataset. The original combined sample covered 76,252 individuals aged 50 or higher. Israel is excluded from the analysis, reducing thus the sample to 74,857 persons.

The objective survival probabilities were obtained from the Human Mortality Database (HMD) (Human Mortality Database n.d.). HMD provides both period and cohort life tables; the latter however are incomplete for persons aged 50 or older. Hence, in this study we used period life tables by country and sex which refer to the 5-year period 2010–2014. As SHARE Wave 6 was undertaken in 2015 and HRS Wave 12 in 2014, we consider these life tables relevant to our study (Post and Hanewald 2010; Balia 2011).

16.2.2 Construction of the Dependent Variable

Subjective survival probabilities (SSPs) are based on Section I of the 'Retirement Plans, Expectations' module of the RAND HRS dataset and in Section I of the 'Retirement and Expectations' module of the RAND SHARE dataset. Respondents asked to report their chances (from 0 to 100) to live up to a target age T (75, 80, 85, etc.). The target age depends on the chronological age at the interview. The reported SSPs correspond to a specific prediction interval, starting from his/her current age up to the target age. Therefore, the cumulative OSPs should cover the same time horizon (Peracchi and Perotti 2010). Hence,

$$OSP_{x,N} = \prod_{t=1}^{N} OSP_{x+t}$$

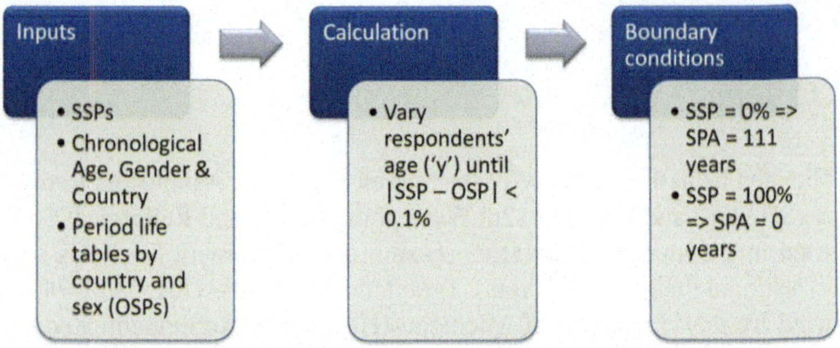

Fig. 16.1 The algorithm for the calculation of 'Self-perceived age'

where 'x' is the Chronological Age ('CA') of the respondent and 'N' is the prediction interval.

'Self-perceived age' ('SPA') is calculated by minimising the absolute difference between SSPs and OSPs. An appropriate algorithm has been developed to estimate 'Self-perceived age'. The code of the algorithm was written in Visual Basic for Applications (VBA) and the structure is presented in Fig. 16.1. Two boundary cases had to be considered explicitly: if SSP = 100% then 'Self-perceived age' is set to 0 years; in contrast, if SSP = 0% then 'Self-perceived age' is set to 111 (the oldest age in the life tables) and the associated OSP is 0%.

16.2.3 Explanatory Variables

16.2.3.1 Demographic Characteristics

This group of variables includes chronological age, gender and the number of parents alive. Country of residence is used as a control variable for the European sample.

16.2.3.2 Physical Health & Cognitive Function

This group of variables includes self-rated health (ranging from 1 = excellent to 5 = poor), a mobility index (ranging from 0 to 5) that shows in how many of the following activities the respondent experiences difficulties: walking one block, walking several blocks, walking across a room, climbing one flight of stairs, and climbing several flights of stairs. Cognitive function is assessed by the respondents total word recall, which is the sum of the immediate and of the delayed word recall scores and ranges from 0 to 20 dependent upon the number of correctly recalled words.

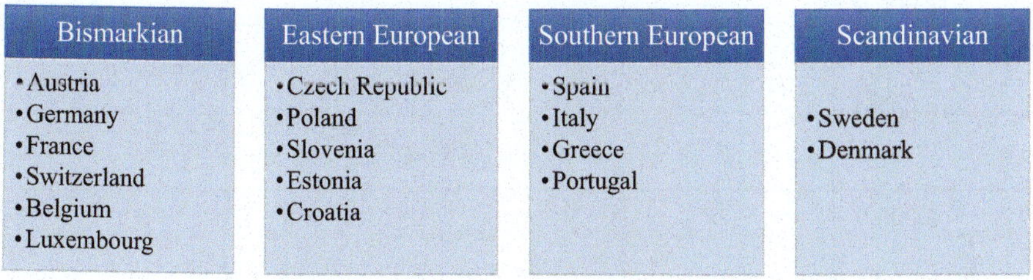

Fig. 16.2 Allocation of European countries in welfare states

16.2.3.3 Socio-economic Status

This group includes total household income in quartiles and educational level in three categories (Lower than upper Secondary, Upper Secondary and Tertiary). The income quartiles were calculated separately for each welfare state (Eikemo et al. 2008). USA is treated as a separate welfare state (Fig. 16.2).

16.2.3.4 Lifestyle & Behavioral Risk Factors

This group of variables includes the frequency of vigorous physical activities within a week (1 = every day to 5 = never) and whether the respondent ever smoked daily.

16.2.3.5 Statistical Modeling

We use Generalised Linear Models with a linear link function to investigate the impact of explanatory variables on 'Self-perceived age'. A different model has been estimated for the total sample, Europe and USA to assess the robustness of our estimates.

16.3 Results

16.3.1 Sample

The distribution of 'Self-perceived age' exhibits some concentration on the value of zero, which corresponds to subjective survival probability of 100%, and on values above 110, which corresponds to subjective survival probability of 0% (Fig. 16.3). These extreme values could distort our estimates; therefore, we excluded them from all subsequent analyses, reducing the total sample to 60,482 persons (Fig. 16.4). The sample characteristics are presented in Tables 16.1 and 16.2.

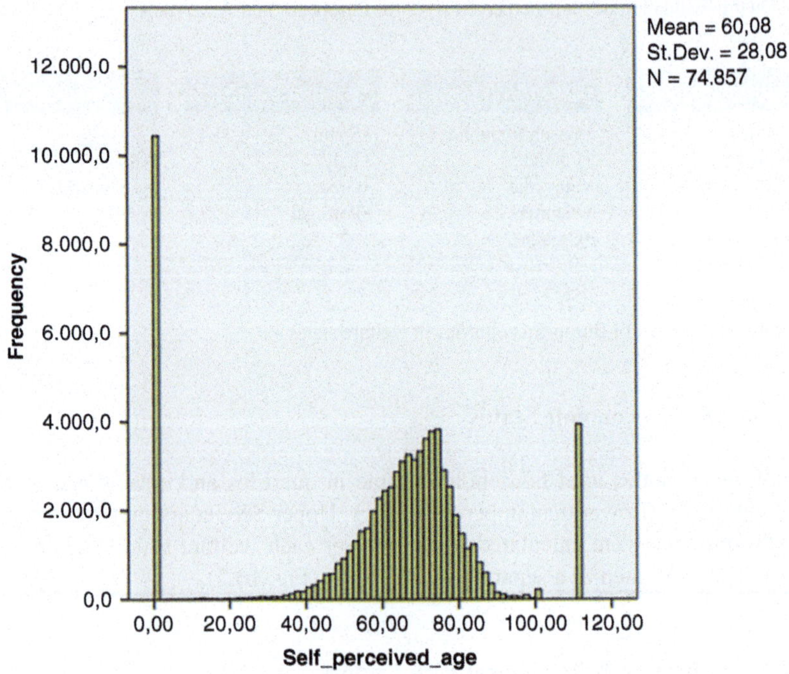

Fig. 16.3 Distribution of 'Self-perceived age'

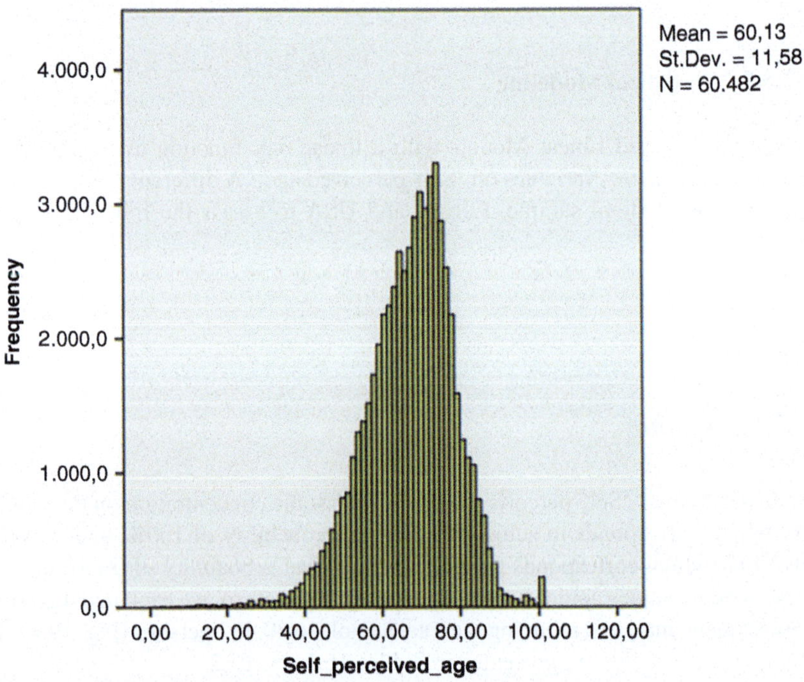

Fig. 16.4 Distribution of 'Self-perceived age' following exclusion of extreme values

Table 16.1 Sample characteristics

Variables	EU & USA	EU	USA
Number of respondents	74,857	58,835	16,022
Number of respondents with SPA = 0	10,437	9,179	1,258
Number of respondents with SPA >110	3,938	1,780	2,158
Number of respondents included in the analyses[a]	60,482	47,876	12,606
Dependent variable[a]			
Self-perceived age (mean)	67.13	67.59	65.42
Independent variables[a]			
Demographic characteristics			
Chronological age (mean)	67.38	67.67	67.01
Male	43.5%	44.4%	42.1%
Parental mortality			
No parents alive	75.9%	76.9%	74.9%
One parent alive	18.4%	17.7%	18.9%
Both parents alive	5.7%	5.5%	6.2%
Physical health & cognitive function			
Self-rated health			
Excellent	7.3%	5.9%	8.4%
Very good	20.6%	17.8%	32.5%
Good	36.2%	37.6%	35.3%
Fair	27.0%	29.4%	19.3%
Poor	9.0%	9.3%	4.5%
Mobility index (mean)	0.65	0.52	1.02
Word recall score (mean)	9.44	9.29	10.13
Lifestyle and behavioral risk factors			
Frequency of vigorous activity			
Less than 1 per week	31.7%	32.0%	26.5%
1 per week	13.8%	14.9%	11.6%
3 per month	10.0%	10.5%	11.0%
Never	44.4%	42.6%	50.9%
Smokers	49.0%	46.8%	55.7%

[a]Descriptive statistics refer only to respondents included in the analyses

Table 16.2 Sample characteristics for socio-economic status

Independent variables for socio economic status				
Household income (mean value by quartile)[a]	**1st quartile**	**2nd quartile**	**3rd quartile**	**4th quartile**
Bismarkian (€)	14,500	28,023	44,660	114,069
Scandinavian (€)	17,127	31,857	48,994	79,959
Southern (€)	3,883	11,517	18,352	38,212
Eastern (€)	2,991	6,563	10,245	21,518
USA ($)	12,156	31,331	60,879	179,839
Education – Frequencies[a]	**EU & USA**	**EU**	**USA**	
Less than upper secondary	35%	40.0%	17.9%	
Upper secondary	40%	37.4%	48.4%	
Tertiary	25%	22.8%	33.7%	

[a]Descriptive statistics refer only to respondents included in the analyses

The average 'Self-perceived age' is 67.13 years and it is similar to Chronological Age. The average 'Self-perceived age' for Americans is 65.42 years, approximately 2 years younger than the average Chronological Age (around 67 years). Approximately seven out of ten respondents report that both their parents have died whereas two out of ten respondents have one parent alive. Regarding health, approximately four out of ten Americans rate their health as excellent or very good. In contrast, only two out of ten Europeans rate their health as excellent or very good. On average, Americans report more mobility difficulties compared to Europeans. Furthermore, the average total recall score is higher for Americans. Approximately three out of ten Europeans do vigorous exercise at least once a week; while this holds for two out of ten Americans. On average, smoking is more prevalent among Americans rather than Europeans.

Household income quartiles were calculated separately for each welfare state (Fig. 16.2). Eastern Europeans have the lowest average household income whereas Scandinavians have the highest average household income in Europe. On average, Bismarkians are the top household income earners in Europe. The range of Americans' household income is wider compared to Europeans. Approximately four out of ten Europeans have lower than upper secondary education. In contrast, two out of ten Americans have lower than upper secondary education. Furthermore, one out of three Americans have tertiary education.

16.3.2 Multivariate Analyses

Table 16.3 shows the findings of the analysis for the total sample and for Europeans and Americans separately. The signs of the coefficients indicate whether a variable has positive or negative contribution to 'Self-perceived age'.

16.3.2.1 Demographic Characteristics

As Chronological Age increases 'Self-perceived age' also increases. The regression coefficients are positive for all models. The impact of Chronological Age is stronger for Americans ($b = 0.762$) than for Europeans ($b = 0.447$). Males have younger 'Self-perceived age' compared to females. The regression coefficients are negative for all models. The impact of gender on 'Self-perceived age' is stronger for Europeans ($b = -4.592$) than Americans ($b = -1.269$). Respondents whose at least one parent has died have older 'Self-perceived age' compared to respondents whose parents are both alive. The impact of parental longevity on 'Self-perceived age' is marginally stronger for Americans than for Europeans.

Table 16.3 Coefficients based on generalised linear models

Independent variables	EU & USA	EU	USA
Intercept	32.961**	39.762**	12.694**
Demographic characteristics			
Chronological age	0.518**	0.447**	0.762**
Male (reference: Female)	−3.948**	−4.592**	−1.269**
Parental mortality (reference: Both parents alive)			
No parents alive	3.463**	3.362**	3.743**
One parent alive	1.851**	1.795**	1.908**
Socio-economic status			
Income quartiles (reference: Q4)			
Income Q1	0.455**	0.619**	−0.137
Income Q2	0.597**	0.613**	0.204
Income Q3	0.494**	0.449**	0.454
Education (reference: Tertiary)			
Less than upper secondary	0.344**	0.235	1.248**
Upper secondary	0.560**	0.435**	0.688**
Physical health & cognitive function			
Self-rated health (reference: Poor)			
Excellent	−7.768**	−7.734**	−7.783**
Very good	−6.151**	−6.527**	−5.182**
Good	−4.141**	−4.472**	−2.742**
Fair	−2.307**	−2.490**	−0.853
Mobility index	0.355**	0.444**	0.179*
Word recall score	−0.035*	−0.079**	0.097**
Lifestyle and behavioral risk factors			
Frequency of vigorous activity	0.173**	0.222**	0.211**
Smoking (reference: Yes)	−0.760**	−0.751**	−0.435*

* $p < 5\%$, ** $p < 1\%$. Controlling for country of residence. The analyses exclude respondents with 'Self-perceived age' = 0 and Self-perceived age' > 110

16.3.2.2 Socio-economic Status

Our regression results suggest that low and medium income earners have older 'Self-perceived age' compared to high income earners. However, the results are conclusive for Europeans only. Respondents at lower and intermediary education levels tend to have older 'Self-perceived age' compared to those at tertiary education level. The impact of educational attainment on 'Self-perceived age' is stronger for Americans ($b = 1.248$ & $b = 0.688$) than for Europeans ($b = 0.235$ & $b = 0.435$).

16.3.2.3 Physical Health & Cognitive Function

Respondents who report excellent, very good, good and fair health have a younger 'Self-perceived age' compared to those who report poor health. The impact of self-rated health on 'Self-perceived age' is marginally stronger for Europeans than for Americans. The number of mobility difficulties is associated with older 'Self-perceived age'. The impact of mobility difficulties on 'Self-perceived age' is stronger for Europeans ($b = 0.444$) than for Americans ($b = 0.179$). Better memory is associated with younger 'Self-perceived age' for Europeans ($b = -0.079$) and older 'Self-perceived age' for Americans ($b = 0.097$).

16.3.2.4 Lifestyle and Behavioral Risk Factors

Physically inactive individuals have older 'Self-perceived age' compared to physically active persons. The regression coefficients are positive for all models and the impact is similar for Americans and Europeans. Our regression results suggest that respondents who never smoked tend to have younger 'Self-perceived age' compared to smokers. The impact of smoking on 'Self-perceived age' is stronger for Europeans ($b = -0.751$) than for Americans ($b = -0.435$).

16.4 Discussion

This study has two objectives: to estimate 'Self-perceived age' by reference to life tables and to evaluate its validity in comparison with actual mortality patterns. To achieve the objectives data from the 6th Wave of the Survey of Health, Ageing and Retirement in Europe (RAND SHARE), the 12th Wave of Health and Retirement Study (RAND HRS) and life tables from the Human Mortality Database (HMD) were used. The original combined sample covered 76,252 individuals aged 50 or higher from 19 countries. The analysis is based on 60,482 individuals after excluding Israel and the extreme values. We estimated three generalised linear models, using 'Self-perceived age' as dependent variable. Below our findings are summarized and discussed in comparison to actual mortality patterns.

16.4.1 Demographic Characteristics

As Chronological Age increases 'Self-perceived age' also increases; this can be explained as a result of the aging process. Our results show that males have younger 'Self-perceived age' compared to females. It is well known that older females have greater average life expectancy as well as more disabilities than males (Newman and

Brach 2001). Furthermore, males usually have on average higher socioeconomic status, level of education, income and tend to report better health status (Mirowsky 1999). Overall we conclude that males are more optimistic about their future survival than females. The impact of gender on the expected future longevity is not captured by 'Self-perceived age'.

Our findings show that respondents who have both parents alive or have one parent alive have younger 'Self-perceived age' compared to respondents whose parents have died. Past research noted that longevity and healthy aging is inherited for three generations (Atzmon et al. 2005). Therefore, we conclude that respondents who have both parents alive may have higher chances of living longer than respondents whose parents have died. The impact of parental longevity on the expected future longevity of the children is captured by 'Self-perceived age'.

16.4.2 Socio-economic Status

Our results indicate that higher income earners and respondents at tertiary education level have younger 'Self-perceived age' compared to respondents with lower socioeconomic status. Past research noted that better socioeconomic status is linked to lower actual mortality (Feinglass et al. 2007; Winkleby and Cubbin 2003). Therefore, we conclude that better socioeconomic status implies lower actual mortality as well as younger 'Self-perceived age'. The impact of socioeconomic status on expected future longevity is captured by 'Self-perceived age'.

16.4.3 Physical Health & Cognitive Function

Our findings show that respondents with poor self-rated health and more mobility difficulties have older 'Self-perceived age' than those with better health. It is worth mentioning that self-rated health as well as mobility impairment are predictors of actual mortality (Idler and Benyamini 1997; Verropoulou 2014; Hirvensalo et al. 2000) .Therefore, we conclude that better health implies lower actual mortality as well as younger 'Self-perceived age'. The impact of health status on expected future longevity is captured by 'Self-perceived age'.

Our findings indicate that Europeans with better memory have younger 'Self-perceived age' than Europeans with poor memory. In contrast, Americans with better memory have older 'Self-perceived age' compared to Americans with poor memory. Past research noted that better memory skills are associated with lower actual mortality (Baker et al. 2008; Shipley et al. 2006). Therefore, we conclude that better memory implies lower actual mortality as well as younger 'Self-perceived age' only for Europeans. The impact of memory skills on 'Self-perceived age' requires further investigation for the American population.

16.4.4 Lifestyle and Behavioral Risk Factors

Our results indicate that physically active individuals and nonsmokers have younger 'Self-perceived age' than smokers and physically inactive persons. It is well known that higher levels of physical activity are related to lower mortality (Gregg et al. 2003) and that smoking increases mortality (Ezzati and Lopez 2003). Therefore, we conclude that adopting a healthier lifestyle implies lower actual mortality as well as younger 'Self-perceived age'. The impact of lifestyle and behavioral risk factors on implied future longevity is captured by 'Self-perceived age'.

16.5 Limitations

Some limitations of this study should be taken into account when considering the findings. First the analysis is based on cross-sectional data and therefore cohort effects have not been accounted for. Second, the calculation of 'Self-perceived age' requires using a life table as input; thus, we cannot use the actual mortality of the respondents. Instead we use HMD period life tables by country and sex, which are estimated for the general population. These life tables reflect average population mortality and do not vary by health status. Therefore, all comparisons are based only on the findings of the relevant literature. Finally, period life tables include several cohorts and reflect the average mortality across these cohorts. Past research noted that period life expectancy is a lagged measure of cohort life expectancy, the lag depending on the pace of mortality improvement (Goldstein and Wachter 2006). On average, we would expect cohort-based 'Self-perceived age' to be marginally younger compared to period-based 'Self-perceived age'.

16.6 Conclusion

This study presents a methodology for estimating individual 'Self-perceived age' by reference to period life tables. The findings of the analysis show that 'Self-perceived age' varies consistently with actual mortality patterns. Health status, frequency of physical activities, socioeconomic and smoking status differentiate the estimated 'Self-perceived age' consistently with actual mortality patterns. The next steps include investigating 'Self-perceived age' variations across sociodemographic and cognition factors and estimating 'Self-perceived age' loadings for life tables by sex and health status.

Acknowledgements This paper uses data from SHARE Wave 6 (DOI: https://doi.org/10.6103/SHARE.w6.600), see Börsch-Supan et al. (2013) for methodological details. The SHARE data collection has been primarily funded by the European Commission through FP5 (QLK6-CT-2001-00360), FP6 (SHARE-I3: RII-CT-2006-062193, COMPARE: CIT5-CT-2005-028857, SHARE-

LIFE: CIT4-CT-2006-028812) and FP7 (SHARE-PREP: N°211909, SHARE-LEAP: N°227822, SHARE M4: N°261982). Additional funding from the German Ministry of Education and Research, the Max Planck Society for the Advancement of Science, the U.S. National Institute on Aging (U01_AG09740-13S2, P01_AG005842, P01_AG08291, P30_AG12815, R21_AG025169, Y1-AG-4553-01, IAG_BSR06-11, OGHA_04-064, HHSN271201300071C) and from various national funding sources is gratefully acknowledged (see www.share-project.org).

This analysis uses data or information from the Harmonized HRS dataset and Codebook, Version A as of February 2018 developed by the Gateway to Global Aging Data. The development of the Harmonized HRS was funded by the National Institute on Aging (R01 AG030153, RC2 AG036619, 1R03AG043052). For more information, please refer to www.g2aging.org.

References

Atzmon, G., Rincon, M., Rabizadeh, P., & Barzilai, N. (2005). Biological evidence for inheritance of exceptional longevity. *Mechanisms of Ageing and Development, 126*(2), 341–345.

Baker, D. W., Wolf, M. S., Feinglass, J., & Thompson, J. A. (2008). Health literacy, cognitive abilities, and mortality among elderly persons. *Journal of General Internal Medicine, 23*(6), 723–726.

Balia, S. (2011). *Survival expectations, subjective health and smoking: Evidence from European countries*. York: University of York, Centre for Health Economics.

Börsch-Supan, A. (2017). *Survey of Health, Ageing and Retirement in Europe (SHARE) Wave 6*. Release version: 6.0.0. SHARE-ERIC. Data set. https://doi.org/10.6103/SHARE.w6.600.

Börsch-Supan, A., Brandt, M., Hunkler, C., Kneip, T., Korbmacher, J., Malter, F., Schaan, B., Stuck, S., & Zuber, S. (2013). Data resource profile: The Survey of Health, Ageing and Retirement in Europe (SHARE). *International Journal of Epidemiology, 42*(4), 992–1001. https://doi.org/10.1093/ije/dyt088.

d'Uva, T. B., O'Donnell, O., & van Doorslaer, E. (2017). Who can predict their own demise? Heterogeneity in the accuracy and value of longevity expectations. *The Journal of the Economics of Ageing*. https://doi.org/10.1016/j.jeoa.2017.10.003.

Eikemo, T. A., Bambra, C., Joyce, K., & Dahl, E. (2008). Welfare state regimes and income-related health inequalities: A comparison of 23 European countries. *The European Journal of Public Health, 18*(6), 593–599.

Elder, T. E. (2007). *Subjective survival probabilities in the health and retirement study: Systematic biases and predictive validity* (Michigan Retirement Research Center Research Paper No. WP 2007-159). Available at SSRN: https://ssrn.com/abstract=1083823; https://doi.org/10.2139/ssrn.1083823.

Ezzati, M., & Lopez, A. D. (2003). Estimates of global mortality attributable to smoking in 2000. *The Lancet, 362*(9387), 847–852.

Feinglass, J., Lin, S., Thompson, J., Sudano, J., Dunlop, D., Song, J., & Baker, D. W. (2007). Baseline health, socioeconomic status, and 10-year mortality among older middle-aged Americans: Findings from the health and retirement study, 1992–2002. *The Journals of Gerontology Series B: Psychological Sciences and Social Sciences, 62*(4), S209–S217.

Gateway to Global Aging Data. (n.d.). Produced by the USC Program on Global Aging, Health & Policy, with funding from the National Institute on Aging (R01 AG030153).

Goldstein, J. R., & Wachter, K. W. (2006). Relationships between period and cohort life expectancy: Gaps and lags. *Population Studies, 60*(3), 257–269.

Gregg, E. W., Cauley, J. A., Stone, K., Thompson, T. J., Bauer, D. C., Cummings, S. R., Ensrud, K. E., & Study of Osteoporotic Fractures Research Group. (2003). Relationship of changes in physical activity and mortality among older women. *JAMA, 289*(18), 2379–2386.

Griffin, B., Loh, V., & Hesketh, B. (2013). A mental model of factors associated with subjective life expectancy. *Social Science & Medicine, 82*, 79–86.

Health and Retirement Study. (n.d.). RAND HRS longitudinal file 2014 v2 public use dataset. Produced and distributed by the University of Michigan with funding from the National Institute on Aging (grant number NIA U01AG009740).

Hirvensalo, M., Rantanen, T., & Heikkinen, E. (2000). Mobility difficulties and physical activity as predictors of mortality and loss of independence in the community-living older population. *Journal of the American Geriatrics Society, 48*(5), 493–498.

Human Mortality Database. (n.d.). University of California, Berkeley (USA), and Max Planck Institute for Demographic Research (Germany). Available at www.mortality.org. (Data downloaded on 26/09/2017).

Hurd, M. D., & McGarry, K. (1995). Evaluation of the subjective probabilities of survival in the health and retirement study. *The Journal of Human Resources, 30*(5), S268–S292.

Hurd, M. D., & McGarry, K. (2002). The predictive validity of subjective probabilities of survival. *The Economic Journal, 112*(482), 966–985.

Idler, E. L., & Benyamini, Y. (1997). Self-rated health and mortality: A review of twenty-seven community studies. *Journal of Health and Social Behavior, 38*(1), 21–37.

Kutlu-Koc, V., & Kalwij, A. (2013). *Individuals' survival expectations and actual mortality* (Discussion Papers (2013–013)).

Liu, J.-T., Tsou, M.-W., & Hammitt, J. K. (2007). Health information and subjective survival probability: Evidence from Taiwan. *Journal of Risk Research, 10*(2), 149–175.

Mirowsky, J. (1999). Subjective life expectancy in the US: Correspondence to actuarial estimates by age, sex and race. *Social Science & Medicine, 49*(7), 967–979.

Newman, A. B., & Brach, J. S. (2001). Gender gap in longevity and disability in older persons. *Epidemiologic Reviews, 23*(2), 343–355.

Peracchi, F., & Perotti, V. (2010). *Subjective survival probabilities and life tables: Evidence from Europe* (Working Paper 10). Einaudi Institute for Economics and Finance (EIEF), p. 16.

Post, T., & Hanewald, K. (2010). *Stochastic mortality, subjective survival expectations, and individual saving behavior* (No. 2010-040. SFB 649 Discussion Paper).

Rappange, D. R., Brouwer, W. B., & Exel, J. (2016). Rational expectations? An explorative study of subjective survival probabilities and lifestyle across Europe. *Health Expectations, 19*(1), 121–137.

Shipley, B. A., Der, G., Taylor, M. D., & Deary, I. J. (2006). Cognition and all-cause mortality across the entire adult age range: Health and lifestyle survey. *Psychosomatic Medicine, 68*(1), 17–24.

van Solinge, H., & Henkens, K. (2018). Subjective life expectancy and actual mortality: Results of a 10-year panel study among older workers. *European Journal of Ageing, 15*(2), 155–164.

Verropoulou, G. (2014). Specific versus general self-reported health indicators predicting mortality among older adults in Europe: Disparities by gender employing SHARE longitudinal data. *International Journal of Public Health, 59*(4), 665–678.

Winkleby, M. A., & Cubbin, C. (2003). Influence of individual and neighbourhood socioeconomic status on mortality among black, Mexican-American, and white women and men in the United States. *Journal of Epidemiology & Community Health, 57*(6), 444–452.

Part IV
Theoretical Issues and Applications

Chapter 17
Spatio-Temporal Aspects of Community Well-Being in Multivariate Functional Data Approach

Włodzimierz Okrasa, Mirosław Krzyśko, and Waldemar Wołyński

17.1 Introduction

The complexity of intertwining methodological and substantive issues involved in analysing the relationship between individual and community well-being while accounting for both temporal and spatial aspects of its cross-level dynamics is addressed here at two stages. First, at the measurement level, the functional data measurement approach is employed in the version of Multivariate Functional Principal Component Analysis (MFPCA) to deal with multidimensionality and temporality of community development/deprivation and of individual (residents') subjective well-being, respectively.

The MFPCA is an extension of the classic principal component analysis PCA from vector data to functional data (Gorecki et al. 2018; 2019) with the procedure of representing data by function or curves (see Ramsay and Silverman 2005) developed on the Besse's (1979) theoretical idea of multivariate data – where random variables take values in general Hilbert space – and its further important developments in different contexts. Of special interest here is an application to factorial methods – principal component analysis, canonical analysis – by Saporta (1981), and by Jacques and Preda (2014), who demonstrated usefulness of combining the MFPCA with cluster analysis.

W. Okrasa (✉)
Cardinal Stefan Wyszynski University in Warsaw, Warsaw, Poland

Statistics Poland, Warsaw, Poland
e-mail: w.okrasa@stat.gov.pl

M. Krzyśko · W. Wołyński
Adam Mickiewicz University, Poznań, Poland
e-mail: krzysko@amu.edu.pl; wolynski@amu.edu.pl

C. H. Skiadas, C. Skiadas (eds.), *Demography of Population Health, Aging and Health Expenditures*, The Springer Series on Demographic Methods and Population Analysis 50, https://doi.org/10.1007/978-3-030-44695-6_17

The advantage of the FPCA over the classic case is to obtain a projection of the analysed units into one or two dimensional subspaces using information for the whole period under study, and to divide them into homogenous groups on the basis of the resulting rankings. Having constructed classifications of both local communities (communes) and their residents for a given period of time (2004–2014, and 2009–2015, respectively), the spatial perspective is involved in the second part of the presentation. The space and place-related effects of the community development (deprivation) on the resulting cross-categorization of individuals are evaluated in terms of spatial patterns (autocorrelation and a tendency to clustering) and spatial dependence, spatial regression (Fischer and Getis 2010). However, no hierarchical *dynamical spatio-temporal model* (DSTM, in the sense of Cressi and Wikle 2011) is attempted here. Instead, the FPCA is assumed to provide an approximate/alternative solution. Nevertheless, further extension of this approach – toward multilevel modelling with spatial effects – is under consideration (e.g. Arcaya et al. 2012; Okrasa and Rozkrut 2018a, b).

The paper is structured along the above topical coordinates as follows. After characterizing fundamental features and advantages of the MFPCA (as compared to the classic approach, using PCA) the outcome of its applications – clusters of communes (gminas) and of their residents according to the multidimensional measures of local deprivation and of subjective self-reported measures of subjective well-being – is treated as an input to further exploration. First, in search for patterns of their co-occurrence (due to the general assumption on interaction between community and individual well-being) using Correspondence Analysis and the subsequent multinomial logistic regression to identify important factors of being classified to higher vs. lower clusters of satisfaction or another measure of subjective well-being. Next, bringing space into the question allows us to check implications of the hypothesis on tendencies to spatial clustering of the resulting categories (clusters) of MFPCA and for the examination of spatial dependence – through autocorrelation and spatial regression – will complete the presentation of the empirical analysis. A brief discussion of results, along with a summary, will conclude the paper.

17.2 Methodological Preliminaries

17.2.1 The Classic Version of the Principal Components Analysis (PCA)

We assume that the analysed objects (voivodships, poviats, gminas) are described by a p-dimensional vector of variables (features) $\boldsymbol{X} = (X_1, \ldots, X_p)^\top$. Suppose further that $\mathbf{E}(\boldsymbol{X}) = \mathbf{0}$.

We look for a random variable (the first main component) U of the form

$$U = \langle \boldsymbol{u}, \boldsymbol{X} \rangle = \boldsymbol{u}^\top \boldsymbol{X} = \sum_{i=1}^{p} u_i X_i,$$

having the maximum variance for all $\boldsymbol{u} \in \mathbb{R}^p$ such that $\langle \boldsymbol{u}, \boldsymbol{u} \rangle = 1$.

Let

$$\lambda_1 = \sup_{\mathbf{u} \in \mathbb{R}^p} \mathrm{Var}\left(\langle \mathbf{u}, \mathbf{X} \rangle\right) = \mathrm{Var}\left(\langle \mathbf{u}_1, \mathbf{X} \rangle\right),$$

at the condition $\langle \mathbf{u}_1, \mathbf{u}_1 \rangle = 1$.

We will call the random variable U_1 the first main component, and the vector $\mathbf{u}_1$ the first weight vector.

Next, we look for the second main component $U_2 = \langle \mathbf{u}_2, \mathbf{X} \rangle = \mathbf{u}_2^\top \mathbf{X}$, maximizing Var($\langle \mathbf{u}, \mathbf{X} \rangle$) and such that $\langle \mathbf{u}_2, \mathbf{u}_2 \rangle = 1$ and is uncorrelated with the first functional main component U_1, i.e. satisfying the condition $\langle \mathbf{u}_1, \mathbf{u}_2 \rangle = 0$.

Generally, k-the main component $U_k = \langle \mathbf{u}_k, \mathbf{X} \rangle$ meets the conditions:

$$\lambda_k = \sup_{\mathbf{u} \in \mathbb{R}^p} \mathrm{Var}\left(\langle \mathbf{u}, \mathbf{X} \rangle\right) = \mathrm{Var}\left(\langle \mathbf{u}_k, \mathbf{X} \rangle\right),$$

$$\left\langle \mathbf{u}_{\kappa_1}, \mathbf{u}_{\kappa_2} \right\rangle = \delta_{\kappa_1 \kappa_2}, \ \kappa_1, \kappa_2 = 1, 2, \ldots, k.$$

For the main components we have:

$$\lambda_1 \geq \lambda_2 \geq \cdots \geq \lambda_p.$$

In addition, for some (usually small) k the following occurs:

$$\lambda_1 + \lambda_2 + \cdots + \lambda_k \approx \lambda_1 + \lambda_2 + \cdots + \lambda_p.$$

This means that the first k major components well explain the total variability of the random vector X and that the other main components contribute little and can be omitted (reduction of the dimension).

The indicator

$$\frac{\lambda_1 + \lambda_2 + \cdots + \lambda_k}{\lambda_1 + \lambda_2 + \cdots + \lambda_p} 100\%$$

is a percentage measure of explaining the variation of a random vector X by the first major k components.

Since

$$\langle \mathbf{u}_k, \mathbf{u}_k \rangle = u_{k1}^2 + u_{k2}^2 + \cdots + u_{kp}^2 = 1,$$

then quantity $|u_{kj}|$ is a measure of the share of the j-th component of the random vector X in the construction k-th main component.

17.2.2 Functional Version of the Principal Components Analysis/FPCA

In addition, we now assume that the analysed objects characterized by variables are observed in many time points (years, months, days). Therefore, an appropriate model describing the examined objects will be p-dimensional random process $X(t) = (X_1(t), \ldots, X_p(t))^\top, t \in I$.

Next let us assume that $X(t) \in L_2^p(I)$, where $L_2(I)$ is a Hilbert space of integrable functions with a square on the interval I, and that the expected value of the process $\mathbf{E}(\mathbf{X}(t)) = \mathbf{0}, t \in I$.

From the above it follows that each component of the process X(t) can be represented in the following form:

$$X_k(t) = \sum_{b=0}^{\infty} \alpha_{kb}\varphi_b(t), t \in I,$$

where in the functions $\varphi_1, \varphi_2, \ldots$ form a base in space $L_2(I)$

The above representation of the X(t) process requires knowledge of an infinite number of α_{kb} coefficients. In practice, we use an approximate representation that uses only a finite number of the first base functions.

Let us assume that the k-th component of the process $\mathbf{X}(t)$ has the following representation:

$$X_k(t) = \sum_{b=0}^{B_k} \alpha_{kb}\varphi_b(t), t \in I,$$

where the number B_k determines the degree of smoothness of the function $X_k(t)$ (the smaller the value B_k, the greater the degree of smoothing).

Similarly to the classical case, we are looking for a random variable (the first functional component) U of the form:

$$U = \langle \boldsymbol{u}, \boldsymbol{X} \rangle = \int_I \boldsymbol{u}(t)^\top \boldsymbol{X}(t)dt$$

having the maximum variance for all $\boldsymbol{u}(t) \in L_2^p(I)$ such that $\langle \boldsymbol{u}, \boldsymbol{u} \rangle = 1$.

In general, the k-th functional master component $U_k = \langle \mathbf{u}_k, \mathbf{X} \rangle$ fulfils the conditions:

$$\lambda_k = \sup_{\mathbf{u} \in L_2^p(I)} \mathrm{Var}\left(\langle \mathbf{u}, \mathbf{X} \rangle\right) = \mathrm{Var}\left(\langle \mathbf{u}_k, \mathbf{X} \rangle\right),$$

$$\left\langle \mathbf{u}_{\kappa_1}, \mathbf{u}_{\kappa_2} \right\rangle = \delta_{\kappa_1 \kappa_2}, \ \kappa_1, \kappa_2 = 1, 2, \ldots, k.$$

Also, the indicator

$$\frac{\lambda_1 + \lambda_2 + \cdots + \lambda_k}{\lambda_1 + \lambda_2 + \cdots + \lambda_p} 100\%$$

is interpreted as a percentage measure of explaining the variability of a random process X(t) by the first k-functional principal components.

In the functional case, we have:

$$\langle \mathbf{u}_k, \mathbf{u}_k \rangle = \int_I u_{k1}^2(t) + u_{k2}^2(t) + \cdots + u_{kp}^2(t) dt = 1.$$

Thus, the quantity $\int_I |u_{kj}(t)| dt$ is a measure of the contribution of *j-th* component of the random process X(t) to the construction *k-th* functional principal component.

Unknown coefficients α_{kb} are estimated on the basis of n independent realizations $\mathbf{x}_1(t)$, $\mathbf{x}_2(t)$, ..., $\mathbf{x}_n(t)$ of the random process $\mathbf{X}(t)$. Since this process is only observed in a finite number of time moments, it is necessary to transform (smooth) discrete data into functional data (for details, see Ramsay and Silverman (2005)).

17.3 Data and Measures

17.3.1 Multi-source Analytical Database (ADB)

Community and individual level data needed to measure local deprivation and subjective well-being, respectively, come from various sources:

measures of local community (communes) development and the relevant covariates are from public statistics, Bank of Local Data, *Statistics Poland*, for the years 2004, 2008, 2010, 2012, 2014, 2016; NUTS5/LAU2; N = 2478 communes/gminas;

subjective well-being measures are based on data from nation-wide surveys:

- *Social Diagnosis (SD)*, carried out in the parallel years (2003–2009–2015), and
- *Time Use Survey* (TUS 2013, *Statistics Poland*).

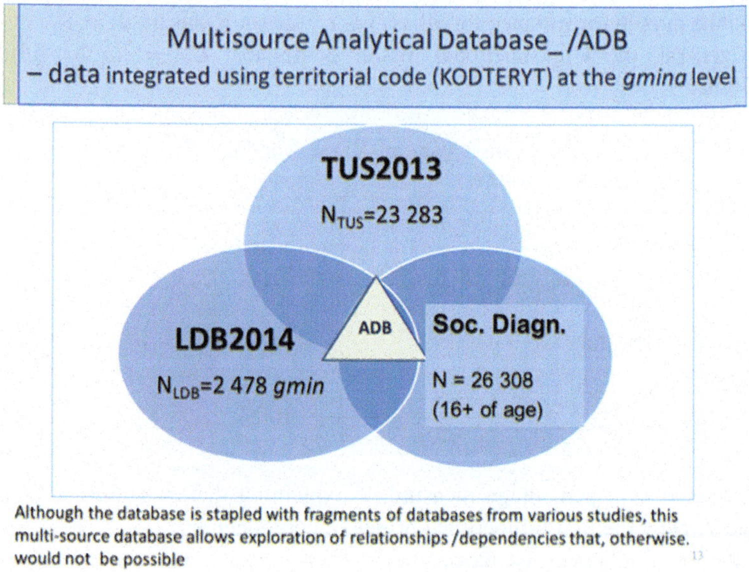

Fig. 17.1 Composition of the multisource analytical database (ADB)

Data are integrated along the matching methodology (e.g. D'Orazio et al. 2006) using territorial code at the NUTS5 *(commune)* level as the key 'integrator' – Fig. 17.1.

17.3.2 Measuring Local Deprivation and Well-Being

- *Multidimensional Index of Local Deprivation (MILD):*

The first measurement task was to construct an appropriate measure of local deprivation covering multidimensional phenomenon of local deprivation by domains. The MILD was built using 'confirmatory' Factor Analysis/PCA. A single-factor version was considered sufficient for the purpose of selecting the relevant items out of the large set of positions used for characterizing the following, customary distinguished 11 domains: *ecology – finance – economy – infrastructure – municipal utilities – culture – housing – social assistance – labour market – education – health* [altogether 65 items were selected]. Since Cronbach's alpha exceeds 0.75, the MILD is considered a reliable composite index of multidimensional deprivation (or under-development),

- *Individual (Subjective) Well-being/ SWB* and *Subjective Community Well-being/SCWB:*

A set of questions asked to respondents of the nation-wide survey *Social Diagnosis* during similar period of time (years 2003–2015) for which the MILD was calculated for the communes embraced five areas of satisfaction – which can be interpreted as an example of operationalization of happiness within a hedonic-evaluative framework of measuring (e.g., CNSTAT 2013; OECD 2013): (i) satisfaction with *living conditions*; (ii)) satisfaction with *living environment*; (iii) satisfaction with *social and family relations*; (iv) satisfaction with *personal situation*, and (v) the degree of *disapproval of antisocial* behaviour. The *synthetic measure of satisfaction (SMS)* – as an indicator of overall subjective well-being attributed to a commune's residents ('compositional' variable, defined in terms of percentage of 'satisfied' or 'very satisfied' on each scale).

- *U-'unpleasant' index of well-being:*

Occasionally, for the purpose of comparison with the above type of self-reported subjective well-being measures another measure is also used. It is a quasi-objective measure of well-being based on data from Time Use Survey (conducted by Statistics Poland in 2013). The idea of developing well-being index with the time-use data was explored at both macro- (Becker 1965; Nordhaus 2009) and micro-level (Kahneman and Krueger 2006). The later, combining econometric and psychometric approaches, is typically based on survey research data with day reconstruction method/DRM (such an approach was also used in TUS2013, conducted on a large national sample, N = 23,000).

Following Krueger, Khaneman et al. (2009) an indicator of well-being is defined in terms of the percentage of time spent on an activity performed weighted by an indicator of emotion – negative/positive- associated with it: 'time of unpleasant state', or U-index:

$$U_i = \Sigma_j I_{ij} h_{ij} / \Sigma_j h_{ij} \quad \text{(in the case of TUS2013 : I} = -1.\ 0.\ +1)$$

and $U = \Sigma_i(\Sigma_j I_{ij} h_{ij}/\Sigma_j h_{ij})$ /N for N-persons /group in population;

For U-binary (−1 & 0 vs. +1), given percentage structure of the indicator, the odds of U (chance of other than 'pleasant' or non-positive state vs. 'pleasant') can be defined as: odds $(U) =_{def} U_i/(1 - U_i)$. Odds U is used instead of the u itself because of its interpretation in the context of community level of development/deprivation in dynamic version, i.e., using Functional Data measures. Besides subjective well-being measures, also introduction of U-index into analysis of community and individual well-being relationship is made to have the opportunity to address some aspects of this relationship in a comparative way, using alternative measures.

17.4 Community Deprivation and Well-Being – Checking the Measurement Approaches

17.4.1 Comparing Static vs. Dynamic Approaches

A comparison of local deprivation by classic (confirmatory) principal component analysis (PCA) and the multivariate functional principal component analysis (MFPCA) is presented in Figs. 17.2a and 17.2b. The overall profile of deprivation of communes (gminas) in terms of values of measures of deprivation in particular domains is very similar, conforming expectations concerning high correlations between the results obtained by these two ways of reducing dimensions of deprivation. It is worthwhile to notice that there are also very small differences between the yearly profiles of deprivation. This means that MFPCA, which is supposed to capture dynamic aspects of local deprivation, preserves the 'centre of gravity' of the total variability among included items over the whole period; in fact, the items are the same in both PCA and MFPCA – therefore, an open question remains, the one about the degree of similarity between such profiles in the case of selecting the items in an independent way.

Another question that arises from the standpoint of the local development analysis concerns the consequences of measuring local deprivation in alternative ways – using PCA and MFPCA, respectively – for policy about allocation of public resources to communes. It is commonly assumed that the distribution of subsidies to them (to gminas) should follow the criterion of 'spatial justice' implied by, for instance, the EU regional conversion and cohesion policies.

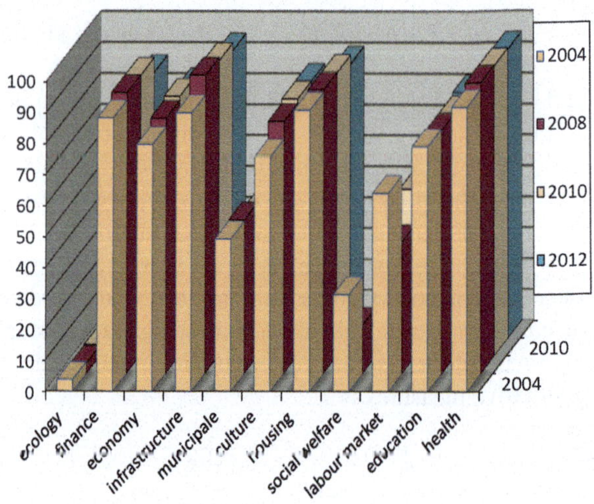

Fig. 17.2 Local deprivation by domains according to PCA, years 2004–2008–2010–2012

A comparison of *actual* distribution of subsidies with the *simulated* one – according to the principle of allocating subsidies proportionally to the value of MILD and the gmina's population share (Okrasa et al. 2006), the so-called *basic allocation formula:*

$$b.a.f. \equiv \underset{r \in S}{\forall r} \left[A(r) \approx \frac{I^r * P^r}{\sum_{i=1}^{S} I_i * P_i} \right]$$

where: I_i and P_i stand for indicator and population size of *i*th commune (1 = 1,. . .,S, and S is a geographic stratum composed of *r* parts, while *r* refers to the stratum for which the allocation is being defined, A(*r*)); (op. cit., p. 1058)

Results of the three systems of distribution of subsidies to communes/gminas are compared in Fig. 17.3: subsidies actually accrued to communes – subsidies simulated according to the above formula using MILD 2016 – and subsidies simulated according to the MFPCA 2004–2016, for all 16 voivodship. Again, there is a big similarity between the amounts allocated according to the two types of measures of local deprivation (based on PCA and MFPCA), showing generally the same direction of deviation from the amount actually obtained by communes in most of the voivodships.

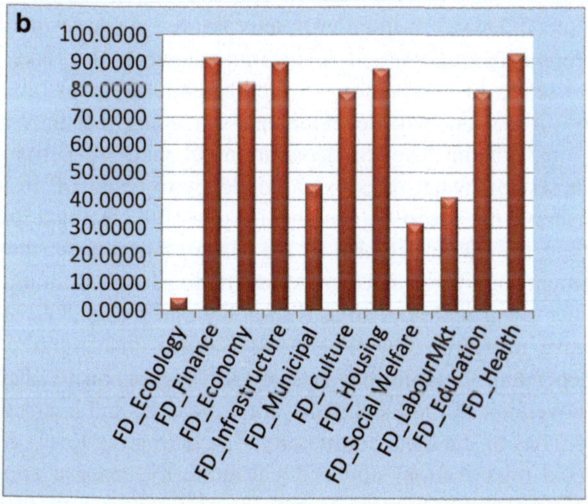

Fig. 17.3 Local deprivation by domains according to MFPCA, years 2004–2016

17.4.2 Initial Exploration of Relationships: Local Risk and Selected Measures of Well-Being

The level of local (under)development or deprivation affects several aspects of individual well-being, which is in general more visible when particular domains of deprivation are explicitly included into analysis. One of the important form in which local deprivation is being manifested is local risk. Therefore, checking its influence on selected aspects (measures) of well-being can serve as introductory part of the needed exploration.

Local Risk is defined as a product of a functional data scale – FD-scale – of local deprivation (‚development") in a domain and the fraction of the commune population (P_k) weighted by the ratio of the domain deprivation (ILD_d) to the total amount of deprivation (in all domains) / MILD_2016, in logarithmic form (Table 17.1):

$$\text{RskFD\_(d-domain)} =_{\text{def}} \text{Ln (FD-deprivation, d-domain)} + \text{Ln } [(P_k) * (ILD_d/MILD)].$$

All the domain-specific risks – associated with local deprivation in particular areas of underdevelopment – remain in negative relation with subjective well-being according to self-reported data of respondents (residents of the selected communes/gminas). However, in order to interpret the above results it is necessary to stress the opposite direction of association between the risks based on the functional data scales of deprivation – which actually are constructed as counter-deprivation, positive version of development – meaning that they also remain in negative relation with the level of deprivation. In fact, subjective well-being in this version shows somewhat paradoxical tendency to better off in less developed communes. It should not surprise given that in spite of the fact that local deprivation is growing along the dwindling size of the place of residence, the feeling about living environment and overall sense of belonging to a local community is generally higher in smaller towns and rural areas than in big cities. (e.g., Okrasa 2017).

An alternative measure of well-being, the above characterized U-index, has served as a dependent variable in a simple OLS regression to check importance both of the FD-version of the local deprivation measure and local risk, along with some characteristics of the local community – including the level of activity of the third sector local organizations and of local authority concern about community revitalization – as potential sources of influence for such a type of the well-being measure. Results are in Table 17.2.

Negative impact of the level of local deprivation and of risk associated with local labour market, as well as of the ratio 'in-work' to 'not-in-work' commune's population, indicates that these factors are conducive to smaller, on average, values of the U-index in a commune, i.e. to assessing their feeling about everyday activities as rather 'pleasant' than 'unpleasant'. On the other hand, factors conducive to

Table 17.1 Subjective well-being according to Functional Data scales by the local risk associated with local deprivation by domains – regression coefficients

Predictors (separate equations):	Functional data scales of subjective well-being selected aspects of satisfaction (with):		
Local risk associated with domains of local deprivation (FD–indicators/'devpt') [N = 386 comm./gminas##]	*All scales* – synthetic measure of satisfaction (SMS)#	Living conditions	Living environment
FD_Risk assoc. w/all domains of local deprivation (2004_16)	−.176** {−3.366)	−.100* (−1.893)	−.107**(−2.022)
FD_Risk assoc. w/depr. ecology	−.188** (−3.583)	−.098* (−1.835)	−202** (−3.852)
FD_Risk assoc. w/depr. finance	−.194** (−3.707)	−.119** (−2.249)	−.096* −1.810)
FD_Risk assoc. w/depr. economy	−.201** (−3 .830)	−.123** (−2.311)	−.096* (−1.812)
FD_Risk ass. w/depr. Infrastruct.	−.204** (−3.895)	−.123** (−2.327)	−.094* (−1.760)
FD_Risk assoc. w/depr. culture	−.190** (−3.621)	−.112** (−2.105)	−.089* (−1.679)
FD_Risk assoc. w/depr. housing	−207** (−3.959)	−.119** (−2.236)	−.107** (−2.015)
FD_Risk ass. w/depr. soc. welfare	−.130** (−2.452)	–	–
FD_Risk assoc. w/depr. labor mkt	−.135** (−2.554)	–	−102** (−1.912)
FD_Risk assoc. w/depr. education	−.105** (−1.975)	–	−.125** {−2.349)
FD_Risk assoc. w/depr. health	−203** (−3.888)	−.123** (−2.310)	−.089* (−1.677)

*$p < 0.05$; **$p < 0.01$
#Some scales are omitted because of low level of association they remain with the local risk by domain of deprivation; 'municipal services' is excluded from the local deprivation scales since it showed no association with any measure of satisfaction. ##Number of gminas reduced due to limited availability of data for respondents of Social Diagnosis/SD present in a sequence of years (years covered by SD: 2009–2011–2013–2019)

the residents' lower well-being include risk associated with counter-deprivation in local economy and fraction of persons temporarily away from home, and also implementation of some program of revitalization/gentrification by the local authorities. Generally, this accords with the previous observations of negative association between subjective well-being according to self-reported data based measures in FD-scales version, and also with growing well-being along the diminishing size of

Table 17.2 Functional data scales of local deprivation and local risk in selected domains, and some characteristics of communes as predictors of individual well-being acc. to U-index

	Unstandardized coefficients		Stand. coefficients		
Model	B	Std. error	Beta	t	Sig.
(Constant)	0.494	0.209		2.364	0.018
FD_Local Deprivation (2004_16)	0.000	0.000	−0.092	−2.204	0.028
Risk assoc. w/depr. labor market	−0.073	0.021	−0.211	−3.542	0.000
Risk assoc. w/depr. loc. economy	0.080	0.027	0.214	2.987	0.003
Temporarily absent/away from home (per 1000 inhabit.)	0.004	0.002	0.071	1.979	0.048
Proportion of 'in-work' to 'not-in-work' in a commune	−0.054	0.010	−0.175	−5.631	0.000
Number of NGOs per 1000 pers	−0.017	0.011	−0.049	−1.534	0.125
Local authority implements a program of revitalization (GPR)	0.069	0.026	0.085	2.715	0.007
F (7, 1012) = 9,7842; p < 0.000					

the place of living. The latter is clearly confirmed by similar tendency in the case of the U-index version of well-being – see Fig. 17.4.

17.5 The Well-Being Equation with Alternative Measures and Hypotheses on Important *Trade-Offs*

Despite that income and subjective well-being relationship is one of the most extensively explored topic in the literature (see Ferrer-i-Carbonell 2005; Clark and D'Ambrosio 2015) the generalizations derived from the empirical studies require relativization to a specific problem context to get an insight into the complex nature of this relation and its determinants. Taking the basic well-being equation as the point of departure:

$$WB = \alpha + \beta\, x_i + \varepsilon_i$$

where x, vector of predictors, and ε_i captures unobserved traits (error term) – different possible specifications are checked in empirical studies being led by hypotheses focusing on specially interesting aspects of this relation. Here, in the context of local community development and individual well-being, two aspects seems to deserve an empirical exploration. One relates to the income vs. work-time *trade-off* – as discussed for instance by Clark (2018) – another to the relative weight of social capital specific to a local community *vis-à-vis* possible changes of income (especially, in view of losing it) – in the version demonstrated by Anand (2018).

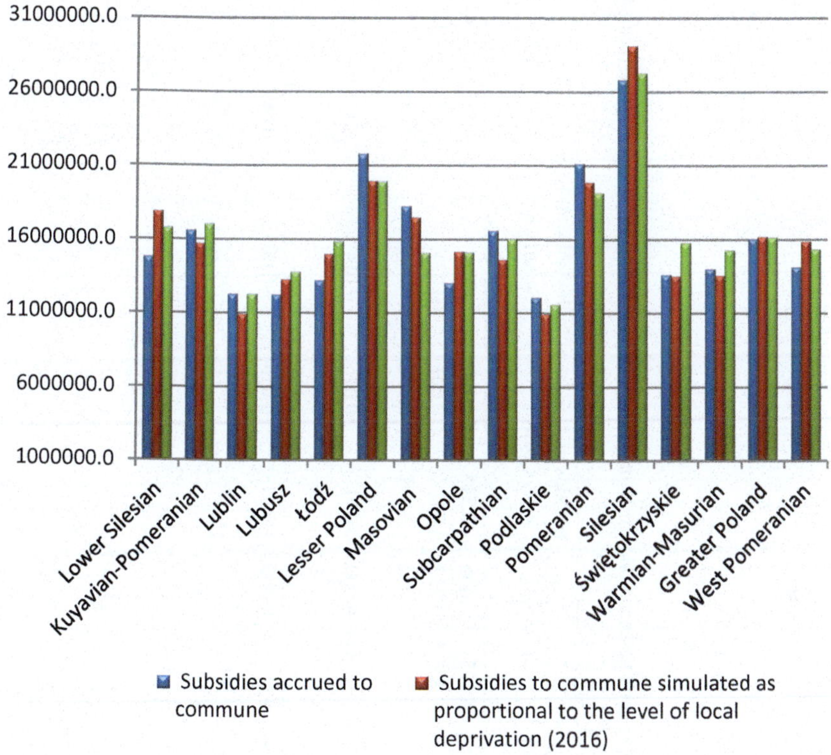

Fig. 17.4 Distribution of public resources (subsidies) to communes (gminas) – amounts obtained compared with amounts simulated as 'fair' – i.e. proportionally to the index of local deprivation MILD_2016, and MFPCA_2004-2016

In a condensed way, both of these issues can be addressed jointly within the OLS regression model with satisfaction with living condition (in the FD-scale version) and U-index as alternative dependent variables, respectively, which are influenced by a set of fundamental independent variables: time-on-job, income (household income per person) local risk associated with deprivation over years 2004–2016 (i.e. MILD in FD-version of the FD-version), and trust in local authority (with emphasize on financial matters). Due to some data limitation – e.g. income should be represented by hourly earnings in the first case, i.e. in income vs. work-time trade-off, rather than by earning – and social capital should be represented by set of items-based scale rather than only by a percentage of residents 'trusting' in local authorities – some caution needs to be stressed about the interpretation of the results presented in Table 17.3.

Indeed, there is a clear trade-off between the time spent on work – which affects in a negative way measures of well-being in both versions, i.e. the self-reported satisfaction with living conditions and the time-based – and income represented

Table 17.3 Subjective well-being acc. to FD-scale of satisfaction with living conditions and odds of U-index (not positive feeling about activities performed last day) by selected factors of influence

Model / predictors	A. FD satisfaction with *living condition*					B. Odds of *U-index* of individual well-being				
	Unstd. Coeff.		Stand. Coeff	t	Sig.	Unstd. Coeff.		Stand. Coeff.	t	Sig.
	B	Std. Error	Beta			B	Std. Error	Beta		
(Constant)	−2.009	0.479		4.195	0.000	−0.315	0.560		−0.563	0.574
Time-on-job	0.002	0.001	−0.023	−2.502	0.012	0.031	0.001	0.221	29.103	0.000
H'hld_Income, per person	3.95E-06	0.000	0.056	6.186	0.000	−3.23E-05	0.000	−0.017	−2.295	0.022
Risk_assoc. w/local depriva-tion/FD_MILD$_{2004–2016}$	−0.131	0.001	−0.053	−5.954	0.000	0.050	0.026	0.014	1.919	0.055
Trust in local authority (financial matters)	0.156	0.029	0.048	5.473	0.000	−0.018	0.026	−0.005	−0.689	0.491
	F (4,13,240) = 22.173; $p < .000$ CV (.48/.56)					*F {4, 17,854) = 223.146; p < .000 CV (.005/.017)*				

here by average household income per person in a commune. It accords with the results reported in the literature (see Clark 2018; Kahneman and Deaton 2010).

As regards the second issue that is touched here in order to check analytical convenience of a way of valuing relative importance of a community's social capital due to possible reduction or loss of the personal income (by a member of household), an approach called *compensating variation (CV)* is used. Its interpretation needs some explanation. Namely, CV is interpreted as the amount of money required to compensate a person for – typically, a price change or inflation – that gives rise to a loss in utility. Following Anand (2018), it is assumed here that the compensating variation for social capital, CV, can be obtained by identifying the utility gain derived from a unit increase in social capital. Formally, a life satisfaction equation can be written as:

$$U^0\left(y^0, SC^0\right) = U^1\left(y^0 + CV, SC^1\right)$$

where y, SC are income and social capital, respectively, and CV stands for compensating variations.

Some transformation, as a result, yields estimation of a ratio of the respective coefficients in the well-being equation, as follows:

$$CV = \beta_{SC}/\beta_y.$$

While the first value in Table 17.3 (for satisfaction with living conditions) is excessively big and unrealistic – mainly because of being approximated on the basis of the average values of relevant variables characterizing communes – the second value (0.29) seems rather quite realistic.

17.6 Patterns of Cross-Level Co-occurrence of the Well-Being Clusters

17.6.1 Correspondence Analysis (CORA) of Categorical Associations

Hypothetically, clusters of communes/gminas and clusters of residents (in fact, also communes/gminas characterized by average values of the relevant variable) derived through MFPCA that was applied to the local deprivation items at the first case and to the measures of well-being at the second, are expected to show a specific pattern of co-occurrence. Such as a pattern given by Correspondence Analysis in Fig. 17.5.

As it might be expected, clusters grouping communes considered higher on the scale of local development occur on similar position (are closer) to the clusters of communes characterized by a higher satisfaction of their residents (all measures are at the FD-types of scales). In other words, local development (counter-deprivation)

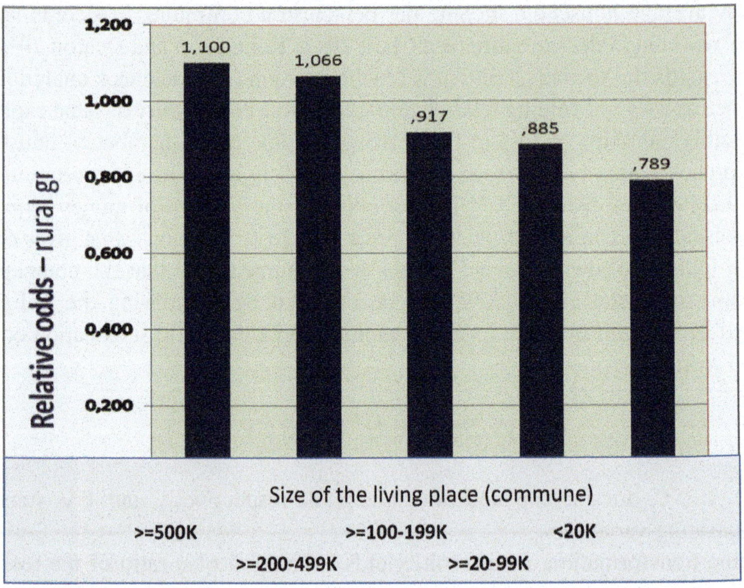

Fig. 17.5 Odds of experiencing 'non-positive' feeling (associated with activities performed last day) – U-index ('unpleasant state') – depending on the size of the place of living

affects positively the overall (multifaceted) satisfaction of residents, since inertia (.074) and chi-square (26.090; .053) values prove significance of such a pattern.

A series of similar graphic results of CORA has been conducted for particular aspects of satisfaction (well-being FD-scales), confirming analytical efficiency of above approach, which allows for better insight into the problem of relationship between quality of a community as a place to live and people's feeling about it (Fig. 17.6). Another example of this is below for satisfaction with overall personal situation and local development (counter-deprivation, without ecology) (Fig. 17.7).

In general, the above results suggest that such an approach may be considered as an appropriate strategy of analysing such a type of problem thanks to employing MFPCA, which is shown to be useful, especially given difficulties encountered with identifying patterns of association with alternative approaches.

17.7 Spatial Aspects of Well-Being Clusters Distribution

17.7.1 Autocorrelation and Spatial Clusters

Having categorized communes by clusters generated through FD-type of scales according, on the one hand, to multidimensional index of local deprivation ($MILD_{2004-16}$) or some domains (index of local deprivation in a particular domain,

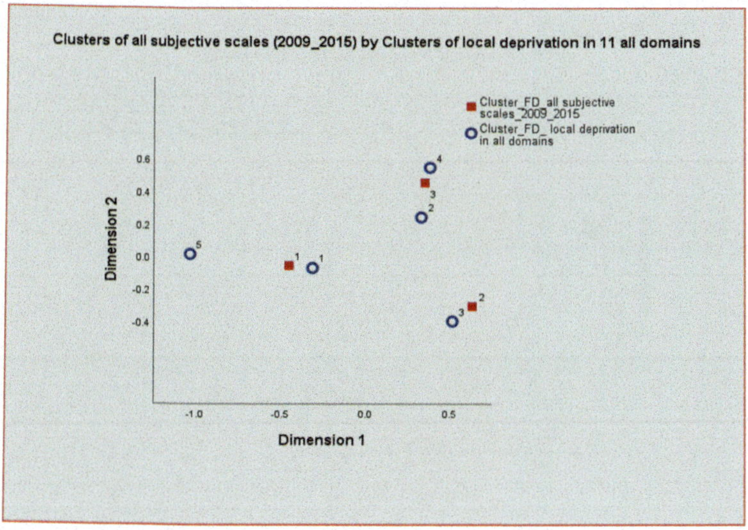

Fig. 17.6 Clusters of communes acc. to the level of local development (counter-deprivation) and clusters of communes with residents characterized by synthetic measures of satisfaction

$ILD_{2004–16}$) and, on the other hand, according to their residents' average subjective well-being, it is reasonable to ask the following question about spatial distribution of these clusters: Is there any tendency among the two types of clusters to behave 'jointly' in a systematic way – especially, along the hypothesis of congruence on the both types of indicators (and actually, of levels – community and individual)? A preliminary answer is being searched by applying LISA/Local Indicators for Spatial Association procedure (e.g. Fischer and Getis 2010), which provides results in terms of scatter plots and cluster maps – see the series of Figs. 17.8, 17.9, and 17.10 below.

Indeed, both the relatively high value of the coefficient of autocorrelation (Moran's I equals 0.36) and the cluster map – see Fig. 17.8 – confirm the first part of the above expectations. Namely, that there is a clear tendency to concentration among the communes/gminas grouped into clusters characterized by, on average, similar level of development. With a clear pattern of less developed communes in eastern part of the country, and those belonging to clusters of more advanced on this scale prevailing in capital metropolitan area and in western part of the country.

Even stronger is the tendency to geographic concentration among communes/gminas with similar developmental characteristics in selected domain – for instance, in local social welfare and labour market, as demonstrated by Figs. 17.9 and 17.10. However, in contrast to overall deprivation/development pattern presented in Fig. 17.8, the domain-specific levels, in both cases, are geographically grouped along north-south axis, rather than east-west – providing useful suggestions for practitioners and decision makers responsible for allocation of public resources in these areas of concern.

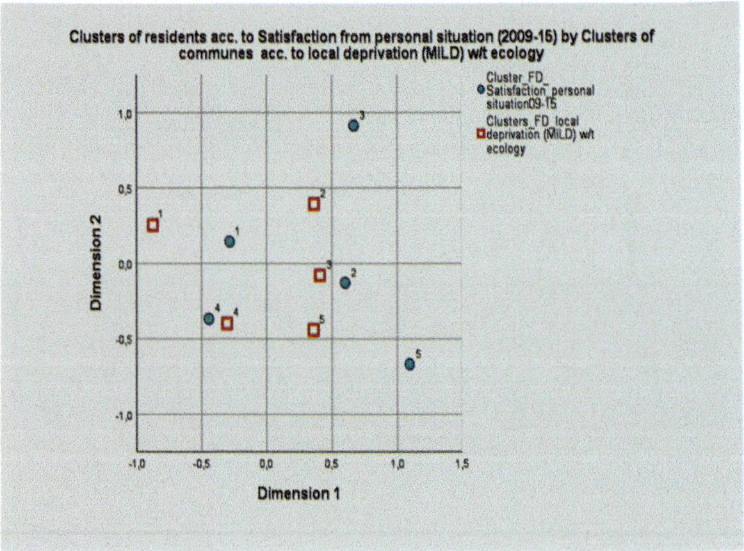

Fig. 17.7 Clusters of communes acc. to the level of local development (counter-deprivation) and clusters of communes with residents' satisfaction with personal situation (*Inertia:* .079; Chi sq. 27.890; .033)

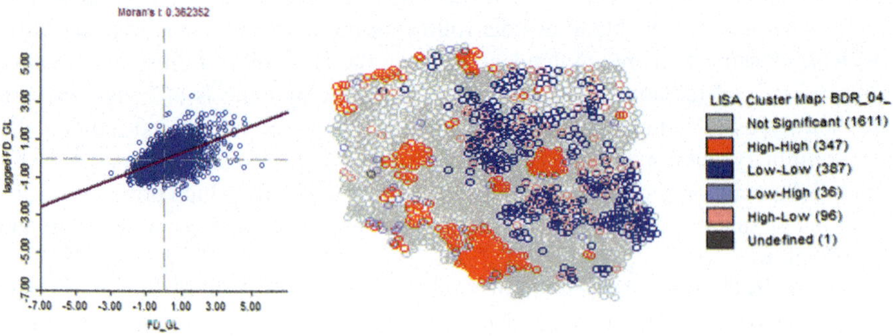

Fig. 17.8 Scatter plot (Moran's I 0.36) and cluster map of local development according to MFPCA measured local deprivation (in reversed, development version) $FD\_MILD_{2004–2016}$

A more complex issue concerning the cross-level relations between community and individual characteristics of development and satisfaction, respectively, can also be considered within the spatial framework. The exemplary problem relates to the situation in the domain of health and health services – here in version of risk associated with deprivation in health (in a positive version, i.e., development) – using FD-type of measure (for years 2004–2016) as key characteristics of the commune's health situation (high autocorrelation, Moran's I 0.26), while subjective well-being is represented in Fig. 17.11 by (a) *depression* (Moran's I 0.15), (b)

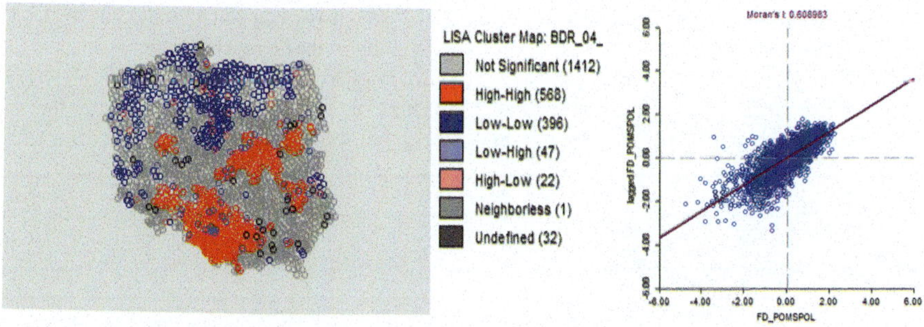

Fig. 17.9 Scatter plot (Moran's I 0. 61) and cluster map of local development in the area of social welfare, according to MFPCA ($FD\_ILD_{2004–2016}$)

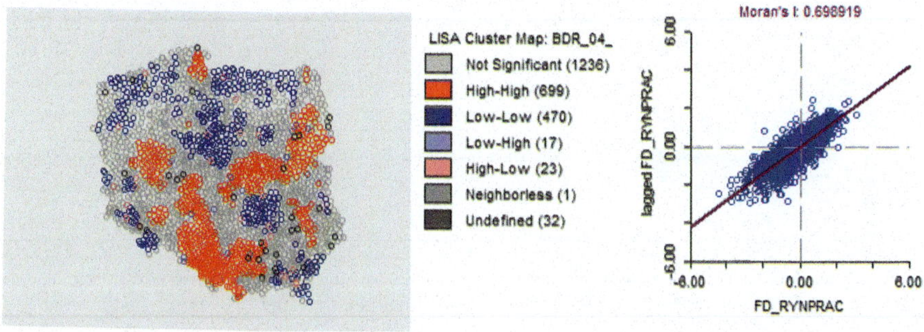

Fig. 17.10 Scatter plot (Moran's I 0.70) and cluster map of local development in the area of labour market, according to MFPCA ($FD\_ILD_{2004–2016}$)
Source: Own calculations

somatic problems (Moran's I 0.14), and (c) *satisfaction with living conditions* (Moran's I 0.16).

The above results – especially the scatter plots and Moran's coefficient (without cluster maps due to very small numbers of the communes/gminas with residents participating in Social Diagnosis survey over sufficiently long period of time) – make it visible that there is a clear propensity in geographic terms of the previously identified associations between community developmental characteristics and residents' subjective well-being (Figs. 17.5 and 17.6). Positive risks associated with health (i.e. chance for improvement) are co-occurring with higher ranked clusters of subjective well-being in all the aspects included: experiencing depression or a somatic problems, and satisfaction with living conditions. To complete the exploration of factors responsible for geographic distribution of the discussed clusters – local deprivation/development and subjective well-being in the functional data framework – it seems desirable to get an insight into the issue of spatial dependence.

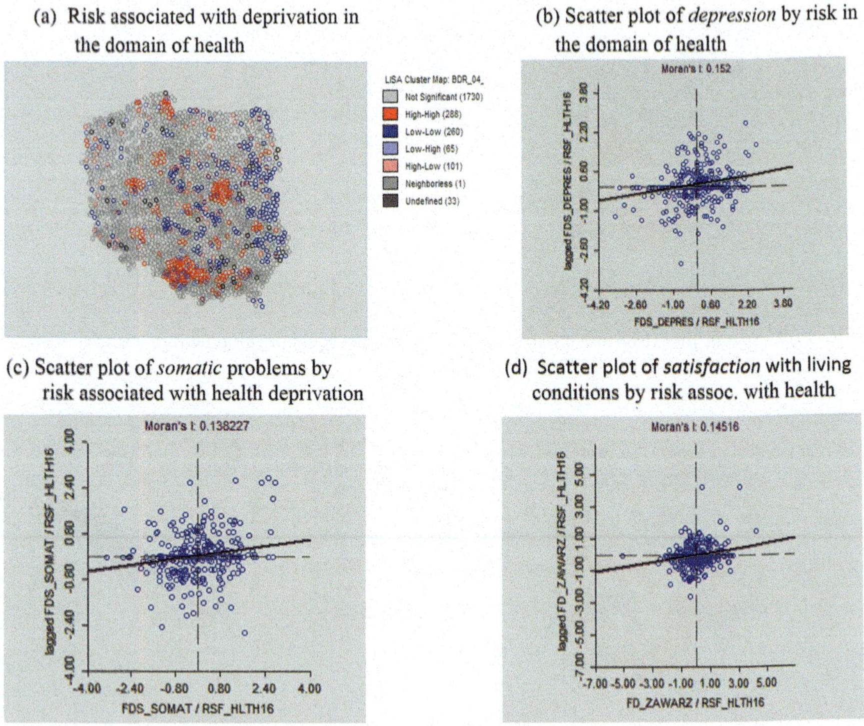

Fig. 17.11 (**a**) Cluster map of risk associated with deprivation in the domain of health (in reversed, positive version) and scatter plots of selected measures of subjective well-being by this risk: (**b**) feeling of depression; (**c**) experiencing somatic problems; and (**d**) satisfaction with living conditions
Source: Own calculations

17.7.2 Spatial Dependence – Spatial Regression

General equation for estimation of parameters of the spatial regression model was used, in the form as below (notation for individual/commune observation *i*):

$$y_i = \rho \sum\nolimits_{j=1}^{n} W_{ij} y_j + \sum\nolimits_{r=1}^{k} X_{ir} \beta_r + \varepsilon_i$$

where: y_i – the dependent variable for observation *i*; X_{ir} *k* – explanatory variables. r = 1. .... k with associated coefficient β_r; ε_i is the disturbance term; *W* is the spatial weights matrix. ρ is parameter of the strength of the average association between the dependent variable values for region/observations and the average of them for their neighbours (e.g. LeSage and Pace 2010. p. 357).

The above specification of the spatial regression model assumes that ε_i is meant as the *spatially lagged* term – versus *spatial error* formulation – for the dependent

Table 17.4 Spatial error model – maximum likelihood estimation

Variable	Coefficient	Std. error	z-value	Probability
CONSTANT	6.58928	4.69413	1.40373	0.16040
RiskFD_LabMkt	1.05379	0.470991	2.2374	0.02526
RiskFD_Economy	−1.4603	0.542753	−2.69055	0.00713
Subsidies FD_2016pc	0.000735	0.001080	0.680927	0.49592
NGOs per 1000 in 2016	−0.46272	0.2308	−2.00488	0.04498
Commune w/program of revitalization	0.18080	0.381625	0.473771	0.63566
Migration_balance	0.04184	0.0414618	1.00924	0.31286
LAMBDA	0.16678	0.0560501	2.97569	0.00292

Dependent – subjective well-being: all scales – SMS/*synthetic measure of satisfaction (N 352)*

REGRESSION DIAGNOSTICS -
DIAGNOSTICS FOR HETEROSKEDASTICITY
RANDOM COEFFICIENTS

TEST	DF	VALUE	PROB
Breusch-Pagan test	6	36.8021	0.00000

SPATIAL ERROR DEPENDENCE FOR WEIGHT MATRIX : BDR_04_16_Juneo5_2019

TEST	DF	VALUE	PROB
Likelihood Ratio Test	1	8.4296	0.00369

variable (which is correlated with the dependent variable), that is:

$$\varepsilon_i = \rho\, W_i . y_i + X_i . \beta + \epsilon_i$$

These two types of models allow us to examine the impact that one observation has on other, proximate observations. Table 17.4 presents results of spatial error model for synthetic measure of satisfaction, the key subjective well-being indicator as dependent variable influenced by community deprivation (development) through FD-based risk associated with deprivation in the domain of *local economy* and in *labour market,* with additional covariates accounting for the presence of the third sector organizations, local authority engagement in revitalization, and the balance (inflow-outflow) of migration.

Moderate size of Lambda coefficient (0.17) still supports the hypothesis that local community, its overall quality and particular areas of deprivation (development) affect significantly the perception and feeling of the residents, determining their attitude and satisfaction with various aspects of their lives.

17.8 Summary and Conclusions

In brief, the functional data-based approach in multivariate functional principal component analysis (MFPCA) is shown to be a useful tool for analysing spatio-temporal aspects of the relationship between the local level community development (deprivation) and individual, subjective well-being. However, due to using available data from different sources, some analyses have had only a demonstrative or illustrative character, without pretending to reach conclusive results. Nevertheless, geographically referred data provide a promising land of opportunities for policy analysis focused on well-being as the ultimate target of the local development. The full gain of combining the functional data approach with geostatistical data would be available to achieve hierarchical/'nested' data structure needed for multi-level spatial modelling – such an attempt is underway towards dynamic spatio-temporal modelling (DSTM).

References

Anand, P., 2018. The Value of Individual and Community Social Resources. The Value Individual and Community Social Resources. https://www.researchgate.net/publication/328701755

Arcaya, M., Brewster, M., Zigler, C. M., & Subremanian, S. V. (2012). Area variations in health. A spatial multilevel modelling approach. *Health Place, 18*(4), 824–831.

Becker, G. S. (1965). A theory of the allocation of time. *Economic Journal, 75*(299), 493–517.

Besse, P. (1979). *Etude descriptive d'un processus*. Ph. D thesis. Universite Paul Sbatier.

Clark, A. E. (2018). Four decades of the economics of happiness: Where next? *Review of Income and Wealth, 64*(2), 245–269.

Clark, A. E., & D'Ambrosio, C. (2015). Attitudes to income inequality: Experimental and survey evidence. In A. Atkinson & F. Bourguignon (Eds.), *Handbook of income distribution* (Vol. 2A, pp. 1147–1208). Amsterdam: Elsevier.

CNSTAT. (2013). Subjective Well-Being: Measuring Happiness, Suffering, and Other Dimensions of Experience. Arthur A. Stone and Christopher Mackie, Editors; Panel on Measuring Subjective Well-Being in a Policy-Relevant Framework; Committee on National Statistics; Division on Behavioral and Social Sciences and Education; National Research Council

Cressi, N., & Wikle, K. C. H. (2011). *Statistics for spatio-temporal data*. Hoboken: Wiley.

D'Orazio, M., Di Zio, M., & Scanu, M. (2006). *Statistical matching. Theory and practice*. Chichester, England: Wiley.

Ferrer-i-Carbonell, A. (2005). Income and well-being: An empirical analysis of the comparison income effect. *Journal of Public Economics, 89*, 997–1019.

Fischer, M. M., & Getis, A. (2010). *Handbook of applied spatial analysis: Software tools, methods and applications*. Berlin: Springer.

Górecki, T., Krzyśko, M., Waszak, & Wołyński, W. (2018). Selected statistical methods of data analysis for multivariate functional data. *Statistical Papers, 59*, 153–182.

Gorecki, T., Krzyśko, M., & Wołyński, W. (2019). Variable selection in multivariate functional data classification. *Statistics in Transition New Series, 17*(3), 449–466.

Jacques, J., & Preda, C. (2014). Model-based clustering for multivariate functional data. *Computational Statistics and Data Analysis, 71*, 92–106.

Kahneman, D., & Deaton, A. (2010). High income improves evaluation of life but not emotional well-being. *PNAS*. https://doi.org/10.1073/pnas.1011492107.

Kahneman, D., & Krueger, A. B. (2006). Developments in the measurement of subjective well-being. *Journal of Economic Perspectives, 20*, 3–24.

Krueger, A. B., Kahneman, D., Schkade, D. A., Schwartz, N., & Stone, A. (2009). National Time Accounting: The currency of life. In A. B. Krueger (Ed.), *Measuring subjective Well-being of nations: National Account of time use and Well-being*. University of Chicago Press.

LeSage, J. P., & Pace, R. K. (2010). Spatial econometrics. In Fischer and Getis (2010).

Nordhaus, W. (2009). *Measuring real income with leisure and household production. In: A. B. Krueger (ed) measuring subjective Well-being of nations*. National Account of Time Use and Well-Being. University of Chicago Press.

OECD. (2013). *OECD guidelines on measuring subjective well-being*. Paris: OECD Publishing.

Okrasa, W. (2017). Community wellbeing, spatial cohesion and individual wellbeing – towards a multilevel spatially integrated framework. In W. Okrasa (Ed.), *Quality of life and spatial cohesion: Development and wellbeing in the local context*. Warsaw: Cardinal Stefan Wyszynski University Press.

Okrasa, W., & Rozkrut, D. (2018a). *The time use data-based measures of the wellbeing effect of community development*. In Proceedings of the 2018 Federal Committee on Statistical Methodology (FCSM) research conference

Okrasa, W., & Rozkrut, D. (2018b). Modelling for improving measurement: Strategies for contextualization of well-being. In IAOS2018_OECD conference better statistics for better lives, Paris, September 1921.

Okrasa, W., Lapins, J., & Vremis, M. (2006). Measuring community deprivation for geographic targeting of public resources – Case of Moldova. *Statistics in Transition, Journal of the Polish Statistical Association, 7*, Nr. 5.

Ramsay, J. O., & Silverman, B. W. (2005). *Functional data analysis* (2nd ed.). New York: Springer.

Saporta, G. (1981). *Methodes exploratoires d'analyse de donnes temporalles*. Cahier du Buro. Ph.D. thesis.

Chapter 18
Properties and Dynamics of the Beta Gompertz Generalized Makeham Distribution

Panagiotis Andreopoulos, Alexandra Tragaki, George Antonopoulos, and Fragkiskos G. Bersimis

18.1 Introduction

18.1.1 Beta Distribution

Beta distribution has been widely used in a variety of scientific fields (Abramowitz and Stegun 2006; Gupta and Nadarajah 2004) due to the fact that it is able to describe a wide range of different data with bounded support. More specifically, beta distribution with two parameters—i.e., left parameter a (shape of beta) and right parameter b (scale of beta)—is used for modeling data, taking values within the interval (0,1). Beta density is expressed as follows:

$$f(x;\alpha,\beta)=\frac{1}{B(\alpha,\beta)}x^{\alpha-1}(1-x)^{\beta-1},\ \ 0<x<1,\qquad \alpha,\beta>0$$

where $B(\alpha,\beta)$ is given by the integral

P. Andreopoulos (✉) · A. Tragaki
Department of Geography, Harokopio University, Athens, Greece
e-mail: pandreop@hua.gr; atragaki@hua.gr

G. Antonopoulos
MSc: Statistics and Operational Research, Athens, Greece

Department of Mathematics, National and Kapodistrian University of Athens, Athens, Greece
e-mail: antono.geo@gmail.com

F. G. Bersimis
Department of Informatics and Telematics, Harokopio University, Tavros, Greece
e-mail: fbersim@hua.gr

C. H. Skiadas, C. Skiadas (eds.), *Demography of Population Health, Aging and Health Expenditures*, The Springer Series on Demographic Methods and Population Analysis 50, https://doi.org/10.1007/978-3-030-44695-6_18

$$B(\alpha, \beta) = \int_0^1 u^{\alpha-1}(1-u)^{\beta-1} du$$

Beta density function may have different shapes (Venter 1983) including left- and right-skewed or the flat shape of uniform density, depending on the combination of its parameters.

In the case that all scores are equally likely, then the parameters could be set as $\alpha = 1$ and $\beta = 1$, as illustrated in Panel B; this gives a "flat" probability density function (pdf).

Some other special cases are:

- If parameters $\alpha = \beta < 1$, the curve of the beta pdf in Panel B takes the shape of a symmetrical bathtub. If $\alpha \neq \beta$, then the shape of a non-symmetrical bathtub appears with a minimum value at $x_0 = \frac{\alpha-1}{\alpha+\beta-2}$.
- If $\alpha = \beta = 1$, the beta distribution B(1, 1) reduces to the uniform distribution $U(0, 1)$ (the pdf is constant and equal to one; Panel B).
- If $\alpha = 1$ and $\beta > 1$, beta pdf is decreasing as x tends to 1, and if $\alpha > 1$ and $\beta = 1$, beta pdf is increasing as x tends to 1.
- If $\alpha < 1$ and $\beta \geq 1$, beta pdf is decreasing as x tends to 1.
- If $\alpha \geq 1$ and $\beta < 1$, beta pdf is increasing as x tends to 1.
- If $\alpha > 1$ and $\beta > 1$, beta pdf (Panel C) is increasing in the domain $(0, x_0)$ and decreasing in the domain $(x_0, 1)$. The peak of the density is in the interior of [0,1] and the mode of the beta distribution is x_0 (Fig. 18.1).

18.1.2 Generated Beta Distribution

In mathematical statistics, the study of the beta distribution is useful in many ways: one can relate the beta distribution to other well-known distributions (uniform, gamma, exponential, normal, etc.) or just present it as an important application of the gamma function. The first distribution of the beta-generated class was the beta-normal distribution, introduced by Eugene et al. (2002). Changing the parent distribution so that it is no longer normal offers more flexibility. Replacing *f(x)* (in Eq. 18.1) by any other parent distribution F yields the general class of beta-generated distributions (Jones 2005).

Beta-generated distributions with more general parent distributions have been studied in the literature, such as the beta Gumbel distribution (Nadarajah and Kotz 2004, 2006), beta-Pareto distribution (Akinsete et al. 2008), generalized beta distribution (Zografos and Balakrishnan 2009) and beta-generalized exponential distribution (Barreto-Souza et al. 2010).

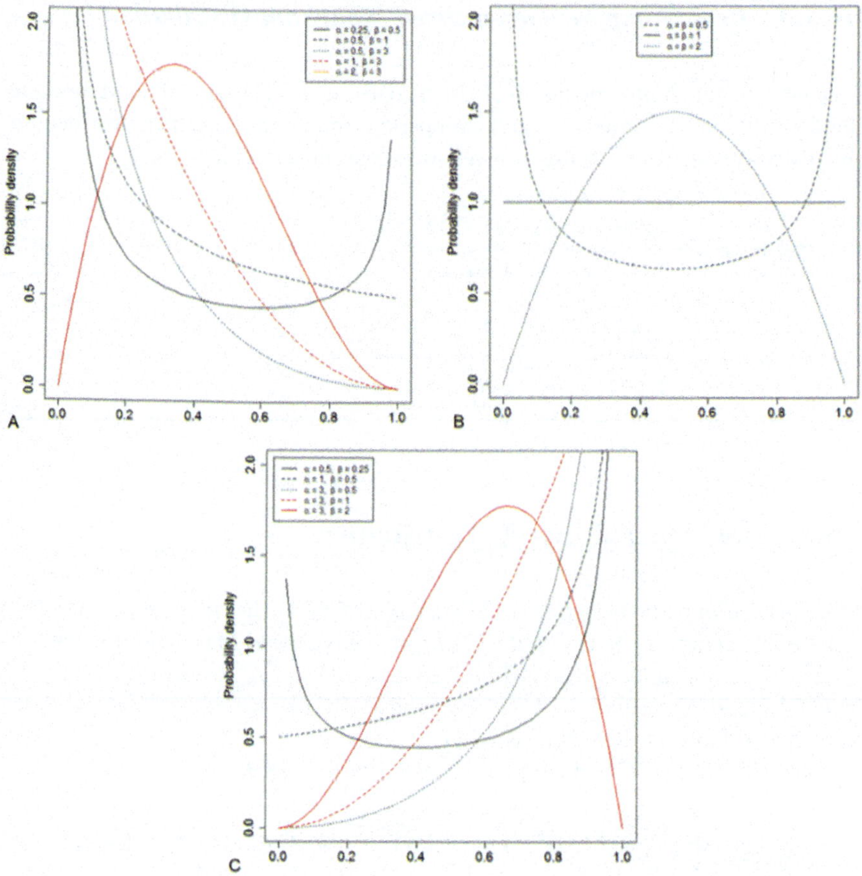

Fig. 18.1 The pdf of the beta distribution for $\alpha < \beta$ (Panel A), $\alpha = \beta$ (Panel B), and $\alpha > \beta$ (Panel C)

An attractive feature of generalized beta distributions is that the generalized beta parameters afford greater control over the weights in both tails and in the center of the generated distribution, compared with the classical beta-normal distribution.

In order to combine beta distribution with the aforementioned mortality models, generated beta distribution is used (Zografos and Balakrishnan 2009). The corresponding pdf is given by Eq. (18.1):

$$g(x) = \frac{f(x)}{B(\alpha, \beta)}[F(x)]^{\alpha-1}[1 - F(x)]^{\beta-1}, \ \ \alpha, \beta > 0 \tag{18.1}$$

18.1.3 Beta Gompertz Generalized Makeham Distribution

The new model (Andreopoulos et al. 2015, 2019) under the name BGGM mortality model is based on mixing beta and Gompertz Generalized Makeham distributions. Its probability density function is given as follows in Eq. (18.2):

$$
\begin{aligned}
f_{BGGM}(x) = {} & \frac{\left(\theta e^{\xi x} + \lambda + \kappa x\right) e^{-\frac{\kappa}{2}x^2 - \lambda x - \frac{\theta}{\xi}\left(e^{\xi x}-1\right)}}{\mathrm{B}\left(\alpha, \beta\right)} \\
& \cdot \left[\left(1 - e^{-\frac{\kappa}{2}x^2 - \lambda x - \frac{\theta}{\xi}\left(e^{\xi x}-1\right)}\right)\right]^{\alpha-1} \cdot \left[e^{-\frac{\kappa}{2}x^2 - \lambda x - \frac{\theta}{\xi}\left(e^{\xi x}-1\right)}\right]^{\beta-1} \\
& \theta, \xi, \lambda, \kappa, \alpha, \beta > 0, x \geq 0
\end{aligned} \tag{18.2}
$$

18.2 The Model of BGGM Distribution

The Beta Gompertz Generalized Makeham (BGGM) distribution is formed by mixing the generated beta with the Gompertz Generalized Makeham distributions. The beta-generated distributions described in Sect. 18.1 are now extended to a more general class of generalized beta-generated (GBG) distributions, given a parent distribution F(x), x ∈ ℝ with density f(x).

The pdf of the BGGM distribution is given by Eq. (18.3):

$$
g_1(x) = \begin{cases} \frac{f(x)}{B(\alpha,\beta)}[F(x)]^{\alpha-1}[1 - F(x)]^{\beta-1}, & \alpha, \beta, \kappa, \lambda, \theta, \xi > 0, x \geq 0 \\ 0 & , \; for \; x < 0 \end{cases} \tag{18.3}
$$

where $f(x) = \left(\theta e^{\xi x} + \lambda + \kappa x\right) e^{-\frac{\kappa}{2}x^2 - \lambda x - \frac{\theta}{\xi}\left(e^{\xi x}-1\right)}$ is the pdf and $F(x) = 1 - e^{-\frac{\kappa}{2}x^2 - \lambda x - \frac{\theta}{\xi}\left(e^{\xi x}-1\right)}$ is the cumulative density function (cdf) of the Generated Gompertz Makeham (GGM) distribution.

If $h(x) = \frac{\kappa}{2}x^2 + \lambda x + \frac{\theta}{\xi}\left(e^{\xi x} - 1\right)$ is the hazard function, then the first derivative of the distribution gives us $h^{'}(x) = \theta e^{\xi x} + \lambda + \kappa x, \; \gamma\iota\alpha \; x > 0$ and $F(x)$ written as $F(x) = 1 - e^{-h(x)}$.

Furthermore, the pdf of BGGM distribution can be rewritten as given in Eq. (18.4)

$$
\begin{aligned}
g_1(x) &= \frac{h^{'}(x)e^{-h(x)}}{\mathrm{B}(\alpha,\beta)}\left(1 - e^{-h(x)}\right)^{\alpha-1}\left(e^{-h(x)}\right)^{\beta-1} \\
&= \frac{h^{'}(x)}{\mathrm{B}(\alpha,\beta)}\left(1 - e^{-h(x)}\right)^{\alpha-1}\left(e^{-h(x)}\right)^{\beta}, x \geq 0
\end{aligned} \tag{18.4}
$$

and the cdf of BGGM distribution can be rewritten as in Eq. (18.5)

$$G_1(x) = \int_{-\infty}^{x} g(t)\,\mathrm{dt} = \frac{1}{\mathrm{B}(\alpha,\beta)} \int_{0}^{x} h'(t)\left(1 - e^{-h(t)}\right)^{\alpha-1}\left(e^{-h(t)}\right)^{\beta}\,\mathrm{dt} \tag{18.5}$$

Considering the transformation of $u = 1 - e^{-h(t)}$, it can be stated that $du = h'(t)e^{-h(t)}dt$ and the limit of integration

- for $t = 0$, because of $h(0) = 0$, is $u = 1 - e^{-h(0)} = 1 - e^0 = 0$
- for $t = x$, is $u = 1 - e^{-h(x)}$
- for $t \to +\infty$, is $u \to 1$, because of $\lim\limits_{t \to +\infty} h(t) \to +\infty$ and

$$\lim_{t \to +\infty} u = \lim_{t \to +\infty}\left(1 - e^{-h(t)}\right) = 1 - 0 = 1. \quad (^{**})$$

$h(x)$ is a continuous, differentiable, and increasing function for $x \geq 0$.
Also, $0 \leq u = 1 - e^{-h(x)} \leq 1$ and finally

$$G_1(x) = \frac{1}{\mathrm{B}(\alpha,\beta)} \int_{0}^{1-e^{-h(x)}} u^{\alpha-1}(1-u)^{\beta-1}\,du = \mathrm{B}\left(1 - e^{-h(x)}, \alpha, \beta\right) \tag{18.6}$$

where $\mathrm{B}(1 - e^{-h(x)}, \alpha, \beta)$ is the incomplete beta function.
The hazard function of the BGGM distribution is given by Eq. (18.7):

$$\begin{aligned} H(x) = \frac{g_1(x)}{1-G_1(x)} &= \frac{\frac{h'(x)}{\mathrm{B}(\alpha,\beta)}\cdot\left(1-e^{-h(x)}\right)^{\alpha-1}\cdot\left(e^{-h(x)}\right)^{\beta}}{1-\mathrm{B}\left(1-e^{-h(x)},\alpha,\beta\right)} \\ &= \frac{h'(x)\cdot\left(1-e^{-h(x)}\right)^{\alpha-1}\cdot\left(e^{-h(x)}\right)^{\beta}}{\mathrm{B}(\alpha,\beta)\cdot\left[1-\mathrm{B}\left(1-e^{-h(x)},\alpha,\beta\right)\right]} \\ &= \frac{h'(x)\cdot\left(1-e^{-h(x)}\right)^{\alpha-1}\cdot\left(e^{-h(x)}\right)^{\beta}}{\mathrm{B}(\alpha,\beta)\cdot\left[1-\frac{1}{\mathrm{B}(\alpha,\beta)}\int_0^x h'(t)\left(1-e^{-h(t)}\right)^{\alpha-1}\left(e^{-h(t)}\right)^{\beta}\,\mathrm{dt}\right]} \\ &= \frac{h'(x)\cdot\left(1-e^{-h(x)}\right)^{\alpha-1}\cdot\left(e^{-h(x)}\right)^{\beta}}{\mathrm{B}(\alpha,\beta)-\int_0^x h'(t)\left(1-e^{\ h(t)}\right)^{\alpha-1}\left(e^{\ h(t)}\right)^{\beta}\,dt} \end{aligned} \tag{18.7}$$

18.3 Special Sub-models of BGGM Distribution

Some well-known distributions arise for specific parameter values of the BGGM distribution. For example, in the case that $\lambda = 0$ and $a = b = 1$, then we get an exponential Gompertz distribution, and, if it also occurs that $\kappa = 0$, then we get a Gompertz distribution (Gompertz 1825).

18.4 Some Statistical Properties of BGGM Distribution

The pdf $f(x)$ should satisfy the following properties (Papaioannou 2001):

1. $f(x) \geq 0$ (*)
2. $\int_{-\infty}^{+\infty} f(x)dx = 1$ (**)

The combination of the beta distribution with the Generalized Gompertz Makeham give us the BGGM distribution with the pdf expression given in Eq. (18.2).

Proof If $h(x) = \frac{\kappa}{2}x^2 + \lambda x + \frac{\theta}{\xi}\left(e^{\xi x} - 1\right)$ is the hazard function, then the first derivative of the distribution gives us $h^{'}(x) = \theta e^{\xi x} + \lambda + \kappa x,\ \gamma\iota\alpha\ x > 0$.

1. It is obvious that $h^{'}(x) \geq 0$ and $h(x) \geq 0$, because all parameters are positive and the functions are increasing on $[0, +\infty)$. The range for $h(x)$ is $[0, +\infty)$, for $x \in [0, +\infty)$, then $e^{-h(x)}$ receives values on $(0, 1]$ and $F(x) = 1 - e^{-h(x)} \in [0, 1)$ for $x \geq 0$ too. Finally, the $f_{BGGM}(x)) \geq 0$ is the product of continuous functions, so $f(x) \geq 0$ (*).
2. The proof of $\int_{-\infty}^{+\infty} f(x)dx = 1$ is the same as (**). However, now the function is written

$$\int_{0}^{\infty} f_{BGGM}(x)dx = \int_{0}^{\infty} \frac{\left(1-e^{-h(x)}\right)^{\alpha-1}.\left[1-\left(1-e^{-h(x)}\right)\right]^{\beta-1}.h'(x)e^{-h(x)}}{B(\alpha,\beta)}dx$$

$$= \int_{0}^{1} \frac{u^{\alpha-1}.(1-u)^{\beta-1}}{\mathrm{B}(\alpha,\beta)}du = \frac{\mathrm{B}(\alpha,\beta)}{\mathrm{B}(\alpha,\beta)} = 1$$

Some possible shapes of $f(x)$, $F(x)$, and the BGGM density function, are given below for the combination of two parameters (Fig. 18.2).

Furthermore, for various values of each parameter obtained, Fig. 18.3 shows the effect of each parameter in the form of the pdf of the BGGM distribution (Currie et al. 2004). The last two graphs show us the difference between the standard deviation of two pdfs, but with similar means. The corresponding distribution functions were conducted by using open code R (www.r-project.org) and corresponding packages of algorithms from the CRAN digital library (Comprehensive R Archive). The estimation of the six parameters was done using minpack.lm (R software).

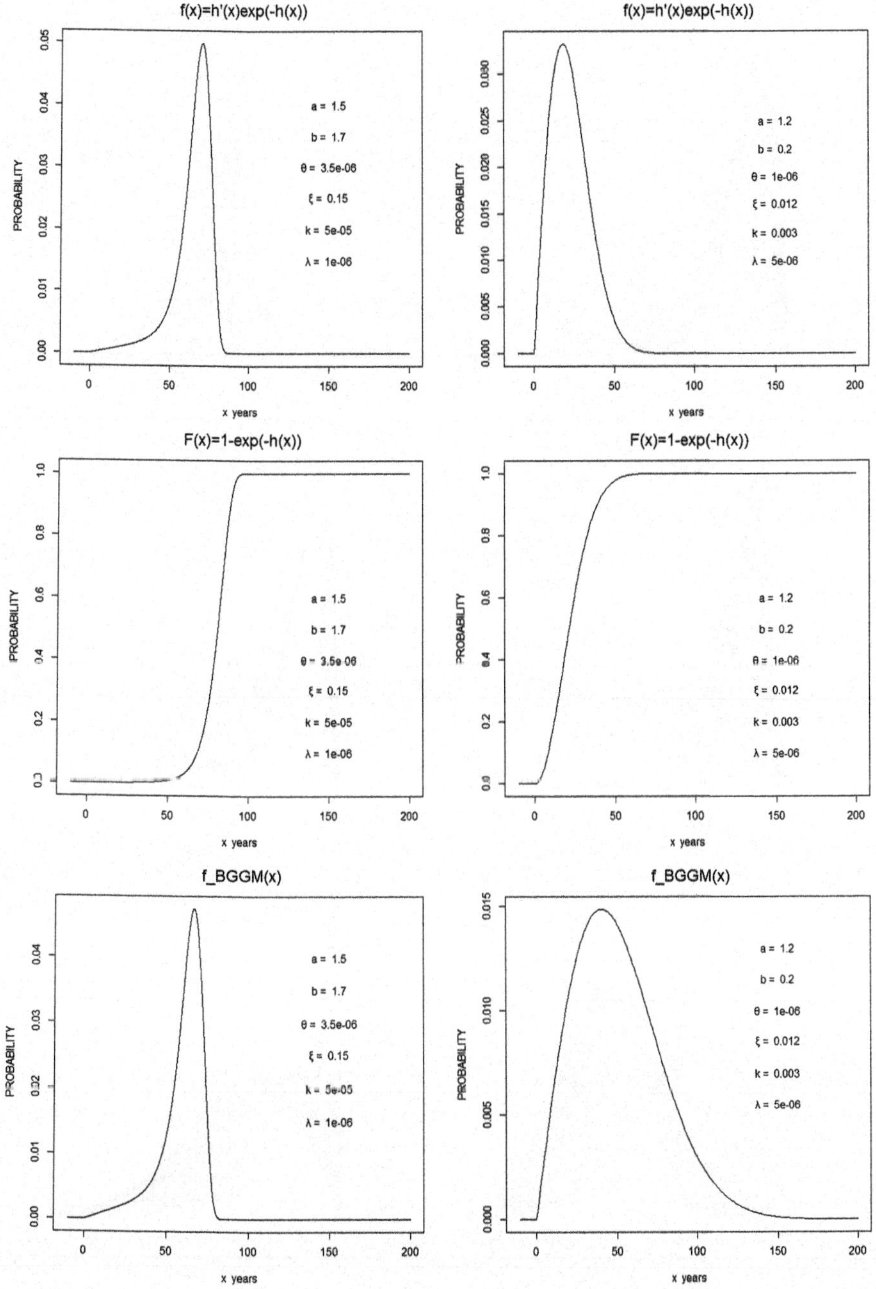

Fig. 18.2 Curves of the BGGM distribution by select values of $f(x)$ and values of cumulative $F(x) = 1 - e^{-h(x)}$. Graphs illustrated using R software

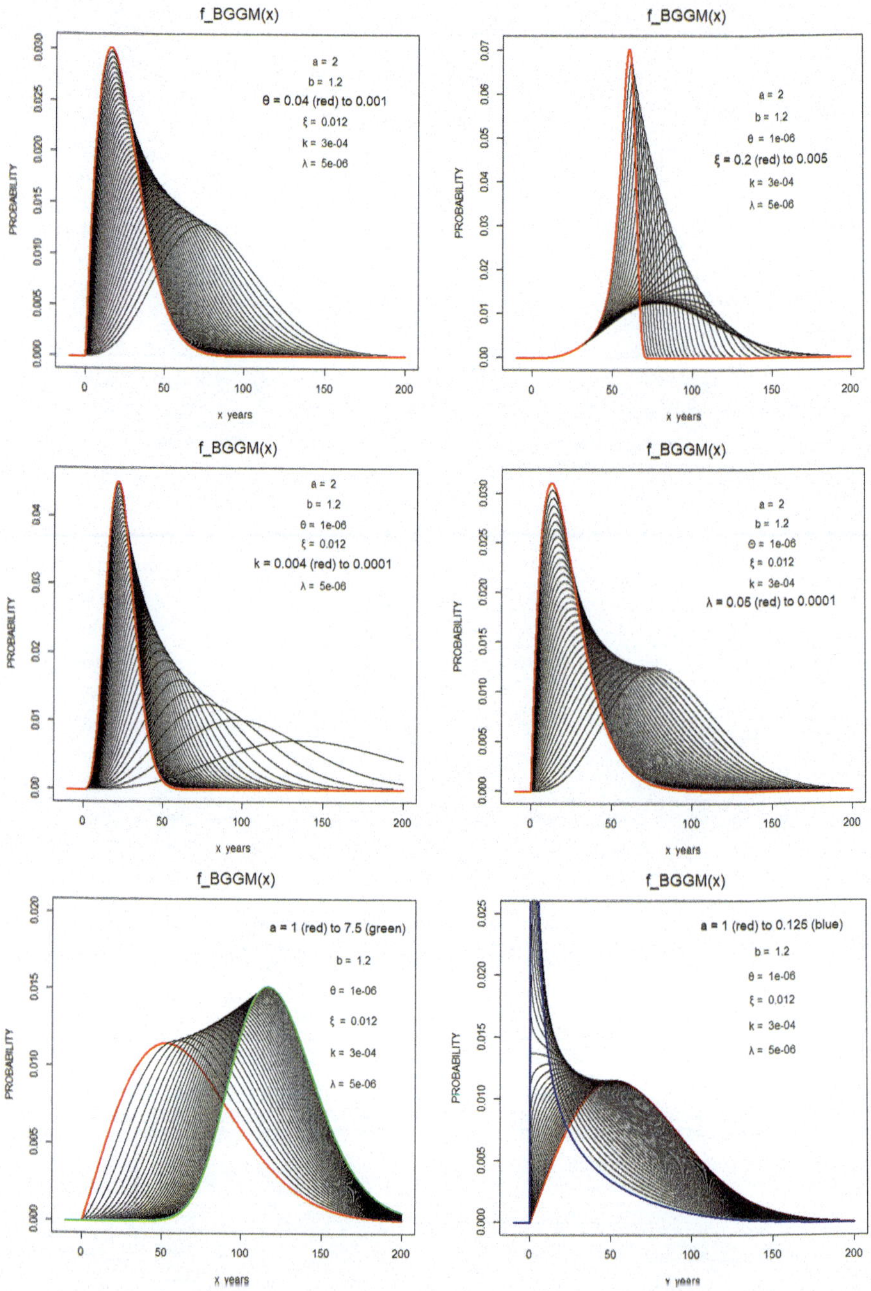

Fig. 18.3 (continued)

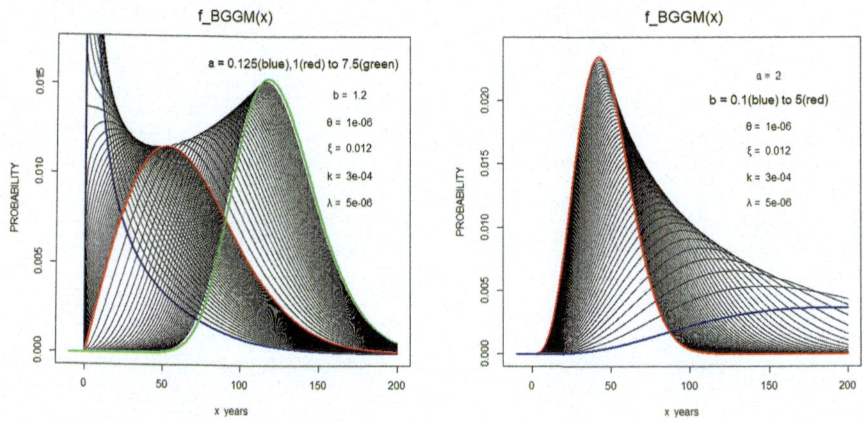

Fig. 18.3 Curves of the pdf of BGGM distribution by select values of parameters. The graphs show the effect on the shape format and scale. Graphs were illustrated using R software

18.5 The Asymptotic Behavior of BGGM Distribution

The asymptotic behavior of mathematical functions shows the convergence of a function for specific parts of its domain or for those values that tend to infinity. We know that for the family of exponential distributions, the following limits apply:

- $\lim\limits_{x \to +\infty} e^x = +\infty$
- $\lim\limits_{x \to -\infty} e^x = 0$
- $\lim\limits_{x \to 0} e^x = 1$

As a sum of polynomial and exponential functions (both emerging functions if they have positive coefficients), the function $h(x) = \frac{\kappa}{2}x^2 + \lambda x + \frac{\theta}{\xi}\left(e^{\xi x} - 1\right)$ is an increasing function. Also, $h(x)$ is positive for $x \geq 0$, with $h(0) = 0$ and $\lim\limits_{x \to +\infty} h(x) \to +\infty$.

The positive derivative $h'(x) = \theta e^{\xi x} + \kappa x + \lambda$ is an increasing function, with $h'(0) = \theta + \lambda$ and $\lim\limits_{x \to +\infty} h'(x) \to +\infty$. Therefore, taking the corresponding limits for each of these cases, we obtain the following results:

- $\mathbf{x} \to 0^+$

$$\lim_{x \to 0^+} f_{BGGM}(x) = \lim_{x \to 0^+} \frac{h'(x)}{B(\alpha,\beta)}\left[1 - e^{-h(x)}\right]^{\alpha-1}\left[e^{-h(x)}\right]^{\beta}$$
$$= \lim_{x \to 0^+} \frac{h'(x)}{B(\alpha,\beta)}\left[1 - e^{-h(x)}\right]^{\alpha-1}\left[e^{-h(x)}\right]^{\beta} = (1)$$

If $\alpha > 1$ $(1) = \frac{\theta+\lambda}{B(\alpha,\beta)}\left[1 - e^{-0}\right]^{\alpha-1}\left[\left(e^{-0}\right)\right]^{\beta} = \frac{\theta+\lambda}{B(\alpha,\beta)}[1-1]^{\alpha-1}[1]^{\beta} = 0$

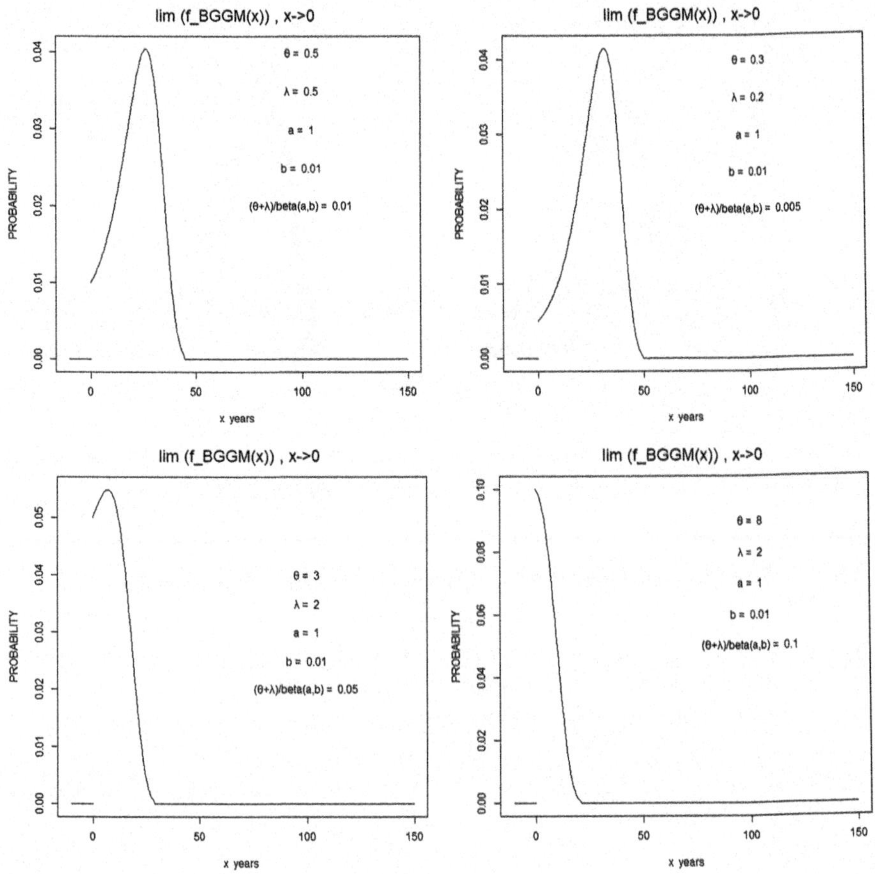

Fig. 18.4 Curves of the pdf of BGGM distribution by select values of parameters when $\alpha = 1$. The graphs show the effect on the shape format, scale and occurrence discontinuity. Graphs illustrated using R software

$$\text{If} \quad \alpha = 1 \ (1) = \frac{\theta+\lambda}{B(\alpha,\beta)}\left[\left(e^{-0}\right)\right]^{\beta} = \frac{\theta+\lambda}{B(\alpha,\beta)}$$

$$\text{If} \quad 0 < \alpha < 1 \ (1) = \frac{\theta+\lambda}{B(\alpha,\beta)}\left(\lim_{x\to 0+}\left[\frac{e^{h(x)}}{e^{h(x)}-1}\right]^{1-\alpha}\right)[1]^{\beta} = +\infty$$

For the case where $\alpha = 1$, discontinuity occurs in the pdf of BGGM distribution to zero, as shown in Fig. 18.4. The limit to zero ($x > 0 \ or \ x \to 0^{+}$) of the value of the distribution is $\frac{\theta+\lambda}{B(\alpha,\beta)}$.

- $\mathbf{x} \to +\infty$

$$\begin{aligned}\lim_{x\to+\infty} f_{BGGM}(x) &= \lim_{x\to+\infty} \frac{h'(x)}{B(\alpha,\beta)} \cdot \left[1-e^{-h(x)}\right]^{\alpha-1} \cdot \left[e^{-h(x)}\right]^{\beta} \\ &= \lim_{x\to+\infty} \frac{h'(x)}{B(\alpha,\beta)\cdot\left[e^{h(x)}\right]^{\beta}} \cdot \left[1-e^{-h(x)}\right]^{\alpha-1} \\ &= \frac{[1-0]^{\alpha-1}}{B(\alpha,\beta)} \cdot \lim_{x\to+\infty} \frac{\left(\theta e^{\xi x}+\lambda+\kappa x\right)}{\left[e^{\frac{\kappa}{2}x^2+\lambda x+\frac{\theta}{\xi}\left(e^{\xi x}-1\right)}\right]^{\beta}} \\ &= \frac{1}{B(\alpha,\beta)} \cdot 0 = 0\end{aligned}$$

18.6 Expansion for the Density Function of BGGM Distribution

It is worth stressing that some important statistical measures, such as, for example, variance, skewness, and kurtosis, can be obtained in terms of moments (Castellares and Lemonte 2019). Therefore, valid expression of the moments is necessary in order to compute amounts such as the aforementioned quantities. In the following, we provide a closed-form expression for the moments of the BGGM distribution. To do that, we make use of the convergent power series expansion

$$(1-z)^b = \sum_{j=0}^{\infty} (-1)^j \binom{b}{j} z^j \quad (*), \ |z| \leq 1, \ b > 0$$

Note that $F(x) = 1 - e^{-\frac{\kappa}{2}x^2-\lambda x-\frac{\theta}{\xi}\left(e^{\xi x}-1\right)} = 1 - e^{-h(x)} := u$ since $0 \leq F(x) = 1 - e^{-h(x)} := u \leq 1$ and the $(*)$, we can infer that the factor $(F(x))^{\alpha-1}$ from the $f_{BGGM}(x)$ can express $(F(x))^{\alpha-1}$ in the form

$$(F(x))^{\alpha-1} = \left(1-e^{-h(x)}\right)^{\alpha-1} = \sum_{j=0}^{\infty} (-1)^j \binom{a-1}{j} \left(e^{-h(x)}\right)^j$$

The pdf is given by

$$\begin{aligned} f_{BGGM}(x) &= \frac{h'(x)}{\mathrm{B}(\alpha,\beta)} \left(1-e^{-h(x)}\right)^{\alpha-1} \left(e^{-h(x)}\right)^{\beta} \\ &= \frac{h'(x)}{\mathrm{B}(\alpha,\beta)} \left[\sum_{j=0}^{\infty} (-1)^j \binom{a-1}{j} \left(e^{-h(x)}\right)^j\right] \left(e^{-h(x)}\right)^{\beta} \\ &= \frac{h'(x)}{\mathrm{B}(\alpha,\beta)} \sum_{j=0}^{\infty} (-1)^j \binom{a-1}{j} e^{-h(x)j} e^{-h(x)\beta} \\ &= \frac{h'(x)}{\mathrm{B}(\alpha,\beta)} \sum_{j=0}^{\infty} (-1)^j \binom{a-1}{j} e^{-h(x)(j+\beta)}. \end{aligned} \tag{18.8}$$

Figure 18.5 presents an approximation of the BGGM probability function by the corresponding sum of Eq. (18.8). The convergence, between black and red lines,

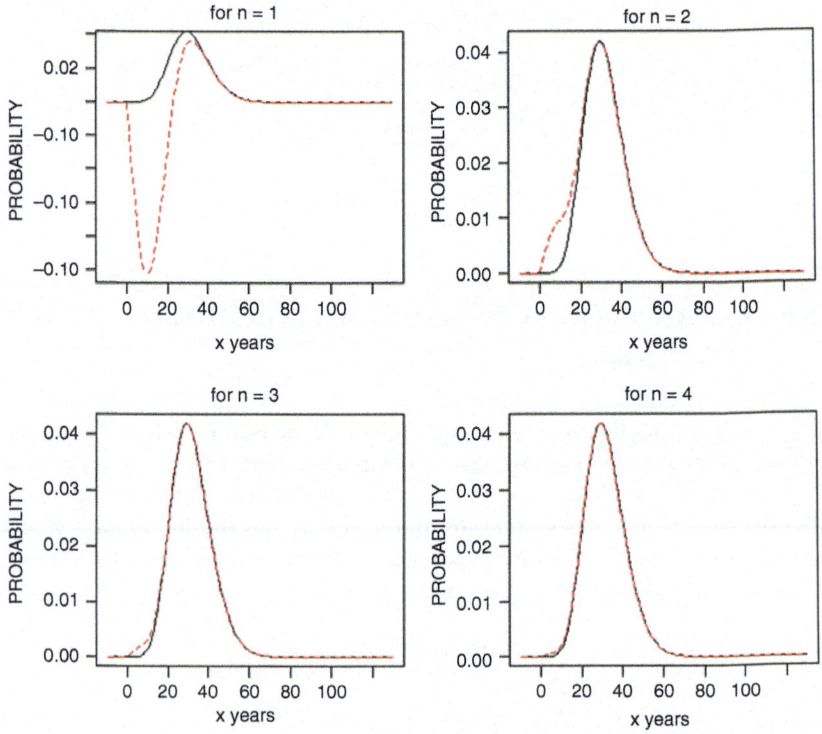

Fig. 18.5 For values $a = 3.2$, $b = 1.2$, $\theta = 0.000001$, $\xi = 0.012$, $k = 0.003$ and $\lambda = 0.000001$, the line with the black font is the pdf of BGGM distribution. The dashed line (red font) is the formula of $\frac{h'(x)}{B(\alpha,\beta)} \sum_{j=0}^{n} (-1)^j \binom{a-1}{j} e^{-h(x)(j+\beta)}$ for $n = 1, 2, 3, 4$. For the other parameters of values, the convergence is satisfactory. Graphs illustrated using R software.

is observed by the first terms of the respective sums and, when n increases, the aforementioned convergence is observed earlier in terms of the sum. The black font line represents the BGGM pdf and the red dotted line represents the sum of Eq. (18.8).

18.7 Some Statistical Measurements of BGGM Distribution

In the diagrams of Fig. 18.6 the shape of the BGGM pdf and BGGM cdf are presented as examples for specific values of parameters a, b, θ, ξ, κ and λ. In addition, Table 18.1 presents calculations of the mean, the standard deviation, and the quantiles of 25% (Q1), 50% (Q2 = median), and 75% (Q3) of the BGGM distribution, for various values of a, b, θ, ξ, κ and λ by using the equations $f(x)$ and $F(x)$. Figure 18.6 shows that even a slight parameter differentiation leads to a

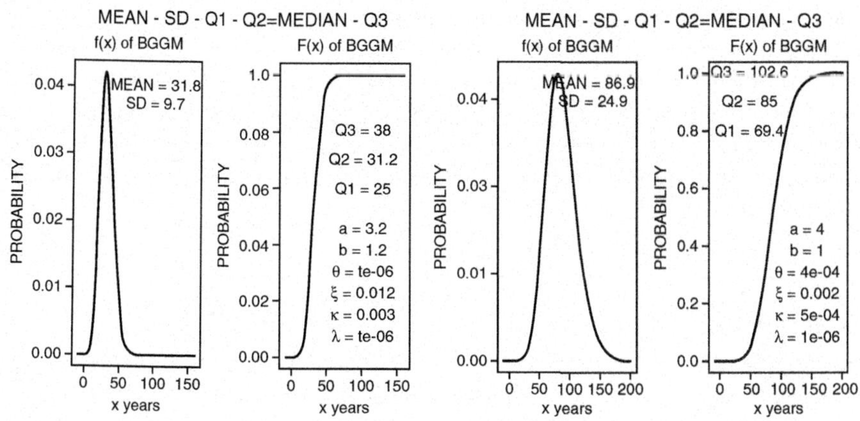

Fig. 18.6 Some statistical measurements of BGGM distribution for various values. Graphs illustrated using R software

Table 18.1 Statistical measurements of BGGM distribution for various values

θ	ξ	k	λ	a	b	MEAN	SD	Q1	Q2 (Median)	Q3
0.000001	0.012	0.003	0.000001	3.2	1.2	31.75	9.67	24.9	31.1	37.9
0.0001	0.012	0.003	0.000001	3.2	1.2	31.71	9.67	24.9	31.0	37.8
0.0004	0.002	0.0005	0.000001	3.2	1.2	76.91	23.67	60.1	75.1	91.8
0.01	0.002	0.0005	0.000001	3.2	1.2	59.61	22.41	43.5	57.7	73.7
0.01	0.002	0.0005	0.000001	4	1.2	64.16	21.83	48.6	62.4	77.8
0.01	0.002	0.0005	0.000001	4	0.6	84.29	30.25	62.7	81.3	103
0.01	0.002	0.0005	0.000001	7	2	62.66	16.22	51.3	61.6	73.0
0.01	0.002	0.0005	0.01	7	2	50.18	15.01	39.5	49.0	59.6
0.01	0.05	0.00005	0.00001	10	0.3	64.18	10.73	57.1	64.4	72.0
0.007	0.05	0.00005	0.00001	10	0.3	71.44	10.91	63.8	71.2	78.9

shape change in the BGGM pdf and BGGM cdf regarding the distribution measures of kurtosis and skewness. This conclusion is reinforced by the numerical values of Table 18.1 where even in the case of the arithmetic value change of a single parameter (i.e., the remaining 5 parameters are stable), the value of central tendency and dispersion measures are sufficiently differentiated.

18.8 Conclusions

A new generalization for the Gompertz Makeham distribution was examined which is called BGGM distribution (Beta Gompertz Generalized Makeham distribution). Some statistical properties of this distribution have been discussed and several measures of this distribution have been derived. We found important results about the asymptotic behavior of BGGM and about expansion for the density function of BGGM distribution. This work's results could be useful in the effort to describe a variety of different data sets and in forecasting their future patterns in several scientific fields, such as demography (Tragaki and Panagiotakos 2018), medicine, education, etc. With regard to extension of this work, future research could involve further exploration of various statistical properties of the proposed BGGM distribution.

References

Abramowitz, M., & Stegun, I. (2006). *Handbook of mathematical functions*. Washington, DC: US Government Printing Office. Eq, 9(3), 366.

Akinsete, A., Famoye, F., & Lee, C. (2008). The beta-Pareto distribution. *Statistics, 42*(6), 547–563.

Andreopoulos, P., & Bersimis, G. F. (2015). Mortality modelling using probability distributions. *SCinTE, 1*, 189.

Andreopoulos, P., Bersimis, G. F., Tragaki, A., & Rovolis, A. (2019). Mortality modeling using probability distributions. Application in Greek mortality data. *Communications in Statistics-Theory and Methods, 48*(1), 127–140.

Barreto-Souza, W., Santos, A. H., & Cordeiro, G. M. (2010). The beta generalized exponential distribution. *Journal of Statistical Computation and Simulation, 80*(2), 159–172.

Castellares, F., & Lemonte, A. J. (2019). On the moments of the generalized Gompertz distribution. *Applied Mathematical Modelling, 72*, 420–424.

Currie, I. D., Durban, M., & Eilers, P. H. (2004). Smoothing and forecasting mortality rates. *Statistical Modelling, 4*(4), 279–298.

Eugene, N., Lee, C., & Famoye, F. (2002). Beta-normal distribution and its applications. *Communications in Statistics-Theory and Methods, 31*(4), 497–512.

Gompertz, B. (1825). On the nature of the function expressive of the law of human mortality, and on a new mode of determining the value of life contingencies. *Philosophical Transactions of the Royal Society of London, 115*, 513–583.

Gupta, A. K., & Nadarajah, S. (2004). *Handbook of beta distribution and its applications*. Boca Raton: CRC Press.

Jones, J. H. (2005). Mathematical hazards models and model life tables, Stanford Summer Short Course, URL: http://www.stanford.edu/~jhj1/teachingdocs/Jones-mortmodel2005.pdf

Nadarajah, S., & Kotz, S. (2004). The beta Gumbel distribution. *Mathematical Problems in Engineering, 2004*(4), 323–332.

Nadarajah, S., & Kotz, S. (2006). The beta exponential distribution. *Reliability Engineering & System Safety, 91*(6), 689–697.

Papaioannou, T. (2001). On distances and measures of information: A case of diversity. *Probability and Statistical Models with applications*, 503–515.

Tragaki, A., & Panagiotakos, D. (2018). Population ageing and cardiovascular health: The case of Greece. *Hellenic Journal of Cardiology, 59*, 360–361.

Venter, G. (1983). Transformed beta and gamma distributions and aggregate losses. *Proceedings of the Casualty Actuarial Society, 70*(133 & 134), 289–308.

Zografos, K., & Balakrishnan, N. (2009). On families of beta- and generalized gamma-generated distributions and associated inference. *Statistical Methodology, 6*, 344–362.

Chapter 19
Increasing Efficiency in the EBT Algorithm

Tin Nwe Aye and Linus Carlsson

19.1 Introduction

Physiologically structured population models (PSPMs) is a natural way to study ecological population dynamics, see for example Metz and Diekmann (1986). A popular model to solve PSPMs is the Escalator Boxcar Train (EBT) method, introduced by de Roos (1988).

In order to specify PSPMs, the *vital rates*, that is, the mortality, growth and death rates for individuals are represented explicitly and depend on their physiological state $x \in \Omega$ where Ω is the set of admissible state (Metz and Diekmann 2014). We assume that the death rate, $\mu(x, E_t)$, the growth rate, $g(x, E_t)$ and the birth rate, $b(x, E_t)$ of individuals depend on the individual state x and on environment state E_t at time t. In a more general setting, the vital rates depend on the entire environment history, but in this paper, we assume that the only influence to the vital rates is the current environment. For simplicity, we will work with a one-dimensional state space which represents the individual size, and offsprings are assumed to have the same birth size x_b. In this article we will consider the case of continuous reproduction. We would like to note that some authors instead investigate pulsed reproduction in the PSPM model, see for example Hartvig et al. (2011), Kelpin et al. (2000), and Persson et al. (1998).

T. Nwe Aye (✉)
Division of Applied Mathematics, Mälardalen University, Västerås, Sweden

University of Mandalay, Mandalay, Myanmar

L. Carlsson
Division of Applied Mathematics, Mälardalen University, Västerås, Sweden
e-mail: linus.carlsson@mdh.se

C. H. Skiadas, C. Skiadas (eds.), *Demography of Population Health, Aging and Health Expenditures*, The Springer Series on Demographic Methods and Population Analysis 50, https://doi.org/10.1007/978-3-030-44695-6_19

The Escalator Boxcar Train is a numerical method designed to solve PSPMs. The basic idea of the EBT model is to follow cohorts of identical individuals throughout their entire life history and account for all life history events of importance, i.e. growth, reproduction, and death of individuals in the population. In a given ecological system, the individuals at time t are represented by a positive measure $d\zeta_t(x) = u(\xi, t)d\xi$, with initial data $d\zeta_0(x)$. The solution is then derived from a weak formulation (see e.g. Brännström et al. 2013) of the following PDE formulation for PSPMs, which is a first order, nonlinear, nonlocal hyperbolic partial differential equation with nonlocal boundary condition

$$\frac{\partial}{\partial t}u(x,t) + \frac{\partial}{\partial x}\big(g(x, E_t)u(x,t)\big) = -\mu(x, E_t)u(x,t) \tag{19.1a}$$

$$g(x_b, E_t)u(x_b, t) = \int_{x_b}^{\infty} b(\xi, E_t)u(\xi, t)d\xi \tag{19.1b}$$

$$u(x, 0) = u_0(x) \tag{19.1c}$$

where $x_b \leq x < \infty$ and $t \geq 0$. In nature, individuals are bounded in size by some number $x_{\max}$, which we will use in the following.
To begin with, the EBT model requires that the initial population is divided into a finite number of cohorts, each cohort is given an index i, where $0 \leq i \leq M$. The EBT continues, approximating the above PDE by tracking the number of individuals, $N_i(t)$, and the mean size, $X_i(t)$ in each cohort. This approximation is done by solving a system of ODEs over a time step of length, Δt, starting from the initial time and is repeated until the final time is reached. The initial distribution is approximated by a sum of Dirac measures

$$d\zeta_0 \approx d\zeta_0^M \equiv \sum_{i=0}^{M} N_i(0)\delta_{X_i(0)}$$

where δ_x is the Dirac measure concentrated at x, see Sect. 19.2 for a more detailed explanation. The approximate solution, $d\zeta_t^M$ is given by the solution of a system of ODEs. The solution $d\zeta_t^M$ converges linearly (see Gwiazda et al. 2014) to $d\zeta_t$ with respect to time steps and the initial approximation.

This paper proves the convergence of a modified EBT model. In addition, we also prove that the computational time of this modified EBT model is proportional to the number of time steps. In the original formulation of the EBT method, the computational time is proportional to the square of the number of time steps. The convergence of the Escalator Boxcar Train has been built in a series of papers, see for examples; Metz and de Roos (1991), Brännström et al. (2013) and Kropielnicka et al. (2018).

This article is outlined as follows: In Sect. 19.2, we present the Escalator Boxcar Train method, including our proposed non-linear merging procedure. In Sect. 19.3, we show that the number of cohorts are bounded when using a general class of merging procedures. In Sect. 19.4, we discuss convergence for a general class of EBT models when allowing merging of internal cohorts. This is proved in two steps, first we prove that the convergence is valid for each time step, with convergence speed $O(\Delta t^2)$, which in turn gives convergence speed of $O(\Delta t)$ for arbitrary long simulation times. In Sect. 19.5, we end with a simulation using the EBT method on a Daphnia colony, where one can see that using the non-linear merging procedure proposed in Sect. 19.4 provides a better result than the naïve linear merging method.

19.2 General Description of the Escalator Boxcar Train

We start with a brief description of the EBT method. We assume that the initial population density ζ_0 is contained in the interval $[x_b, x_{max})$ which we divide up into small intervals $\Omega_i = [x_{i-1}, x_i)$, $i = 1, 2, \ldots, M$ where $x_i = x_b + i\Delta x$, $i = 0, 1, 2, \ldots, M$, $\Delta x > 0$ is a constant, and $x_M = x_{\max}$. The total number of individuals in cohort i is written $N_i(t)$ with sizes in Ω_i. As the sizes of individuals in each cohort are not completely identical, the mean state will be used and is denoted by $X_i(t)$. We divide the initial population density into groups, which are characterized by pairs $(X_i(t), N_i(t))$. Thus, the initial values could be defined as

$$N_i(0) = \int_{x_{l-1}}^{x_i} d\zeta(x)$$

$$X_i(0) = \frac{1}{N_i(0)} \int_{x_{i-1}}^{x_i} x d\zeta(x) \tag{19.2}$$

for $i = 1, 2, \ldots, M$ and we set $N_0(0) = 0$ and $X_0(0) = x_b$.

Once individuals are allocated in the particular cohort, they remain in the same cohort till the moment of death. Evolution of the cohort's characteristics is described by ODEs (see Sect. 19.2.1), in particular, the amount of individuals, $N_i(t)$, changes its value due to the process of birth or death, while the size, $X_i(t)$, evolves by *e.g.* nutrient resources. The cohort, i, with the smallest mean size, that is $X_0(t)$ is called the *boundary cohort*, all other cohorts are called *internal cohorts*.

19.2.1 The Dynamics of Internal and Boundary Cohorts

The equations composing the EBT scheme are as follows. The dynamics of the internal cohorts are derived as

$$\frac{dN_i}{dt} = -\mu(X_i, E_t)N_i(t) \tag{19.3}$$

$$\frac{dX_i}{dt} = g(X_i, E_t) \tag{19.4}$$

where $i = 1, 2, \ldots, M$.

In this paper, we assume that the mortality function and growth rate function has derivatives that are uniformly continuous.

Using the variant of the EBT model given in Brännström et al. (2013), the boundary cohort follows the same dynamics as the internal cohorts, with the additional term of the total population fecundity, $\sum_{i=0}^{M} b(X_i, E_t)N_i(t)$. The dynamics of the boundary cohort is thus

$$\frac{dN_0}{dt} = -\mu(X_0, E_t)N_0(t) + \sum_{i=0}^{M} b(X_i, E_t)N_i(t) \tag{19.5}$$

$$\frac{dX_0}{dt} = g(X_0, E_t) \tag{19.6}$$

The system of equations, (19.3)–(19.6) with initial data (19.2) together with the environmental ODE, is solved from the starting time, $t = 0$ to the final time $t = t_{\max}$. This is done numerically by taking the partition $t_i = i\Delta t$, and solve the system from t_0 to t_1, and then from t_i to t_{i+1}, $i = 1, 2, \ldots$ using the solution from the previous step as initial data until the final time is reached. The environmental ODE is specified by the eco-system that is modeled.

In the following, we refer to cohorts that contain a small number[1] of individuals, as *a small cohort*.

19.2.2 Process of Internalizing the Boundary Cohort

The number and size of individuals in the boundary cohort increase in the course of time due to the reproduction of individuals and the environment. As a consequence, an inapplicable large approximation error will occur unless something is done about it. Therefore, we need to introduce a new boundary cohort at a suitable time.
At each time step, if the current boundary cohort's density is zero, we reset the size of individuals in the boundary cohort to x_b, otherwise, a new boundary cohort is introduced, still with index 0. The old boundary cohort is transformed into an internal cohort. When a boundary cohort is internalized, it is given the index 1 and all old internal cohorts indices are increased by 1. For this reason, the number of internal cohorts will increase[2] over time, which is inconvenient for computational purposes. In fact, the computational cost soon becomes too high, because only small simulations can be done. To overcome an increase of internal cohorts, we introduce a merging procedure for small internal cohorts.

[1]By "a small number" we mean, depending on the model, that the number of individuals in that cohort is less than ten per mille of the total number of individuals in the population.

[2]This means that the number M in Sect. 19.2.1 changes depending on time.

19.2.3 Merging Procedure of Internal Cohorts

Accounting for internalization, the classical EBT scheme gives rise to an expanding system of ODEs, resulting in a polynomial calculation time with respect to the total number of time steps. To overcome this problem, different authors solve this by simply deleting small cohorts (Persson et al. 1998; Claessen et al. 2002; Zhang et al. 2017; Rinke and Vijverberg 2005; Pascual and Caswell 1997; Tuljapurkar and Caswell 2012) or merging small cohorts in a linear way (Kelpin et al. 2000). Both these approaches result in an uncontrollable error compared to the analytical solution. In contrast, we propose a procedure of merging internal cohorts in order to reduce the system of ODEs, while keeping the order of convergence. We will consider a merging procedure when the number of individuals in an internal cohort falls below a certain threshold, $N_{\min}$, this is checked for all cohorts in each time step. When this is the case, merging is conducted of this cohort with a cohort of *similar size*, meaning that the nearest cohort in size should be closer than some threshold value called, Δx_{merge}. This is done in such a way, so that the expected number of offsprings for the merged cohort stays the same as if no merging was performed, see Sect. 19.4. If there is no cohort with similar size, merging is not conducted.

Remark 1 The total number of cohorts, M, is not a constant, due to internalization of the boundary cohorts, and merging of small internal cohorts.

Remark 2 Note that the constant Δx defined in Sect. 19.2 is the size of the initial cohorts, whereas Δx_{merge} is the threshold value of the difference between mean sizes of two cohorts, that is, we allow merging at time t of two cohorts a and b provided that $|X_a(t) - X_b(t)| < \Delta x_{\text{merge}}$.

Remark 3 When merging any two internal cohorts of similar size, the convergence rate will stay the same as the original EBT formulation has when no merging is allowed.

19.3 Boundedness of Cohorts When Merging Is Allowed

In this section we show that the number of cohorts is bounded under the following natural assumptions

(i) The size of individuals are bounded from below by x_b and from above by $x_{\max}$. We denote the minimum weight of an individual by m_{x_b}.
(ii) We assume that the total biomass of individuals is bounded by T_b.
(iii) There are numbers $N_{\min} > 0$ and $\Delta x_{\text{merge}} > 0$ such that, at each time step, merging of internal cohort a and internal cohort b is performed if $N_a(t) < N_{\min}$ and that the closest cohort to a is cohort b satisfying

$$|X_a(t) - X_b(t)| < \Delta x_{\text{merge}}$$

Theorem 1 *Under assumptions (i)–(iii) above, the total number of cohorts in the EBT model is bounded by a constant.*

Proof. From Assumption (i) and (iii) it follows that the amount of small cohorts, M_s, must satisfy

$$M_s(t) \leq \frac{x_{\max} - x_b}{\Delta x_{\text{merge}}}$$

Denoting the number of large cohorts by M_l, and using assumptions (i)–(iii) gives

$$M_l(t)\, m_{x_b}\, N_{\min} \leq m_{x_b} \sum_{\text{largecohorts}} N_i(t) \leq T_b$$

Thus, the total number of cohorts are bounded by

$$M(t) = M_s(t) + M_l(t) \leq \frac{x_{\max} - x_b}{\Delta x_{\text{merge}}} + \frac{T_b}{m_{x_b}\, N_{\min}}$$

Thus for all times t, the total amount of cohorts are bounded by a constant.

19.4 Convergence of Merging Internal Cohorts

In PSPMs, different kind of reproduction functions appear depending on what species is modelled (see e.g. De Roos et al. 1992; Claessen et al. 2002). We show that for a general class of reproduction functions, the convergence rate of the EBT model is not affected when merging is done, using a specified weighted mean dependent on the reproduction function. To proceed, we consider two cohorts $(X_a(t), N_a(t))$ and $(X_b(t), N_b(t))$, with similar mean size. In this chapter, we use the assumptions (i)–(iii) given in Sect. 19.3. For notational purpose, we denote the initial data with a sub-index zero, e.g., $X_{a_0} = X_a(0)$, $N_{a_0} = N_a(0)$ and so on.

The dynamics of the number of individuals and mean size in the internal cohort is defined by Equations (19.3) and (19.4).

The class of models we consider, have the reproduction rate per individual of the form

$$b(X, F) = \begin{cases} c(F)\, X^n & \text{if } X > x_j \\ 0 & \text{(otherwise)} \end{cases} \tag{19.7}$$

where $b(X, F)$ denotes the birth rate of adult individuals, $n \geq 1$ is a constant, $c(F) > 0$ is a bounded function depending on the food density F. This class of reproduction models cover for example the models used in Nwe Aye and Calrsson (2017) and Ryabov et al. (2017).

When merging the two cohorts $(X_a(t), N_a(t))$ and $(X_b(t), N_b(t))$ into one cohort $(X_m(t), N_m(t))$ at time t, we naturally add the number of individuals in both cohorts, *i.e.*

$$N_m(t) = N_a(t) + N_b(t) \tag{19.8}$$

And to get the expected number of offsprings, we use Equation (19.7) to initialize the merged cohort size as

$$X_m = \left(\frac{N_a(t)X_a^n(t) + N_b(t)X_b^n(t)}{N_a(t) + N_b(t)} \right)^{\frac{1}{n}} \tag{19.9}$$

For notational purpose we denote $b_m(t) = b(X_m(t), F(t))$ for the merged cohort and $b_w(t) = b(X_a(t), F(t)) + b(X_b(t), F(t))$ for the sum of the two original cohorts.

Remark 4 If n is replaced by 1 in Equation (19.9), the merging is called *linear*.

Theorem 2 *Under the above assumptions we get*

$$b_m(t + \Delta t) = b_w(t + \Delta t) + O(\Delta x_{\text{merge}} \cdot \Delta t)$$

Here b_m is the resulting cohort when merging was done at time t.

The proof relies on the fact that $(s + k\xi)^r = s^r + O(\xi)$ and Taylor's expansion in Δt. Due to the length and technical matter, the proof is postponed to Appendix B.

As an immediate consequence we get

Corollary 1 *If the maximum difference in sizes when merging two cohorts satisfies $\Delta x_{\max} = O(\Delta t)$ in the EBT model, then*

$$b_m(t + \Delta t) = b_w(t + \Delta t) + O(\Delta t^2)$$

Theorem 3 *Under the same assumptions of Theorem 2 and Corollary 1, it follows that*

$$b_m(s) = b_w(s) + O(\Delta t)$$

for each $s > t$. Here b_m is the resulting cohort when merging was done at time t.

The proof follows the same ideas as the proof of Theorem 2. We first use Corollary 1 for one small time step, with $O(\Delta t^2)$ convergence rate, then add time steps together, loosing one order of convergence. This is done by the principle of mathematical induction. The proof is postponed to Appendix C.

19.5 Simulation of the Daphnia Model

In this section, we model the life of Daphnia's. In the simulations we show that the best result, when allowing merging, is gained when using the merging proposed in Sect. 19.4, cf. Equation (19.10). In particular, this non-linear merging is shown to produce better results than the naïve way of linear merging or simply deleting small cohorts. The Daphnia pulex is a species of the Daphnia, known as the water flea because of its unique swimming style. The behaviors of Daphnia can be seen in a series of papers, see for example; Ebert (2005), Dodson and Ramcharan (1991), and Hebert (1978).

19.5.1 The Daphnia's Model Specification

Daphnia behavior is extremely influenced by the size of an individual. An important assumption for this model is that Daphnia individuals can shrink when they can't get enough food, see de Roos (1988). We denote the length of an individual Daphnia by x and the food density by F. The reproduction of adult Daphnia is described by the function

$$b(x, F) = \begin{cases} r_{max} x^2 \frac{F}{f_h + F} & \text{if } x > x_j \\ 0 & \text{if } x \leqslant x_j \end{cases} \tag{19.10}$$

where $b(x, F)$ denotes the birth rate of adult Daphnia per unit of time, f_h is the half-saturation food density, x_j is the barrier for when an individual transits from juvenile to adult, and r_{max} is the maximum reproduction rate per unit of surface area.

When this is solved with the EBT model, we denote the mean size of individuals in cohort i by X_i and the number of individuals by N_i. Comparing Equation (19.10) with the merging procedure stated in Sect. 19.4, the merging of two cohorts $(X_a(t), N_a(t))$ and $(X_b(t), N_b(t))$ into one will be done by using Equation (19.8), thus

$$X_m(t) = \left(\frac{N_a(t) X_a^2(t) + N_b(t) X_b^2(t)}{N_a(t) + N_b(t)} \right)^{1/2} \tag{19.11}$$

which we will call *square-square root* merging.

This will later be compared to the *linear* merging, given by

$$X_m(t) = \frac{N_a(t) X_a(t) + N_b(t) X_b(t)}{N_a(t) + N_b(t)} \tag{19.12}$$

The Daphnia model assumes a constant mortality rate

$$\mu(x, F) = \mu$$

The growth rate of Daphnia is represented by

$$\frac{dx}{dt} = r_g \left(x_{max} \frac{F}{f_h + F} - x(t) \right)$$

where r_g is the growth rate constant and x_{max} is the maximum length that Daphnia reaches under abundance of food.

The feeding rate, I, of an individual Daphnia with length x at food density F can be calculated by:

$$I(x, F) = \nu x^2 \frac{F}{F_h + F}$$

The main nutrition for the Daphnia pulex is the algae Chlamydomonas rheinhardii. Therefore, the food density will be used as the environmental factor for this EBT model. The food density equation is:

$$\frac{dF}{dt} = \rho(K - F(t)) - \sum_{i=0}^{M} I(X_i(t), F(t))N_i(t)$$

We will use the EBT model, with the constants in the simulation, as given in Table 19.1.

Table 19.1 Interpretation of constants used for the Daphnia's life history is explained by e.g. De Roos and Persson, see Kooijman (1986), Kooijman and Metz (1984). The values are developed from practical experiments. Milligram of carbon (mgC), millimeter (mm), liter(L), and day (d) are used in units

Symbol	Value	Unit	Interpretation
ν	0.007	mgC/mm^2	Maximum ingestion rate per surface area
f_h	0.164	mgC/L	Half saturation food density
x_b	0.6	mm	Length at birth
x_j	1.4	mm	Length at maturity
x_{max}	3.5	mm	Maximum length
r_g	0.11	d^{-1}	Growth rate
r_{max}	1.0	mm^2	Maximum reproduction rate
μ	0.05	d^{-1}	Mortality rate
ρ	0.5	d^{-1}	Resource regrowth rate
k	0.25	mgC/L	Maximum resource density

19.5.2 Comparing Results Between Linear and Non-linear Merging

The original EBT method, adapted to the Daphnia's model solves a system of ODEs in each time step. We have run the simulations in Matlab where we solve the ODEs using the built in package ode45, where we have used the standard options with relative error of 10^{-5}.

When internalizing the boundary cohort, the system of ODEs is increased. We modify the original EBT method, to include merging[3] of internal cohorts, this reduces the number of ODEs, so that the maximum number of cohorts is bounded (cf. Theorem 1), which result in shorter simulation time. After each time step, the boundary cohort is internalized and merging is carried out.

The merging is done by checking whether the cohort with the smallest amount of individuals is less than N_{min}, and if this is the case, we find the closest cohort to this one and merge if the distance is less than Δx_{merge}. In this case we merge the two cohorts according to Equations (19.8) and (19.11). This is repeated until no more cohort satisfy these criterions. To demonstrate that the square-square root merging is the merging procedure to use, we also run simulation with linear merging using Equation (19.12) as well as discarding small cohorts. We then compare all these data with simulations of the EBT model without any merging or discarding small cohorts.

The EBT model for the Daphnia population is very robust, so to show that the square-square root merging outperforms the linear merging, we choose time steps ranging from one eighth of a day to one day.

The results in Table 19.2 present a convincing evidence that the square-square root merging is the right merging to use. In this table the relative error is smaller for the square-square root method compared to linear merging for all different time steps. We have used a moving average of 50 days,[4] this to avoid the time lag that occurs for any model involving merging or discarding of cohorts.

We also find that when using merging, compared to keeping all internal cohorts, the simulation time is decreased from quadratic to linear dependence on the total number of time steps, see Appendix A Tables 19.3, 19.4, 19.5, 19.6, 19.7, 19.8, and 19.9 (see also Nwe Aye and Calrsson 2017 for a similar table of time consumption).

[3]Some modellers discard "negligible" cohorts, but because of the non-linearity of the EBT model, this might give rise to an arbitrarily large error in the solution. We later see that this indeed is the case.

[4]The absolute value of the relative change of the moving average over 50 days. It is computed for j_s as

$$j_s = \frac{|\sum_{i=0}^{49}(m_{j_s}(t - i\Delta t) - m_j(t - i\Delta t))|}{\sum_{i=0}^{49} m_j(t - i\Delta t)}$$

where $m_j(t)$ is the biomass of juvenile individual at time t for non-merging simulation and m_{j_s} for square-square root merging.

Table 19.2 Data of the absolute value of the relative change for biomass of juvenile (j), adult (a) and total biomass (tb) with respect to the non-merging EBT model values for j, a, and tb. The rows correspond to different time steps. The indices s and l stands for square-square root respectively linear merging. The simulation was run for 3000 days. The biomass was compared on the last time step of the simulations

Time step (days)	j_l(‰)	j_s(‰)	a_l(‰)	a_s(‰)	tb_l(‰)	tb_s(‰)
1	4.865	3.414	12.417	1.399	3.345	2.991
1/2	3.426	2.637	25.297	19.688	0.242	0.213
1/4	0.017	0.001	0.051	0.014	0.009	0.003
1/8	0.215	0.150	1.943	1.700	0.062	0.019

The constants which are used in the simulation are; $\Delta x_{\text{merge}} = 0.4$ mm, $N_{\min} = 0.01$, and the initial populations was divided up into $M = 32$ cohorts, and the simulation is assumed to be in a unit volume tank with unit surface area.[5]

This resulted in 6133 cohorts when no merging was allowed, and dropped down to 203 cohorts when we allowed merging.

More explanations and simulation results is given in Appendix A, where also the underlying data is provided.

By comparing the values of biomass in the above tables, the relative error between the biomass for non-merging cohorts and linear merging cohorts is bigger than the relative error between the biomass for non-merging cohorts and square-square root merging cohorts. This collaborates the results in Sect. 19.4, and hence, when modelling Daphnia's, the square-square root merging of cohorts is the correct merging to use.

19.6 Results and Discussion

The Escalator Boxcar Train is a powerful and favorable mathematical model to use when solving physiologically structured population models and is frequently used by biologists and the research community. The major disadvantage of the EBT model is that in its original formulation the number of ODEs to solve increases at each internalization points. To ensure that the solution from EBT method does not diverge from the exact solution of the PDE formulation of PSPMs, these internalizations have to be done in sufficiently small time steps. This makes longer simulations impossible.

To overcome this disadvantage, different authors suggest that small cohorts are either discarded or merged with other cohorts in a linear way. Both these methods result in errors which can not be controlled. We suggest that merging should be done in such a way, so that the expected number of offsprings stay the same as if

[5]With these chosen constants and vital rate functions, all the assumptions and hypothesis in Theorems 1 and 3 are fulfilled

no merging was done. Using this idea, we have proved that the computational time is reduced from quadratic to linear time when this merging procedure is used, at the same time the error of the solutions is not affected compared to the exact solution to the PDE formulation of PSPMs.

Even though the linear merging provides similar convergence rate in the numerical simulations as the square square-root merging, it has not to our knowledge, been proved to possess the same convergence rate in general EBT models. For example, this might result in wrong solutions when studying catastrophic behavior in a population. We, therefore, suggest that the non-linear merging should be used when the EBT model is used.

The convergence results for the non-linear merging has been shown analytically in Sects. 19.3 and 19.4 and collaborated numerically in Sect. 19.5. Theorems 2 and 3 are proved under the assumption that the birth rate is given in the form of Equation (19.7). It should be noted that the proofs of these theorems can be modified to EBT models in which the birth rates are given by more general functions, provided that the birth rates are smooth enough.

Appendices

A Detailed Description and Numerical Results of Simulations of Daphnia

In this section, we recount results from our simulations of a Daphnia colony. In Sect. 19.5.2, we used these results to show that our proposed square-square root merging-procedure produces the best results as well as it reduces the computational time from quadratic to linear proportional to the number of time steps.

The simulations for a Daphnia colony was done by using the original EBT model by de Roos (1988) with the boundary dynamics given in Brännström et al. (2013). Table 19.3 shows simulations without any merging, which we will use as "the correct solutions". The data in Table 19.4 (linear merging) and Table 19.5 (square-square root merging) are the results when we used merging in our simulations, see Sect. 19.5 Equations (19.11) and (19.12). Each simulation was ran over a total of 3000 days with different time steps[6] (first column) and the second column shows the elapsed time.[7] In the third column, we present the maximum number of internal cohorts in the simulation. Column 4–6 show the biomass average taken over the last 50 days[8] of the simulation for juveniles, adults, total biomass, respectively.

[6]Due to Theorem 3 we let $\Delta x_{\text{merge}} = 0.4\Delta t$ for each different time step Δt.

[7]We used a MacBook Pro 2.5 GHz intel core i5 with 4Gb of ram memory and Matlab R2015b to simulate the EBT model for Daphnia.

[8]The reason why we used an average over 50 days is that this time span covers at least two full generations of Daphnia.

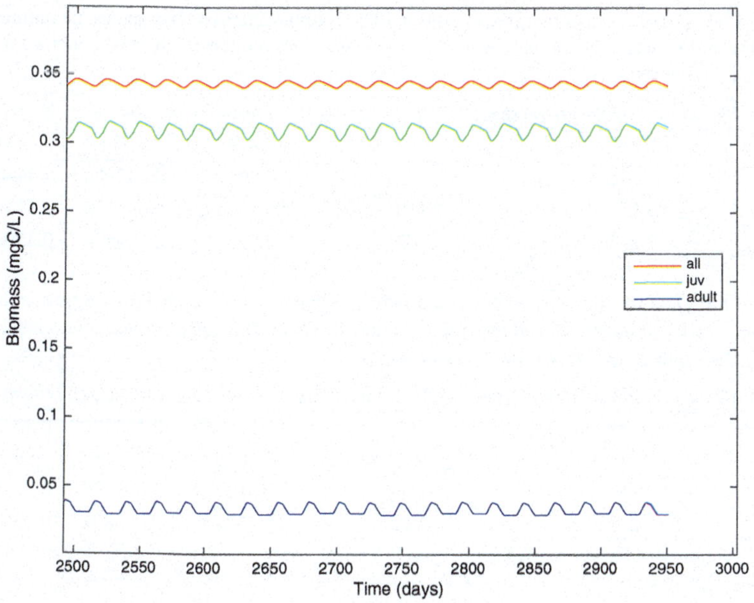

Fig. 19.1 A graphical representation of a simulation with merging of cohorts. It represents the moving average of juvenile biomass, mature biomass and total biomass of juvenile and adult. The moving average was taken over 50 days

Table 19.3 Data for non-merging of cohorts. The column, Elapsed time, shows the running time for simulations, and the column, Internal cohorts, the total number of internal cohorts when no merging is conducted

Time step(days)	Elapsed time(seconds)	Internal cohorts	$j(mgC/L)$	$a(mgC/L)$	$tb(mgC/L)$
1	52	1367	0.314052	0.030286	0.344338
1/2	109	1681	0.300732	0.044019	0.344750
1/4	398	3165	0.319454	0.044176	0.363629
1/8	1864	6133	0.325488	0.024779	0.350268

To see that the square-square root outperforms the linear merging with respect to convergence, we present the results from Tables 19.3, 19.4, and 19.5 in Sect. 19.5 with absolute relative change[9] of the biomasses of linear and square-square root merging compared to the results for the non-merging simulation (Fig. 19.1).

[9]The absolute relative change for the biomass of linear merging compared to non-merging is calculated by

$$\text{relative change} = \frac{|\text{biomass of non}-\text{merging} - \text{biomass of linear merging}|}{\text{biomass of non}-\text{merging}}$$

Table 19.4 Data for linear merging of cohorts. The column, Elapsed time, shows the running time for simulations, and the column, Internal cohorts, the total number of internal cohorts for linear merging of cohorts

Time step(days)	Elapsed time(seconds)	Internal cohorts	$j(mgC/L)$	$a(mgC/L)$	$tb(mgC/L)$
1	28	56	0.312524	0.030662	0.343186
1/2	51	72	0.301762	0.042905	0.344667
1/4	101	119	0.319449	0.044178	0.363627
1/8	214	203	0.325419	0.024828	0.350246

Table 19.5 Data for square-square root merging of cohorts. The column, Elapsed time, shows the running time for simulations, and the column, Internal cohorts, the total number of internal cohorts, when using square-square root merging of cohorts

Time step(days)	Elapsed time(seconds)	Internal cohorts	$j(mgC/L)$	$a(mgC/L)$	$tb(mgC/L)$
1	28	56	0.312979	0.030328	0.343308
1/2	46	72	0.301525	0.043152	0.344677
1/4	91	119	0.319454	0.044175	0.363629
1/8	219	203	0.325439	0.024821	0.350261

We make the ansatz that the running time, T, for the simulations given in Tables 19.3, 19.4, and 19.5 is given on the form $T(\Delta t) = C(\Delta t)^p$, where C and p are constants.

For the non-merging procedure, the best time consumption function of the above form is given by the constants $C = 19.98$ and $p = -2.181$ The 95% confidence intervals for C is (4.084, 35.89) and for p the confidence interval is (−2.569, −1.793). The time consumption is proportional to the square of the number of time steps in the simulation when merging is not allowed. The simulations run with one of the merging procedures gives the best time consumption function with the constants $C = 24.63$ and $p = -1.038$ with 95% confidence intervals (18.23, 31.04) for C and (−1.173, −0.9021) for p, that is, the time consumption is directly proportional to the number of time steps when merging is allowed. We calculated above these values, using the built in function fittype and fit in Matlab.

In Tables 19.6 and 19.7, simulation was taken over a period of 300 days with different time steps. Note that the elapsed time when $\Delta t = 1/128$ is very large, we believe that this is caused by memory constraints in the computer.

In Tables 19.8 and 19.9, the simulation was taken 1 day for each time step to run for different simulation length.

B Proof of Theorem 2

We will, for reading convenient, write Δx instead of Δx_{merge} in the proofs given in Appendices B and C.

Table 19.6 Simulation of a Daphnia colony using the original EBT method, over a period of 300 days without any merging procedure. The best time consumption function of the form $T(\Delta t) = C(\Delta t)^p$ is given by the constants $C = 0.0067$ and $p = -2.759$ with 95% confidence intervals (0.0037, 0.0097) for C and (−2.85, −2.667) for p. When the last row in the above data is excluded, the value for p is −2.264 with a 95% confidence interval (−2.425, −2.103)

Time step	Elapsed time	Internal cohort
1	2	137
1/2	3.4	203
1/4	6.7	363
1/8	13.6	685
1/16	35.3	1330
1/32	127.5	2617
1/64	634.4	5194
1/128	4369.5	10346

Table 19.7 Simulation of a Daphnia colony using the original EBT method, over a period of 300 days with merging procedure. The best time consumption function of the form $T(\Delta t) = C(\Delta t)^p$ is given by the constants $C = 0.5277$ and $p = -1.34$ with 95% confidence intervals (0.3994, 0.6559) for C and (−1.391, −1.288) for p

Time step	Elapsed time	Internal cohort
1	2	55
1/2	3.2	72
1/4	6.1	118
1/8	11.6	203
1/16	23.9	360
1/32	51.2	568
1/64	139.2	957
1/128	351	1376

Table 19.8 Data for non-merging of cohorts, i.e. using the original EBT formulation. The column, Elapsed time, shows the running time for simulations, and the column, Internal cohorts, the total number of internal cohorts

Simulation length (days)	Elapsed time (seconds)	Internal cohorts
100	0.7	66
200	1.4	95
400	2.5	181
800	5	366
1600	11.2	732
3200	29.5	1462
6400	110	2922

The proof below shows that the merged cohort $(X_m(t), N_m(t))$ differ at most with $C\Delta t^2$ compared to the number of offsprings from two internal cohorts $(X_a(t), N_a(t))$ and $(X_b(t), N_b(t))$ over a small time step Δt, where C is a constant.

Table 19.9 Results from simulation, using square-square root merging of cohorts, i.e. our proposed modified version of the original EBT formulation. The column, Elapsed time, shows the running time for simulations, and the column, Internal cohorts, the total number of internal cohorts

Simulation length (days)	Elapsed time(seconds)	Internal cohorts
100	0.8	55
200	1.4	55
1000	5.5	56
2000	11	56
4000	21	56
8000	41	56

Proof. *(Proof of Theorem* 2 *in Sect.* 19.4). The following proof relies heavily on the fact that

$$(s + k\xi)^r = s^r + O(\xi) \tag{19.13}$$

In the case $n = 1$, Equations (19.7) and (19.9) is linear, and the proof becomes trivial.

For $n > 1$, we get an expression for the newborn individuals of non-merging cohorts, $b_w = b(X_a, F) + b(X_b, F)$, by using Taylor's expansion in Δt.

$$\begin{aligned} b_w(t + \Delta t) &= b_w(t) + b'_w(t)\Delta t + O(\Delta t^2) \\ &= c(N_a X_a^n + N_b X_b^n) \\ &\quad + c(N'_a X_a^n + nN_a X_a^{n-1} X'_a + N'_b X_b^n + nN_b X_b^{n-1} X'_b)\Delta t \\ &\quad + O(\Delta t^2) \end{aligned}$$

where Equation (19.7) has been used. In the following, we sometimes suppress the variable t. Using Equations (19.3) and (19.4) in the last equality gives

$$\begin{aligned} &b_w(t + \Delta t) \\ &= c(N_a X_a^n + N_b X_b^n) \\ &\quad + c\left(-\mu N_a X_a^n + nN_a X_a^{n-1} g(X_a, E_t)\right) \Delta t \\ &\quad + c\left(-\mu N_b X_b^n + nN_b X_b^{n-1} g(X_b, E_t)\right) \Delta t + O(\Delta t^2) \\ &= c(N_a X_a^n + N_b X_b^n) - \mu c\left(N_a X_a^n + N_b X_b^n\right) \Delta t \\ &\quad + nc\left(N_a X_a^{n-1} g(X_a, E_t) + N_b X_b^{n-1} g(X_b, E_t)\right) \Delta t + O(\Delta t^2) \end{aligned} \tag{19.14}$$

When we merge cohorts, using similar calculations as above, we get the following expression for the newborn individuals,

$$
\begin{aligned}
& b_m(t + \Delta t) \\
&= b_m(t) + b'_m(t)\Delta t + O(\Delta t^2) \\
&= cN_m X_m^n + c(N'_m X_m^n + nN_m X_m^{n-1} X'_m)\Delta t + O(\Delta t^2)
\end{aligned}
$$

We substitute Equations (19.3) and (19.4) in the above equation.

$$
\begin{aligned}
& b_m(t + \Delta t) \\
&= cN_m X_m^n + c(-\mu N_m X_m^n + nN_m X_m^{n-1} g(X_m, E_t))\Delta t + O(\Delta t^2)
\end{aligned}
$$

To proceed, we substitute Equation (19.9) in the above equation.

$$
\begin{aligned}
& b_m(t + \Delta t) \\
&= cN_m \frac{N_a X_a^n + N_b X_b^n}{N_a + N_b} - c\mu N_m \frac{N_a X_a^n + N_b X_b^n}{N_a + N_b} \Delta t \\
&\quad + cnN_m \frac{(N_a X_a^n + N_b X_b^n)^{1-\frac{1}{n}}}{(N_a + N_b)^{1-\frac{1}{n}}} g(X_m, E_t)\Delta t + O(\Delta t^2)
\end{aligned}
$$

Since $N_m = N_a + N_b$ (Equation (19.8)), we get

$$
\begin{aligned}
& b_m(t + \Delta t) \\
&= c(N_a X_a^n + N_b X_b^n) - c\mu(N_a X_a^n + N_b X_b^n)\Delta t \\
&\quad + cn(N_a X_a^n + N_b X_b^n)^{1-\frac{1}{n}} (N_a + N_b)^{\frac{1}{n}} g(X_m, E_t)\Delta t + O(\Delta t^2)
\end{aligned}
\tag{19.15}
$$

To show that Equation (19.15) converges to Equation (19.14) we use Taylor expansions

$$
g(X_m, E_t) = g(X_a, E_t) + O(\Delta x) \tag{19.16}
$$

and

$$
g(X_b, E_t) = g(X_a, E_t) + O(\Delta x) \tag{19.17}
$$

where $X_b = X_a + \Delta x$.

Subtracting Equations (19.15) from (19.14) gives

$$
\begin{aligned}
b_w(t+\Delta t) - b_m(t+\Delta t) &= nc\bigg(N_a X_a^{n-1} g(X_a, E_t) + N_b X_b^{n-1} g(X_b, E_t) \\
&\quad - (N_a X_a^n + N_b X_b^n)^{1-\frac{1}{n}} (N_a + N_b)^{\frac{1}{n}} g(X_m, E_t)\bigg)\Delta t \\
&\quad + O(\Delta t^2) \\
&= nc\bigg(N_a X_a^{n-1} g(X_a, E_t) + N_b X_b^{n-1} g(X_b, E_t) \\
&\quad - (N_a X_a^{n-1} + N_b X_a^{n-1}) g(X_m, E_t) \\
&\quad + (N_a X_a^{n-1} + N_b X_a^{n-1}) g(X_m, E_t) \\
&\quad - (N_a X_a^n + N_b X_b^n)^{1-\frac{1}{n}} (N_a + N_b)^{\frac{1}{n}} g(X_m, E_t)\bigg)\Delta t \\
&\quad + O(\Delta t^2)
\end{aligned}
$$

By using Equations (19.16) and (19.17) in the above, we get

$$
\begin{aligned}
& b_w(t+\Delta t) - b_m(t+\Delta t) \\
& = nc\bigg(N_a X_a^{n-1} g(X_a, E_t) + N_b X_b^{n-1}(g(X_a, E_t) + O(\Delta x)) \\
& - (N_a X_a^{n-1} + N_b X_a^{n-1})(g(X_a, E_t) + O(\Delta x))\bigg)\Delta t \\
& + nc\bigg((N_a X_a^{n-1} + N_b X_a^{n-1}) g(X_m, E_t) \\
& - (N_a X_a^n + N_b X_b^n)^{1-\frac{1}{n}} (N_a + N_b)^{\frac{1}{n}} g(X_m, E_t)\bigg)\Delta t \\
& + O(\Delta t^2)
\end{aligned}
$$

which simplifies to

$$
\begin{aligned}
& b_w(t+\Delta t) - b_m(t+\Delta t) \\
& = nc\bigg((N_a X_a^{n-1} + N_b X_b^{n-1}) - (N_a X_a^{n-1} + N_b X_a^{n-1})\bigg) g(X_a, E_t)\Delta t \\
& \quad + nc\bigg((N_a X_a^{n-1} + N_b X_a^{n-1}) - (N_a X_a^n + N_b X_b^n)^{1-\frac{1}{n}} (N_a + N_b)^{\frac{1}{n}}\bigg) g(X_m, E_t)\Delta t \\
& \quad + O(\Delta x \cdot \Delta t) + O(\Delta t^2)
\end{aligned}
\tag{19.18}
$$

Studying the first term of the right hand side in Equation (19.18), and using $X_b = X_a + \Delta x$, we get

$$N_a X_a^{n-1} + N_b X_b^{n-1} = N_a X_a^{n-1} + N_b (X_a + \Delta x)^{n-1}$$
$$= N_a X_a^{n-1} + N_b X_a^{n-1} (1 + \frac{\Delta x}{X_a})^{n-1}$$

By using the fact that $(s + k\xi)^r = s^r + O(\xi)$, the above expression becomes

$$(N_a X_a^{n-1} + N_b X_b^{n-1}) - (N_a X_a^{n-1} + N_b X_a^{n-1}) = O(\Delta x) \qquad (19.19)$$

The second term in Equation (19.18) is evaluated as

$$(N_a X_a^n + N_b X_b^n)^{1-\frac{1}{n}} (N_a + N_b)^{\frac{1}{n}}$$
$$= (N_a X_a^n + N_b (X_a + \Delta x)^n)^{1-\frac{1}{n}} (N_a + N_b)^{\frac{1}{n}}$$
$$= \left(N_a X_a^n + N_b \left(1 + \frac{\Delta x}{X_a} \right)^n X_a^n \right)^{1-\frac{1}{n}} (N_a + N_b)^{\frac{1}{n}}$$
$$= X_a^{n-1} \left(N_a + N_b \left(1 + \frac{\Delta x}{X_a} \right)^n \right)^{\frac{n-1}{n}} (N_a + N_b)^{\frac{1}{n}}$$

Again, using $(s + k\xi)^r = s^r + O(\xi)$, the above expression becomes

$$(N_a X_a^n + N_b X_b^n)^{1-\frac{1}{n}} (N_a + N_b)^{\frac{1}{n}} - (N_a X_a^{n-1} + N_b X_a^{n-1}) = O(\Delta x) \qquad (19.20)$$

By using Equations (19.20) and (19.19), Equation (19.18) becomes

$$b_w(t + \Delta t) - b_m(t + \Delta t) = nc\, g(X_a, E_t) O(\Delta x) \cdot \Delta t + nc\, g(X_m, E_t) O(\Delta x) \cdot \Delta t$$
$$+ O(\Delta x \cdot \Delta t) + O(\Delta t^2)$$
$$= O(\Delta x \cdot \Delta t)$$

where we have used that the function c is bounded.

This completes the proof.

C Proof of Theorem 3

The proof below shows that if merging is done at the time t then the number of offsprings from the merged cohort $(X_m(s), N_m(s))$ differ at most with $(s - t)C\Delta t$ compared to two internal non merged cohorts $(X_a(s), N_a(s))$ and $(X_b(s), N_b(s))$

for any future time $s \geq t$, where Δt is the time step and C is a constant independent of s and Δt. That is, we prove that the convergence when merging internal cohorts has a convergence rate of $O(\Delta t)$ for arbitrary long simulation times.

Since the assumptions in Corollary 1 is met we will in the following, without notice, change $O(\Delta x)$ to $O(\Delta t)$. In addition to Equation (19.13) we will also use the second order Taylor expansion

$$(s + k\xi)^r = s^r + rs^{r-1}k\xi + O(\xi^2) \tag{19.21}$$

Proof. *(Proof of Theorem* 3 *in Sect.* 19.4*)*. This proof uses the principle of mathematical induction, we will start with the assumption and the induction step.

Let $\Delta t > 0$ be fixed.

After inspecting the proof of Theorem 2, we may assume that

$$N_m(\zeta) = N_a(\zeta) + N_b(\zeta) + O(\Delta t^2) \tag{19.22}$$

and

$$X_m(\zeta) = \left(\frac{X_a^n(\zeta)N_a(\zeta) + X_b^n(\zeta)N_b(\zeta)}{N_a(\zeta) + N_b(\zeta)} \right)^{\frac{1}{n}} + O(\Delta t^2) \tag{19.23}$$

for some $\zeta \geq t$.

Firstly we want to prove that

$$N_m(\zeta + \Delta t) = N_a(\zeta + \Delta t) + N_b(\zeta + \Delta t) + O(\Delta t^2)$$

and

$$X_m(\zeta+\Delta t) = \left(\frac{X_a^n(\zeta + \Delta t)N_a(\zeta + \Delta t) + X_b^n(\zeta + \Delta t)N_b(\zeta + \Delta t)}{N_a(\zeta + \Delta t) + N_b(\zeta + \Delta t)} \right)^{\frac{1}{n}} + O(\Delta t^2)$$

To prove this, we use the fact that

$$\mu(X_m, E_\zeta) = \mu(X_a, E_\zeta) + O(\Delta x) \tag{19.24}$$

and

$$\mu(X_b, E_\zeta) = \mu(X_a, E_\zeta) + O(\Delta x) \tag{19.25}$$

For merging cohort, we get an expression for the number of individuals by using Taylor's expansion in Δt

$$N_m(\zeta + \Delta t) = N_m(\zeta) + N_m'(\zeta)\Delta t + O(\Delta t^2) \tag{19.26}$$

By using Equation (19.3), we get

$$N_m(\zeta + \Delta t) = N_m(\zeta) - \mu(X_m, E_\zeta)N_m(\zeta)\Delta t + O(\Delta t^2)$$

By using Equations (19.22) and (19.24), the above equation becomes

$$\begin{aligned} N_m(\zeta + \Delta t) = {} & N_a(\zeta) + N_b(\zeta) + O(\Delta t^2) \\ & - \mu(X_a, E_\zeta)(N_a(\zeta) + N_b(\zeta) + O(\Delta t^2))\Delta t \\ & + O(\Delta x \cdot \Delta t) + O(\Delta t^2) \end{aligned} \tag{19.27}$$

In the same way, we get

$$N_a(\zeta + \Delta t) = N_a(\zeta) - \mu(X_a, E_\zeta)N_a(\zeta)\Delta t + O(\Delta t^2) \tag{19.28}$$

and

$$N_b(\zeta + \Delta t) = N_b(\zeta) - \mu(X_b, E_\zeta)N_b(\zeta)\Delta t + O(\Delta t^2) \tag{19.29}$$

By subtracting Equations (19.28) and (19.29) from Equation (19.27), we get

$$\begin{aligned} N_m(\zeta + \Delta t) - (N_a(\zeta + \Delta t) + N_b(\zeta + \Delta t)) &= O(\Delta x \cdot \Delta t) + O(\Delta t^2) \\ &= O(\Delta t^2) \end{aligned} \tag{19.30}$$

Again, using Taylor expansion, we get an expression for the size of individuals

$$X_m(\zeta + \Delta t) = X_m(\zeta) + X'_m(\zeta)\Delta t + O(\Delta t^2)$$

By using Equation (19.4), we get

$$X_m(\zeta + \Delta t) = X_m(\zeta) + g(X_m, E_\zeta)\Delta t + O(\Delta t^2)$$

This, together with Equations (19.16) and (19.23), gives

$$\begin{aligned} X_m(\zeta + \Delta t) = {} & \left(\frac{X_a^n(\zeta)N_a(\zeta) + X_b^n(\zeta)N_b(\zeta)}{N_a(\zeta) + N_b(\zeta)}\right)^{\frac{1}{n}} + O(\Delta t^2) \\ & + g(X_a, E_\zeta)\Delta t + O(\Delta x \cdot \Delta t) + O(\Delta t^2) \end{aligned} \tag{19.31}$$

similarly, we get

$$X_a(\zeta + \Delta t) = X_a(\zeta) + g(X_a, E_\zeta)\Delta t + O(\Delta t^2) \tag{19.32}$$

and

$$X_b(\zeta + \Delta t) = X_b(\zeta) + g(X_b, E_\zeta)\Delta t + O(\Delta t^2) \tag{19.33}$$

By using Equations (19.28) and (19.29) in the above, gives

$$\begin{aligned} N_a(\zeta + \Delta t) + N_b(\zeta + \Delta t) &= N_a(\zeta) - \mu(X_a, E_\zeta)N_a(\zeta)\Delta t + O(\Delta t^2) \\ &\quad + N_b(\zeta) - \mu(X_b, E_\zeta)N_b(\zeta)\Delta t + O(\Delta t^2) \end{aligned}$$

By using Equation (19.25), we get

$$\begin{aligned} N_a(\zeta + \Delta t) + N_b(\zeta + \Delta t) &= N_a(\zeta) - \mu(X_a, E_\zeta)N_a(\zeta)\Delta t + O(\Delta t^2) \\ &\quad + N_b(\zeta) - (\mu(X_a, E_\zeta) + O(\Delta x))N_b(\zeta)\Delta t + O(\Delta t^2) \end{aligned}$$

simplifying the above equation yields

$$\begin{aligned} N_a(\zeta + \Delta t) + N_b(\zeta + \Delta t) &= (N_a(\zeta) + N_b(\zeta))(1 - \mu(X_a, E_\zeta)\Delta t) + O(\Delta x \cdot \Delta t) \\ &\quad + O(\Delta t^2) \end{aligned} \tag{19.34}$$

By using Equations (19.28), (19.29), (19.32) and (19.33), we get

$$\begin{aligned} &X_a^n(\zeta + \Delta t)N_a(\zeta + \Delta t) + X_b^n(\zeta + \Delta t)N_b(\zeta + \Delta t) \\ &= (X_a(\zeta) + g(X_a, E_\zeta)\Delta t + O(\Delta t^2))^n(N_a(\zeta) - \mu(X_a, E_\zeta)N_a(\zeta)\Delta t + O(\Delta t^2)) \\ &\quad + (X_b(\zeta) + g(X_b, E_\zeta)\Delta t + O(\Delta t^2))^n(N_b(\zeta) - \mu(X_b, E_\zeta)N_b(\zeta)\Delta t + O(\Delta t^2)) \end{aligned}$$

Now we use Equations (19.13), (19.17), and (19.25) in the above to get

$$\begin{aligned} &X_a^n(\zeta + \Delta t)N_a(\zeta + \Delta t) + X_b^n(\zeta + \Delta t)N_b(\zeta + \Delta t) \\ &= \left((X_a(\zeta) + g(X_a, E_\zeta)\Delta t)^n + O(\Delta t^2)\right)(N_a(\zeta) - \mu(X_a, E_\zeta)N_a(\zeta)\Delta t + O(\Delta t^2)) \\ &\quad + \left((X_b(\zeta) + g(X_a, E_\zeta)\Delta t)^n + O(\Delta t^2)\right)(N_b(\zeta) - \mu(X_a, E_\zeta)N_b(\zeta)\Delta t + O(\Delta t^2)) \end{aligned}$$

With $X_b = X_a + \Delta x$, we get

$$\begin{aligned} &X_a^n(\zeta + \Delta t)N_a(\zeta + \Delta t) + X_b^n(\zeta + \Delta t)N_b(\zeta + \Delta t) \\ &= (X_a(\zeta) + g(X_a, E_\zeta)\Delta t)^n(N_a(\zeta) - \mu(X_a, E_\zeta)N_a(\zeta)\Delta t) \\ &\quad + \left(X_a(\zeta) + \Delta x + g(X_a, E_\zeta)\Delta t\right)^n\left(N_b(\zeta) - \mu(X_a, E_\zeta)N_b(\zeta)\Delta t\right) \\ &\quad + O(\Delta t^2) \end{aligned}$$

By using Equation (19.21), we get

$$
\begin{aligned}
&X_a^n(\zeta+\Delta t)N_a(\zeta+\Delta t)+X_b^n(\zeta+\Delta t)N_b(\zeta+\Delta t)\\
&=(X_a^n(\zeta)+nX_a^{n-1}(\zeta)g(X_a,E_\zeta)\Delta t)(N_a(\zeta)-\mu(X_a,E_\zeta)N_a(\zeta)\Delta t)\\
&\quad+((X_a(\zeta)+\Delta x)^n+n(X_a(\zeta)+\Delta x)^{n-1}g(X_a,E_\zeta)\Delta t)(N_b(\zeta)-\mu(X_a,E_\zeta)N_b(\zeta)\Delta t)\\
&\quad+O(\Delta t^2)
\end{aligned}
$$

Using the same expansion again, gives

$$
\begin{aligned}
&X_a^n(\zeta+\Delta t)N_a(\zeta+\Delta t)+X_b^n(\zeta+\Delta t)N_b(\zeta+\Delta t)\\
&=\Big(X_a^n(\zeta)+nX_a^{n-1}(\zeta)g(X_a,E_\zeta)\Delta t\Big)\Big(N_a(\zeta)-\mu(X_a,E_\zeta)N_a(\zeta)\Delta t\Big)\\
&\quad+\Big(X_a^n(\zeta)+nX_a^{n-1}\Delta x+nX_a^{n-1}(\zeta)g(X_a,E_\zeta)\Delta t\Big)\Big(N_b(\zeta)-\mu(X_a,E_\zeta)N_b(\zeta)\Delta t\Big)\\
&\quad+O(\Delta x\cdot\Delta t)+O(\Delta x^2)+O(\Delta t^2)
\end{aligned}
$$

We simply the above equation to

$$
\begin{aligned}
&X_a^n(\zeta+\Delta t)N_a(\zeta+\Delta t)+X_b^n(\zeta+\Delta t)N_b(\zeta+\Delta t)\\
&=X_a^n(\zeta)(N_a(\zeta)+N_b(\zeta))-\mu(X_a,E_\zeta)X_a^n(\zeta)(N_a(\zeta)+N_b(\zeta))\Delta t\\
&\quad+nX_a^{n-1}(\zeta)g(X_a,E_\zeta)(N_a(\zeta)+N_b(\zeta))\Delta t+nX_a^{n-1}N_b(\zeta)\Delta x+O(\Delta t^2)
\end{aligned}
\tag{19.35}
$$

Observing that

$$
\frac{1}{N_a(\zeta+\Delta t)+N_b(\zeta+\Delta t)}=\frac{1}{(N_a(\zeta)+N_b(\zeta))(1-\mu(X_a,E_\zeta)\Delta t)}+O(\Delta t^2)
$$

By using Equations (19.34) and (19.35), we get

$$
\begin{aligned}
&\frac{X_a^n(\zeta+\Delta t)N_a(\zeta+\Delta t)+X_b^n(\zeta+\Delta t)N_b(\zeta+\Delta t)}{N_a(\zeta+\Delta t)+N_b(\zeta+\Delta t)}\\
&=\frac{X_a^n(\zeta)-\mu(X_a,E_\zeta)X_a^n(\zeta)\Delta t+nX_a^{n-1}(\zeta)g(X_a,E_\zeta)\Delta t}{(1-\mu(X_a,E_\zeta)\Delta t)}\\
&\quad+\frac{nX_a^{n-1}(\zeta)N_b(\zeta)\Delta x}{(N_a(\zeta)+N_b(\zeta))(1-\mu(X_a,E_\zeta)\Delta t)}+O(\Delta t^2)
\end{aligned}
\tag{19.36}
$$

At time ζ we get, by similar calculations, and using $X_b=X_a+\Delta x$

$$\frac{X_a^n(\zeta)N_a(\zeta)+X_b^n(\zeta)N_b(\zeta)}{N_a(\zeta)+N_b(\zeta)}=\frac{X_a^n(\zeta)N_a(\zeta)+(X_a(\zeta)+\Delta x)^n N_b(\zeta)}{N_a(\zeta)+N_b(\zeta)}$$

Using Equation (19.21), the above equation becomes

$$\frac{X_a^n(\zeta)N_a(\zeta)+X_b^n(\zeta)N_b(\zeta)}{N_a(\zeta)+N_b(\zeta)}$$
$$=\frac{X_a^n(\zeta)N_a(\zeta)+\left(X_a^n(\zeta)+nX_a^{n-1}(\zeta)\Delta x+O(\Delta x^2)\right)N_b(\zeta)}{N_a(\zeta)+N_b(\zeta)}$$

simplifying the above equation gives

$$\frac{X_a^n(\zeta)N_a(\zeta)+X_b^n(\zeta)N_b(\zeta)}{N_a(\zeta)+N_b(\zeta)}=X_a^n(\zeta)+\frac{nX_a^{n-1}(\zeta)N_b(\zeta)\Delta x}{N_a(\zeta)+N_b(\zeta)}+O(\Delta x^2) \tag{19.37}$$

By using Equations (19.31), (19.36) and (19.37), we get

$$X_m(\zeta+\Delta t)-\left(\frac{X_a^n(\zeta+\Delta t)N_a(\zeta+\Delta t)+X_b^n(\zeta+\Delta t)N_b(\zeta+\Delta t)}{N_a(\zeta+\Delta t)+N_b(\zeta+\Delta t)}\right)^{\frac{1}{n}}$$
$$=\left(X_a^n(\zeta)+\frac{nX_a^{n-1}(\zeta)N_b(\zeta)\Delta x}{N_a(\zeta)+N_b(\zeta)}+O(\Delta x^2)\right)^{\frac{1}{n}}$$
$$-\left(\frac{X_a^n(\zeta)-\mu(X_a,E_\zeta)X_a^n(\zeta)\Delta t+nX_a^{n-1}(\zeta)g(X_a,E_\zeta)\Delta t}{1-\mu(X_a,E_\zeta)\Delta t}\right.$$
$$\left.+\frac{nX_a^{n-1}(\zeta)N_b(\zeta)\Delta x}{(N_a(\zeta)+N_b(\zeta))(1-\mu(X_a,E_\zeta)\Delta t)}+O(\Delta t^2)\right)^{\frac{1}{n}}$$
$$+g(X_a,E_\zeta)\Delta t+O(\Delta t^2)$$

By using Equation (19.13), we get

$$X_m(\zeta+\Delta t)-\left(\frac{X_a^n(\zeta+\Delta t)N_a(\zeta+\Delta t)+X_b^n(\zeta+\Delta t)N_b(\zeta+\Delta t)}{N_a(\zeta+\Delta t)+N_b(\zeta+\Delta t)}\right)^{\frac{1}{n}}$$
$$=\left(X_a^n(\zeta)+\frac{nX_a^{n-1}(\zeta)N_b(\zeta)\Delta x}{N_a(\zeta)+N_b(\zeta)}\right)^{\frac{1}{n}}$$
$$-\left(X_a^n(\zeta)+\frac{nX_a^{n-1}(\zeta)g(X_a,E_\zeta)\Delta t}{1-\mu(X_a,E_\zeta)\Delta t}+\frac{nX_a^{n-1}(\zeta)N_b(\zeta)\Delta x}{(N_a(\zeta)+N_b(\zeta))(1-\mu(X_a,E_\zeta)\Delta t)}\right)^{\frac{1}{n}}$$
$$+g(X_a,E_\zeta)\Delta t+O(\Delta t^2)$$

By using Equation (19.21), we get

$$X_m(\zeta+\Delta t)-\left(\frac{X_a^n(\zeta+\Delta t)N_a(\zeta+\Delta t)+X_b^n(\zeta+\Delta t)N_b(\zeta+\Delta t)}{N_a(\zeta+\Delta t)+N_b(\zeta+\Delta t)}\right)^{\frac{1}{n}}$$

$$=\left(X_a^n(\zeta)+\frac{nX_a^{n-1}(\zeta)N_b(\zeta)\Delta x}{N_a(\zeta)+N_b(\zeta)}\right)^{\frac{1}{n}}$$

$$-\left(X_a^n(\zeta)+\frac{nX_a^{n-1}(\zeta)N_b(\zeta)\Delta x}{(N_a(\zeta)+N_b(\zeta))(1-\mu(X_a,E_\zeta)\Delta t)}\right)^{\frac{1}{n}}$$

$$-\frac{1}{n}\left(X_a^n(\zeta)+\frac{nX_a^{n-1}(\zeta)N_b(\zeta)\Delta x}{(N_a(\zeta)+N_b(\zeta))(1-\mu(X_a,E_\zeta)\Delta t)}\right)^{\frac{1}{n}-1}\frac{nX_a^{n-1}(\zeta)g(X_a,E_\zeta)\Delta t}{1-\mu(X_a,E_\zeta)\Delta t}$$

$$+g(X_a,E_\zeta)\Delta t+O(\Delta t^2)$$

Again, using Equation (19.13), we get

$$X_m(\zeta+\Delta t)-\left(\frac{X_a^n(\zeta+\Delta t)N_a(\zeta+\Delta t)+X_b^n(\zeta+\Delta t)N_b(\zeta+\Delta t)}{N_a(\zeta+\Delta t)+N_b(\zeta+\Delta t)}\right)^{\frac{1}{n}}$$

$$=\left(X_a^n(\zeta)+\frac{nX_a^{n-1}(\zeta)N_b(\zeta)\Delta x}{N_a(\zeta)+N_b(\zeta)}\right)^{\frac{1}{n}}$$

$$-\left(X_a^n(\zeta)+\frac{nX_a^{n-1}(\zeta)N_b(\zeta)\Delta x}{(N_a(\zeta)+N_b(\zeta))(1-\mu(X_a,E_\zeta)\Delta t)}\right)^{\frac{1}{n}}$$

$$-\left(X_a^n(\zeta)\right)^{\frac{1}{n}-1}\frac{X_a^{n-1}(\zeta)g(X_a,E_\zeta)\Delta t}{1-\mu(X_a,E_\zeta)\Delta t}+g(X_a,E_\zeta)\Delta t+O(\Delta t^2)$$

Canceling a lot of powers in the last row, gives

$$\begin{aligned}&X_m(\zeta+\Delta t)-\left(\frac{X_a^n(\zeta+\Delta t)N_a(\zeta+\Delta t)+X_b^n(\zeta+\Delta t)N_b(\zeta+\Delta t)}{N_a(\zeta+\Delta t)+N_b(\zeta+\Delta t)}\right)^{\frac{1}{n}}\\&=\left(X_a^n(\zeta)+\frac{nX_a^{n-1}(\zeta)N_b(\zeta)\Delta x}{N_a(\zeta)+N_b(\zeta)}\right)^{\frac{1}{n}}\\&-\left(X_a^n(\zeta)+\frac{nX_a^{n-1}(\zeta)N_b(\zeta)\Delta x}{(N_a(\zeta)+N_b(\zeta))(1-\mu(X_a,E_\zeta)\Delta t)}\right)^{\frac{1}{n}}\\&-\frac{g(X_a,E_\zeta)\Delta t}{1-\mu(X_a,E_\zeta)\Delta t}+g(X_a,E_\zeta)\Delta t+O(\Delta t^2)\end{aligned}\tag{19.38}$$

Where Equation (19.21), has been used to prove

$$\left(X_a^n(\zeta)+\frac{nX_a^{n-1}(\zeta)N_b(\zeta)\Delta x}{N_a(\zeta)+N_b(\zeta)}\right)^{\frac{1}{n}}-\left(X_a^n(\zeta)+\frac{nX_a^{n-1}(\zeta)N_b(\zeta)\Delta x}{(N_a(\zeta)+N_b(\zeta))(1-\mu(X_a,E_\zeta)\Delta t)}\right)^{\frac{1}{n}}$$
$$= O(\Delta t^2) \tag{19.39}$$

A simple Taylor expansion gives

$$g(X_a, E_\zeta)\Delta t - \frac{g(X_a, E_\zeta)\Delta t}{1-\mu(X_a, E_\zeta)\Delta t} = O(\Delta t^2) \tag{19.40}$$

By using Equations (19.39), (19.40) and Corollary 1, Equation (19.38) becomes

$$X_m(\zeta+\Delta t) = \tag{19.41}$$
$$\left(\frac{X_a^n(\zeta+\Delta t)N_a(\zeta+\Delta t)+X_b^n(\zeta+\Delta t)N_b(\zeta+\Delta t)}{N_a(\zeta+\Delta t)+N_b(\zeta+\Delta t)}\right)^{\frac{1}{n}} + O(\Delta t^2)$$

According to the Equation (19.7), the expression for the newborn individuals of non-merging cohorts are

$$b_w(\zeta+\Delta t) = cN_a(\zeta+\Delta t)X_a^n(\zeta+\Delta t) + cN_b(\zeta+\Delta t)X_b^n(\zeta+\Delta t)$$

By using Equation (19.35), we get

$$b_w(\zeta+\Delta t) \tag{19.42}$$
$$= c(N_a(\zeta)+N_b(\zeta))\big(X_a^n(\zeta) - \mu(X_a,E_\zeta)X_a^n(\zeta)\Delta t + nX_a^{n-1}(\zeta)g(X_a,E_\zeta)\Delta t\big)$$
$$+ cnX_a^{n-1}(\zeta)N_b(\zeta)\Delta x + O(\Delta t^2)$$

For the newborn individuals of merging cohort, we get

$$b_m(\zeta+\Delta t) = cN_m(\zeta+\Delta t)X_m^n(\zeta+\Delta t)$$

By using Equations (19.27) and (19.41), we have

$$b_m(\zeta+\Delta t)$$
$$= c\Big(N_a(\zeta)+N_b(\zeta) - \mu(X_a,E_\zeta)\,(N_a(\zeta)+N_b(\zeta))\,\Delta t + O(\Delta t^2)\Big)\cdot$$
$$\cdot\left(\left(\frac{X_a^n(\zeta+\Delta t)N_a(\zeta+\Delta t)+X_b^n(\zeta+\Delta t)N_b(\zeta+\Delta t)}{N_a(\zeta+\Delta t)+N_b(\zeta+\Delta t)}\right)^{\frac{1}{n}} + O(\Delta t^2)\right)^n$$

By using Equations (19.36), we get

$$
\begin{aligned}
&b_m(\zeta + \Delta t) \\
&= c(N_a(\zeta) + N_b(\zeta))\Big(X_a^n(\zeta) - \mu(X_a, E_\zeta)X_a^n(\zeta)\Delta t + nX_a^{n-1}(\zeta)g(X_a, E_\zeta)\Delta t\Big) \\
&\quad + cnX_a^{n-1}N_b(\zeta)\Delta x + O(\Delta t^2)
\end{aligned}
\tag{19.43}
$$

By subtracting Equation (19.42) from Equation (19.43), we get, by using Corollary 1

$$
|b_m(\zeta + \Delta t) - b_w(\zeta + \Delta t)| = O(\Delta t^2) \tag{19.44}
$$

Note that, since the derivative of all the functions above are uniformly continuous, the constant involved in the big-O notation does not depend on ζ.

This concludes the induction step.

By Corollary (1), the induction base step is trivial. By using the principle of mathematical induction, the above calculation results in

$$
|b_m(t + n\Delta t) - b_w(t + n\Delta t)| \leq Cn\Delta t^2 \tag{19.45}
$$

where C is a constant.

This means that for any $s > t$, we may divide this interval into n equidistant subintervals, and since Equation (19.45) holds for any fixed $\Delta t > 0$ we choose $\Delta t = \frac{s-t}{n}$ and finally get

$$
\begin{aligned}
|b_m(s) - b_w(s)| &= |b_m(t + n\Delta t) - b_w(t + n\Delta t)| \\
&\leq C\frac{s-t}{\Delta t}\Delta t^2 = (s-t)C\Delta t
\end{aligned}
$$

That is

$$
b_m(s) = b_w(s) + O(\Delta t)
$$

This completes the proof.

Acknowledgments This research was supported by International Science Programme (ISP) in collaboration with South-East Asia Mathematical Network (SEAMaN). Tin Nwe Aye is grateful to the research environment Mathematics and Applied Mathematics (MAM), Division of Applied Mathematics, Mälardalen University for providing an excellent and inspiring environment for research education and research.

References

Brännström, Å., Carlsson, L., & Simpson, D. (2013). On the convergence of the escalator boxcar train. *SIAM Journal on Numerical Analysis, 51*(6), 3213–3231.

Claessen, D., Van Oss, C., de Roos, A. M., & Persson, L. (2002). The impact of size-dependent predation on population dynamics and individual life history. *Ecology, 83*(6), 1660–1675.

de Roos, A. M. (1988). Numerical methods for structured population models: the escalator boxcar train. *Numerical Methods for Partial Differential Equations 4*(3), 173–195.

de Roos, A. M., Diekmann, O., & Metz, J. A. J. (1992). Studying the dynamics of structured population models: A versatile technique and its application to Daphnia. *The American Naturalist, 139*(1), 123–147.

Dodson, S., & Ramcharan, C. (1991). Size-specific swimming behavior of daphnia pulex. *Journal of plankton research, 13*(6), 1367–1379.

Ebert, D. (2005). *Ecology, epidemiology, and evolution of parasitism in Daphnia*. National Library of Medicine.

Gwiazda, P., Jablonski, J., Marciniak-Czochra, A., & Ulikowska, A. (2014). Analysis of particle methods for structured population models with nonlocal boundary term in the framework of bounded Lipschitz distance. *Numerical Methods for Partial Differential Equations, 30*(6), 1797–1820.

Hartvig, M., Andersen, K. H., & Beyer, J. E. (2011). Food web framework for size-structured populations. *Journal of theoretical Biology, 272*(1), 113–122.

Hebert, P. (1978). The population bilogy of daphnia (crustacea, daphnidae). *Biological Reviews, 53*(3), 387–426.

Kelpin, F. D. L., Kirkilionis, M. A., & Kooi, B. W. (2000). Numerical methods and parameter estimation of a structured population model with discrete events in the life history. *Journal of Theoretical Biology, 207*(2), 217–230.

Kooijman, S., & Metz, J. A. J. (1984). On the dynamics of chemically stressed populations: the deduction of population consequences from effects on individuals. *Ecotoxicology and environmental safety, 8*(3), 254–274.

Kooijman, S. (1986). Energy budgets can explain body size relations. *Journal of Theoretical Biology, 121*(3), 269–282.

Kropielnicka, K., Carrillo, J., Gwiazda, P., & Marciniak-Czochra, A. (2018). The escalator boxcar train method for a system of aged-structured equations in the space of measures.

Metz, J. A. J., & de Roos, A. M. (1991). Towards a numerical analysis of the escalator boxcar train. *Differential Equations with Applications in Biology, Physics, and Engineering, 133*, 91.

Metz, J. A. J., & Diekmann, O. (1986). Formulating models for structured populations. In *The dynamics of physiologically structured populations* (pp. 78–135). Springer.

Metz, J. A. J., & Diekmann, O. (2014). *The dynamics of physiologically structured populations* (Vol. 68). Springer.

Nwe Aye, T., & Calrsson, L. (2017). Numerical stability of the escalator boxcar train under reducing system of ordinary differential equations. In C. H. Skiadas (Ed.), *ASMDA2017* (pp. 83–94). London, UK: De Morgan House.

Pascual, M., & Caswell, H. (1997). From the cell cycle to population cycles in phytoplankton–nutrient interactions. *Ecology, 78*(3), 897–912.

Persson, L., Leonardsson, K., De Roos, A. M., Gyllenberg, M., & Christensen, B. (1998). Ontogenetic scaling of foraging rates and the dynamics of a size-structured consumer-resource model. *Theoretical population Biology, 54*(3), 270–293.

Rinke, K., & Vijverberg, J. (2005). A model approach to evaluate the effect of temperature and food concentration on individual life-history and population dynamics of Daphnia. *Ecological Modelling, 186*(3), 326–344.

Ryabov, A. B., de Roos, A. M., Meyer, B., Kawaguchi, S., & Blasius, B. (2017). Competition-induced starvation drives large-scale population cycles in antarctic krill. *Nature Ecology & Evolution*, 1(7):0177.

Tuljapurkar, S., & Caswell, H. (2012). *Structured-population models in marine, terrestrial, and freshwater systems* (Vol. 18). Springer Science & Business Media.

Zhang, L., Dieckmann, U., & Brännström, Å. (2017). On the performance of four methods for the numerical solution of ecologically realistic size-structured population models. *Methods in Ecology and Evolution, 8*(8), 948–956.

Chapter 20
Psychometric Validation of Constructs Defined by Ordinal-Valued Items

Anastasia Charalampi, Catherine Michalopoulou, and Clive Richardson

20.1 Introduction

Attitude scaling methods are employed extensively in sample survey investigations in social science, educational, medical and health research. The development of attitude scales is based on the principle that a single question provides a poor indicator of a person's general attitude (or cognitive and non-cognitive skills or personality traits) and therefore, in order to measure an attitude more accurately, "a sample of beliefs [opinion-questions] covering a range of aspects of the attitude" Moser and Kalton (1975: 351) has to be defined. The respondent's attitude is usually measured by summing (or averaging) his or her responses for each of the items into a total score. In order to allow for the computation of respondents' scores, each item is assigned the same number (and labeling) of response categories. In Likert (or Likert-type) scaling, items are worded alternately as positive or negative to control for acquiescence response. Therefore, before carrying out an analysis, the values of all negatively (or positively) worded items must be reversed in order to achieve correspondence between the ordering of response categories so that in constructing the overall scale (or subscales) low and high scores would indicate negative and positive attitudes, respectively.

A prerequisite of scaling theory is to investigate the scale's structure (dimensionality) and provide evidence on the psychometric properties of the scaleor subscales

A. Charalampi (✉) · C. Michalopoulou · C. Richardson
Panteion University of Social and Political Sciences, Athens, Greece
e-mail: acharalampi@panteion.gr; kmichal@panteion.gr; crichard@panteion.gr

C. H. Skiadas, C. Skiadas (eds.), *Demography of Population Health, Aging and Health Expenditures*, The Springer Series on Demographic Methods and Population Analysis 50, https://doi.org/10.1007/978-3-030-44695-6_20

by ascertaining their reliability and validity before computing the respondents' scores. This investigation depends on whether the goal is theory development — subscales are not predetermined as dimensions by theory — or theory testing — subscales are predetermined as dimensions by theory Tabachnick and Fidell (2007), Thompson (2005). In the case of theory development, principal components analysis (PCA) or exploratory factor analysis (EFA) is performed to define components or factors as subscales and component or factor loadings are reported. Though PCA is exploratory in nature it differs from common factor analyses defined as EFA Bartholomew et al. (2008), Fabrigar et al. (1999), Thompson (2005). The choice between PCA and EFA depends on whether the goal is to reduce the correlated observed variables into a smaller set of independent composite variables (components) or to test a theoretical model of latent factors causing the observed variables Bartholomew et al. (2008), Fabrigar et al. (1999). In the case of theory testing Charalampi (2018), Charalampi et al. (2018, 2019), Michalopoulou (2017)), first a sample of adequate size is randomly split into two halves and EFA is performed on one half-sample. Then the structure is investigated by carrying out confirmatory factor analysis (CFA) on the second half-sample. Applying this approach, the structure of the attitude scale identified by EFA is validated by performing CFA. In both theory development and theory testing, based on the EFA and the CFA results for the total sample, the validity and reliability of the resulting subscales (or overall scale) and their distributional properties are assessed.

The first and most important consideration in any statistical analysis — whether univariate, bivariate or multivariate — is to ascertain the level of measurement of the input variables which guides the correct choice of the methods to be used. The items of an attitude scale can be categorical or continuous Tabachnick and Fidell (2007). Furthermore, following Stevens (2002), we distinguish four levels of measurement: nominal, ordinal, interval and ratio. In all empirical research, nominal and ordinal items are considered as categorical, and interval and ratio items as scale — treated in statistical analyses as continuous Blalock (1979). In the literature, the level of measurement of the items of most attitude scales is ordinal.

In this paper, to demonstrate the importance of ascertaining the level of measurement of the items in choosing the methods to be used, we carry out the investigation and assessment of the 2012 European Social Survey (ESS) short eight-item version of the Center for Epidemiologic Studies Depression scale (CES-D 8) for Italy and Spain when items are considered as ordinal. The CES-D 8 serves as a good example because the only evidence on its structure that has been reported to our knowledge results from measurement invariance tests Van de Velde et al. (2009, 2010). Apart from such tests, only Karim et al. (2015) validated the CES-D 8, for the older adult (60–90) population in 23 European countries based on samples selected from the 2012 ESS. However, the methods used were appropriate for interval scale items and not ordinal.

20.1.1 The CES-D 8 Scale

Depression is "a common mental disorder that presents with depressed mood, loss of interest or pleasure, decreased energy, feelings of guilt or low self-worth, disturbed sleep or appetite, and poor concentration" Marcus et al. (2012: 6). It is a common public health outcome and significantly decreases quality of life especially in older adults Blazer (2009), Irwin et al. (1999), O'Halloran (2014), Beekman et al. (1997).

Effective screening for depression has become a crucial public health issue as it is often difficult to detect and as a result remains untreated Cheung et al. (2007). In the past decades many screening tests were developed in order to improve recognition of depression Beekman et al. (1997).

Radloff (1977) developed the CES-D as a short self-report instrument designed to measure symptomatology of depression in clinical and non-clinical settings Andresen et al. (1994), Carleton et al. (2013), Cole et al. (2000), Karim et al. (2015). The 20 items of the CES-D represent 20 descriptive statements of depressed mood Andresen et al. (1994). It is a popular screening tool widely used in many countries and studies Beekman et al. (1997), Cheung et al. (2007), Cole et al. (2000), Irwin et al. (1999). However, it should not be used as a clinical diagnostic tool Radloff (1977), Van de Velde et al. (2009).

Although the CES-D has proven a reliable and valid instrument Karim et al. (2015) and convenient to use in most settings Irwin et al. (1999), its administration appeared to be problematic among older adults. It may be exhausting and difficult for elderly respondents to remain focused for all the time required to complete the instrument Karim et al. (2015), O'Halloran et al. (2014) and they may find the questions emotionally stressful Irwin et al. (1999).

Researchers have made efforts to build more parsimonious versions of the CES-D to manage the problems associated with administration O'Halloran et al. (2014), Karim et al. (2015). The eight-item version of CES-D (CES-D 8) scale proposed by Van de Velde et al. (2009, 2010) was first embedded in Round 3 of the ESS. It is included in section D of the rotating module questionnaire of the ESS as part of personal and social wellbeing. As mentioned in our previous work Charalampi (2018), Charalampi et al. (2018, 2019), the personal and social wellbeing module was first included in the questionnaire of Round 3 (2006) of the ESS and was repeated with certain changes in Round 6 (2012) of the survey European Social Survey (2015), Jeffrey et al. (2015). Combining theoretical models and evidence from statistical analyses, six key dimensions were defined for the 2012 ESS measurement of personal and social wellbeing as follows European Social Survey (2015), Jeffrey et al. (2015): evaluative wellbeing, emotional wellbeing, functioning, vitality, community and supportive relationships. The CES-D 8 is comprised of four variables from emotional wellbeing, three items from vitality and one from the supportive relationships dimension.

20.2 Method

20.2.1 Participants

The analysis was based on the ESS Round 6 Data (2012) for Italy and Spain. The ESS implements all the strict methodological prerequisites for comparability over time and cross-nationally Kish (1994), Carey (2000) by applying probability sampling, minimum effective achieved sample sizes in all participating countries and a maximum target non-response rate of 30% The ESS Sampling Expert Panel (2016). Face-to-face interviewing is used for data collection. The survey population is defined as all persons aged 15 and over residing within private households in each country, regardless of their nationality, citizenship or language; this definition applies to all rounds of the survey. In Table 20.1, the demographic and social characteristics of the participants are presented.

As shown, gender was almost equally distributed in the Italian and Spanish samples, with slightly more women than men. The mean age was around 47 years, more than 53.6% of the participants were married, the majority had completed secondary education or lower and at least 43.7% were in paid work.

20.2.2 Instrument

Table 20.2 presents the CES-D 8 scale items in the ESS questionnaire. The scale is comprised of six negatively (D5-D7, D9, D11 and D12) and two positively (D8 and D10) worded items, respectively, so as to reduce acquiescence response biases. The theoretical definition of the scale requires the scoring of the positively worded items to be reversed before analysis so that low and high scores on the scale (or subscales) will indicate a lower and higher frequency of depressive complaints, respectively Van de Velde et al. (2009: 19). The response categories range from 1 to 4 and are defined as follows: 1 (none or almost none of the time); 2 (some of the time); 3 (most of the time); 4 (all or almost all of the time). Therefore, the level of measurement is ordinal.

Table 20.1 Participants' demographic and social characteristics: European Social Survey, 2012

Country	*N*	Men (%)	Women (%)	Age Mean (SD)	Married (%)	Secondary education or lower (%)	In paid work* (%)
Italy	913	48.3	51.7	47.2 (18.3)	53.6	77.7	43.8
Spain	1746	48.2	51.8	47.6 (18.0)	54.9	79.2	43.7

*The reference period for the respondent's main activity was defined as during the last 7 days

Table 20.2 The eight-item version of the Center for Epidemiologic Studies-Depression Scale (CES-D 8): European Social Survey, 2012

Depression items	ESS questionnaire	Aligned scale
Felt depressed, how often past week	D5	1–4
Felt that everything you did was an effort, how often past week	D6	1–4
Sleep was restless, how often past week	D7	1–4
Were happy, how often past week	D8	1–4 (R)
Felt lonely, how often past week	D9	1–4
Enjoyed life, how often past week	D10	1–4 (R)
Felt sad, how often past week	D11	1–4
Could not get going, how often past week	D12	1–4

R = these items were reversed before analysis
1 = none or almost none of the time; 2 = some of the time; 3 = most of the time; 4 = all or almost all of the time

20.2.3 Statistical Analyses

In the first step of the analysis, the sample in each country was randomly split into two halves using an SPSS script taken from Raynald's SPSS Tools Levesque (2012). EFA was performed on the first half in order to assess the construct validity of the scale and the suggested structure was subsequently validated by carrying out CFA on the second half. Statistical analyses were performed using Mplus Version 7.4 and IBM SPSS Statistics Version 20.

According to Tabachnick and Fidell (2007), a sample size of 300 cases or more is adequate for performing factor analysis. Since the sample sizes were 960 (Italy) and 1889 (Spain), the half-samples were 480 (Italy) and 944 (Spain) and were therefore considered large enough for carrying out factor analyses separately in each country.

Initially, missing data analysis and data screening for outliers and unengaged responses were performed Charalampi (2018), Charalampi et al. (2018, 2019), Michalopoulou (2017). Only cases with missing values on all items were excluded automatically from the analysis Muthén and Muthén (1998–2012). In addition, cases were eliminated if they exhibited low standard deviation (<0.5), i.e. no variance in the responses Gaskin (2016). Data screening for outliers was based on background variables e.g., gender (dichotomy), age (ratio) and education (pseudo-interval). Cases were eliminated if they were shown in the boxplots as outliers Gaskin (2016); see also Brown (2015), Tabachnick and Fidell (2007), Thompson (2005).

20.2.3.1 Exploratory Factor Analysis (EFA)

In performing EFA, as in our previous work Charalampi (2018), Charalampi et al. (2018, 2019), Michalopoulou (2017), the following sequence of decisions was required Brown (2015), Cabrera-Nguyen (2010), Tabachnick and Fidell (2007), Thompson (2005):

1. Initially, the items' frequency distributions were inspected for floor and ceiling effects, bearing in mind that percentages of responses in the range 1–15 are normally deemed acceptable Anagnostopoulos et al. (2013).
2. Decision on the matrix of association coefficients to analyze: as the items were considered as ordinal, the polychoric correlation matrix was employed as the appropriate matrix of associations Brown (2015).
3. Factor extraction method: robust weighted least squares was applied as the appropriate method of factor extraction Brown (2015).
4. Decision on the number of factor to be extracted: considering the theoretical definition of the scale Van de Velde et al. (2009), one to two factors were tested for each country. Model fit was considered adequate if $\chi^2/df < 3$, comparative fit index (CFI) and Tucker-Lewis index (TLI) values were greater than or close to .95 and the root-mean-square error approximation (RMSEA) $\leq .06$ with the 90% confidence interval (CI) upper limit $\leq .06$ Bollen (1989), Brown (2015), Hu and Bentler (1999), Schmitt (2011), Tabachnick and Fidell (2007), Thompson (2005). Model fit was considered acceptable if $\chi^2/df < 3$, CFI and TLI values were >.90 and RMSEA <.08 with the 90% CI upper limit <.08 Hu and Bentler (1999), Marsh et al. (2004).
5. Factor rotation method: based on the correlations between factors and the simple structure criterion, geomin rotation was applied as the appropriate oblique factor rotation method Brown (2015).
6. Items were considered salient if their factor loadings were >.30 and therefore the meaning of each dimension was inferred from these items Thompson (2005). Items with loadings >.30 on one factor and >.22 on another factor were considered as "cross-loading" items, i.e. items that loaded on multiple factors Stevens (2002). Items with loadings <.30 on all factors (i.e., low communalities) were excluded from the analysis. Also, factors with salient loadings for only two items were eliminated from the analysis as poorly defined Brown (2015).

20.2.3.2 Confirmatory Factor Analysis (CFA)

In performing CFA, as in our previous work Charalampi (2018), Charalampi et al. (2018, 2019), Michalopoulou (2017), the following sequence of decisions was required Brown (2015), Cabrera-Nguyen (2010), Tabachnick and Fidell (2007), Thompson (2005):

1. Model estimation: the model indicated as best by the EFA performed on the first half-sample was tested. CFA of the polychoric correlation matrix was performed using robust weighted least squares.

2. Model evaluation: model fit was considered adequate or acceptable as in EFA, i.e. adequate if $\chi^2/df < 3$, CFI and TLI values were greater than or close to .95 and RMSEA $\leq$ .06 with the 90% CI upper limit $\leq$ .06, or acceptable if $\chi^2/df < 3$, the CFI and TLI values were >90 and RMSEA < .08 with the 90% CI upper limit < .08.
3. Model misspecification searches: searches for modification indices and further specifications were performed and, where necessary, correlations between error variances were introduced Brown (2015), Thompson (2005).

20.2.3.3 Scale Construction and Assessment

First, all items were rescaled into a 0–3 scale Van de Velde et al. (2009). The scale (or subscales) was constructed for the full sample by summing up the defining items and descriptive statistics were computed. Based on the CFA results for the full sample, as in our previous work Charalampi (2018), Charalampi et al. (2018, 2019), Michalopoulou (2017), the average variance extracted (AVE) was computed for the scale (or subscales) by averaging the sum of all squared standardized factor loadings in order to assess the convergent validity of the respective construct. Convergent validity was considered adequate if the AVE was above or around .50, i.e. a relaxed version of the Fornell and Larcker (1981) criterion for AVE $\geq$ .50 (see also Anagnostopoulos et al. (2013). Average inter-item correlations in the recommended range of .15–.5 that cluster near their mean value were used as an indication of the unidimensionality of the scale Clark and Watson (1995). Furthermore, based on the CFA results for the full sample, the composite reliability coefficient Raykov (2007) was computed using the calculator provided by Colwell (2016); this is more appropriate than the commonly used Cronbach's alpha coefficient Brown (2015), Raykov (2007). A scale (or subscales) was considered reliable if the composite reliability coefficient was above or around .70, i.e. using a more relaxed version of the Nunnally and Bernstein (1994) criterion for Cronbach's alpha coefficients $\geq$.70.

20.3 Results

The screening of both half-samples identified 45 unengaged responses in Italy and 143 in Spain and it was decided to exclude them in the analysis. In the Italian sample, four outlying cases with a Higher Education degree were detected and it was decided not to reject them from the analysis. There were no cases with missing values on all items in the sample of either country. Therefore, the analysis was based on 915 cases of Italy and 1746 cases of Spain.

20.3.1 EFA Results

First, frequency distributions and mode and median values of the items based on the first half-sample were inspected for Italy and Spain (Table 20.3). The full range of possible responses was used for all items. However, strong floor effects were detected for all items in both countries except for one item in the Italian half-sample that exhibited strong ceiling effects (D10). As shown, in both countries, the proportion of missing values was negligible, exceeding 1.1% only for two items of the Italian sample (D10 and D12).

EFA was performed with robust weighted least squares of the polychoric matrix of associations computed from the first half-samples of each country and geomin rotation was applied. Two different models were tested with one and two factors, respectively Van de Velde et al. (2009). The one-factor solution demonstrated inadequate model fit for both Italy (χ^2/df = 9.49, CFI = .889, TLI = .846, RMSEA = .136 with the 90% CI upper limit = .154) and Spain (χ^2/df = 14.71, CFI = .952, TLI = .933, RMSEA = .126 with the 90% CI upper limit = .139). The two-factor solution also exhibited inadequate model fit for Italy: χ^2/df = 3.54, CFI = 0.979, TLI = 0.954, RMSEA (90% CI) = .074 (.052–.098). Furthermore, this solution indicated the presence of method effects Brown (2015), Marsh (1996), Van de Velde et al. (2009) as the first factor consisted of the six negatively worded items

Table 20.3 Item analysis of the CES-D 8 scale for Italy and Spain based on the first half-samples: European Social Survey, 2012

Country/item	Mode	Median	Frequency percent of response categories				
			0	1	2	3	NA
Italy (n = 461)							
D5	0	0	60.5	33.6	3.3	2.2	0.4
D6	0	1	45.2	43.4	7.4	3.3	0.7
D7	0	1	49.1	32.8	10.9	6.1	1.1
D8	1	1	20.7	44.3	30.6	4.1	0.2
D9	0	0	57.9	31.7	6.8	3.1	0.7
D10	2	2	15.7	29.9	36.0	16.4	2.0
D11	1	1	36.5	55.7	4.4	2.4	1.1
D12	1	1	39.1	49.8	6.3	2.8	2.0
Spain (n = 865)							
D5	0	0	50.7	37.2	8.6	3.0	0.6
D6	0	1	47.0	38.5	9.6	4.6	0.2
D7	0	1	42.1	41.4	11.8	4.7	0.1
D8	1	1	21.3	43.8	29.3	5.4	0.2
D9	0	0	66.7	24.6	4.8	3.6	0.3
D10	1	1	16.5	37.1	34.4	10.9	1.1
D11	1	1	37.6	50.2	7.9	4.0	0.4
D12	1	1	44.5	44.6	7.6	3.0	0.3

NA = no answer (missing values). Data weighted by the design weight (dweight)

Table 20.4 Exploratory factor analysis of the CES-D 8 scale performed with robust weighted least squares of the polychoric correlation matrix on the first half-samples of Italy and Spain: European Social Survey, 2012

Item	CES-D 8 Italy ($n = 461$)	CES-D 8 Spain ($n = 865$)
D5	.820	.859
D6	.623	.691
D7	.502	.612
D8	.580	.648
D9	.577	.518
D10	.410	.518
D11	.756	.815
D12	.664	.837

Goodness of fit indices for the Italian model: $\chi^2/df = 9.49$, CFI = .889, TLI = .846, RMSEA (90% CI) = .136 (.118–.154)
Goodness of fit indices for the Spanish model: $\chi^2/df = 14.71$, CFI = .952, TLI = .933, RMSEA (90% CI) = .126 (.113–.139)

and the second factor the remaining two positively worded items and was therefore not considered in further analysis. In the case of Spain, there was no solution as the estimation procedure did not converge for a two-factor model. In Table 20.4, the factor loadings of the one-factor model are presented for Italy and Spain. In both cases, all items exhibited strong factor loadings (greater than .50). As each factor was defined by all the items of the CES-D 8 scale, this name was retained.

20.3.2 *CFA Results*

In both countries, the one first-order factor model indicated as preferable by the EFA results was tested by performing CFA of the polychoric correlation matrix using robust weighted least squares on the second half-samples. Correlated errors were introduced for the two reversed (positively worded) items (D8 and D10) in order to rule out methods effects Brown (2015), Marsh (1996), Van de Velde et al. (2009). Modification searches were conducted. The CFA results for Italy (Fig. 20.1) showed acceptable model fit for the solution with one first-order factor: $\chi^2/df = 2.13$, CFI = .988, TLI = .981, RMSEA (90% CI) = .050 (.028–.072). The CFA results for Spain (Fig. 20.2) also showed adequate model fit for the one first-order factor solution: $\chi^2/df = 2.55$, CFI = .995, TLI = .992, RMSEA (90% CI) = .042 (.027–.057). Therefore, the CFA findings supported the unidimensional structure of the CES-D 8 for both Italy and Spain.

20.3.3 *Scale Construction and Assessment*

Scales were constructed by summing up their (rescaled) defining items. In Table 20.5, descriptive statistics, composite reliability and convergent validity based on

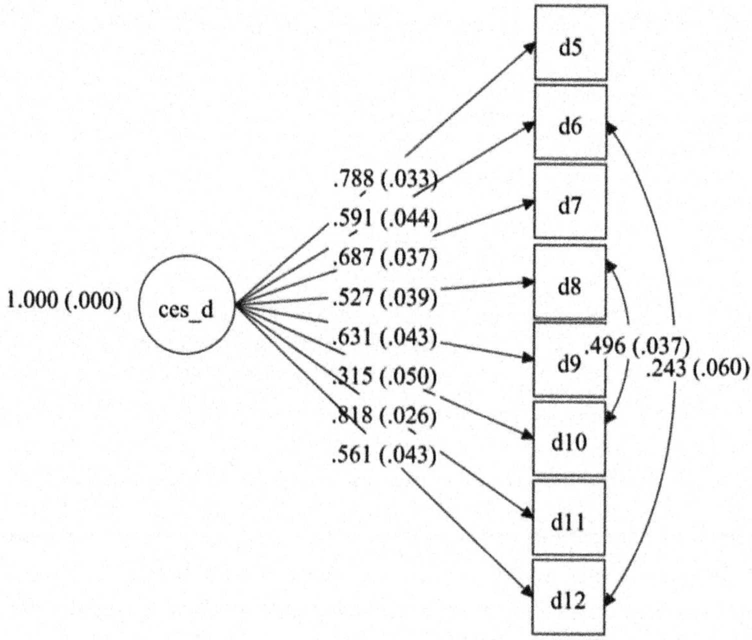

Fig. 20.1 Standardized solution for the model with one first-order factor based on CFA analysis performed on the second half-sample of Italy (n = 454). Observed variables are represented by squares and the latent variable by a circle. Goodness of fit indices for this model: $\chi^2/df = 2.13$, CFI = .988, TLI = .981, RMSEA (90% CI) = .050 (.028–.072)

the full samples are presented for Italy and Spain. The AVE was computed for each scale based on the CFA analysis repeated for the full samples of Italy (Fig. 20.3: $\chi^2/df = 2.62$, CFI = .991, TLI = .985, RMSEA = .042 with the 90% CI = .027–.057) and Spain (Fig. 20.4: $\chi^2/df = 3.58$, CFI = .996, TLI = .994, RMSEA = .038 with the 90% CI = .028–.049). The CES-D 8 demonstrated adequate convergent validity in Spain (AVE above or around .50). However, in Italy the CES-D 8 exhibited problematic convergent validity. The average inter-item correlations in both countries were within the recommended range for unidimensionality (.15–.5). In both Italy and Spain, the CES-D 8 was reliable with composite reliability values .822 and .878 ($\geq$.70), respectively.

20.4 Conclusions

In this paper, the importance of ascertaining the level of measurement of the items in the choice of the methods to be applied when validating a construct was demonstrated using the 2012 ESS measurement of the CES-D 8 scale for Italy and Spain, where items were considered as ordinal.

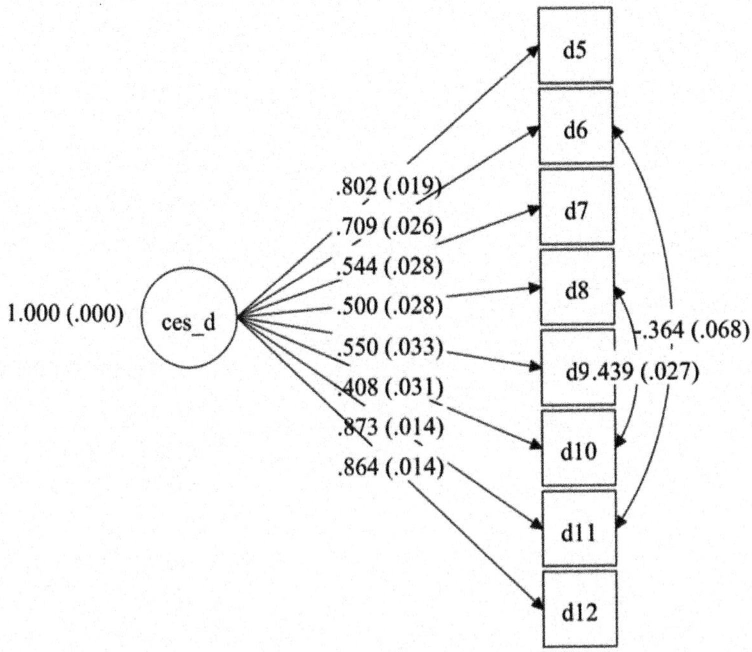

Fig. 20.2 Standardized solution for the model with one first-order factor based on CFA analysis performed on the second half-sample of Spain ($n = 887$). Observed variables are represented by squares and the latent variable by a circle. Goodness of fit indices for this model: $\chi^2/df = 2.55$, CFI = .995, TLI = .992, RMSEA (90% CI) = .042 (.027–.057)

Table 20.5 Descriptive statistics, convergent validity, composite reliability and internal consistencies of the depression scale based on the full samples of Italy and Spain: European Social Survey, 2012

	CES-D 8 Italy ($N = 915$)	CES-D 8 Spain ($N = 1746$)
Number of items	8	8
Mean (standard error)	6.88 (0.136)	6.49 (0.102)
95% confidence interval	6.62–7.15	6.29–6.69
Standard deviation	4.005	4.258
Skewness	1.068	1.027
Kurtosis	1.411	1.038
Convergent validity	.380	.486
Composite reliability	.822	.878
Average inter-item correlation	.315	.381
Min.-max. correlations	.115–.499	.191–.647
Range of correlations	.384	.456

All items were rescaled into a 0–3 scale. Standard errors for skewness and kurtosis of the Italian scale were 0.082 and 0.164, respectively. Standard errors for skewness and kurtosis of the Spanish scale were 0.059 and 0.118, respectively

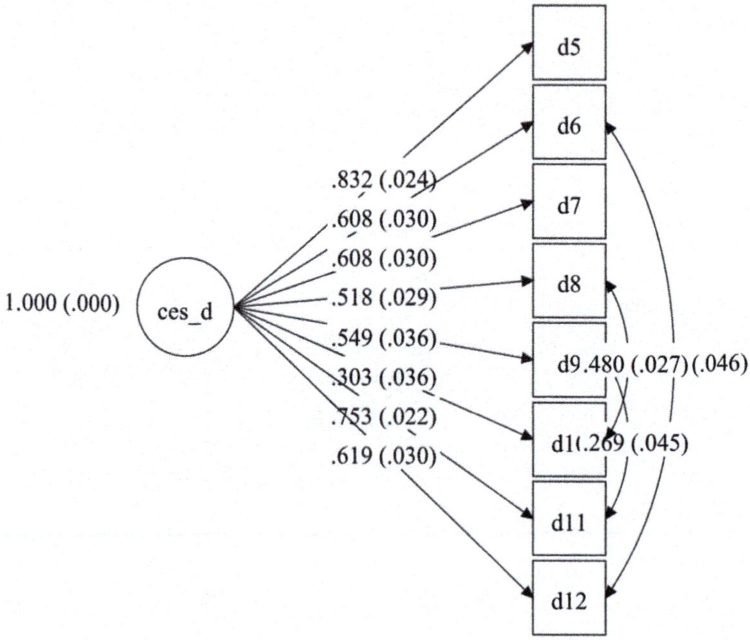

Fig. 20.3 Standardized solution for the model with one first-order factor based on CFA analysis performed on the full sample of Italy ($N = 915$). Observed variables are represented by squares and the latent variable by a circle. Goodness of fit indices: $\chi^2/df = 2.62$, CFI = .991, TLI = .985, RMSEA (90% CI) = .042 (.027–.057)

The investigation of the structure (dimensionality) of the 2012 ESS measurement of CES-D 8 scale by applying the traditional approaches of EFA and CFA to randomly split half-samples resulted in both countries in a unidimensional structure defined by all the eight items of the scale.

The demonstration of the complex sequence of decisions required in performing EFA and CFA based on current theory and practice should be noted among the strengths of the study. In both countries, item analysis carried out on the first-half samples indicated floor effects for all items except for a single item in the Italian half-sample that exhibited ceiling effects. In each case, two models were tested by applying the appropriate method of factor extraction (robust weighted least squares) because it "performs well … for variables with floor or ceiling effects" Brown (2015: 335). In both countries, EFA resulted in a one-factor solution defined by all eight items of the CES-D 8 scale but demonstrated inadequate fit to the data. The two-factor model of Italy indicted the presence of methods effects Brown (2015), Marsh (1996), Van de Velde et al. (2009). Therefore, in performing CFA, correlated errors for the two reversed items of the CES-D 8 were introduced to rule out methods effects Brown (2015), Marsh (1996), Van de Velde et al. (2009)). The one first-order factor model was tested by performing CFA which resulted in acceptable and adequate model fits for Italy and Spain, respectively. The analysis for the full sample

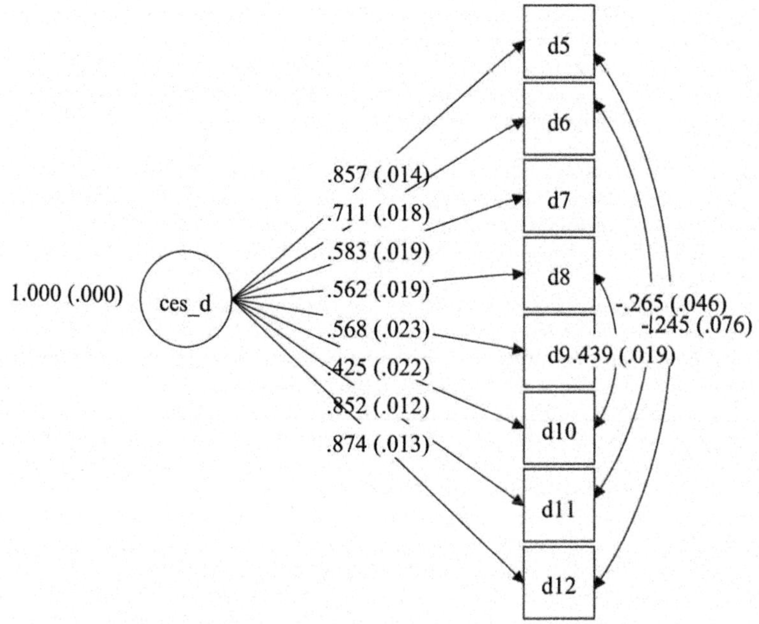

Fig. 20.4 Standardized solution for the model with one first-order factor based on CFA analysis performed on the full sample of Spain ($N = 1746$). Observed variables are represented by squares and the latent variable by a circle. Goodness of fit indices: $\chi^2/df = 3.58$, CFI = .996, TLI = .994, RMSEA (90% CI) = .038 (.028–.049)

demonstrated that the CES-D 8 was reliable for both countries but of problematic convergent validity for Italy.

In this paper, the unidimensionality of the CES-D 8 scale was confirmed in both countries and results on its psychometric properties were provided. Further research is necessary in every country and both rounds of the ESS that included the CES-D 8 scale (2006 and 2012) in order to establish that it is suitable for use in analyses. The methodology presented may be easily applied to other scales comprised of ordinal items and defined as multidimensional by theory (theory testing). In the case of theory development, the preliminary considerations and the sequence of decisions for performing EFA may be applied with the appropriate modifications.

References

Anagnostopoulos, F., Yfantopoulos, J., Moustaki, I., & Niakas, D. (2013). Psychometric and factor analytic evaluation of the 15D health-related quality of life instrument: The case of Greece. *Quality of Life Research, 22*(8), 1973–1986.

Andresen, E. M., Malmgren, J. A., Carter, W. B., & Patrick, D. L. (1994). Screening for depression in well older adults: Evaluation of a short form of the CES-D. *American Journal of Preventive Medicine, 10*(2), 77–84.

Bartholomew, D. J., Steele, F., Moustaki, I., & Galbraith, J. (2008). *Analysis of multivariate social science data*. London: Chapman & Hall/CRC.

Beekman, A. T. F., Deeg, D. J. H., Van Limbeek, J., Braam, A. W., De Vries, M. Z., & Van Tilburg, W. (1997). Criterion validity of the Center for Epidemiologic Studies Depression scale (CES-D): Results from a community-based sample of older subjects in the Netherlands. *Psychological Medicine, 27*, 231–235.

Blalock, H. M., Jr. (1979). *Social statistics (revised second edition)*. New York: McGraw-Hill.

Blazer, D. G. (2009). Depression in late life: Review and commentary. *Focus, 7*(1), 118–136.

Bollen, K. A. (1989). *Structural equations with latent variables*. New York: Wiley.

Brown, T. A. (2015). *Confirmatory factor analysis for applied research* (2nd ed.). New York: The Guilford Press.

Cabrera-Nguyen, P. (2010). Author guidelines for reporting scale development and validation result. *Journal of the Society for Social Work and Research, 1*(2), 99–103.

Carey, S. (Ed.). (2000). *Measuring adult literacy: The International Adult Literacy Survey (IALS) in the European context*. London: Office for National Statistics.

Carleton, R. N., Thibodeau, M. A., Teale, M. J. N., Welch, P. G., Abrams, M. P., Robinson, T., & Asmundson, G. J. G. (2013). The Center for Epidemiologic Studies Depression scale: A review with a theoretical and empirical examination of item content and factor structure. *PLoS One, 8*, 3. https://doi.org/10.1371/journal.pone.0058067.

Charalampi, A. (2018). *The importance of items' level of measurement in investigating the structure and assessing the psychometric properties of multidimensional constructs (Doctoral dissertation)*. Athens: Panteion University of Social and Political Sciences.

Charalampi, A., Michalopoulou, C., & Richardson, C. (2018). Validation of the 2012 European Social Survey measurement of wellbeing in seventeen European countries. *Applied Research in Quality of Life, Advance Online Publication*. https://doi.org/10.1007/s11482-018-9666-4.

Charalampi, A., Michalopoulou, C., & Richardson, C. (2019). Determining the structure and assessing the psychometric properties of multidimensional scales constructed from ordinal and pseudo-interval items, Communications in Statistics – Case studies. *Data Analysis and Applications, Advance Online Publication*. https://doi.org/10.1080/23737484.2019.1579683.

Cheung, Y. B., Liu, K. Y., Phil, M., & Yip, P. S. F. (2007). Performance of the CES-D and its short forms in screening suicidality and hopelessness in the community. *The American Association of Suicidology, 37*(1), 79–88.

Clark, L. A., & Watson, D. (1995). Constructing validity: Basic issues in objective scale development. *Psychological Assessment, 7*(3), 309–319.

Cole, S. R., Kawachi, I., Maller, S. J., & Berkman, L. F. (2000). Test of item-response bias in the CES-D scale: Experience from the New Haven EPESE study. *Journal of Clinical Epidemiology, 53*, 285–289.

Colwell, S. R. (2016). *The composite reliability calculator, Technical Report*. https://doi.org/10.13140/RG.2.1.4298.088.

ESS ERIC. (2015). *European social survey, measuring and reporting on Europeans' wellbeing: Findings from the European Social Survey*. London: ESS ERIC. Retrieved from: https://www.europeansocialsurvey.org/docs/findings/ESS1-6_measuring_and_reporting_europeans_wellbeing.pdf.

Fabrigar, L. R., Wegener, D. T., MacCallum, R. C., & Strahan, E. J. (1999). Evaluating the use of exploratory factor analysis in psychological research. *Psychological Methods, 4*(3), 272–299.

Fornell, C., & Larcker, D. F. (1981). Evaluating structural equation models with unobservable variables and measurement error. *Journal of Marketing Research, 18*(1), 39–50.

Gaskin, J. (2016). Data screening. In: *Gaskination's StatWiki*. http://statwiki.kolobkreations.com/index.php?title=Data_screening. Accessed 30 June 2016.

Hu, L., & Bentler, P. M. (1999). Cutoff criteria for fit indexes in covariance structure analysis: Conventional criteria versus new alternatives. *Structural Equation Modeling, 6*(1), 1–55.

Irwin, M., Artin, K. H., & Oxman, M. N. (1999). Screening for depression in the older adults: Criterion validity of the 10-item Center for Epidemiological Studies Depression scale (CES-D). *Archives of Internal Medicine, 159*(15), 1701–1704.

Jeffrey, K., Abdallah, S., & Quick, A. (2015). Europeans' personal and social wellbeing: Topline results from Round 6 of the European Social Survey, *ESS Topline Result Series – Issue 5*. http://www.europeansocialsurvey.org/docs/findings/ESS6_toplines_issue_5_personal_and_social_wellbeing.pdf. Accessed 2 June 2016.

Karim, J., Weisz, R., Bibi, Z., & ur Rehman, S. (2015). Validation of the eight-item Center for Epidemiologic Studies Depression scale (CES-D) among older adults. *Current Psychology, 34*(4), 681–692. https://doi.org/10.1007/s12144-014-9281-y.

Kish, L. (1994). Multi-population survey designs: Five types with seven shared aspects. *International Statistical Review, 62*(2), 167–186.

Levesque, R. (2012). *Raynald's SPSS Tools [computer software]*. http://www.spsstools.net. Accessed 5 February 2012.

Marcus, M., Yasamy, T. Y., Ommeren, M., Chisholm, D., & Saxena, S. (2012). Depression: A global health concern. In *Depression: A global crisis* (pp. 6–8). Occoquan: World Federation for Mental Health. Retrieved from https://www.who.int/mental_health/management/depression/wfmh_paper_depression_wmhd_2012.pdf.

Marsh, H. W. (1996). Positive and negative global self-esteem: A substantively meaningful distinction or artifactors? *Journal of Personality and Social Psychology, 70*(4), 810–819. https://doi.org/10.1037/0022-3514.70.4.810.

Marsh, H. W., Hau, K. T., & Wen, Z. (2004). In search of golden rules: Comment on hypotheses-testing approaches to setting cutoff values for fit indexes and dangers in overgeneralizing Hu and Bentler's (1999) findings. *Structural Equation Modeling, 11*(3), 320–341.

Michalopoulou, C. (2017). Likert scales require validation before application – Another cautionary tale. *BMS Bulletin de Méthodologie Sociologique, 134*, 5–23.

Moser, C., & Kalton, G. (1975). *Survey methods in social investigation*. London: Heinemann Educational Books.

Muthén, L. K., & Muthén, B. O. (1998–2012). *Mplus user's guide* (7th ed.). Los Angeles: Muthén & Muthén.

Nunnally, J. C., & Bernstein, I. H. (1994). *Psychometric theory*. New York: McGraw-Hill.

O'Halloran, A. M., Kenny, R. A., & King-Kallimanis, B. L. (2014). The latent factors for depression from the short forms of the CES-D are consistent, reliable and valid in community-living older adults. *European Geriatric Medicine, 5*, 97–102.

Radloff, L. S. (1977). The CES-D scale: A self-report depression scale for research in the general population. *Applied Psychological Measurements, 1*(3), 385–401.

Raykov, T. (2007). Reliability if deleted, not "alpha if deleted": Evaluation of scale reliability following component deletion. *British Journal of Mathematical and Statistical Psychology, 60*(2), 201–216.

Schmitt, T. A. (2011). Current methodological considerations in exploratory and confirmatory factor analysis. *Journal of Psychoeducational Assessment, 29*, 304–322.

Stevens, J. (2002). *Applied multivariate statistics for the social sciences*. Mahwah: Lawrence Erlbaum Associates.

Tabachnick, B. G., & Fidell, L. S. (2007). *Using multivariate statistics*. Upper Saddle River: Pearson Allyn & Bacon.

The ESS. (2016). *Sampling expert panel, sampling guidelines: Principles and implementation for the European Social Survey*. London: ESS ERIC Headquarters. http://www.europeansocialsurvey.org/. Accessed 2 June 2016.

Thompson, B. (2005). *Exploratory and confirmatory factor analysis: Understanding concepts and applications*. Washington DC: American Psychological Association.

Van de Velde, S., Levecque, K., & Bracke, P. (2009). Measurement equivalence of the CES-D 8 in the general population in Belgium: A gender perspective. *Archives of Public Health, 67*, 15–29. https://doi.org/10.1186/0778-7367-67-1-15.

Van de Velde, S., Bracke, P., Levecque, K., & Meuleman, B. (2010). Gender differences in depression in 25 European countries after eliminating measurement bias in the CES-D 8. *Social Science Research, 39*, 396–404.

Chapter 21
Robust Minimal Markov Model for Dengue Virus Type 3

Jesús E. García and V. A. González-López

21.1 Introduction

The Dengue virus (DENV) is very effective in attacking populations, since by means of its four types consolidates its effective and cyclic contamination, alternating its types for years. The four types produce the same affections and potential clinical fatalities causes by *dengue hemorrhagic fever* (DHF) and *dengue shock syndrome* (DSS). Here, we analyze data coming from the first dengue fever (DF) outbreak in the central of Henan province, China, northern temperate regions, in late autumn, 2013. The strains detected in Henan Province were genotype II of DENV type 3. The epidemiological investigation showed that all cases of dengue had no travel history or collective activities. However, the business activities in Henan are associated with a flow of people coming and going to Laos, Guangzhou, and Yunnan, for work reasons. Also, in those places, the dengue prevailed during all 2013, see Sang et al. (2016). In the present work, we consider nine complete genomic sequences of DENV type 3, genotype II, registered in Henan during 2013 (fasta format). On them, we impose a structure of discrete stochastic processes in a finite alphabet with finite memory. We seek a representation of the genomic structure of these sequences and to achieve this goal, producing a robust representation, we first classify the nine sequences in order of representativeness. To do this, we apply the classifier introduced in Fernández et al. (2019). Secondly, and with the sequences classified as more representative, we build the hypothetical profile of a sequence of that group. We apply the partition Markov models, introduced in García and González-López (2017). In parallel, we check the conjecture of some possible connections between the genomic sequences of Henan and other coming from a province of China. So we

J. E. García (✉) · V. A. González-López
Department of Statistics, University of Campinas, Campinas, SP, Brazil
e-mail: jg@ime.unicamp.br; veronica@ime.unicamp.br

C. H. Skiadas, C. Skiadas (eds.), *Demography of Population Health, Aging and Health Expenditures*, The Springer Series on Demographic Methods and Population Analysis 50, https://doi.org/10.1007/978-3-030-44695-6_21

compare the Henan sequences with two other complete sequences available in the NCBI database to identify whether they are distant or close to the native ones, for this comparison we use the metric introduced in García et al. (2018). We emphasize that in this work, we have used complete genomic sequences in order to guarantee the validity of the stochastic techniques described in Sect. 21.2.
In Sect. 21.2, we present the notions that we use in modeling, in Sect. 21.3, we describe the data and its source. In Sect. 21.4, we present the results. The conclusions follow in Sect. 21.5.

21.2 Preliminaries

Let (X_t) be a discrete time, order o (with $o < \infty$) Markov chain on a finite alphabet A. Let us call $\mathcal{S} = A^o$ the state space, denote the string $a_m a_{m+1} \ldots a_n$ by a_m^n where $a_i \in A$, $m \le i \le n$. For each $a \in A$ and $s \in \mathcal{S}$ define the conditional probability $P(a|s) = \text{Prob}(X_t = a|X_{t-o}^{t-1} = s)$. In a given sample x_1^n, coming from the stochastic process, the number of occurrences of s in the sample x_1^n is denoted by $N_n(s)$ and the number of occurrences of s followed by a in the sample x_1^n is denoted by $N_n(s, a)$. In this way, $\frac{N_n(s,a)}{N_n(s)}$ is the estimator of $P(a|s)$.
Consider now, two Markov chains $(X_{1,t})$ and $(X_{2,t})$, of order o, arranged on the finite alphabet A with state space $\mathcal{S}$. Given $s \in \mathcal{S}$ denote by $\{P(a|s)\}_{a \in A}$ and $\{Q(a|s)\}_{a \in A}$ the sets of conditional probabilities of $(X_{1,t})$ and $(X_{2,t})$ respectively. We give a local metric d_s (introduced by García et al. 2018) that, when evaluated in a given string s, allows us to define how far or near the processes are.

Definition 2.1 Consider two Markov chains $(X_{1,t})$ and $(X_{2,t})$, of order o, with finite alphabet A, state space $\mathcal{S} = A^o$ and independent samples $x_{1,1}^{n_1}$, $x_{2,1}^{n_2}$ respectively.

i. Define for a string $s \in \mathcal{S}$,

$$d_s(x_{1,1}^{n_1}, x_{2,1}^{n_2}) = \frac{\alpha}{(|A|-1)\ln(n_1+n_2)} \sum_{a \in A} \Big\{ N_{n_1}(s,a) \ln\left(\frac{N_{n_1}(s,a)}{N_{n_1}(s)}\right) + N_{n_2}(s,a) \ln\left(\frac{N_{n_2}(s,a)}{N_{n_2}(s)}\right) - N_{n_1+n_2}(s,a) \ln\left(\frac{N_{n_1+n_2}(s,a)}{N_{n_1+n_2}(s)}\right) \Big\},$$

ii.

$$\text{dmax}(x_{1,1}^{n_1}, x_{2,1}^{n_2}) = \max_{s \in \mathcal{S}} \{d_s(x_{1,1}^{n_1}, x_{2,1}^{n_2})\},$$

with $N_{n_1+n_2}(s,a) = N_{n_1}(s,a) + N_{n_2}(s,a)$, $N_{n_1+n_2}(s) = N_{n_1}(s) + N_{n_2}(s)$, where N_{n_1} and N_{n_2} are given as usual, computed from the samples $x_{1,1}^{n_1}$ and $x_{2,1}^{n_2}$ respectively. With α a real and positive value.

Both notions (Definition 2.1: i. and ii.) are consistent, so when the sample sizes increase, if the underlying law generating both samples is the same, i. (and ii.) goes to zero, and if the underlying law generating the samples are different, i. (and ii.) goes to infinity. Depending on the circumstances we can apply one or the other notion, since i. is local (for each state $s \in \mathcal{S}$), and ii. is global (for the whole state space $\mathcal{S}$). On the other hand, i. is a metric in the strict sense of the term, because it is non-negative, symmetric and it verifies the triangular inequality. Already, option ii barely loses the last property cited for i.
The metric i. is derived from the Bayesian Information Criterion (BIC) computed for two circumstances: (a) assuming that the samples come from the same stochastic law, (b) supposing that the samples do not come from the same stochastic law. Under the formulation of the BIC criterion and when the sample sizes grow (a) is pointed out as true if the BIC in such case adopts a higher value than when computing it in the case (b). When comparing these BIC values, the metric i. results (see García et al. 2018).
The BIC criterion was originally formulated with the constant $\alpha = 2$ (see Schwarz 1978), but it is still valid for any other positive α value. If $\alpha = 2$ it is possible to obtain a cutoff value for the metric i. so that if $d_s < 1$, the samples can be considered as generated by the same law for that state s (see García et al. 2018). Next, the Definition 2.1 ii. is used to formulate the sequence classifier.

Definition 2.2 Given a finite collection $\{x_{j,1}^{n_j}\}_{j=1}^m$ of samples from the processes $\{X_{j,t}\}_{j=1}^m$ with probabilities $\{P_j\}_{j=1}^m$, over the finite alphabet A, with state space $\mathcal{S} = A^o$ ($o < \infty$). For a fixed $i \in \{1, 2, \ldots, m\}$ define

$$V(x_{i,1}^{n_i}) = \text{median}\{\text{dmax}(x_{i,1}^{n_i}, x_{j,1}^{n_j}) : j \neq i, 1 \leq j \leq m\}.$$

Where, given a sequence $\{z_j\}_{j=1}^l$, $\text{median}\{z_j, 1 \leq j \leq l\} = z_{(k+1)}$ if $l = 2k+1$ and $\text{median}\{z_j, 1 \leq j \leq l\} = \frac{z_{(k)}+z_{(k+1)}}{2}$ if $l = 2k$, for k an integer and $z_{(j)}$ denoting the jth order statistic of the collection $\{z_j\}_{j=1}^l$.

The next results, proved in Fernández et al. (2019), give us an adequate tool to classify sequences, according to their underlying laws.
Under the assumptions of Definition 2.2, for each i, $1 \leq i \leq m$, set $\xi_i = |\{j : 1 \leq j \leq m, P_j = P_i\}|$,

a.

$$V(x_{i,1}^{n_i}) \underset{\min\{n_1,\cdots,n_m\}\to\infty}{\longrightarrow} \infty \text{ if, and only if, } \xi_i \leq \lceil \frac{m}{2} \rceil.$$

b.

$$V(x_{i,1}^{n_i}) \underset{\min\{n_1,\cdots,n_m\}\to\infty}{\longrightarrow} 0, \text{ if, and only if, } \xi_i > \lceil\frac{m}{2}\rceil.$$

V is a classifier with the ability to identify the most representative sequences of the collection with the lowest values, attributing high values to the least representative ones. Moreover, V takes advantage of the properties of the Definition 2.1, being the value 1 still a referential, since if, in the computation of d_s is used $\alpha = 2$, the properties mentioned above will remain. A relevant aspect in V is that its efficiency as a classifier depends on the existence of a law of the majority and that such law prevails in more than 50% of the collection. V is consistent if more than half of the sequences are generated by identical law, as shown by the result b. and it fails if it is not the case, as the result a. shows.
As anticipated in the introduction, once the sequences are classified, in order of representativeness, we can use a subset of sequences to define a global profile of the performance of a hypothetical sequence, following the law of the majority. To achieve this purpose we use a model introduced in García and González-López (2017) and defined as follows.

Definition 2.3 Let (X_t) be a discrete time order o Markov chain on a finite alphabet A, with state space $\mathcal{S} = A^o$,

i. $s, r \in \mathcal{S}$ are equivalent (denoted by $s \sim_p r$) if $P(a|s) = P(a|r)\ \forall a \in A$.
ii. (X_t) is a Markov chain with partition $\mathcal{L} = \{L_1, L_2, \ldots, L_{|\mathcal{L}|}\}$ if this partition is the one defined by the equivalence relationship $\sim_p$ introduced by item i.

This model aims to reduce the parameters to be estimated, which are the transition probabilities. By identifying states that share the same probabilities, it is possible to use the occurrence of all the states that make up the same part L_i to estimate a single conditional probability.
Under this model, it is necessary to estimate the partition $\mathcal{L}$ and then the transition probabilities conditioned to each part of the partition. The estimation process can be carried out consistently, through the BIC criterion, as shown in García and González-López (2017), since it allows defining a metric in the state space to decide which states make up each part L_i. Given x_1^n a sample of (X_t), if $\hat{\mathcal{L}} = \{L_1, \cdots, L_k\}$ is the estimated partition, the conditional probability is estimated by $\hat{P}(a|L) = \frac{N_n(L,a)}{N_n(L)}$ with $N_n(L) = \sum_{s\in L} N_n(s)$ and $N_n(L, a) = \sum_{s\in L} N_n(s, a)$, for $a \in A$.
Suppose that there are several realizations that can be used to define the model – Definition 2.3. In that case, we consider the sequences as independent to (a) construct the parts (b) compute the estimators of each conditional probability.

Table 21.1 Complete sequences of Dengue Virus Type 3, genotype II, from the outbreak in Henan, China during 2013

Sequence	Accession number	Size
HN/2013/110	KJ622199	10710
HN/2013/108	KJ622198	10710
HN/2013/107	KJ622197	10710
HN/2013/92	KJ622196	10710
HN/2013/50	KJ622195	10711
HN/2013/22	KJ622194	10711
HN/2013/21	KJ622193	10706
HN/2013/20	KJ622192	10710
HN/2013/11	KJ622191	10710

21.3 Dengue Virus Type 3

The sequences (see Table 21.1) were obtained from http://www.ncbi.nlm.nih.gov/ (NCBI – National Center for Biotechnology Information). The nine sequences are complete sequences of dengue virus type 3 genotype II, isolated in Henan, China, during the Dengue Fever (DF) outbreak in 2013. Coming from the Henan Center for Disease Control and Prevention, No. 105 Nongyedonglu, Zhengzhou, Henan 450016, China. The alphabet is $A = \{a, c, g, t\}$ and the data follows the *FASTA* format, for example, the sequence HN/2013/110 begins with

$$agttgttagtctacgtggaccgacaagaacagtttcgactcggaagcttgcttaacgtag\ldots$$

The Henan sequences (Table 21.1) are part of a study carried out in Sang et al. (2016), in which are included several regions of China to investigate the transmission dynamics of the Dengue considering the geographical distribution, in the Dengue outbreak (types: 1 and 3) happened in China during 2013. We also compare the 9 Henan complete sequences with other two complete sequences of DENV type 3, genotype II: accesion numbers: KF824902 (size 10707), KF824903 (size 10707) from Xishuangbanna obtained in 2013 (see Sang et al. 2016), also available in the NCBI base.

According to the sample sizes reported in Table 21.1, the memory to be used must follow $o < 5 = \lfloor \log_{|A|}(10700) \rfloor - 1$, with $|A| - 4$. It is given that the bases of A are organized in triples, $o = 3$. In this way, the state space $\mathcal{S}$ is composed by concatenations $a_1a_2a_3$ with $a_i \in \{a, c, g, t\}$.

21.4 Results

Tables 21.2 and 21.3 expose the dmax values calculated from the Definition 2.1, ii. It is confirmed from these values that the sequences of Henan are closer to each other as well as the sequences of Xishuangbanna are also close to each other, relative to

Table 21.2 *dmax* values (see Definition 2.1-ii.) between sequences of Henan set, y = 62219

dmax	KJy2	KJy3	KJy4	KJy5	KJy6	KJy7	KJy8	KJy9
KJy1	0.001649	0.003179	0.002572	0.003269	0.002007	0.001649	0.002020	0.001649
KJy2	–	0.001660	0.001440	0.001087	0.002007	0.000600	0.000858	0.000697
KJy3	–	–	0.001660	0.002826	0.002007	0.001660	0.001660	0.001660
KJy4	–	–	–	0.001440	0.002608	0.001440	0.001440	0.001440
KJy5	–	–	–	–	0.002007	0.001075	0.001489	0.001075
KJy6	–	–	–	–	–	0.002007	0.002007	0.002007
KJy7	–	–	–	–	–	–	0.000838	0.000564
KJy8	–	–	–	–	–	–	–	0.001489

Table 21.3 *dmax* values (see Definition 2.1-ii.) between the Xishuangbanna sequences and the Henan sequences. dmax(KFx2,KFx3) = 0.003141, x = 82490 and y = 62219

	KJyi, i = 1, 2, 4, 5, 6, 7, 8, 9	KJy3
KFx2	0.048113	0.048114
KFx3	0.048113	0.048114

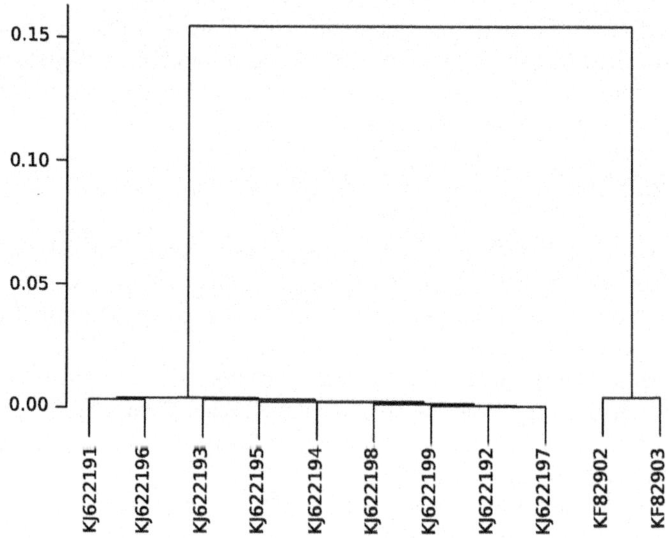

Fig. 21.1 Dendrogram by *average* criterion built from the *dmax* values, reported in Tables 21.2 and 21.3

the distance between the members of the Henan and Xishuangbanna groups, from Table 21.3. Figure 21.1 shows a dendrogram built using the *dmax* values for each pair of sequences (see values in Tables 21.2 and 21.3). As we can visualize the sequences of Xishuangbanna can be considered as another cluster.

Tables 21.4 and 21.5 show the results of the classification (see Definition 2.2) in two situations (a) considering the set of 9 complete sequences of Henan, (b) considering

Table 21.4 Classification of Henan Sequences

Sequence	V
KJ622197	0.001258
KJ622192	0.001263
KJ622194	0.001440
KJ622195	0.001465
KJ622199	0.001465
KJ622198	0.001489
KJ622193	0.001660
KJ622196	0.002007
KJ622191	0.002013

Table 21.5 Classification including Foreing Sequences

Sequence	V
KJ622192	0.001544
KJ622197	0.001544
KJ622194	0.001550
KJ622199	0.001569
KJ622198	0.001575
KJ622195	0.001748
KJ622193	0.001834
KJ622196	0.002007
KJ622191	0.002296
KF824902	0.048113
KF824903	0.048113

the set of 11 sequences, that includes the complete sequences of Henan and those of Xishuangbanna.

According to our results, the sequences that best represent Henan's set are KJ622197 and KJ622192. And those two that least represent the Henan's set are KJ622196 and KJ622191. Moreover, for the second comparison, the sequences that best represent the set are KJ622192 and KJ622197. And those two that least represent the set are KF824902 and KF824903. That is, although the original set (only formed by Henan sequences) has undergone alterations with the inclusion of two sequences that appear to be quite different from the others, the representation was maintained by the same sequences indicated by Table 21.4. And as expected, the less representative sequences are the foreign sequences.

Note how the inclusion of the two Xishuangbanna sequences has affected the classification of the most representative sequences, an alteration that can be perceived when comparing the classifications reported in Tables 21.4 and 21.5. The variability introduced by the sequences of Xishuangbanna generates an increase in the value of the classifier V.

In Table 21.6 we show the estimated partition from the most representative sequence (lowest V), when considering the set of Henan sequences. In that case, the partition

Table 21.6 Estimated partition, from the sequence KJ622197 ($V = 0.001258$), see Table 21.4

Part	States
L_1	*aaa, gcg, tcg, cag, gca, aga, gaa, tag, aag, gag, gga, tgg, agg, ggg*
L_2	*aac, gcc, atc, cac, cgc, ttc, agc, ctc, tac, gac, acc, ggc, gtc, ccc, tcc, tgc*
L_3	*aat, cat, gtg, ata, gct, ctg, ttg, atg*
L_4	*aca, cga, cgg, cca, cta, tta, tct, caa, tca, ccg, tga, taa*
L_5	*acg, ctt, gta, act, agt, ttt, tat, ggt, tgt, att, gat, cct, gtt, cgt*

Table 21.7 Estimated conditional probabilites $\hat{P}(\cdot|L_i), i = 1, \cdots, 5, \cdot \in A$, from the sequence KJ622197 ($V = 0.001258$), see Tables 21.4 and 21.6. In bold type the highest values per line

i	a	c	g	t
1	**0.385772**	0.182475	0.270388	0.161365
2	**0.418481**	0.234562	0.103510	0.243447
3	0.238727	0.168435	**0.407162**	0.185676
4	**0.303811**	0.233494	0.227590	0.235105
5	0.165637	0.226277	**0.340258**	0.267827

Table 21.8 Estimated partition, from the two most representative sequences KJ62219x, x = 2, 7, see Tables 21.4 and 21.5

Part	States
L_1	*aaa, gcg, tcg, cag, gca, aga, gaa, tag*
L_2	*aac, gcc, atc, cac, cgc, ttc, agc, ctc, tac, gac*
L_3	*aag, gag, gga, tgg, agg, ggg*
L_4	*aat, cat, gtg, ata, gct, ctg, ttg, atg*
L_5	*aca, cga, cgg, cca, cta, tta, tct, caa, tca, ccg, tga, taa*
L_6	*acc, ggc, gtc, ccc, tcc, tgc*
L_7	*acg, ctt, gta*
L_8	*act, agt, ttt, tat, ggt, tgt, att, gat, cct, gtt, cgt*

is composed by 5 parts. Table 21.7 reports the resulting transition probabilities for the model detailed in Table 21.6.

We see that when incorporating the second most representative sequence according to V (Table 21.4, Henan set) the parts originally pointed by the sequence of minor V (KJ622197), are separated into others, giving space to the partition reported in Table 21.8. See the relationship between the parts in Table 21.10. The effect of this separation of parts is evident from the comparison of Tables 21.7 and 21.9. For example, the parts that remained equally constituted maintain the probabilities (compare line 3 (4) – Table 21.7 and line 4 (5) – Table 21.9). Already, in relation to the parts that have been separated from one case to another, there are several patterns that are revealed, for example, part 1 of Table 21.6 reveals conditional probabilities relatively similar to parts 1 and 3 of Table 21.8. The reason being that this pattern does not seem so clear when comparing line 5 of Table 21.7 and lines 7 and 8 of Table 21.9.

Table 21.9 Estimated conditional probabilites $\hat{P}(\cdot|L_i), i = 1, \cdots, \cdot \in A$, from the two most representative sequences KJ62219x, x = 2, 7, see Tables 21.4, 21.5 and 21.8. In bold type the highest values per line

i	a	c	g	t
1	**0.349872**	0.192187	0.282863	0.175078
2	**0.426441**	0.240000	0.116610	0.216949
3	**0.422874**	0.172434	0.257478	0.147214
4	0.238806	0.168159	**0.407297**	0.185738
5	**0.303356**	0.233557	0.227919	0.235168
6	**0.403351**	0.224227	0.078608	0.293814
7	0.176849	**0.308682**	0.286174	0.228296
8	0.163210	0.209113	**0.351581**	0.276097

Table 21.10 Relation between partitions: from the sequence KJ622197 (left) and from the sequences KJ62219x, x = 2, 7 (right)

Part from Table 21.6	Parts from Table 21.8
1	1 & 3
2	2 & 6
3	4
4	5
5	7 & 8

Table 21.11 From left to right: (1) Sequences used to fit the model (according to Table 21.4); (2) The number of parts in the partition $|\mathcal{L}|$

Sequences KJ62219x	Total of parts $|\mathcal{L}|$
$x = 7$	5
$x = 7, 2$	8
$x = 7, 2, 4$	9
$x = 7, 2, 4, 5$	11
$x = 7, 2, 4, 5, 9$	13
$x = 7, 2, 4, 5, 9, 8$	14
$x = 7, 2, 4, 5, 9, 8, 3$	17
$x = 7, 2, 4, 5, 9, 8, 3, 6$	18
$x = 7, 2, 4, 5, 9, 8, 3, 6, 1$	18

The following table shows how the number of parts of the partition of the model (see Definition 2.3) is affected, by incorporating more genomic sequences into the model: first line using the most representative sequence, second line using the two most representative, …kth line using the k most representative sequences, for k = 1, …, 9, where the situations of the 1st and 2nd lines were already shown and the last represents the model using all the sequences (Table 21.11).
As expected, the relationship between the parts defined by the most representative sequence and the parts from the six most representative sequences follow the pattern of the comparison made with the two most representative sequences. In summary, what is perceived is that the parts, in Tables 21.6 and 21.7 that have a probability of 40%, have preserved the magnitudes of the conditional probabilities in the new partition, defined by the six most representative sequences (see Tables 21.12 and 21.13). For example, see the cases reported in lines 1, 2, and 3 of Table 21.14.

Table 21.12 Estimated partition, from the six most representative sequences KJ62219x, x = 2, 4, 5, 7, 8, 9, see Tables 21.4 and 21.5

Part	States
L_1	$aaa, gcg, tcg, cag, gca, aga, gaa, tag$
L_2	$aac, gcc, atc, cac, cgc, ttc$
L_3	aag, gag, gga, tgg
L_4	$aat, cat, gtg, ata, gct, ctg, ttg$
L_5	$aca, cga, cgg, cca, cta, tta$
L_6	$acc, ggc, gtc, ccc, tcc, tgc$
L_7	acg, ctt, gta
L_8	$act, agt, ttt, tat, ggt, tgt, att, gat, cct, gtt$
L_9	agc, ctc, tac, gac
L_{10}	agg, ggg
L_{11}	atg
L_{12}	caa, tca, ccg, tga, taa
L_{13}	cgt
L_{14}	tct

Table 21.13 Estimated conditional probabilites $\hat{P}(\cdot|L_i), i = 1, \cdots, 5, \cdot \in A$, from the two most representative sequences KJ62219x, x = 2, 4, 5, 7, 8, 9, see Tables 21.4, 21.5 and 21.12. In bold type the highest values per line

i	a	c	g	t
1	**0.350147**	0.192078	0.282702	0.175074
2	**0.441852**	0.213587	0.123259	0.221302
3	**0.406225**	0.179569	0.272812	0.141394
4	0.235984	0.176011	**0.394744**	0.193262
5	**0.283128**	0.227324	0.258865	0.230683
6	**0.403479**	0.224179	0.078590	0.293751
7	0.176754	**0.308516**	0.286020	0.228709
8	0.165528	0.209472	**0.345782**	0.279218
9	**0.402432**	0.280260	0.106900	0.210407
10	**0.469200**	0.152711	0.214681	0.163408
11	0.251077	0.132923	**0.464615**	0.151385
12	**0.335936**	0.239899	0.198321	0.225844
13	0.107143	0.196429	**0.500000**	0.196429
14	0.217391	0.234783	0.208696	**0.339130**

Table 21.14 Relation between partitions: from the sequence KJ622197 (left) and from the sequences KJ62219x, x = 2, 4, 5, 7, 8, 9 (right)

Part from Table 21.6	Parts from Table 21.12
1	1, 3 & 10
2	2, 6 & 9
3	4 & 11
4	5, 12 & 14
5	7, 8 & 13

21.5 Conclusions

In this paper, the stochastic profile of genomic sequences of DENV 3 – genotype II, coming from the 2013 outbreak in Henan Province, China is described. We combine a series of stochastic tools from the latest developments in the investigation of discrete stochastic processes with finite memory and finite alphabets. We determine the distance between the genomic sequences and based on these distances, an indicator of representativeness is established for each sequence, which orders the list of sequences in order of representativeness of the set. The most representative receives the lowest indicator, and the least representative receives the highest indicator. Through the classifier, it is possible to establish which sequences best represent the set and from that information, it is possible to select a subset of the set of sequences to establish a robust profile of the set. Under the theoretical assumption of the classifier (see Definition 2.2 and Fernández et al. 2019) if more than 50% of the sequences follow the same stochastic law, the classifier is capable of attributing the best classifications to the sequences that follow the law of the majority. That is, under this assumption the model would be built with the sequences that actually come from the law of the majority because the sequences with higher classifications (lower indices of representativeness) are excluded from the model. From the total of 9 Henan sequences, we have selected the six most representative, based on the classifier. The profile found through the Partition Markov Model (Definition 2.3 – García and González-López 2017) shows the stochastic performance of the sequences that can be described by 14 basic units (called parts). The states (from the state space $\mathcal{S}$) that make up each of these basic units share the same transition probability for any element of the genomic alphabet $A = \{a, c, g, t\}$. We note that 6 of these units: 1, 2, 3, 6, 9 and 10 show a predilection for choosing the element a as next element, and units 4 and 11 choose the element g as the next element. These units have shown consistency in establishing these preferences for doing the transition. That is, in the process of incorporation into the model, of sequences in order of representativeness, these preferences have been maintained (see Tables 21.6–21.7, 21.8–21.9 and 21.12–21.13). In relation to the comparison of the Henan sequences and those from Xishuangbanna, we find that the groups are shown separately. Although the values of $dmax$ are all less than 1, which indicates their similarity, we see that there is less homogeneity between the groups than within them, which makes us suspect that they could be strains of different origins. This finding raises the possibility that the origin of the outbreak in Henan does not necessarily come from Xishuangbanna.

In the present paper, we apply a series of tools from stochastic processes, for the construction of a profile that represents the stochastic behavior of genomic sequences collected in the outbreak in Henan during 2013. This could be a way of building a genomic mosaic useful in comparisons with other genomic mosaics. For example, coming from other regions, from which it is speculated that the Henan outbreak came.

References

Fernández, M., García Jesús, E., Gholizadeh, R., & González-López, V. A. (2019). Sample selection procedure in daily trading volume processes. *Mathematical Methods in the Applied Sciences*. https://doi.org/10.1002/mma.5705.

García Jesús, E., Gholizadeh, R., & González-López, V. A. (2018). A BIC-based consistent metric between Markovian processes. *Applied Stochastic Models in Business and Industry, 34*(6), 868–878.

García Jesús, E., & González-López, V. A. (2017). Consistent estimation of partition Markov models. *Entropy, 19*(4), 160.

Sang, S., Wang, S., Bi, P., Lv, M., & Liu, Q. (2016). The epidemiological characteristics and dynamic transmission of dengue in China, 2013. *PLoS Neglected Tropical Diseases, 10*(11), e0005095.

Schwarz, G. (1978). Estimating the dimension of a model. *The Annals of Statistics, 6*(2), 461–464.

Chapter 22
Determining Influential Factors in Spatio-temporal Models

Rebecca Nalule Muhumuza, Olha Bodnar, Joseph Nzabanita, and Rebecca N. Nsubuga

22.1 Introduction

Influenza is an airborne disease caused by influenza viruses, Type A, Type B, Type C in humans. Symptoms of influenza include high fever, muscle pains, sore throat, headache, sneezing, feeling tired. It is among the emerging and re-emerging pathogens. As a consequence there has been a growing interest in human epidemiology to gain insight into the disease dynamics using stochastic epidemic modeling.

Influenza epidemic is usually present in winter season when households are in close contact with the infected, but it can as well occur in public places and work places. Different types of stochastic models have been suggested in the literature to capture the dynamics in the influenza epidemic. One type of models is based on the theory of stochastic processes and have been intensively discussed in applied

R. N. Muhumuza (✉)
Division of Applied Mathematics, School of Education, Culture and Communication (UKK), Mälardalen University, Västerås, Sweden

Department of Mathematics, Busitema University, Tororo, Uganda, East Africa

Department of Mathematics, Makerere University, Kampala, Uganda
e-mail: rebecca.nalule@mdh.se

O. Bodnar
Unit of Statistics, School of Business, Örebro University, Örebro, Sweden
e-mail: olha.bodnar@oru.se

J. Nzabanita
Department of Mathematics, CST-University of Rwanda, Kigali, Rwanda

R. N. Nsubuga
MRC/UVRI and LSHTM Uganda Research Unit, Entebbe, Uganda

C. H. Skiadas, C. Skiadas (eds.), *Demography of Population Health, Aging and Health Expenditures*, The Springer Series on Demographic Methods and Population Analysis 50, https://doi.org/10.1007/978-3-030-44695-6_22

probability recently (cf. Keeling and Rohani 2011 and Zarebski et al. 2017). The other type of the stochastic models are coming from the theory of multivariate statistics with the aim to capture both the spatial and the temporal dynamics in the observed data (cf. Serfling 1963, Ozonoff et al. 2006, Paul and Held 2011, and Held et al. 2017).

In this paper we extend Serfling's model in several directions. First, a latent variable is included in the model which is used to describe the unobservable number of influenza cases in the whole region. Second, this latent variable is considered as a state variable in the proposed Kalman filter whose dynamics are captured by the state equation. Third, the application of the Kalman filter representation of the model equation allows to incorporate temporal dependence structure into the model. These three generalizations lead to the consideration of a general spatio-temporal model which is widely used in environmental statistics (see, e.g. Fassò et al. (2007), Bodnar and Schmid (2010), among others).

Spatio-temporal processes provide a very useful tool for analyzing complicated data showing both temporal and spatial dependence. Although the theory is widely used in many areas of research and it is well presented in statistical literature, the recent developments of computer industry provide the possibility for further developments of new stochastic methods of data analysis and their practical applications which can be used to deal with challenging problems observed in modern big data analysis. In the present paper we contribute to the methods of construction of the spatial interpolation and temporal prediction in the case of influenza data.

The prediction of the values of a spatio-temporal process is a goal in different areas of application in statistics. The classical approach is based on the linear kriging predictor, which is obtained by minimizing the mean squared error (cf., Zimmerman 2006, Genton 2007, and Cressie and Wikle 2015). The linear predictor, however, could be too restrictive, especially, when the spatio-temporal process is not a Gaussian process (see, e.g. Stein 2012). As a possible solution, Bodnar and Schmid (2010) extended the linear kriging predictor by introducing the locally weighted scatterplot smoothing (LOESS) predictor which is one of many modern modeling methods based on classical approaches and applied usually in regression analysis as a generalization of both linear and nonlinear least squares regressions. The idea behind the LOESS approach is to fit simple models to localized subsets of the data to build up a function that describes the deterministic part of the variation in the data, point by point (see, e.g., Cleveland 1979). The length of the subset could be chosen differently, for example by cross-validation (Givens and Hoeting 2012).

The rest of the paper is organized as follows. The spatio-temporal model for influenza data is introduced in the next section. The estimation of the model parameters is discussed in Sect. 22.3. In Sect. 22.4, we present the LOESS predictor and provide the spatial prediction and the temporal forecast related to the proposed model, while the choice of the sub-sample via cross-validation is provided in Sect. 22.5. The application of the theoretical findings to real data is described in Sect. 22.6.

22.2 Spatio-temporal Model for Weekly Influenza Data

In this section we develop a non-linear extension of Serfling's model used to capture the dynamics in the weekly influenza data which is implemented by the Centers for Disease Control and Prevention (CDC). Let Z_t denote observed weekly influenza at time t. Then Serfling's model (see, Serfling 1963) is given by

$$Z_t = \alpha_0 + \alpha_1 t + \beta_1 \sin\left(\frac{2\pi t}{52}\right) + \beta_2 \cos\left(\frac{2\pi t}{52}\right) + \varepsilon_t, \tag{22.1}$$

where a pair of harmonic terms is used to capture the underlying sinusoidal behavior of seasonal influenza. The model residuals $\{\varepsilon_t\}$ are assumed to be independent and normally distributed with zero mean and variance σ_ε^2. The parameters α_0, α_1, β_1, and β_2 are estimated by fitting the model to real data by the maximum likelihood approach or by the ordinary least squares. Several extensions of this model are discussed in Ozonoff et al. (2006).

We consider a spatio-temporal extension of Serfling's model which could be used to model the influenza data measured in several locations simultaneously. Moreover, the LOESS (locally weighted scatterplot) will be used for a more flexible modeling of the nonlinear dynamics observed in data as well as to construct a nonlinear interpolation of process values where no measurement is available and to make a prediction of future realizations.

Let $Z(\mathbf{s}, t)$ denote the number of influenza cases observed in the region $\mathbf{s}$ (determined by the coordinates of its center) at time t and define by $\mathbf{Z}_t = (Z(\mathbf{s}_0, t), Z(\mathbf{s}_1, t), \ldots, Z(\mathbf{s}_n, t))'$ the observation vector of all cases available at time point t. We model the dynamics in $\mathbf{Z}_t$ by applying a generalization of the spatio-temporal process suggested by Fassò et al. (2007), Fassò and Cameletti (2010) which is given by

$$\mathbf{Z}_t = \mathbf{U}_t + \boldsymbol{\varepsilon}_t\,, \tag{22.2}$$

$$\mathbf{U}_t = \mathbf{f}(\mathbf{X}_t, Y_t) + \boldsymbol{\omega}_t\,, \tag{22.3}$$

$$Y_t = gY_{t-1} + \eta_t\,, \tag{22.4}$$

$$Y_0 \sim \mathcal{N}(\mu_0, \sigma_0^2)\,, \tag{22.5}$$

where the observable spatio-temporal process $\{\mathbf{Z}_t\}$ is related to the "true" spatio-temporal process $\{\mathbf{U}_t\}$, which is not observable, in the observation equation (22.2), while the dynamics of $\{\mathbf{U}_t\}$ are captured in the model equation (22.3) by using the covariates vector $\mathbf{X}_t$ and the latent process Y_t described in (22.4) and (22.5). The latent process $\{Y_t\}$ is assumed to be space constant and is used to model the time series behaviour of the influenza cases in the whole region.

The three error processes $\{\boldsymbol{\varepsilon}_t\}$, $\{\boldsymbol{\omega}_t\}$, and $\{\eta_t\}$ are assumed to be mutually independent and normally distributed with $\boldsymbol{\varepsilon}_t \sim \mathcal{N}_{n+1}(\mathbf{0}, \boldsymbol{\Sigma}_\varepsilon)$, $\boldsymbol{\Sigma}_\varepsilon = \sigma_\varepsilon^2\mathbf{I}$, and $\boldsymbol{\eta}_t \sim \mathcal{N}(\mathbf{0}, \sigma_\eta^2)$. The process $\{\boldsymbol{\omega}_t\}$ is assumed to be pure spatial Gaussian with mean zero and a time constant covariance function

$$Cov(\omega(\mathbf{s}, t), \omega(\tilde{\mathbf{s}}, t)) = \sigma_\omega^2 C_\theta(||\mathbf{s} - \tilde{\mathbf{s}}||),$$

where θ is a parameter and $||\mathbf{s} - \tilde{\mathbf{s}}||$ denotes the distance between the two regions $\mathbf{s}$ and $\tilde{\mathbf{s}}$ from which the observations are taken. $C_\theta(\cdot)$ stands for the covariogram which is assumed to be isotropic.

In the model equation (22.3) the unobservable spatio-temporal process $\{\mathbf{U}_t\}$ is modeled as a measurable function of the $(n+1) \times d$-dimensional matrix $\mathbf{X}_t$ of covariates observed at time t and the latent temporal process $\{Y_t\}$. Several choices of the function $\mathbf{f}(\cdot, \cdot)$ are considered in the literature. If $\mathbf{f}(\cdot, \cdot)$ is a linear function, i.e.

$$\mathbf{f}(\mathbf{X}_t, Y_t) = \mathbf{X}_t \boldsymbol{\beta} + \mathbf{F}_t Y_t, \tag{22.6}$$

the model (22.2), (22.3), (22.4), and (22.5) becomes the general spatio-temporal process as defined by Fassò et al. (2007). Under this model assumption, the parameter vector $\boldsymbol{\beta}$ is the same for all regions and is used to capture the influence of $\mathbf{X}_t$ on $\{\mathbf{U}_t\}$. The $(n+1)$-dimensional vector $\mathbf{F}_t$ is assumed to be known and it specifies the influence of the latent process Y_t in each spacial point. If no specific information is provided, then the common choice is to set $\mathbf{F}_t = \mathbf{1}_{n+1}$ (see, Fassò and Cameletti 2010) where the symbol $\mathbf{1}_{n+1}$ stands for the $(n+1)$-dimensional vector of ones. Another possibility is to use the loadings of a principal component decomposition (see, e.g., Wikle and Cressie 1999 and Fassò et al. 2007). Finally, following Serfling's model we set $\boldsymbol{\beta} = (\alpha_0, \alpha_1, \beta_1, \beta_2)'$ and

$$\mathbf{X}_t = \mathbf{1}_{n+1} \left[1\ t\ \sin\left(\frac{2\pi t}{52}\right)\ \cos\left(\frac{2\pi t}{52}\right) \right].$$

Another possibility of the definition of $\mathbf{f}(\cdot, \cdot)$ is based on the semi-parametric approach. In particular, Bodnar and Schmid (2010) suggested to approximate $\mathbf{f}(\cdot, \cdot)$ locally by linear functions using the data from the nearest l regions. Several l can be considered with the most appropriate one to be chosen by cross-validation. We will follow this approach and will discuss the problems like parameter estimation, spatial interpolation, and temporal prediction in Sects. 22.3 and 22.4.

Finally, we point out that the general model (22.2) (22.3), (22.4), and (22.5) can be presented in the state-space framework. Namely, substituting (22.3) into (22.2) leads to

$$\mathbf{Z}_t = \mathbf{f}(\mathbf{X}_t, Y_t) + \mathbf{e}_t, \tag{22.7}$$

$$Y_t = g Y_{t-1} + \eta_t, \tag{22.8}$$

where $\{\mathbf{e}_t\} = \{\boldsymbol{\varepsilon}_t + \boldsymbol{\omega}_t\}$ is a zero-mean Gaussian process with the covariance matrix

$$\boldsymbol{\Sigma}_e = \sigma_\omega^2 \left(\Gamma(||\mathbf{s}_i - \mathbf{s}_j||)\right)_{i,j=0,\ldots,n} \tag{22.9}$$

with

$$\Gamma(h) = \begin{cases} 1 + \gamma & \text{for } h = 0 \\ C_\theta(h) & \text{for } h > 0 \end{cases}$$

and $\gamma = \sigma_\varepsilon^2/\sigma_\omega^2$. The equations (22.7) and (22.8) define a classical state-space model (see, e.g., Durbin and Koopman 2012) when $\mathbf{f}(.,.)$ is a linear function. In the empirical part of the paper $C_\theta(h) = \exp(-\theta h)$ is used following Fassò and Cameletti (2010).

22.3 Parameter Estimation

For given region $\mathbf{s}_0$ and l nearest regions $\mathbf{s}_{i_1}(\mathbf{s}_0), \mathbf{s}_{i_2}(\mathbf{s}_0), \ldots, \mathbf{s}_{i_l}(\mathbf{s}_0)$, the unknown model parameters are presented by $\boldsymbol{\Psi} = (\boldsymbol{\beta}', g, \mu_0, \sigma_0^2, \sigma_\eta^2, \sigma_\omega^2, \gamma, \theta)'$. Following Fassò and Cameletti (2010) we estimate $\boldsymbol{\Psi}$ by employing the maximum likelihood method.

The likelihood function of the model (22.2), (22.3), (22.4), and (22.5) depends non-linearly on the parameter vector $\boldsymbol{\Psi}$. As a result, an analytical expression of the estimated parameters cannot be derived and numerical procedure should be used instead. The first possibility is the application of Newton-Raphson's method (cf. Durbin and Koopman 2012), while we follow the approach considered by Fassò et al. (2007) that is based on the expectation-maximization (EM) algorithm.

Using the results presented in the appendix of Bodnar and Schmid (2010), the closed-form formulas for some components of the vector $\boldsymbol{\Psi}$ can be obtained. Let the estimator of $\boldsymbol{\Psi}$ after the k-th iteration of the EM algorithm be denoted by $\boldsymbol{\Psi}^{(k)} = (\boldsymbol{\beta}^{(k)\,\prime}, g^{(k)}, \mu_0^{(k)}, \sigma_0^{2\,(k)}, \sigma_\eta^{2\,(k)}, \sigma_\omega^{2\,(k)}, \gamma^{(k)}, \theta^{(k)})'$. Then, given the observed number of influenza cases $\tilde{\mathbf{Z}}_1(\mathbf{s}_0), \tilde{\mathbf{Z}}_2(\mathbf{s}_0), \ldots, \tilde{\mathbf{Z}}_T(\mathbf{s}_0)$ consisting of data from l nearest regions $\tilde{\mathbf{Z}}_t(\mathbf{s}_0) = (Z(\mathbf{s}_{i_1}(\mathbf{s}_0), t), Z(\mathbf{s}_{i_2}(\mathbf{s}_0), t), \ldots, Z(\mathbf{s}_{i_l}(\mathbf{s}_0), t))'$, it holds for the $(k+1)$-th iteration that

$$\boldsymbol{\beta}^{(k+1)} = \left(\sum_{t=1}^{T} \mathbf{X}_t' \boldsymbol{\Sigma}_e^{-1} \mathbf{X}_t\right)^{-1} \sum_{t=1}^{T} \mathbf{X}_t' \boldsymbol{\Sigma}_e^{-1} (\tilde{\mathbf{Z}}_t(\mathbf{s}_0) - y_t^T \mathbf{1}), \tag{22.10}$$

$$\sigma_w^{2\,(k+1)} = \frac{\sigma_w^{2\,(k)}}{Tl} tr(\boldsymbol{\Sigma}_e^{-1}\mathbf{W}), \tag{22.11}$$

$$g^{(k+1)} = \frac{S_{10}}{S_{00}}, \tag{22.12}$$

$$\sigma_\eta^{2\,(k+1)} = S_{11} - \frac{S_{10}^2}{S_{00}}, \tag{22.13}$$

$$\mu_0^{(k+1)} = y_0^T, \tag{22.14}$$

$$\sigma_0^{2\,(k+1)} = P_0^T, \tag{22.15}$$

with $S_{00}^k = (1/T)\sum_{t=1}^T((y_{t-1}^T)^2 + P_{t-1}^T)$, $S_{10}^k = (1/T)\sum_{t=1}^T(y_t^T y_{t-1}^T + P_{t,t-1}^T)$, $S_{11}^k = (1/T)\sum_{t=1}^T((y_t^T)^2 + P_t^T)$, $\mathbf{\Sigma}_e = \mathbf{\Sigma}_e^{(k)}$, and

$$\mathbf{W} = \sum_{t=1}^{T}(\tilde{\mathbf{Z}}_t(\mathrm{s}_0) - \mathbf{X}_t\boldsymbol{\beta} - \mathbf{1}y_t^T)(\tilde{\mathbf{Z}}_t(\mathrm{s}_0) - \mathbf{X}_t\boldsymbol{\beta} - \mathbf{1}y_t^T)' + P_t^T\mathbf{1}\mathbf{1}'), \tag{22.16}$$

where the quantities y_t^T, P_t^T, and $P_{t,t-1}^T$ are the Kalman smoother outputs which are calculated recursively using $\hat{\mathbf{\Psi}}^{(k)}$ (see, Fassò and Cameletti 2009). First, the predicted values are computed by

$$y_t^{t-1} = g^{(k)}y_{t-1}^{t-1} \qquad \text{and} \qquad P_t^{t-1} = (g^{(k)})^2 P_{t-1}^{t-1} + \sigma_\eta^{2\,(k)}$$

with the initial values $y_0^0 = \mu_0^{(k)}$ and $P_0^0 = \sigma_0^{2\,(k)}$. Second, the filtered values are given by

$$y_t^t = y_t^{t-1} + \mathbf{A}_t(\tilde{\mathbf{Z}}_t - \mathbf{X}_t\boldsymbol{\beta}^{(k)} - y_t^{t-1}\mathbf{1}) \qquad \text{and} \qquad P_t^t = P_t^{t-1} - P_t^{t-1}\mathbf{A}_t\mathbf{1},$$

where $\mathbf{A}_t = P_t^{t-1}\mathbf{1}'(P_t^{t-1}\mathbf{1}\mathbf{1}' + \mathbf{\Sigma}_e)^{-1}$. Finally, the smoothed values are obtained by the following Kalman smoother backward recursion for $t = T, \ldots, 1$

$$y_{t-1}^T = y_{t-1}^{t-1} + B_{t-1}(y_t^T - y_t^{t-1}) \qquad \text{and} \qquad P_{t-1}^T = P_{t-1}^{t-1} + B_{t-1}^2(P_t^T - P_t^{t-1}),$$

where $B_{t-1} = P_{t-1}^{t-1}g^{(k)}/P_t^{t-1}$. The initial values y_T^T and P_T^T are the output of the previously defined Kalman filter equations. The smoothed lag-one covariance is computed by

$$P_{t,t-1}^T = B_{t-1}P_t^t + B_{t-1}B_t(P_{t+1,t}^T - g^{(k)}P_t^t) \quad \text{for} \quad t = T-1, \ldots, 1$$

with $P_{T,T-1}^T = (I - \mathbf{A}_T\mathbf{1})g^{(k)}P_{T-1}^{T-1}$.

The remaining two parameters, namely $(\theta^{(k+1)}, \gamma^{(k+1)})'$ are obtained by numerical maximization of

$$T\log(|\mathbf{\Sigma}_e|) + tr(\mathbf{\Sigma}_e^{-1}\mathbf{W}),$$

where $\mathbf{\Sigma}_e$ is a function of $\theta^{(k+1)}$ and $\gamma^{(k+1)}$ as given in (22.9). This optimization can be performed by applying, for example, the Newton-Raphson algorithm (Givens and Hoeting 2012).

22.4 Spatial Interpolation and Temporal Prediction

We start with the discussion of the LOESS (locally weighted scatterplot smoothing) predictor as suggested in Bodnar and Schmid (2010). The aim is to construct a linear interpolation of $U(\mathbf{s}_0, t)$ using the observation of the nearest l regions, namely using $\tilde{\mathbf{Z}}_t(\mathbf{s}_0) = (Z(\mathbf{s}_{i_1}(\mathbf{s}_0), t), Z(\mathbf{s}_{i_2}(\mathbf{s}_0), t), \ldots, Z(\mathbf{s}_{i_l}(\mathbf{s}_0), t))'$, where $\mathbf{s}_{i_1}(\mathbf{s}_0), \mathbf{s}_{i_2}(\mathbf{s}_0), \ldots, \mathbf{s}_{i_l}(\mathbf{s}_0)$ correspond the nearest l regions to s_0. If $l = n$, then the LOESS predictor coincides with the linear predictor.

By construction, the LOESS kriging predictor is nonlinear and, thus, provides a flexible way of the spatial interpolation of the spatio-temporal process. Instead of using the measurement results from all regions, only the nearest l regions have an influence on the interpolated value. As a result, it also reduces the computation time needed to calculate the inverse of large-dimensional covariance matrix and avoids problems when this matrix is ill-conditioned. From the other side the LOESS predictor possesses locally all properties of the linear predictor. As a result, it is optimal in terms of minimizing the mean squared error between the observed and predicted values.

Mathematically, the LOESS kriging predictor is defined by $\tilde{\lambda}_{0,t} + \tilde{\boldsymbol{\lambda}}_t'\tilde{\mathbf{Z}}_t(\mathbf{s}_0)$ and it is obtained by minimizing the mean squared error given by

$$\begin{aligned} & E((U(\mathbf{s}_0, t) - \tilde{\lambda}_{0,t} - \tilde{\boldsymbol{\lambda}}_t'\tilde{\mathbf{Z}}_t(\mathbf{s}_0))^2) \\ &= (\mu_{0,t} - \tilde{\lambda}_{0,t} - \tilde{\boldsymbol{\lambda}}_t'\boldsymbol{\mu}_{\mathbf{s}_0,t})^2 + k_{0,t} - 2\tilde{\boldsymbol{\lambda}}_t'\mathbf{k}_{\mathbf{s}_0,t} + \tilde{\boldsymbol{\lambda}}_t'\mathbf{K}_{\mathbf{s}_0,t}\tilde{\boldsymbol{\lambda}}_t \,, \end{aligned}$$

where

$$\begin{aligned} \boldsymbol{\mu}_{\mathbf{s}_0,t} &= E(\tilde{\mathbf{Z}}_t(\mathbf{s}_0)), \\ \mathbf{k}_{\mathbf{s}_0,t} &= Cov(U(\mathbf{s}_0, t), \tilde{\mathbf{Z}}_t(\mathbf{s}_0)) = E((U(\mathbf{s}_0, t) - \mu_{0,t})(\tilde{\mathbf{Z}}_t(\mathbf{s}_0) - \boldsymbol{\mu}_{\mathbf{s}_0,t})), \\ \mathbf{K}_{\mathbf{s}_0,t} &= Cov(\tilde{\mathbf{Z}}_t(\mathbf{s}_0), \tilde{\mathbf{Z}}_t(\mathbf{s}_0)) = E((\tilde{\mathbf{Z}}_t(\mathbf{s}_0) - \boldsymbol{\mu}_{\mathbf{s}_0,t})(\tilde{\mathbf{Z}}_t(\mathbf{s}_0) - \boldsymbol{\mu}_{\mathbf{s}_0,t})'). \end{aligned}$$

The solution is expressed as

$$\hat{U}(\mathbf{s}_0, t; l) = \mu_{\mathbf{s}_0,t} + \mathbf{k}_{\mathbf{s}_0,t}'\mathbf{K}_{\mathbf{s}_0,t}^{-1}(\tilde{\mathbf{Z}}_t(\mathbf{s}_0) - \boldsymbol{\mu}_{\mathbf{s}_0,t}) \,. \tag{22.17}$$

Using the Kalman filter representation (22.7)–(22.8) of the considered spatio-temporal process, the prediction of future values of the process $\mathbf{Z}_t$ is constructed by using the observed data $\mathbf{Z}_1, \ldots, \mathbf{Z}_T$. For example, the predictor at time point $T+1$ of $\mathbf{U}_{T+1}$ is given by

$$\hat{\mathbf{U}}_{T+1} = \mathbf{f}(\mathbf{X}_{T+1}, \hat{Y}_{T+1}) \quad \text{with} \quad \hat{Y}_{T+1} = gY_T \,. \tag{22.18}$$

22.5 Choice of l Nearest Regions: Cross-Validation

An important question in the construction of the LOESS predictor is the determination of the number of the nearest regions. We propose to solve this problem by using cross-validation which is an important numerical procedure in statistics (see, Givens and Hoeting 2012), especially, when a kernel density estimator is constructed. The number of nearest region is chosen as the solution of the following minimization problem:

$$l_0 = \min_l CV_l \quad \text{with} \quad CV_l = \frac{1}{t}\sum_{t=1}^{T} E((U(\mathbf{s}_0, t) - \hat{U}(\mathbf{s}_0, t; l))^2) \tag{22.19}$$

The empirical counterpart of (22.19) is given by

$$\hat{l}_0 = \min_l \widehat{CV}_l$$

with

$$\widehat{CV}_l = \frac{1}{t(n+1)}\sum_{t=1}^{T}\sum_{i=0}^{n}((Z(\mathbf{s}_i, t) - \hat{U}(\mathbf{s}_i, t; l))^2) \tag{22.20}$$

which is used to compute the value of l in the empirical illustration of Sect. 22.6.

22.6 Empirical Study

In this section we apply the developed theoretical results to weekly data of confirmed influenza cases observed in the two southern states of Germany, namely Baden-Württemberg and Bavaria. The data were taken in the period from 2001 to 2007 and consist of 364×140 observations measured in each of 140 regions of these two states.

In Fig. 22.1 we present the values of the cross-validation criteria defined in (22.19) with the empirical counterpart given in (22.20) when the number of nearest regions to the region s_0 is equal to $l \in \{20, 25, \ldots, 130, 135, 139\}$. Large values of the $\widehat{CV}_l$ are shown in the figure when l is relatively small, that is $l \leq 50$. As l increases, a considerable improvement in the spatial interpolation is observed, especially for the values of l larger than 110 with the global minimum observed for $l = 125$. As a result, we will use $l = 125$ as the optimal number of nearest regions for which the contribution to the spatial interpolation of the weekly influenza cases appears to have be the most impact.

In Fig. 22.2 the original data of registered influenza cases in the two southern states of Germany, Baden-Württemberg and Bavaria, for the seventh week in year

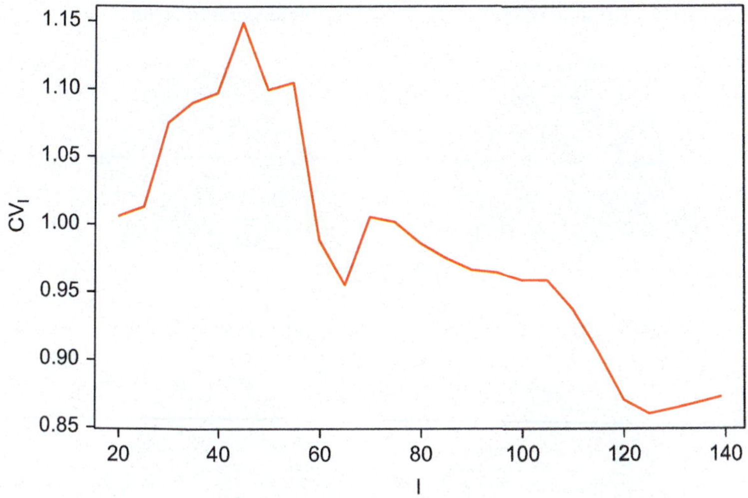

Fig. 22.1 Empirical cross-validation criteria $\widehat{CV}_l$ as a function of the number of nearest regions $l \in \{20, 25, \ldots, 130, 135, 139\}$

2007 are displayed in the top left plot together with three interpolations computed for $l = 65$ (top right plot), $l = 125$ (bottom left plot), and $l = 139$ (bottom right plot). The number $l = 65$ of the nearest regions is chosen due to the results presented in Fig. 22.1 where a local minimum of the cross-validation criteria takes place. The choice $l = 125$ corresponds to the global minimum observed in Fig. 22.1, while $l = 139$ is equal to the number of all regions in the considered two south states of Germany except the one for which the interpolation is constructed. The latter case corresponds to the linear interpolation, while the first two values of l present nonlinear interpolation based on the LOESS approach.

In Fig. 22.2 we observe that the map with interpolated values of influenza cases obtained for $l = 125$ provides the best fit to the original data of registered influenza cases. Interestingly, all three regions with the largest number of the registered influenza cases are detected. It does not hold for two other maps consisting of the interpolated values when $l = 65$ and $l = 139$. Moreover, the application of all 139 regions leads to the worst results by being too conservative with respect to the influence of the regions with small number of registered influenza cases. In contrast, the application of $l = 65$ nearest regions leads to the situation where the number of influenza cases is often overestimated.

Acknowledgments This research was supported by the Sida bilateral programme Capacity Building in Mathematics and its Applications (pr. nr. 316), Swedish International Development Cooperation Agency (Sida) and International Science Programme in Mathematical Sciences (IPMS). The authors are also grateful to the research environment Mathematics and Applied Mathematics (MAM), Division of Applied Mathematics, Mälardalen University for providing an excellent and inspiring environment for research education and research.

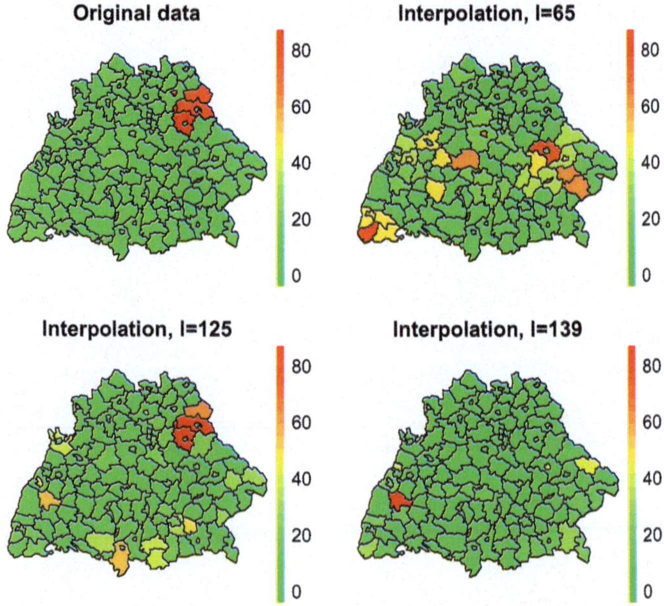

Fig. 22.2 Observed and interpolated number of influenza cases in the seventh week of 2007. The original data are displayed in the top left figure, while interpolated values are given in the top right figure ($l = 65$), in the bottom left figure ($l = 125$, optimal value following the cross-validation criteria), in the bottom right figure ($l = 139$, all regions of Baden-Württemberg and Bavaria states with the exception of the region for which the interpolation is computed)

References

Bodnar, O., & Schmid, W. (2010). Nonlinear locally weighted kriging prediction for spatio-temporal environmental processes. *Environmetrics, 21*(3–4), 365–381.

Cleveland, W. S. (1979). Robust locally weighted regression and smoothing scatterplots. *Journal of the American Statistical Association, 74*(368), 829–836.

Cressie, N., & Wikle, C. K. (2015). *Statistics for spatio-temporal data*. Wiley, New Jersey.

Durbin, J., & Koopman, S. J. (2012). *Time series analysis by state space methods* (Vol. 38). Oxford University Press, Oxford.

Fassò, A., & Cameletti, M. (2009). The EM algorithm in a distributed computing environment for modelling environmental space–time data. *Environmental Modelling and Software, 24*(9), 1027–1035.

Fassò, A., & Cameletti, M. (2010). A unified statistical approach for simulation, modeling, analysis and mapping of environmental data. *Simulation, 86*(3), 139–153.

Fassò, A., Cameletti, M., & Nicolis, O. (2007). Air quality monitoring using heterogeneous networks. *Environmetrics, 18*(3), 245–264.

Genton, M. G. (2007). Separable approximations of space-time covariance matrices. *Environmetrics: The Official Journal of the International Environmetrics Society, 18*(7), 681–695.

Givens, G. H., & Hoeting, J. A. (2012). *Computational statistics* (Vol. 710). Wiley, New Jersey.

Held, L., Meyer, S., & Bracher, J. (2017). Probabilistic forecasting in infectious disease epidemiology: The 13th armitage lecture. *Statistics in Medicine, 36*(22), 3443–3460.

Keeling, M. J., & Rohani, P. (2011). *Modeling infectious diseases in humans and animals.* Princeton University Press, New Jersey.
Ozonoff, A., Sukpraprut, S., & Sebastiani, P. (2006). Modeling seasonality of influenza with hidden Markov models. In *Proceedings of the American Statistical Association.*
Paul, M., & Held, L. (2011). Predictive assessment of a non-linear random effects model for multivariate time series of infectious disease counts. *Statistics in Medicine, 30*(10), 1118–1136.
Serfling, R. E. (1963). Methods for current statistical analysis of excess pneumonia-influenza deaths. *Public Health Reports, 78*(6), 494.
Stein, M. L. (2012). *Interpolation of spatial data: Some theory for kriging.* Springer Science & Business Media, New York.
Wikle, C. K., & Cressie, N. (1999). A dimension-reduced approach to space-time Kalman filtering. *Biometrika, 86*(4), 815–829.
Zarebski, A. E., Dawson, P., McCaw, J. M., & Moss, R. (2017). Model selection for seasonal influenza forecasting. *Infectious Disease Modelling, 2*(1), 56–70.
Zimmerman, D. L. (2006). Optimal network design for spatial prediction, covariance parameter estimation, and empirical prediction. *Environmetrics: The Official Journal of the International Environmetrics Society, 17*(6), 635–652.

Chapter 23
Describing Labour Market Dynamics Through Non Homogeneous Markov System Theory

Maria Symeonaki and Glykeria Stamatopoulou

23.1 Introduction

In recent years, there has been an increasing interest in the study of the underline mechanisms behind school-to-work transitions. The aim of the present paper is to estimate the transition probabilities between labour market states and the input probabilities to these states in order to determine the differences, if any, between transitions flows among southern European countries. We use the theory of non homogeneous Markov systems (NHMS) introduced in Vassiliou (1982). The theory of Markov systems provides a powerful tool for describing a dynamic population system that evolves over time, according to probabilistic laws and it is used in different research areas such as sociology and social policy, operation research and manpower planning, ecology, psychology, etc. Since 1982, extensive literature has been published on NHMS (see for example Dimitriou et al. 2013; Dimitriou and Tsantas 2010; Malefaki et al. 2014; Symeonaki 2015, 2018; Symeonaki and Stamou 2004; Symeonaki et al. 2002; Tsaklidis 1999, 2000; Tsantas and Vassiliou 1993; Vassiliou 2013, 2014; Vassiliou and Georgiou 1990; Vassiliou and Symeonaki 1998, 1999; Vassiliou and Tsakiridou 2004; Vassiliou and Tsaklidis 1989). Apparently, the model of a NHMS has not been applied systematically to labour market research and to the estimation of transition probabilities between labour market states, except for Symeonaki, Karamessini and Stamatopoulou (2019a, b), Symeonaki and Stamatopoulou (2015), Karamessini, Symeonaki and Stamatopoulou (2016a), Karamessini, Symeonaki, Stamatopoulou and Papazachariou (2016b). However, in these studies the labour market transitions have not been theoretically adapted to the model of a NHMS which is most suitable for this purpose since it models

M. Symeonaki (✉) · G. Stamatopoulou
Department of Social Policy, Panteion University of Social and Political Sciences, Athens, Greece
e-mail: msymeon@panteion.gr

C. H. Skiadas, C. Skiadas (eds.), *Demography of Population Health, Aging and Health Expenditures*, The Springer Series on Demographic Methods and Population Analysis 50, https://doi.org/10.1007/978-3-030-44695-6_23

simultaneously, transitions between labour market states, entry to these states and exit from the system.

Different approaches to a similar end can be found in the literature. For example, in Flek and Mysíková (2015), labour market flows are estimated for central Europe. Likewise, Ward-Warmedinge et al. (2013) use Markov analysis in order to examine the main flows that affect the changes in unemployment rates in European countries.

Our approach reports that the situation in the labour market system has worsened during the years of the economic crisis for the countries under study, but to a different extend for each country. A significant change to the school to-work transition probabilities is detected for all countries, but changes to the transition flows between employment, unemployment, inactivity and vice versa differ for each country.

The paper begins by laying out the theoretical dimensions of the research and a brief overview of the data source and the methodology used in the analysis. It then goes on to Sect. 23.3 presenting the results and the comparative analysis between the transition flows by country, while Sect. 23.4 concludes.

23.2 Data and Methods

One of the challenges in studying the evolution of the labour market systems and transitions between labour market states is to find the adequate data sources that can provide the necessary information in order to reveal the dynamics of the transition processes. The distinction between the national and international surveys can be found in the literature, as they offer different kinds of advantages and allow the study of different aspects in the research of labour market transitions. More specifically, while national panels offer longitudinal micro-data and contain a larger number of variables concerning labour market, the large-scale cross-national surveys cover a broad set of European countries, use larger sample sizes and enable the comparability across countries and over time. Considering that we aim to compare transitions between different European countries, we use comparable data drawn from the European Labour Force Survey (EU-LFS), for the years 2006–2013.

The EU-LFS[1] is a cross-sectional household research survey, aiming to collect detailed information on the trends of European labour markets in order to develop a series of structural indicators harmonised across European Union member states, such as the employment and unemployment rates. The data are collected on quarterly and annual basis since 1983 by the national statistical institutes of all 28 EU member states, three EFTA countries (Iceland, Norway and Switzerland) and two candidate countries (North Macedonia and Turkey). The target population is all the private households as well as their members aged 15 and over regardless of whether they participate or not in the labour force.

[1] http://ec.europa.eu/eurostat/web/microdata/european-union-labour-force-survey

The current labour status[2] and the working situation one year before the survey[3] are the variables that will be used in the present analysis. The individual states whether he/she:

1. Has a job or profession, including unpaid work for a family business or holding, including an apprenticeship or paid traineeship, etc.,
2. Unemployed,
3. Pupil, student, further training, unpaid work experience,
4. In retirement or early retirement or has given up business,
5. Permanently disabled,
6. In compulsory military service,
7. Fulfilling domestic tasks,
8. Other inactive person,

For the purpose of the analysis we combine these categories and define the following labour market states:

1 → Employment,
2 → Unemployment, and
3 → Inactivity.

Apparently, the entrance to the labour market system is represented by the transition from the third category (in Education or Training) to either one of the three labour market states of employment, unemployment or inactivity. On the contrary, the loss probabilities are defined as the probability of a member leaving the labour market states (e.g. retire).

More specifically, the transition probabilities are the conditional probabilities:

$p_{ij}(t)$=prob{an individual moves to state *j* at time *t* | he/she was in state *i* at time *t-1*}, $\forall_i = 1, 2, 3$.

The input probabilities are the conditional probabilities:

$p_{oj}(t)$=prob{an individual moves to state *j* at time *t* | he/she was a pupil, a student, in further training or unpaid work experience at time *t-1*}, $\forall_i = 1, 2, 3$.

The loss probabilities are the conditional probabilities:

$p_{i,k+1}(t)$=prob{an individual leaves the system at time *t* / he or she was in state *i* at time *t-1*}, $\forall i = 1,2,3$.

Clearly the set space includes three categories, $S = \{1, 2,3\}$. Figure 23.1 provides us with all the possible transitions into the system.

Let us now provide a short description of a NHMS. Consider a population that is stratified into categories according to a specific characteristic. Let also $S = \{1, 2, \ldots, k\}$ be the state space of a NHMS, i.e. the set of all categories. The basic parameters of a NHMS are the following:

[2]MAINSTAT.
[3]WSTAT1Y.

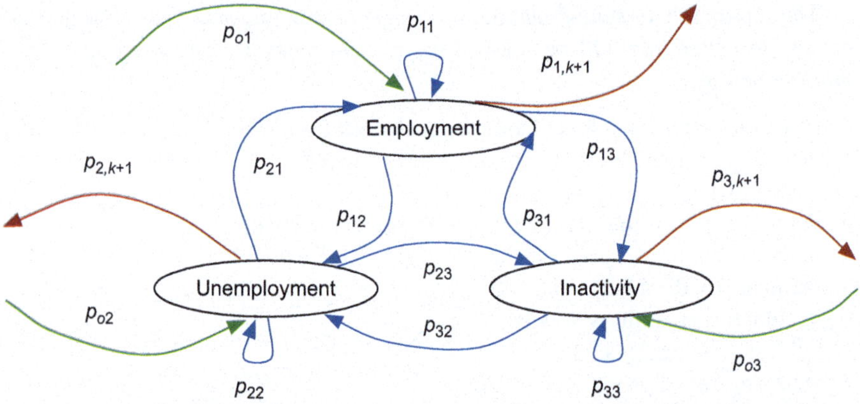

Fig. 23.1 The transition diagram of the labour market system

t	the parameter of time,
$N_r(t)$	the number of members in state r at time t, $r = 1, 2, \ldots, k$, $t = 0, 1, 2\ldots$,
$p_{ij}(t)$	the transition probability from state i to state j at time t,
$p_{oj}(t)$	the input probability,
$p_{i,\,k+1}(t)$	the loss probability,
$q_{ij}(t)$	the total transition probability from state i to state j at time t, i.e. $q_{ij}(t) = p_{ij}(t) + p_{i,\,k+1}(t)p_{oj}(t)$,
$T(t)$	the total number of members serving the system at time t,
$\Delta T(t)$	$=T(t) - T(t-1)$,
$\{\mathbf{P}(t)\}_{t=0}^{\infty}$	$= \left\{p_{ij}(t)\right\}_{t=0}^{\infty}, i, j = 1, 2, \ldots, k,$
$\{\mathbf{p}_o(t)\}_{t=0}^{\infty}$	$= \left\{p_{oj}(t)\right\}_{t=0}^{\infty}, j = 1, 2, \ldots, k,$
$\{\mathbf{p}_{k+1}(t)\}_{t=0}^{\infty}$	$= \left\{p_{i,k+1}(t)\right\}_{t=0}^{\infty}, i = 1, 2, \ldots, k,$
$\{\mathbf{Q}(t)\}_{t=0}^{\infty}$	$= \left\{q_{ij}(t)\right\}_{t=0}^{\infty}, i, j = 1, 2, \ldots, k,$
$\{\mathbf{N}(t)\}_{t=0}^{\infty}$	$= \{N_i(t)\}_{t=0}^{\infty}, i = 1, 2, \ldots, k,$
$\{\mathbf{q}(t)\}_{t=0}^{\infty}$	$= \left\{\frac{N_i(t)}{T(t)}\right\}_{t=0}^{\infty}, i = 1, 2, \ldots, k.$

Apparently, the expected population structure of the NHMS at time $t + 1$, $\mathbf{N}(t + 1)$, that provides the number of members in each state at time $t + 1$ is:

$$\mathbf{N}(t+1) = \mathbf{N}(t)\mathbf{Q}(t) + \Delta T(t)\mathbf{p}_o(t). \tag{23.1}$$

The NHMS just described will be used to model the labour market transitions of an individual for Greece, Spain, Italy and Portugal, providing the necessary information for estimating the transition probabilities between labour market states and the probabilities of entry to the labour market. Moreover, we focus on four different relative mobility indices, which are calculated in order to reveal the mobility of individuals within the labour market system. More specifically, we compute the following relative mobility indices:

$$\text{Prais} - \text{Shorrocks mobility index}: \quad M(PS) = \frac{1}{k\text{-}1}(k - tr\,(\mathbf{Q})) \tag{23.2}$$

$$\text{Immobility index}: \quad IM = \frac{tr\,(\mathbf{Q})}{k} \tag{23.3}$$

$$\text{Bartholomew mobility index}: \quad M_B = \frac{1}{k}\sum\nolimits_{i=1}^{k}\sum\nolimits_{j=1}^{k} q_{ij}\,|i - j| \tag{23.4}$$

$$\text{Prais} - \text{Bibby mobility index}: \quad M_T = 1 - \frac{tr\,(\mathbf{Q})}{k}. \tag{23.5}$$

Furthermore, we use the well-established concept of a distance between two matrices provided by Eqs. (23.6) and (23.7), in order to capture the differences that occurred in the transitions between the labour market states from 2006 to 2013 in Southern Europe.

$$d_1\,(\mathbf{A}, \mathbf{B}) = \sum_{j=1}^{k}\sum_{i=1}^{k}\left|a_{ij} - b_{ij}\right| \tag{23.6}$$

$$d_2\,(\mathbf{A}, \mathbf{B}) = \left(\sum_{j=1}^{k}\sum_{i=1}^{k}\left(a_{ij} - b_{ij}\right)^2\right)^{1/2} \tag{23.7}$$

23.3 Results

We now proceed to the estimation of the transition probabilities from each labour market state to the other, the input probabilities to the different labour market states as well the loss probabilities by country for 2006 to 2013.

It is known that $\mathbf{Q}(t) = \mathbf{P}(t) + \mathbf{p}'_{k+1}(t) \cdot \mathbf{p}_o(t)$ and it is important to note that $\mathbf{Q}(t)$ is a stochastic matrix, where $\mathbf{P}(t)$ is sub-stochastic. The elements of the matrix of total transition probabilities represent the total gain of state j, since it captures the probability that either a member moves from state i to state j or a new member goes to state j, when he/she enters the system and an old member that was in state i leaves the system.

Table 23.1 Input and loss probabilities and transition probability matrices, Greece, 2006–2013

LFS	N	P	p_0, p_{k+1}	Q
2006	262,884	$\begin{bmatrix} 0.947 & 0.021 & 0.011 \\ 0.231 & 0.716 & 0.037 \\ 0.020 & 0.017 & 0.947 \end{bmatrix}$	$p_0 = [0.395\ \ 0.372\ \ 0.233]$ $p_{k+1} = \begin{bmatrix} 0.021 & 0.016 & 0.016 \end{bmatrix}$	$\begin{bmatrix} 0.955 & 0.029 & 0.016 \\ 0.237 & 0.722 & 0.041 \\ 0.026 & 0.023 & 0.951 \end{bmatrix}$
2007	257,228	$\begin{bmatrix} 0.949 & 0.019 & 0.010 \\ 0.226 & 0.718 & 0.035 \\ 0.019 & 0.016 & 0.951 \end{bmatrix}$	$p_0 = \begin{bmatrix} 0.367 & 0.357 & 0.276 \end{bmatrix}$ $p_{k+1} = \begin{bmatrix} 0.022 & 0.021 & 0.014 \end{bmatrix}$	$\begin{bmatrix} 0.957 & 0.027 & 0.016 \\ 0.234 & 0.726 & 0.040 \\ 0.024 & 0.020 & 0.956 \end{bmatrix}$
2008	255,701	$\begin{bmatrix} 0.949 & 0.018 & 0.010 \\ 0.241 & 0.702 & 0.037 \\ 0.018 & 0.015 & 0.950 \end{bmatrix}$	$p_0 = \begin{bmatrix} 0.343 & 0.325 & 0.305 \end{bmatrix}$ $p_{k+1} = \begin{bmatrix} 0.023 & 0.020 & 0.017 \end{bmatrix}$	$\begin{bmatrix} 0.957 & 0.026 & 0.017 \\ 0.248 & 0.709 & 0.043 \\ 0.023 & 0.021 & 0.956 \end{bmatrix}$
2009	258,428	$\begin{bmatrix} 0.938 & 0.031 & 0.011 \\ 0.201 & 0.739 & 0.039 \\ 0.018 & 0.021 & 0.942 \end{bmatrix}$	$p_0 = \begin{bmatrix} 0.309 & 0.418 & 0.273 \end{bmatrix}$ $p_{k+1} = \begin{bmatrix} 0.020 & 0.021 & 0.019 \end{bmatrix}$	$\begin{bmatrix} 0.944 & 0.040 & 0.016 \\ 0.208 & 0.748 & 0.044 \\ 0.024 & 0.029 & 0.947 \end{bmatrix}$
2010	262,172	$\begin{bmatrix} 0.929 & 0.041 & 0.010 \\ 0.166 & 0.779 & 0.036 \\ 0.017 & 0.025 & 0.938 \end{bmatrix}$	$p_0 = \begin{bmatrix} 0.257 & 0.431 & 0.312 \end{bmatrix}$ $p_{k+1} = \begin{bmatrix} 0.020 & 0.019 & 0.020 \end{bmatrix}$	$\begin{bmatrix} 0.934 & 0.050 & 0.016 \\ 0.171 & 0.787 & 0.042 \\ 0.022 & 0.034 & 0.944 \end{bmatrix}$
2011	236,551	$\begin{bmatrix} 0.912 & 0.054 & 0.010 \\ 0.113 & 0.838 & 0.030 \\ 0.011 & 0.029 & 0.940 \end{bmatrix}$	$p_0 = \begin{bmatrix} 0.195 & 0.537 & 0.268 \end{bmatrix}$ $p_{k+1} = \begin{bmatrix} 0.024 & 0.019 & 0.020 \end{bmatrix}$	$\begin{bmatrix} 0.917 & 0.067 & 0.016 \\ 0.117 & 0.848 & 0.035 \\ 0.015 & 0.034 & 0.951 \end{bmatrix}$
2012	212,307	$\begin{bmatrix} 0.912 & 0.054 & 0.010 \\ 0.113 & 0.838 & 0.030 \\ 0.011 & 0.029 & 0.940 \end{bmatrix}$	$p_0 = \begin{bmatrix} 0.154 & 0.555 & 0.291 \end{bmatrix}$ $p_{k+1} = \begin{bmatrix} 0.025 & 0.020 & 0.026 \end{bmatrix}$	$\begin{bmatrix} 0.903 & 0.078 & 0.019 \\ 0.092 & 0.876 & 0.032 \\ 0.012 & 0.045 & 0.943 \end{bmatrix}$
2013	215,244	$\begin{bmatrix} 0.900 & 0.062 & 0.012 \\ 0.097 & 0.862 & 0.023 \\ 0.008 & 0.026 & 0.938 \end{bmatrix}$	$p_0 = \begin{bmatrix} 0.172 & 0.552 & 0.276 \end{bmatrix}$ $p_{k+1} = \begin{bmatrix} 0.026 & 0.018 & 0.028 \end{bmatrix}$	$\begin{bmatrix} 0.904 & 0.076 & 0.020 \\ 0.100 & 0.872 & 0.028 \\ 0.013 & 0.041 & 0.946 \end{bmatrix}$

Source: EU-LFS, 2006–2013; own calculations

Table 23.1 presents the results of all estimations for the case of Greece. It is clear that the situation in the labour market in Greece has worsened substantially during the years of the economic crisis. Apparently, the probabilities of a new member that was a pupil, a student or in training, to enter the labour market and become employed have dropped significantly from 0.395 in 2006 to 0.172 in 2013. On the other hand, the probability of an unemployed individual to stay unemployed the next year has become higher, from 0.716 in 2006 to 0.862 in 2013. Analogously, individuals have less probabilities of leaving the state of inactivity and successfully enter the state of employment (0.020 in 2006 and 0.008 in 2013).

Furthermore, we estimate the distances between the transition probability matrices according to Eqs. (23.6) and (23.7).

Table 23.2 Input and loss probabilities and transition probability matrices, Spain, 2006–2013

LFS	N	P	p_0, p_{k+1}	Q
2006	84,252	$\begin{bmatrix} 0.927 & 0.030 & 0.029 \\ 0.394 & 0.373 & 0.194 \\ 0.038 & 0.029 & 0.870 \end{bmatrix}$	$p_0 = \begin{bmatrix} 0.613 & 0.306 & 0.081 \end{bmatrix}$ $p'_{k+1} = \begin{bmatrix} 0.014 & 0.039 & 0.063 \end{bmatrix}$	$\begin{bmatrix} 0.935 & 0.034 & 0.031 \\ 0.418 & 0.385 & 0.197 \\ 0.076 & 0.048 & 0.876 \end{bmatrix}$
2007	85,686	$\begin{bmatrix} 0.931 & 0.028 & 0.025 \\ 0.409 & 0.387 & 0.167 \\ 0.043 & 0.030 & 0.864 \end{bmatrix}$	$p_0 = \begin{bmatrix} 0.632 & 0.278 & 0.090 \end{bmatrix}$ $p'_{k+1} = \begin{bmatrix} 0.016 & 0.037 & 0.063 \end{bmatrix}$	$\begin{bmatrix} 0.941 & 0.032 & 0.027 \\ 0.432 & 0.397 & 0.171 \\ 0.083 & 0.047 & 0.870 \end{bmatrix}$
2008	86,543	$\begin{bmatrix} 0.917 & 0.045 & 0.024 \\ 0.371 & 0.425 & 0.169 \\ 0.034 & 0.031 & 0.876 \end{bmatrix}$	$p_0 = \begin{bmatrix} 0.597 & 0.377 & 0.026 \end{bmatrix}$ $p'_{k+1} = \begin{bmatrix} 0.014 & 0.035 & 0.026 \end{bmatrix}$	$\begin{bmatrix} 0.925 & 0.050 & 0.025 \\ 0.391 & 0.438 & 0.171 \\ 0.069 & 0.053 & 0.878 \end{bmatrix}$
2009	90,715	$\begin{bmatrix} 0.885 & 0.074 & 0.025 \\ 0.275 & 0.555 & 0.140 \\ 0.027 & 0.038 & 0.879 \end{bmatrix}$	$p_0 = \begin{bmatrix} 0.440 & 0.444 & 0.116 \end{bmatrix}$ $p'_{k+1} = \begin{bmatrix} 0.016 & 0.030 & 0.056 \end{bmatrix}$	$\begin{bmatrix} 0.892 & 0.081 & 0.027 \\ 0.288 & 0.568 & 0.144 \\ 0.052 & 0.063 & 0.885 \end{bmatrix}$
2010	91,375	$\begin{bmatrix} 0.904 & 0.060 & 0.021 \\ 0.256 & 0.582 & 0.133 \\ 0.022 & 0.031 & 0.890 \end{bmatrix}$	$p_0 = \begin{bmatrix} 0.383 & 0.498 & 0.119 \end{bmatrix}$ $p'_{k+1} = \begin{bmatrix} 0.015 & 0.029 & 0.057 \end{bmatrix}$	$\begin{bmatrix} 0.910 & 0.068 & 0.022 \\ 0.267 & 0.596 & 0.137 \\ 0.044 & 0.059 & 0.897 \end{bmatrix}$
2011	87,343	$\begin{bmatrix} 0.908 & 0.057 & 0.021 \\ 0.239 & 0.620 & 0.135 \\ 0.041 & 0.033 & 0.871 \end{bmatrix}$	$p_0 = \begin{bmatrix} 0.366 & 0.516 & 0.118 \end{bmatrix}$ $p'_{k+1} = \begin{bmatrix} 0.014 & 0.026 & 0.055 \end{bmatrix}$	$\begin{bmatrix} 0.913 & 0.064 & 0.023 \\ 0.248 & 0.634 & 0.118 \\ 0.061 & 0.061 & 0.878 \end{bmatrix}$
2012	90,634	$\begin{bmatrix} 0.895 & 0.071 & 0.020 \\ 0.198 & 0.670 & 0.108 \\ 0.026 & 0.034 & 0.886 \end{bmatrix}$	$p_0 = \begin{bmatrix} 0.330 & 0.562 & 0.108 \end{bmatrix}$ $p'_{k+1} = \begin{bmatrix} 0.014 & 0.024 & 0.054 \end{bmatrix}$	$\begin{bmatrix} 0.899 & 0.078 & 0.022 \\ 0.205 & 0.683 & 0.110 \\ 0.043 & 0.064 & 0.892 \end{bmatrix}$
2013	89,241	$\begin{bmatrix} 0.905 & 0.062 & 0.019 \\ 0.187 & 0.678 & 0.114 \\ 0.021 & 0.032 & 0.891 \end{bmatrix}$	$p_0 = \begin{bmatrix} 0.234 & 0.496 & 0.270 \end{bmatrix}$ $p'_{k+1} = \begin{bmatrix} 0.014 & 0.021 & 0.056 \end{bmatrix}$	$\begin{bmatrix} 0.908 & 0.068 & 0.022 \\ 0.191 & 0.688 & 0.119 \\ 0.034 & 0.059 & 0.906 \end{bmatrix}$

Source: EU-LFS, 2006–2013; own calculations

$$d_1\left(\mathbf{P}(2006), \mathbf{P}(2013)\right) = 0.413 \quad d_2\left(\mathbf{P}(2006), \mathbf{P}(2013)\right) = 0.209$$

The outcomes state that there is a difference between the transition probability matrices in years 2006 and 2013.

Table 23.2 provides the respective results for Spain. There is strong evidence that the situation in the labour market in Spain has also deteriorated during the years of the economic crisis. Clearly, the probabilities of a new member entering the labour market and becoming employed have dropped considerably from 0.613 in 2006 to

0.234 in 2013. The comparison between Greece and Spain shows that Greece is in an inferior situation, but it was also in a worse position to begin with. Moreover, the probability of an unemployed individual to remain unemployed the next year has become higher, from 0.373 in 2006 to 0.678 in 2013. If we now consider the probability of an individual to leave the state of inactivity and successfully enter the state of employment, this has dropped from 0.038 in 2006 to 0.021 in 2013. If we estimate the distances between the transition probability matrices we will get that:

$$d_1\left(\mathbf{P}(2006), \mathbf{P}(2013)\right) = 0.697 \quad d_2\left(\mathbf{P}(2006), \mathbf{P}(2013)\right) = 0.380$$

The outcomes reveal that the changes in the transition probability matrices for Spain between the years 2006 and 2013 are larger than those for Greece.

Apparently, the results for Italy reveal a quite similar situation (Table 23.3). Firstly, the school-to-work probabilities have dropped from 0.456 in 2006 to 0.290, whereas the school-to-unemployment probability increased from 0.474 to 0.607 for the same years. However, we notice that the transition probabilities between labour market states did not change considerably from 2006 to 2013. This is also evident from the distances between the transition probability matrices that report lower values, i.e. the observed difference between **P**(2006) and **P**(2013) in Italy is not significant:

$$d_1\left(\mathbf{P}(2006), \mathbf{P}(2013)\right) = 0.198 \quad d_2\left(\mathbf{P}(2006), \mathbf{P}(2013)\right) = 0.010.$$

Finally, Table 23.4 provides information concerning Portugal. In the Portuguese databases from 2006 to 2010, the variable concerning the main labour status one year before the time of the survey ('WSTAT1Y') poses limitations in the analysis. Particularly, the information on the individuals being in retirement or permanently disabled is missing and it seems that they are grouped under the label 'other inactive persons', not allowing the separation of these states. Therefore, only outcomes from 2011 to 2013 are presented. Through these years, one can detect only a small decrease in the school-to-work transition from 2011 to 2013 and an increase in the probability of someone entering the system to be unemployed. In addition, the transition probabilities between labour market states did not change considerably from 2011 to 2013.

Figure 23.2 summarises the input probabilities from school to labour market states for the observed countries. It is evident that there is a negative trend through the years for all countries in the probabilities of a young person entering the labour market and becoming employed, with Greece scoring worse between all countries. For the case of Spain, the drop through years seems to be larger, since the probabilities to employment drop significantly from 0.613 in 2006 to 0.234, implying that Spain is affected to a greater degree than the other countries (see Appendix, Table 23.A5).

On the other hand, the transition probabilities from school-to-unemployment are by far the highest probability after the economic crisis, with the highest value found

Table 23.3 Input and loss probabilities and transition probability matrices, Italy, 2006–2013

LFS	N	P	p_0, p_{k+1}	Q
2006	586,642	$\begin{bmatrix} 0.948 & 0.025 & 0.012 \\ 0.237 & 0.708 & 0.040 \\ 0.021 & 0.017 & 0.936 \end{bmatrix}$	$p_0 = \begin{bmatrix} 0.456 & 0.474 & 0.070 \end{bmatrix}$ $p'_{k+1} = \begin{bmatrix} 0.015 & 0.015 & 0.026 \end{bmatrix}$	$\begin{bmatrix} 0.955 & 0.032 & 0.013 \\ 0.244 & 0.715 & 0.041 \\ 0.032 & 0.029 & 0.939 \end{bmatrix}$
2007	584,637	$\begin{bmatrix} 0.949 & 0.025 & 0.012 \\ 0.248 & 0.702 & 0.038 \\ 0.020 & 0.015 & 0.938 \end{bmatrix}$	$p_0 = \begin{bmatrix} 0.632 & 0.278 & 0.090 \end{bmatrix}$ $p'_{k+1} = \begin{bmatrix} 0.014 & 0.012 & 0.027 \end{bmatrix}$	$\begin{bmatrix} 0.955 & 0.032 & 0.013 \\ 0.256 & 0.707 & 0.037 \\ 0.032 & 0.028 & 0.940 \end{bmatrix}$
2008	579,602	$\begin{bmatrix} 0.944 & 0.027 & 0.012 \\ 0.231 & 0.721 & 0.037 \\ 0.019 & 0.016 & 0.941 \end{bmatrix}$	$p_0 = \begin{bmatrix} 0.478 & 0.460 & 0.062 \end{bmatrix}$ $p'_{k+1} = \begin{bmatrix} 0.017 & 0.011 & 0.024 \end{bmatrix}$	$\begin{bmatrix} 0.952 & 0.035 & 0.013 \\ 0.236 & 0.726 & 0.038 \\ 0.030 & 0.027 & 0.943 \end{bmatrix}$
2009	570,662	$\begin{bmatrix} 0.937 & 0.034 & 0.014 \\ 0.197 & 0.758 & 0.034 \\ 0.015 & 0.015 & 0.944 \end{bmatrix}$	$p_0 = \begin{bmatrix} 0.392 & 0.519 & 0.089 \end{bmatrix}$ $p'_{k+1} = \begin{bmatrix} 0.015 & 0.011 & 0.026 \end{bmatrix}$	$\begin{bmatrix} 0.943 & 0.042 & 0.015 \\ 0.202 & 0.763 & 0.035 \\ 0.025 & 0.029 & 0.946 \end{bmatrix}$
2010	573,984	$\begin{bmatrix} 0.937 & 0.034 & 0.014 \\ 0.204 & 0.739 & 0.045 \\ 0.019 & 0.018 & 0.938 \end{bmatrix}$	$p_0 = \begin{bmatrix} 0.358 & 0.556 & 0.086 \end{bmatrix}$ $p'_{k+1} = \begin{bmatrix} 0.017 & 0.012 & 0.025 \end{bmatrix}$	$\begin{bmatrix} 0.943 & 0.044 & 0.013 \\ 0.208 & 0.745 & 0.047 \\ 0.028 & 0.032 & 0.940 \end{bmatrix}$
2011	570,490	$\begin{bmatrix} 0.939 & 0.035 & 0.011 \\ 0.214 & 0.724 & 0.049 \\ 0.019 & 0.020 & 0.935 \end{bmatrix}$	$p_0 = \begin{bmatrix} 0.360 & 0.546 & 0.094 \end{bmatrix}$ $p'_{k+1} = \begin{bmatrix} 0.015 & 0.013 & 0.026 \end{bmatrix}$	$\begin{bmatrix} 0.944 & 0.043 & 0.013 \\ 0.219 & 0.731 & 0.050 \\ 0.028 & 0.034 & 0.938 \end{bmatrix}$
2012	525,991	$\begin{bmatrix} 0.933 & 0.043 & 0.012 \\ 0.189 & 0.749 & 0.049 \\ 0.017 & 0.024 & 0.938 \end{bmatrix}$	$p_0 = \begin{bmatrix} 0.330 & 0.574 & 0.096 \end{bmatrix}$ $p'_{k+1} = \begin{bmatrix} 0.012 & 0.013 & 0.096 \end{bmatrix}$	$\begin{bmatrix} 0.937 & 0.050 & 0.013 \\ 0.193 & 0.756 & 0.051 \\ 0.024 & 0.036 & 0.940 \end{bmatrix}$
2013	530,173	$\begin{bmatrix} 0.929 & 0.048 & 0.011 \\ 0.167 & 0.778 & 0.041 \\ 0.015 & 0.022 & 0.939 \end{bmatrix}$	$p_0 = \begin{bmatrix} 0.290 & 0.607 & 0.103 \end{bmatrix}$ $p'_{k+1} = \begin{bmatrix} 0.012 & 0.014 & 0.024 \end{bmatrix}$	$\begin{bmatrix} 0.932 & 0.055 & 0.012 \\ 0.171 & 0.786 & 0.042 \\ 0.022 & 0.036 & 0.941 \end{bmatrix}$

Source: EU-LFS, 2006–2013; own calculations

in Italy, in 2013. Concerning the probabilities to become inactive after entering the labour market, we notice low values in general, with Greece being clear in a more negative position. A striking point in this figure is that a sudden increase in the probabilities to the inactivity in Spain has appeared from 2012 to 2013.

Turning now to the mobility and immobility indices for the surveyed countries, the results are presented in Fig. 23.3. It seems that there is generally a drop in the mobility rates, since all the indices' values gradually decreased and respectively a

Table 23.4 Input and loss probabilities and transition probability matrices, Portugal, 2011–2013

LFS	N	P	p_0, p_{k+1}	Q
2011	139,634	$\begin{bmatrix} 0.833 & 0.064 & 0.029 \\ 0.225 & 0.711 & 0.026 \\ 0.011 & 0.015 & 0.908 \end{bmatrix}$	$\mathbf{p_0} = \begin{bmatrix} 0.449 & 0.489 & 0.062 \end{bmatrix}$ $\mathbf{p_{k+1}} = \begin{bmatrix} 0.074 & 0.038 & 0.066 \end{bmatrix}$	$\begin{bmatrix} 0.866 & 0.100 & 0.034 \\ 0.242 & 0.730 & 0.028 \\ 0.041 & 0.047 & 0.912 \end{bmatrix}$
2012	140,210	$\begin{bmatrix} 0.803 & 0.082 & 0.032 \\ 0.178 & 0.749 & 0.032 \\ 0.010 & 0.014 & 0.926 \end{bmatrix}$	$\mathbf{p_0} = \begin{bmatrix} 0.350 & 0.578 & 0.072 \end{bmatrix}$ $\mathbf{p_{k+1}} = \begin{bmatrix} 0.083 & 0.041 & 0.050 \end{bmatrix}$	$\begin{bmatrix} 0.832 & 0.130 & 0.038 \\ 0.192 & 0.773 & 0.035 \\ 0.027 & 0.043 & 0.930 \end{bmatrix}$
2013	139,939	$\begin{bmatrix} 0.819 & 0.072 & 0.032 \\ 0.186 & 0.744 & 0.031 \\ 0.009 & 0.013 & 0.939 \end{bmatrix}$	$\mathbf{p_0} = \begin{bmatrix} 0.385 & 0.538 & 0.077 \end{bmatrix}$ $\mathbf{p_{k+1}} = \begin{bmatrix} 0.077 & 0.113 & 0.038 \end{bmatrix}$	$\begin{bmatrix} 0.849 & 0.113 & 0.038 \\ 0.201 & 0.765 & 0.034 \\ 0.024 & 0.034 & 0.942 \end{bmatrix}$

Source: EU-LFS, 2006–2013; own calculations

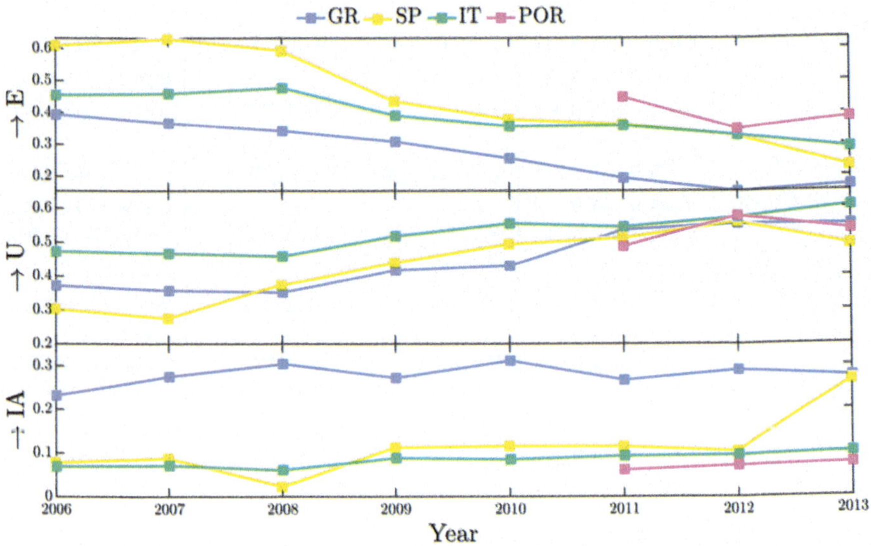

Fig. 23.2 The input probabilities, Southern Europe, 2006–2013*Source*: EU-LFS, 2006–2013; own calculations

simultaneous increase in immobility occurs. Comparison between countries shows that Greece and Italy are the least mobile countries, as their rates take the lowest values through the years. Noteworthy is the case of Spain, where a sudden drop in all mobility rates is detected in 2012.

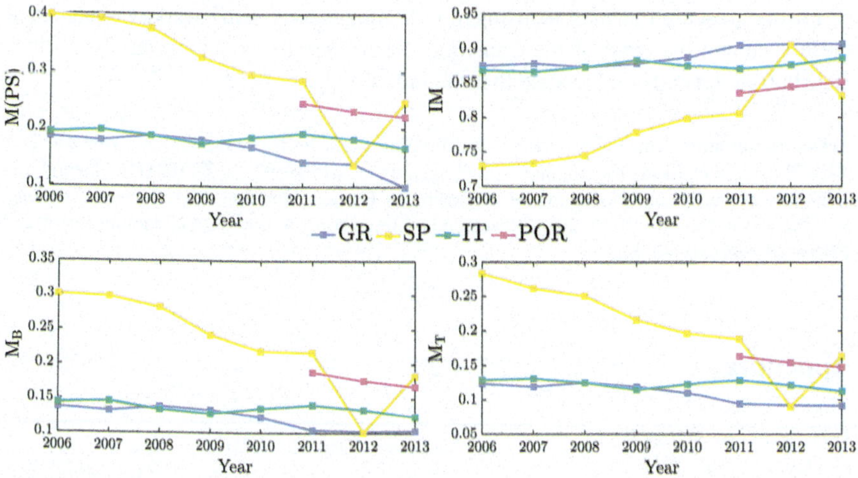

Fig. 23.3 The (im)mobility rates, Southern Europe, 2006–2013*Source*: EU-LFS, 2006–2013; own calculations

23.4 Conclusions and Future Work

In the present paper, the theory of non-homogeneous Markov systems is used to model the transitions between labour market states in the four Southern European countries. With data drawn from the EU-LFS from 2006 until 2013, we compute transition probability matrices that show all the transitions that occur in the system. We also compute mobility and immobility indices, as well as distance measures to gain more information about these transitions.

Through the analysis it is proved that the probabilities to remain in the same category prevail in all countries and years, meaning that individuals are more likely not to change working conditions from previous years. However, the crisis' impact on these probabilities is clear for every country studied, with an evident impact on school-to-work transitions.

More specifically, the transition probabilities within the labour market in Spain are affected to a greater degree than in other countries, while Italy seems to be affected the least. On the other hand, a decrease in the probabilities of a young person entering the labour market and becoming employed is reported though the years, with Greece scoring the lowest among all countries. Moreover, the probabilities to inactivity seem to be higher in Greece than in other southern European countries. Nevertheless, we should keep in mind that limitations in the dataset prevent us from having a clearer picture of the situation in Portugal.

As far as the results from the mobility indices are concerned, Greece and Italy seem to have lower levels of mobility, while Spain is proved to be the most mobile, with a decrease in the mobility rates detected across years. More detailed analysis will be needed to fully understand the apparent sudden drop that is noticed in the mobility rates for 2012.

Further research relates to using non-homogeneous semi-Markov systems. In these systems, the time of the transition occurrence is a random variable, whose distribution depends on the state that the individual will move to.

Acknowledgments This study has received funding from the European Union's Horizon 2020 research and innovation program under grant agreement No 649395 (NEGOTIATE – Negotiating early job-insecurity and labour market exclusion in Europe, Horizon 2020, Societal Challenge 6, H2020-YOUNG-SOCIETY-2014, Research and Innovation Action (RIA), Duration: 01 March 2015 – 28 February 2018).

Appendix

Table A.1 $M(PS)$, IM, M_B, M_T, Greece, 2006–2013

Greece	$M(PS)$	IM	M_B	M_T
2006	0.186	0.876	0.138	0.124
2007	0.180	0.879	0.133	0.120
2008	0.189	0.874	0.139	0.126
2009	0.181	0.879	0.134	0.120
2010	0.168	0.888	0.124	0.111
2011	0.142	0.905	0.105	0.095
2012	0.139	0.907	0.103	0.093
2013	0.098	0.907	0.104	0.092

Source: EU-LFS, 2006–2013; own calculations

Table A.2 $M(PS)$, IM, M_B, M_T, Spain, 2006–2013

Spain	$M(PS)$	IM	M_B	M_T
2006	0.402	0.732	0.303	0.286
2007	0.396	0.736	0.300	0.264
2008	.379	0.747	0.284	0.253
2009	0.328	0.781	0.244	0.218
2010	0.298	0.801	0.221	0.199
2011	0.287	0.808	0.219	0.191
2012	0.139	0.907	0.103	0.093
2013	0.249	0.834	0.185	0.166

Source: EU-LFS, 2006–2013; own calculations

Table A.3 $M(PS)$, IM, M_B, M_T, Italy, 2006–2013

Italy	$M(PS)$	IM	M_B	M_T
2006	0.195	0.869	0.145	0.130
2007	0.199	0.867	0.147	0.132
2008	0.189	0.874	0.135	0.126
2009	0.174	0.884	0.129	0.116
2010	0.186	0.876	0.137	0.124
2011	0.193	0.871	0.142	0.129
2012	0.183	0.877	0.135	0.122
2013	0.167	0.887	0.125	0.113

Source: EU-LFS, 2006–2013; own calculations

Table A.4 $M(PS)$, IM, M_B, M_T, Portugal, 2011–2013

Portugal	$M(PS)$	IM	M_B	M_T
2011	0.246	0.836	0.189	0.164
2012	0.232	0.845	0.177	0.155
2013	0.222	0.852	0.168	0.148

Source: EU-LFS, 2006–2013; own calculations

Table A.5 The input probabilities to employment, unemployment and inactivity, Southern Europe, 2006–2013

Greece	→ E	→ U	→ IA
2006	**0.395**	**0.372**	**0.233**
2007	0.367	0.357	0.276
2008	0.343	0.352	0.305
2009	0.309	0.418	0.273
2010	0.257	0.431	0.312
2011	0.195	0.537	0.268
2012	0.154	0.555	0.291
2013	**0.172**	**0.552**	**0.276**
Spain	→ E	→ U	→ IA
2006	**0.613**	**0.306**	**0.081**
2007	0.632	0.278	0.090
2008	0.597	0.377	0.026
2009	0.440	0.444	0.116
2010	0.383	0.498	0.119
2011	0.366	0.516	0.118
2012	0.330	0.562	0.107
2013	**0.234**	**0.496**	**0.269**
Italy	→ E	→ U	→ IA
2006	**0.456**	**0.474**	**0.070**
2007	0.460	0.468	0.072
2008	0.478	0.460	0.062
2009	0.392	0.519	0.089
2010	0.358	0.556	0.086
2011	0.360	0.546	0.094
2012	0.330	0.574	0.096
2013	**0.290**	**0.607**	**0.103**
Portugal	→ E	→ U	→ IA
2010	**0.449**	**0.489**	**0.062**
2011	0.350	0.578	0.072
2012	**0.385**	**0.538**	**0.077**

Source: EU-LFS, 2006–2013; own calculations

References

Dimitriou, V. A., & Tsantas, N. (2010). Evolution of a time dependent Markov model for training and recruitment decisions in manpower planning. *Linear Algebra and its Applications, 433*(11–12), 1950–1972.

Dimitriou, V. A., Georgiou, A. C., & Tsantas, N. (2013). The multivariate non-homogeneous Markov manpower system in a departmental mobility framework. *European Journal of Operational Research, 228*, 112–121.

Flek, V., & Mysíková, M. (2015). Unemployment dynamics in Central Europe: A labour flow approach. *Prague Economic Papers, 24*(1), 73–87.

Karamessini, M., Symeonaki, M., & Stamatopoulou, G. (2016a). *The role of the economic crisis in determining the degree of early job insecurity in Europe*, https://blogg.hioa.no/negotiate/files/2015/04/NEGOTIATE-working-paper-D3.3.pdf

Karamessini, M., Symeonaki, M., Stamatopoulou, G., & Papazachariou, A. (2016b). *The careers of young people in Europe during the economic crisis: Identifying risk factors* (Negotiate working papers), https://blogg.hioa.no/negotiate/files/2015/04/NEGOTIATE-working-paper-no-D3.2-The-careers-of-young-people-in-Eurpa-during-the-economic-crisis.pdf

Malefaki, S., Limnios, N., & Dersin, P. (2014). Reliability of maintained systems under a semi-Markov setting. *Reliability Engineering and System Safety, 131*, 282–290.

Symeonaki, M. (2015). Theory of Fuzzy non-homogeneous Markov systems with fuzzy states. *Quality and Quantity, 49*(6), 2369–2385.

Symeonaki, M. (2018). Rate of convergence in fuzzy non-homogeneous Markov systems. *Communications in Statistics – Theory and Methods*. https://doi.org/10.1080/03610926.2017.1395044.

Symeonaki, M., & Stamatopoulou, G. (2015). *A Markov system analysis application on labour market dynamics: The case of Greece*. In IWPLMS, 22–24 June, Athens.

Symeonaki, M., & Stamou, G. (2004). Theory of Markov systems with fuzzy states. *Fuzzy Sets and Systems, 143*, 427–445.

Symeonaki, M., Stamou, G., & Tzafestas, S. (2002). Fuzzy non-homogeneous Markov systems. *Applied Intelligence, 17*(2), 203–214.

Symeonaki, M., Karamessini, M., & Stamatopoulou, G. (2019a). Gender-based differences on the impact of the economic crisis on labour market flows in Southern Europe. In J. Bozeman & C. Skiadas (Eds.), *Data analysis and applications: New and classical approaches* (pp. 107–120). London: ISTE Science Publishing LTD.

Symeonaki, M., Karamessini, M., & Stamatopoulou, G. (2019b). Measuring school-to-work transition probabilities in Europe with evidence from the EU-SILC. In J. Bozeman & C. Skiadas (Eds.), *Data analysis and applications: New and classical approaches* (pp. 121–136). London: ISTE Science Publishing LTD.

Tsaklidis, G. (1999). The stress tensor and the energy of a continuous time homogeneous Markov system with fixed size. *Journal of Applied Probability, 36*, 21–29.

Tsaklidis, G. (2000). The closed continuous time homogeneous semi-Markov system as a non-Newtonian fluid. *Applied Stochastic Models in Business and Industry, 16*, 73–83.

Tsantas, N., & Vassiliou, P.-C. G. (1993). The non-homogenous Markov system in a stochastic environment. *Journal of Applied Probability, 30*, 285–301.

Vassiliou, P.-C. G. (1982). Asymptotic behaviour of Markov systems. *Journal of Applied Probability, 19*, 851–857.

Vassiliou, P.-C. G. (2013). Fuzzy semi-Markov migration process in credit risk. *Fuzzy Sets and Systems, 223*(0), 39–58.

Vassiliou, P.-C. G. (2014). Semi-Markov migration process in a stochastic market in credit risk. *Linear Algebra and its Applications, 450*, 13–43.

Vassiliou, P.-C. G., & Georgiou, A. K. (1990). Asymptotically attainable structures in non-homogeneous Markov systems. *The Journal of the Operational Research Society, 38*, 537–545.

Vassiliou, P.-C. G., & Symeonaki, M. (1998). The perturbed non-homogeneous Markov system in continuous time. *Applied Stochastic Models and Data Analysis, 13*(3–4), 207–216.

Vassiliou, P.-C. G., & Symeonaki, M. (1999). The perturbed non-homogeneous Markov system. *Linear Algebra and Applications, 289*(1–3), 319–332.

Vassiliou, P.-C. G., & Tsakiridou, H. (2004). Asymptotic behaviour of first passage probabilities in the perturbed non-homogeneous semi-Markov systems. *Communications in Statistics – Theory and Methods, 33*(3), 651–679.

Vassiliou, P.-C. G., & Tsaklidis, G. (1989). The rate of convergence of the vector of variances and covariances in non-homogeneous Markov systems. *Journal of Applied Probability, 26*, 776–783.

Ward-Warmedinge, M., Melanie, E., & Macchiarelli, C. (2013). Transitions in labour market status in the European Union, LEQS Paper 69.

Part V
Life-Time, Survival, Pension, Labor Force and Further Estimates

Chapter 24
The Wide Variety of Regression Models for Lifetime Data

Chrys Caroni

Introduction

This paper examines various approaches that have been proposed for the analysis of lifetimes (times-to-event) in relation to covariates, for example, to examine how the survival of a group of patients in a medical study depends on their gender, age and health-related markers. The models that are considered are regression-type models in which the combined effect of the covariates appears through a linear predictor of the form $\boldsymbol{\beta}'\mathbf{x}$, where $\boldsymbol{\beta}$ is a vector of coefficients and $\mathbf{x}$ is a vector of covariate values.

In general, suppose an outcome variable Y follows some distribution with probability density function $f(y|\theta)$ indexed by parameters θ. A strong tradition in statistical modelling is to allow covariates $\mathbf{X}$ that influence the value of Y to do so by modifying the parameters of the distribution, although not its functional form. Thus

$$f(y|\mathbf{x}, \theta) = f(y|\theta(\mathbf{x})).$$

The basic example is the standard linear regression model $Y = \boldsymbol{\beta}'\mathbf{x} + \epsilon$ with $\epsilon \sim N(0, \sigma^2)$ which implies $Y \sim N(\boldsymbol{\beta}'\mathbf{x}, \sigma^2)$. Thus $Y|\mathbf{x}$ follows the normal distribution with mean parameter $\mu = \mu(\mathbf{x}) = \boldsymbol{\beta}'\mathbf{x}$.

Modelling the outcome variable's distribution as remaining within the same family but with different parameters, is familiar from generalised linear modelling (in which the mean parameter depends on $\mathbf{x}$ as in the regression example above) and GAMLSS (in which up to three parameters of a distribution may depend on covariates: Rigby et al. 2017). It has not customarily been adopted in modelling the

C. Caroni (✉)
Department of Mathematics, National Technical University of Athens, Athens, Greece
e-mail: ccar@math.ntua.gr

C. H. Skiadas, C. Skiadas (eds.), *Demography of Population Health, Aging and Health Expenditures*, The Springer Series on Demographic Methods and Population Analysis 50, https://doi.org/10.1007/978-3-030-44695-6_24

dependence of lifetime distributions on covariates in survival analysis and reliability modelling (but see Burke and MacKenzie (2017), below). Consider, however, the Weibull distribution, widely used in modelling lifetime data. The survival function is

$$S(t) = 1 - F(t) = P(T > t) = \exp\left\{-(t/\alpha)^{\eta}\right\},\ t > 0,\ \alpha > 0,\ \eta > 0$$

and the hazard function $h(t) = f(t)/S(t) = \eta\alpha^{-\eta}t^{\eta-1}$. If the scale parameter α depends on covariates $\mathbf{X}$, so that $\alpha(\mathbf{x}) = e^{\boldsymbol{\beta}'\mathbf{x}}$, then it follows that the ratio of the hazards for two units with different covariate vectors $\mathbf{x}_1$ and $\mathbf{x}_2$ is $e^{-\eta\boldsymbol{\beta}'(\mathbf{x}_1-\mathbf{x}_2)}$ which is independent of time t. This is the well-known proportional hazards (PH) property. Alternatively, instead of arising from how the values of parameters are changed by covariates, this property can be taken as *defining* how covariates affect the outcome variable. This was done most famously in Cox's semiparametric PH regression model

$$h(t|\mathbf{x}) = e^{\boldsymbol{\beta}'\mathbf{x}}h_0(t)$$

in which the underlying distribution – and hence the baseline hazard $h_0(t|\mathbf{x})$ – is not even specified parametrically.

The great popularity of the PH model has led to proposals of various other "proportional" models in the literature. The purpose of the present paper is to review these briefly, compare some of their properties and comment on their differences. Other regression models, not possessing a proportionality property, will also be mentioned for completeness.

Regression Models with Proportionality Properties

Proportional Odds

After PH, the best-known regression model with a proportionality property for lifetime data is the proportional odds (PO) model (Bennett 1983a,b). The *odds* of the occurrence of the event by time t:

$$\theta(t) = \frac{F(t)}{1 - F(t)} = \frac{1 - S(t)}{S(t)}.$$

is related to covariates in the PO model by

$$\theta(t|\mathbf{x}) = \theta_0(t)g(\mathbf{x}),$$

where $\theta_0(t)$ is the baseline odds and $g(\mathbf{x})$ is a suitable non-negative function, usually $g(\mathbf{x}) = e^{\boldsymbol{\beta}'\mathbf{x}}$, which gives the model

$$\ln \theta(t|\mathbf{x}) = \ln \theta_0(t) + \boldsymbol{\beta}'\mathbf{x}.$$

This is *logistic regression*, the most widely used regression model for binary data. However, PO models are used infrequently. With the exception of the regression model based on the log-logistic distribution, which also fits into the accelerated failure time framework, they are mathematically and computationally more difficult to handle. Thus they have not been included in most of the computing packages, although the R package "timereg" (Scheike 2019) allows the fitting of a semiparametric PO model. Detailed discussions of PO models can be found in Collett (2014).

From the above equation for $\theta(t|\mathbf{x})$, it follows that

$$S(t|\mathbf{x}) = \{1 + \theta_0(t)g(\mathbf{x})\}^{-1}$$

in a PO model, from which the ratio of the hazard functions for two units with different covariate values is

$$\frac{h(t|\mathbf{x}_1)}{h(t|\mathbf{x}_2)} = \frac{g(\mathbf{x}_1)}{1 + \theta_0(t)g(\mathbf{x}_1)} \Big/ \frac{g(\mathbf{x}_2)}{1 + \theta_0(t)g(\mathbf{x}_2)}$$

$$\to 1 \ (t \to \infty)$$

because $\theta_0(t) \to \infty$ as $t \to \infty$ (since $S_0(t)$ tends to zero) whereas $g(.)$ does not change with t. In other words, the PO model implies that initial differences between units disappear (in terms of the hazard), whereas the PH model demands that they never change. Depending on the context, either might be more appropriate. A badly-made unit remains badly made, so the PO model seems unreasonable in such situations. But if a patient's treatment has only a temporary effect that wears off with time, the PH property – that initial differences in the covariates continue to have the same effect for ever and ever – seems unrealistic.

An obvious implication of the above result on hazard rates is that it is impossible for any distribution to possess both the PO and PH properties.

Proportional Mean Residual Life

The mean residual life (MRL) of a unit, conditional on the unit's current age t, is defined as the expected remaining lifetime beyond t,

$$\mu(t) = E[T - t|T \geq t] = \frac{\int_t^\infty S(u)du}{S(t)}$$

(if it exists). This is familiar in demography under the name life expectancy. Under the proportional mean residual life (PMRL)model,

$$\mu(t|\mathbf{x}) = g(\mathbf{x})\mu_0(t),$$

where $\mu_0(t)$ is the baseline MRL function.

Oakes and Dasu (1990, 2003) suggest the MRL function as a basis for modelling lifetime data because it summarizes the entire remaining life distribution and not just the immediate risk of failure. They claim that this is likely to be the important information for the design of maintenance and repair strategies. On the other hand, Hougaard (1999) states that, in contrast to these industrial and demographic applications, the evaluation of mean lifetime in biostatistics is considered unacceptable. One reason is the difficulty of estimating the right tail of the residual lifetime distribution (exacerbated by censoring), which – as acknowledged by Oakes and Dasu (2003) – can have a strong influence on the mean; secondly, there may be a tendency for readers to think in terms of the normal distribution when they are presented with means, which could be very misleading; thirdly, there may sometimes be a proportion of the population (a "cured fraction") that will never experience the event, which makes it impossible to calculate a mean. The last of these objections obviously does not apply in industrial and demographic applications, because units will always fail and individuals will always die eventually.

This model was developed further by Maguluri and Zhang (1994) and subsequently by others; see Chen and Cheng (2005). The PMRL model does not seem to have entered general use.

Proportional Reversed Hazards

Another "proportional" model is the proportional reversed hazard rate (PRHR) model (see Gupta and Gupta 2007). The reversed hazard rate $r(t)$ is related to the conditional probability that an event occurred in the interval of length δt *before* time t, whereas the hazard is related to the occurrence of the event in the interval of length δt *after* time t. Thus

$$r(t)\delta t = P(t - \delta t < T \leq t|T \leq t) = f(t)\delta t/F(t),$$

so $r(t) = (d/dt) \ln F(t)$. In the PRHR model, covariates act multiplicatively on the baseline function $r_0(t)$,

$$r(t|\mathbf{x}) = g(\mathbf{x})r_0(t).$$

Equivalently, $F(t|\mathbf{x}) = [F_0(t)]^{g(\mathbf{x})}$ similar to the definition $S(t|\mathbf{x}) = [S_0(t)]^{g(\mathbf{x})}$ which can be derived for the proportional hazards model. A PRHR model was first suggested, although not studied, by Kalbfleisch and Lawless (1989).

An example of a family of distributions with the PRHR property is the exponentiated Weibull which has baseline distribution function $F(t|\alpha) = [1 - \exp(-t^\alpha)]^\theta$.

Relations Between Proportionality Models

Given a lifetime distribution with distribution function F, survival function S and probability density function f, denote the hazard function, reversed hazard function and odds by h, r and θ, respectively. Then, by definition,

$$h = \frac{f}{S} = \left(\frac{f}{F}\right)\left(\frac{F}{S}\right) = r\theta \,.$$

Write h_1 and h_2 for the hazard function corresponding to two different parameter values, and similarly for the other functions. Then

$$\frac{h_2}{h_1} = \left(\frac{r_2}{r_1}\right)\left(\frac{\theta_2}{\theta_1}\right).$$

This expression relates the ratios of the hazard, reversed hazard and odds functions. Notice that if any two of them have the proportionality property – that is, they are independent of time – then so must the third. But, as shown above, no lifetime distribution can simultaneously possess the proportional hazards and proportional odds properties. Consequently, no distribution can have more than one of these properties. Thus, the properties of proportional hazards, reversed proportional hazards and proportional odds are mutually exclusive. The same equation holds some obvious implications for the behaviour of the various ratios. If the distribution has proportional odds, then ratios of hazards must be in constant ratio to ratios of reversed hazards. Similarly, proportional reversed hazards implies constant ratios of hazard ratios to odds ratios.

Multi-parameter Regression Models

Threshold (or First Hitting Time) Regression

The Weibull PH and log-logistic PO models described above are single-parameter regression models, in which the dependence on covariates is introduced into one parameter while the other remains invariant. It is possible to allow dependence in further parameters. One survival analysis model which is becoming well known and includes this feature is threshold regression (or first hitting time regression) based on the inverse Gaussian (IG) distribution (Lee and Whitmore 2006; Caroni 2017). Parameterising the distribution as

$$f(t|y_0, \mu, \sigma^2) = \frac{y_0}{\left(2\pi\sigma^2 t^3\right)^{1/2}} \exp\left[-\frac{(y_0 + \mu t)^2}{2\sigma^2 t}\right],$$

dependence on covariates is modelled by

$$\mu = \boldsymbol{\beta}'\mathbf{u}$$
$$\ln y_0 = \boldsymbol{\gamma}'\mathbf{v},$$

where the parameters $\boldsymbol{\beta}$ and $\boldsymbol{\gamma}$ are vectors of regression coefficients and $\mathbf{u}$ and $\mathbf{v}$ are vectors of covariates (possibly the same vector).

This distribution arises if the lifetime is conceptualised as a realisation of an underlying Wiener process with drift μ and variance σ^2 which will usually be taken to be equal to one. For example, a patient's survival depends on a latent health status $Y(t)$ and death occurs when this falls to the threshold zero for the first time, having started at $Y(0) = y_0 > 0$. The IG distribution given above is the distribution of the first passage time from y_0 to zero. If only the drift depends on covariates whereas the starting level is constant across units, this is a single-parameter IG regression which can be fitted as a generalised linear model.

General Frameworks

Burke and MacKenzie (2017) extended single-parameter regression to either two or three parameters (depending on the distribution). Their R package 'mpr' provides fitting for a small number of survival distributions, including the Weibull and log-logistic but not the IG. The GAMLSS framework is a general approach to regression modelling, in which all the parameters of the univariate response variable can be modelled as additive functions of the explanatory variables in a very flexible manner. The distributions currently available include the Weibull and the IG (Rigby et al. 2017). However, the IG parameterisation is not the one employed in the first hitting time regression model.

References

Bennett, S. (1983a). Analysis of survival data by the proportional odds model. *Statistics in Medicine, 2*(2), 273–277.

Bennett, S. (1983b). Log-logistic regression models for survival data. *Applied Statistics, 32*(2), 165–171.

Burke, K., & MacKenzie, G. (2017). Multi-parameter regression survival modeling: An alternative to proportional hazards. *Biometrics, 73*(2), 678–686.

Caroni, C. (2017). *First hitting time regression models*. ISTE-Wiley, London.

Chen, Y. Q., & Cheng, S. (2005). Semiparametric regression analysis of mean residual life with censored survival data. *Biometrika, 92*(1), 19–29.

Collett, D. (2014). *Modelling survival data in medical research*, 3rd edn. Chapman and Hall, Boca Raton.

Gupta, R. C., & Gupta, R. D. (2007). Proportional reversed hazard rate model and its applications. *Journal of Statistical Planning and Inference, 137*(11), 3525–3536.

Hougaard, P. (1999). Fundamentals of survival data. *Biometrics, 55*(1), 13–22.

Kalbfleisch, J. D., & Lawless, J. F. (1989). Inference based on retrospective ascertainment: An analysis of the data on transfusion-related AIDS. *Journal of the American Statistical Association, 84*(406), 360–372.

Lee, M.-L. T., & Whitmore, G. A. (2006). Threshold regression for survival analysis: Modeling event times by a stochastic process reaching a boundary. *Statistical Science, 21*(4), 501–513.

Maguluri, G., & Zhang, C. H. (1994). Estimation in the mean residual life regression model. *Journal of the Royal Statistical Society, Series B, 56*(3), 477–489.

Oakes, D., & Dasu, T. (1990). A note on residual life. *Biometrika, 77*(2), 409–410.

Oakes, D., & Dasu, T. (2003). Inference for the proportional mean residual life model. Institute of Mathematical Statistics Lecture Notes – Monograph Series. 43, 105–116.

Rigby, R., Stasinopoulos, M., Heller, G., & De Bastiani, F. (2017). Distributions for Modelling Location, Scale and Shape: Using GAMLSS in R. Draft version available from https://www.gamlss.com/wp-content/uploads/2018/01/DistributionsForModellingLocationScaleandShape.pdf. Accessed 21 Apr 2020.

Scheike, T. (2019). Package "timereg". https://cran.r-project.org/web/packages/timereg/timereg.pdf.

Chapter 25
Analysing the Risk of Bankruptcy of Firms: Survival Analysis, Competing Risks and Multistate Models

Francesca Pierri and Chrys Caroni

Introduction

Quantitative methods have long been applied to contribute to controlling credit risk, the possibility that a bank borrower or counter-party will fail to meet its obligations in accordance with agreed terms. There is particular interest in such models during an economic downturn, when firms with an inadequate financial base may be forced to close down. Multivariate discriminant analysis (MDA) and logistic regression were the first techniques employed in building credit risk models (Balcaen and Ooghe 2006). J.W. Beaver first used MDA in 1966 and J.A. Ohlson introduced logistic regression to predict company failure in 1980; the latter subsequently became the dominant method.

The above methods analyse the occurrence of the event of interest, but not its timing. Models of survival and reliability for analyzing times-to-event data were already familiar in fields such as medicine and engineering. The idea of employing survival analysis for building credit-scoring models was introduced by Narain 1992. In previous papers (Pierri and Caroni 2017; Caroni and Pierri 2020), we used techniques from survival analysis to study the closure of Small Business Enterprises in Italy. Our first analysis employed the well-known Cox semi-parametric proportional hazards regression model, extended to include time-dependent covariates (Pierri and Caroni 2017). In the second analysis, recognising that the event of closure in fact represented three distinct endpoints – voluntary

F. Pierri
Department of Economics, University of Perugia, Perugia, Italy
e-mail: francesca.pierri@unipg.it

C. Caroni (✉)
Department of Mathematics, National Technical University of Athens, Athens, Greece
e-mail: ccar@math.ntua.gr

C. H. Skiadas, C. Skiadas (eds.), *Demography of Population Health, Aging and Health Expenditures*, The Springer Series on Demographic Methods and Population Analysis 50, https://doi.org/10.1007/978-3-030-44695-6_25

winding-up, closure not due to any action by creditors or the Court, and formal bankruptcy – we extended the first analysis by using *competing risks* techniques to examine how the occurrence of each potential mode of a firm's failure was related to covariates (Caroni and Pierri 2020). Two techniques were compared: the first modelled the *cause-specific hazards*, while the second used *subdistribution hazards*.

A competing risks model can be thought of as representing a situation in which the underlying process (here, the life course of the firm) starts from the initial state (the firm is in active operation) and may make the transition to one of several other states, each of which represents an alternative endpoint. The latter are mutually exclusive absorbing states of the underlying stochastic process; the process terminates upon entry to one of them. This simple structure is a special case of a *multistate model* (see, for example, Putter et al. 2007). In a general multistate model, further transitions between some or all of the states are possible. This is what happens in our data on Italian firms. For example, since voluntary liquidation requires the firm to be solvent, if it is subsequently unable to meet its obligations, or if the liquidator does not complete the process, legal processes take over. Figure 25.1 shows the structure of the model. As movements cannot be retraced, this is a unidirectional multistate model.

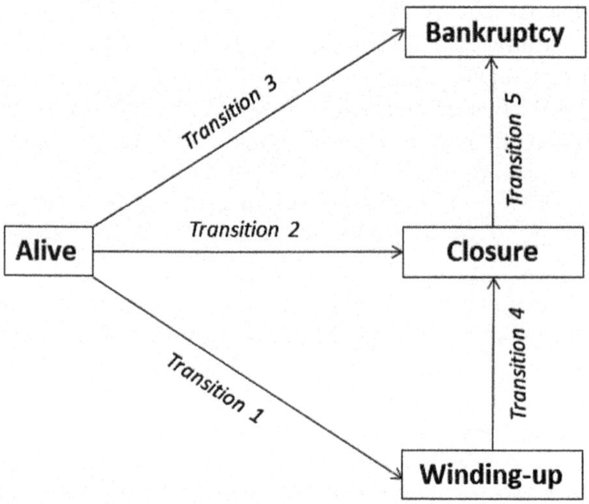

Fig. 25.1 Possible transitions between states in the model of lifetimes of Small Business Enterprises

Methodology

Competing Risks

Let T denote the random time until the event and $\boldsymbol{\beta}$ a vector of coefficients that expresses the effects of covariates $\mathbf{x}$ upon T. Covariate values are assumed to be measured at time zero and to be independent of time. Define the survival function

$$S(t|\mathbf{x}) = P(T > t|\mathbf{x}), \tag{25.1}$$

which is the probability that a unit (firm) will remain in operation until at least time t. Its complement $F(t|\mathbf{x}) = P(T \leq t|\mathbf{x})$ is the *cumulative incidence function*, the probability of failure up to time t. The hazard function $h(t)$ expresses the instantaneous rate of failure at time t among units that still survive at that time:

$$h(t) = \lim_{\delta t \to 0} \frac{P[t < T \leq t + \delta t | T > t]}{\delta t} = \frac{f(t)}{S(t)}. \tag{25.2}$$

The two basic functions h and S are linked by

$$S(t|\mathbf{x}) = e^{-H(t|\mathbf{x})}, \tag{25.3}$$

where the cumulative hazard $H(t|\mathbf{x})$ is the integral of $h(\tau|\mathbf{x})$ from time zero to t. This shows that h and S provide equivalent information, because one can be derived from the other. In the proportional hazards model, which includes Cox regression,

$$h(t|\mathbf{x}) = h_0(t)\exp\left\{\boldsymbol{\beta}'\mathbf{x}\right\}, \tag{25.4}$$

(in the usual specification). From equation (25.3),

$$S(t|\mathbf{x}) = S_0(t)^{e^{\boldsymbol{\beta}'\mathbf{x}}}, \tag{25.5}$$

where $S_0(t)$ is the baseline survival function corresponding to the baseline hazard $h_0(t)$. The effect of covariate x_i is expressed by the quantity e^{β_i}: an increase of one unit in the value x_i leads, from (25.4), to the multiplication of the hazard by this amount while, from (25.5), the survival function is raised to this power. Therefore, the coefficients $\boldsymbol{\beta}$ show the effect of each covariate on both the rate (hazard function) and risk (one minus the survival function) of failure.

The above outline did not consider the situation of competing risks, in which there are two principal alternative approaches to the analysis of the data. This is because the basic identity between hazard functions and survival functions given above does not apply. It is necessary to concentrate either on hazards or cumulative incidence. In the present paper, we present results for the *cause-specific hazards*. By adding a subscript denoting the model of failure to the definition of hazard, we define the cause-specific hazard for the kth of K causes as

$$h_k(t) = \lim_{\delta t \to 0} \frac{P[t < T \leq t + \delta t \bigcap D = k | T > t]}{\delta t}, \tag{25.6}$$

where D denotes the cause of failure. This is the probability of failure from cause k in the interval $(t, t + \delta t]$ given that the unit has survived until time t. It does not lead directly to the cumulative incidence of failure from this cause, because survival to time t depends on the hazards from all the causes.

Alternatively, in order to examine cumulative incidences, we would work with the *subdistribution hazard* function (Fine and Gray 1999)

$$h_k^s(t) = \lim_{\delta t \to 0} \frac{P[t < T \leq t + \delta t \bigcap D = k | T > t \bigcup (T \leq t \cap D \neq k)]}{\delta t}, \tag{25.7}$$

which gives the instantaneous rate of failure in units that have not yet experienced an event of type k, although they may have experienced a different event. In contrast to the risk set at time t for the cause-specific hazard – which consists of the units still surviving at that time – the risk set for the subdistribution hazard must also include units that have already failed before time t, but from causes other than cause k (even though, in reality, they cannot in fact fail again). Results from this approach are given by Caroni and Pierri (2020). The choice between approaches is discussed by, for example, Lau et al. (2009).

Both versions of the hazard function under competing risks, equations (25.6) and (25.7), can be extended to include the multiplicative effects of covariates exactly as in the proportional hazards models defined by equation (25.4). The definition of cause-specific hazard shows that a standard program for fitting the Cox model can always be employed in order to calculate it, by treating failures from causes other than the one of interest as censored at the failure time.

Multistate Models

Following Putter et al. (2007), we define a hazard rate or transition intensity for the transition from state i to j:

$$\lambda_{ij}(t) = \lim_{\delta t \to 0} \frac{P[t < T \leq t + \delta t | T > t]}{\delta t}, \tag{25.8}$$

analogous to a cause-specific hazard in competing risks analysis. The time t is measured from entry into the initial state. This is known as the $clock\ forward$ approach. This multistate model is assumed to be a Markov model, so that transitions depend only on the present state and not on the history. Each of the hazard rates is modelled using Cox's semi-parametric proportional hazards regression model. In order to fit this set of models, it is convenient to rearrange the data file so that each firm is represented by several rows of data, one for each transition for which it is at risk at some time. For example, suppose a firm enters state W after 30 months and remains in that state until the study terminates 72 months after the start. Then it is represented by four rows of data. There is one row for each of the three possible transitions from the initial state A (active) to states W, C or B, all with starting time zero and finishing time 30 months. The transition to W ends with an event, but those to C and B end with right censoring because no such transition took place. The fourth row for this firm is for the transition W→C, with starting time 30 and finishing with right censoring at time 72. There is no row of data for this firm for the remaining possible transition C→B, because the firm never entered C and therefore was not at risk for this transition.

This data structure is demonstrated fully in Table 25.1, showing four firms with different types of transition histories: the first firm experienced bankruptcy directly (transition = 3 with status = 1), the second both closure and bankruptcy (transition = 2 and transition = 5 with status = 1), the third voluntary winding-up followed by closure (transition = 1 and transition = 4 with status = 1), and the fourth voluntary winding-up, then closure and finally bankruptcy (transition = 1, transition = 4 and transition = 5 with status = 1).

Table 25.1 Example of data structure for possible transitions between the states of the model

id	From	To	Transition	Label	Tstart	Tstop	Time	Status
1	1	2	1	A→W	0	35	35	0
1	1	3	2	A→C	0	35	35	0
1	1	4	3	A→B	0	35	35	1
2	1	2	1	A→W	0	30	30	0
2	1	3	2	A→C	0	30	30	1
2	1	4	3	A→B	0	30	30	0
2	3	4	5	C→B	30	47	17	1
3	1	2	1	A→W	0	46	46	1
3	1	3	2	A→C	0	46	46	0
3	1	4	3	A→B	0	46	46	0
3	2	3	4	W→C	46	49	3	1
3	3	4	5	C→B	49	72	23	0
4	1	2	1	A→W	0	58	58	1
4	1	3	2	A→C	0	58	58	0
4	1	4	3	A→B	0	58	58	0
4	2	3	4	W→C	58	64	6	1
4	3	4	5	C→B	64	70	6	1

Application

Data

The data employed in the analysis presented here were extracted from the same sources as were used in earlier analyses and described in detail by Pierri and Caroni (2017). In brief, for the present analysis, data on 15184 enterprises were compiled from files made available by the Chamber of Commerce of Perugia in the Umbria Region of Italy. These businesses were active at the start of 2008, at which time the financial data listed in Table 25.2 were recorded. Together with the sector of activity, legal form and geographical location, these data were used as covariates in modelling whether and when the enterprise moved from the active state to one of the three states of inactivity described above in the Introduction. In addition, any transitions between the three inactivity states were recorded. Data were updated to the end of 2013. Table 25.3 shows the recorded numbers of transitions.

Table 25.2 Covariates investigated in the survival analyses. In addition, the firm's Activity Sector (6 categories), Legal Form (3 categories) and Geographical Location (2 categories) were represented by indicator variables

Abbreviation	Variable
CR	Current Ratio
QR	Quick Ratio
L	Leverage
IRR	Investment Rigidity Ratio
TAR	Tangible Assets Ratio
IAR	Intangible Assets Ratio
FFAR	Financial Fixed Assets Ratio
OFAR	Other Fixed Assets Ratio
IER	Investment Elasticity Ratio
IIR	Inventories Impact Ratio
LR	Liquidity Ratio
STLR	Short Term Liquidity Ratio
LTLR	Long Term Liquidity Ratio
ER	Equity Ratio
DR	Debt Ratio
PDR	Permanent Debt Ratio
CDR	Current Debt Ratio
ROA	Return on Assets
ROE	Return on Equity
ROT	Return on Turnover
ROS	Return on Sales
TUR	Turnover

Table 25.3 Observed transitions between the four states of the model

	Still active in 2013	Winding-up (W)	Closure (C)	Bankruptcy (B)	Total
First transition From active in 2008	12135 (79.9%)	1808 (11.9%)	1005 (6.6%)	236 (1.6%)	15184 (100%)
Subsequent transitions From	W	1194[a] (66.0%)	614 (34.0%)	0	1808 (100%)
	C	0	987[a] (98.2%)	18 (1.8%)	1005 (100%)
	B	0	0	236[a] (100%)	236 (100%)

[a]These firms did not make a second transition within the duration of the study

Results: Competing Risks

Table 25.4 shows the statistically significant ($P < 0.05$) effects of covariates (estimated hazard ratios, HR) after a backward elimination procedure. The importance of applying the competing risks approach is demonstrated by the large differences between results for the different causes. Debt ratios are significant for all causes but with very different HR; a higher debt ratio increases the hazard for every cause but most of all for bankruptcy. For another example, the firm's sector of economic activity affects only the hazard of bankruptcy, for which every sector has an increased HR compared to the reference sector, Agriculture, in which bankruptcy was a rare event (only 2.6% of the failures in this sector). The covariates describing type of firm acted in opposite directions for different causes. A Cox regression for failure from any cause gives an HR of 1.01 for cooperatives and 1.15 for limited liability companies, compared to publicly-traded companies, which would misleadingly suggest that the type of firm had little effect on the failure rate.

Results: Multistate Model

Table 25.5 presents results from fitting the multistate model. For each transition, the statistically significant covariate ($P < 0.05$) effects (estimated hazard ratios, HR) after a backward elimination procedure are reported. First of all, we note the absence of significant effects for transition 5, attributable to both the rarity of the event (1.8%) and its nature. In fact, during a voluntary winding-up, if the liquidator is not able to pay all the debts, the firm's activity is stopped and the insolvency procedure will start. Moreover, transition 4 (winding-up to closure) is strongly increased by a high value of Short Term Liquidity Ratio and slowed down by the Permanent Debt

Table 25.4 Cause-specific hazard ratios (with 95% confidence intervals) of significant covariates in competing risks analysis

	Winding-up	Closure	Bankruptcy
Activity sector[a]			
Construction			4.14 (1.01–17.0)
Manufacturing			4.33 (1.06–17.7)
Type of firm[b]			
Cooperative	2.51 (1.62–3.89)	0.48 (0.33–0.70)	
Limited liability	1.61 (1.06–2.44)	0.50 (0.38–0.67)	
Location: Terni[c]	1.22 (1.10–1.35)		
IRR	2.01 (1.73–2.33)		
TAR	0.14 (0.10–0.20)		
IAR		1.71 (1.22–2.41)	
FFAR		4.00 (2.89–5.52)	
IIR	0.68 (0.53–0.86)		
STLR	1.82 (1.35–2.45)		
DR	1.85 (1.67–2.06)	1.22 (1.03–1.44)	3.43 (2.71–4.34)
PDR	0.52 (0.42–0.64)	0.56 (0.42–0.74)	0.23 (0.13–0.44)
ROA	0.995 (0.993–0.998)		
TUR		0.98 (0.96–0.99)	
Age	0.970 (0.964–0.976)	0.963 (0.955–0.971)	0.979 (0.964–0.994)

Reference categories: [a]Agriculture; [b]Publicly-traded; [c]Perugia

Table 25.5 Transition-specific hazard ratios (with 95% confidence intervals) of significant covariates in multistate analysis. There were no significant effects for transition 5 from closure to bankruptcy

	Transition 1 winding-up	Transition 2 closure	Transition 3 bankruptcy	Transition 4 W→C
Location: Terni[a]	1.22 (1.10–1.35)			
TAR	0.29 (0.22–0.38)		0.43 (0.20–0.92)	
IAR	0.28 (0.23–0.35)	1.74 (1.19–2.55)	3.83 (1.66–8.85)	
FFAR	1.63 (1.15–2.32)	4.13 (2.95–5.76)		
IIR			2.15 (1.25–3.69)	
LR	1.42 (1.21–1.66)		1.83 (1.18–2.83)	
STLR			0.30 (0.10–0.89)	2.62 (1.67–4.19)
ER	0.87 (0.84–0.91)		0.82 (0.74–0.92)	
PDR	0.69 (0.55–0.85)		0.39 (0.20–0.75)	0.69 (0.49–0.98)
CDR	0.87 (0.84–0.91)		0.82 (0.74–0.92)	
ROA	0.994 (0.991–0.998)			
ROT	0.88 (0.82–0.93)			
TUR	1.15 (1.08–1.22)			
Age	0.967 (0.962–0.974)	0.966 (0.958–0.974)	0.978 (0.963–0.993)	0.979 (0.968–0.991)

Reference category: [a]Perugia

Ratio. Mainly, two financial indexes (IAR and FFAR) affect the closure event, as in competing risks analysis.

Moreover we notice a larger number of significant effects for the winding-up cause compared to others. These sometimes follow the same direction but with different intensity (FFAR), sometimes with the same strength (CDR, ER). The geographical district affects only the hazard of voluntary winding-up, while the firm's age, when significant, has quite similar intensity (close to 1) for all the transitions. Profitability indexes affect only the winding-up event while the activity sector and the type of firm are not significant at all in multistate analysis.

Conclusions

This work present two different methodologies, *competing risks* and *multistate*, applied to financial data. The object of interest was to investigate whether the same factors influence different events and which model would provide better prediction of the future state of a firm.

The two approaches underline the importance of the same group of indexes with the same sign for comparable events; in some cases multistate results are more concise and in other cases the opposite. For example, the multistate results highlight that for the closure event the analysis of financial indexes (IAR, FFAR) gains a dominant position, while for the winding-up to closure transition the same role is played by the liquidity and indebtedness indexes (STLR, PDR). In contrast, these two aspects are merged in competing risk for the closure event. However, when we consider bankruptcy, the competing risk analysis mainly focuses its attention on indebtedness indexes (DR, PDR), besides the activity sector, whereas in multistate modeling the group of liquidity ratios (IIR, LR, STLR), indebtedness (PDR, CDR) and financial indexes (TAR IAR) seem to play an important role.

Even though transition 5, that is, proceeding from closure to bankruptcy, has no significant covariates, probably due to the small number of cases, looking at multistate results, the analysis confirms what happens in practice: if the liquidator is able to increase the Short Term Liquidity Ratio, he will be able to drive the company from voluntary winding-up to closure, otherwise not.

Acknowledgments Francesca Pierri received the support of "Fondo Ricerca di Base, 2018" from Universitá degli Studi di Perugia for the project "Survival Analysis and Competing Risks: a study on firms in Umbria Region".

References

Balcaen, S., & Ooghe, H. (2006). 35 years of studies on business failure: An overview of the classic statistical methodologies and their related problems. *British Accounting Review, 38*(1), 63–93.

Caroni, C., & Pierri F. (2020). Different causes of closure of Small Business Enterprises: Survival analysis with alternative models for competing risks. *Electronic Journal of Applied Statistical Analysis, 13*(1), 211–228.

Fine, J. P., & Gray, R. J. (1999). A proportional hazards model for the subdistribution of a competing risk. *Journal of the American Statistical Association, 94*(446), 496–509.

Lau, B., Cole, S. R., & Gange, S. J. (2009). Competing risk regression models for epidemiologic data. *American Journal of Epidemiology, 170*(2), 244–256.

Narain, B. (1992). *Survival analysis and the credit-granting decision.* In L. C. Thomas J. N. Crook, & D. B. Edelman (Eds.), *Credit scoring and credit control* (pp. 109–122). Oxford University Press.

Pierri, F., & Caroni, C. (2017). Bankruptcy prediction by survival models based on current and lagged values of time-varying financial data. *Communications in Statistics: Case Studies, Data Analysis and Applications, 3*(3–4), 62–70.

Putter, H., Fiocco, M., Geskus, R. B. (2007). Tutorial in biostatistics: Competing risks and multi-state models. *Statistics in Medicine, 26*(11), 2389–2430.

Chapter 26
A Bayesian Modeling Approach to Private Preparedness Behavior Against Flood Hazards

Pedro Araújo, Gilvan Guedes, and Rosangela Loschi

Introduction

Floods are one of the major causes of health and economic losses among natural disasters worldwide (Kellenberg and Mobarak 2011). Preparedness behavior against flood hazards however is low in many parts of the world, including regions with high levels of education, environmental awareness and economic development (Lindell and Perry 2012).

Some studies in social psychology and insurance economics studies have devoted some theoretical and empirical efforts to explain persons' flood preparedness behavior (Terpstra and Lindell 2013; Alary et al. 2013). The Protective Action Decision Model (PADM) is a well-established conceptual approach representing these efforts. PADM was proposed in social psychology (Lindell and Perry 1992, 2012) to explain the way people make decisions with regard to the adoption of protective actions against imminent disasters or environmental hazards. The main goal is to identify the factors or attributes that influence the individual's decision. Among the factors, PADM assumes the hazard-related (perception of effectiveness) and the resource-related (opportunity costs) factors. Three hazard-related factors, named perceived efficacy for (i) protect person and (ii) protect property, and (iii) perceived utility of the actions for other purposes, are considered. Perceived requirement for money, equipments, knowledge, time and effort to implement the protective action are among the resource-related factors. The desirable behavior of

P. Araújo · R. Loschi
Department of Statistics, Universidade Federal de Minas Gerais, Belo Horizonte, Brazil
e-mail: loschi@est.ufmg.br

G. Guedes (✉)
Demography Department, UFMG, Belo Horizonte, Brazil
e-mail: grguedes@cedeplar.ufmg.br

C. H. Skiadas, C. Skiadas (eds.), *Demography of Population Health, Aging and Health Expenditures*, The Springer Series on Demographic Methods and Population Analysis 50, https://doi.org/10.1007/978-3-030-44695-6_26

PADM should indicate that, fixing the level of risk aversion, actions against floods are positively associated with the perceived effectiveness of the protective action and negatively related to their direct and indirect (opportunity) costs. Although PADM allows to identify the effects of the perception of effectiveness and opportunity costs of private measures in protective action adoption, it omits the risk aversion that is a key factor in any private insurance model. That may possibly induce a omission bias.

This work is inspired by real situation, detailed presented in Sect. 26, where we like to evaluate the behavior of people that live under risk of frequent floods. Our data is collected in Governador Valadares, a cite located in Minas Gerais state in Brazil, covering part of Rio Doce basin, and evaluate if people preparedness behavior in that region has the same features as those shown in the study presented by Terpstra and Lindell (2013). Rio Doce basin plays an important role in the economical development of Eastern Minas Gerais and North Espírito Santo regions. Despite its importance to the region economy, flooding of Rio Doce has also imposing great loses to cities in its board, not only from the economic point-of view. Governador Valadares has large share of the population under risk of river floods. The three big floods occurring in 1979, 1997 and 2012 destroyed houses, environment and killed many people.

As assumed in PADM, the intention of the individual adopting each protection action depends on her/his perception of effectiveness and opportunity costs of the action as well as of some personal features as her/his risk aversion, sex and socio-economic condition. The effects of which can vary through the protective measures.

Risk aversion and the perceptions of effectiveness and opportunity costs are latent traits of the individual that are indirectly measured. Besides, the individual responses to the flood hazard actions previously presented are naturally correlated as they are individual dependent.

Our goal is to introduce a statistical model to accommodate all these data features providing an alternative model to properly analyze such kind of data. We introduce a two stage model. In the first stage, an item response theory (IRT) model (Rasch 1960; Birnbaum 1968; Lord 1980) is fitted to estimate the perceptions of effectiveness and opportunity costs allowing us to appropriately quantify the uncertainty inherent to such quantities. In the second stage, a mixed logistic regression model relating the probability of adopting protective measures against floods to covariates directly measured from individuals and to the latent traits representing risk-aversion and perceptions about the effectiveness (PE) and the opportunity cost (PCO) of those measures is built. We assume a Bayesian approach for inference and investigate if the hypotheses supporting PADM are satisfied in a study involving individuals under risk of river floods in Governador Valadares. A random effect reflecting unmeasured individuals features is included into the model to correlate the individual responses to the different protective measures considered in our study.

Our Data

The survey data used in this study come from a pioneering research project in Brazil addressing environmental attitude, awareness, and behavior at the local level, with detailed questions on climate change perception and adaptive measures under risk of flood hazards. Data are part of the research project entitled "Migration, Vulnerability, and Environmental Change in the Rio Doce Valley". The project was approved by the Research Ethics Committee at the Universidade Federal de Minas Gerais (Protocol CAAE 12650413.0.0000.5149).

Face-to-face interviews were conducted in the urban area of Governador Valadares between 2014 and 2015, based on a questionnaire successfully applied in other countries (Kievik and Gutteling 2011; Zaalberg et al. 2009) and for other hazards (Terpstra and Lindell 2013). The survey was based on a multi-stage sampling design. The first stage used clusters of neighborhoods, with clustering based on spatial contiguity and socioeconomic status of the neighborhood. Within each cluster, sample was stratified by sex and age groups, and within each stratum households were randomly selected. A minimum sample size was estimated as 1,069 households, based on a significance level of 5% and a tolerance of 3% for sample proportions. Variance estimate was 0.25, yielding the most conservative minimum sample size (Groves et al. 2011). Because of additional budget resources granted, sample size was increased to 1,226 households. Due to missing information on some characteristics selected for this analysis, our analytical sample reduced to 1,164 households.

Data used for modeling preparedness behavior were collected following the Protective Action Decision Model framework proposed in Lindell and Perry (2012), with selected dimensions transformed into an structured questionnaire as used in Terpstra and Lindell (2013). The six response variables of interest measures the adoption intentions in relation to the following flood hazard actions: emergency kit (KIT), search for information about flood consequences, evacuation routs to safe places in neighbor (INFO), a list telling what to do in case of evacuation or flood (LIST), agreement with family or friends on how to help each other during evacuation (COOP), use of sandbags or flood skirts (SAB) and acquisition of contractual flood insurance (INSUR).

Covariates representing determinants of preparedness behavior were classified in two groups. Group 1 comprises individual-level characteristics: gender, risk aversion, risk perception, and household income. Risk aversion and risk perception were created as a combination of other variables in the questionnaire, as explained in Sect. 26. Although the questionnarie contained a direct question on household income in interval categories, we simulated its continuous value applying the

Integral Transform Theorem[1]. The same method was applied to risk aversion (ordinal scale) (Angus 1994).

Group 2 includes the covariates that depend on both individuals characteristics and the protective actions. Such covariates are: effectiveness to protect people (EPa), effectiveness to protect properties (EPr), usefulness for other purposes (UP), cost to implement the action (COST), time to implement the action (TIME), effort needed in the action implementation (Effort), knowledge and skills required in the action implementation (KS) and the need for cooperation during action implementation (CoIA).

The response variables and covariates in Group 2, originally measured in the Likert scale with 5 categories, are transformed into binary variables assuming 0 if the response ranges from1 to 3 (representing low to intermediate levels of effectiveness, opportunity cost or adoption intention) and 1 otherwise.

The Proposed Model

Assume that a random sample of n individuals is independently interviewed to obtain their opinion about J protective actions. Let Y_{ij}, $i = 1, \ldots, n$, and $j = 1, \ldots, J$, be a binary response variable assuming 1 if individual i has high intention of adopting the protective action j and 0, otherwise. Let $\boldsymbol{X}_i = (X_{1i}, \ldots, X_{pi})$ be the vector of observable covariates which responses represent individuals' features exogenous to the protective actions. In this study we use gender and household income as elements of $\mathbf{X_i}$ measured without any transformation. Let $\boldsymbol{W}_i = (W_{1i}, \ldots, W_{pi})$ be the vector of observable indicators used as proxies of latent covariates. We use risk aversion and risk perception as the two elements of $\boldsymbol{W}_i$. Such random variables are transformed to proxy their related latent analogs as discussed in Sect. 26.

Some covariates of interest are latent traits of the individuals. These latent covariates measure the perception θ_{ij} of subject i about the effectiveness of protective action j and the perception λ_{ij} of individual i about the opportunity cost of protective action j. Such latent covariates are indirectly measured through some observable covariates as discussed in Sect. 26.

Let $\beta_{0j} \in \mathcal{R}$ be the model intercept. Denote by $\beta_{\theta j}$, $\beta_{\lambda j}$, $\boldsymbol{\beta}^*_j = (\beta^*_{1j}, \ldots, \beta^*_{pj})$ and $\boldsymbol{\beta}_j = (\beta_{1j}, \ldots, \beta_{pj})$ is the real fixed effect related to covariates θ_{ij}, λ_{ij}, $\boldsymbol{X}_i$ and $\boldsymbol{W}_i$, respectively. Assume the linear predictor η_{ij} given by

[1]Given a latent continuous random variable X with its cumulative strictly increasing distribution function F_X, the Probability Integral Transform Theorem states that the variable $Y = F_X(X)$ follows a $U[0, 1]$ distribution. The observable variable $\tilde{X}$ is discrete and we aim to simulate a sample of X using an observable sample of $\tilde{X}$. Therefore, generating an i.i.d. sample $\{y_n\}_{nN}$ of Y and calculating $x_n = F^1 X(y_n)$, we obtain a sample $\{x_n\}_{nN}$ of a random variable with the same distribution of X. For the quality of the simulation for income using the same dataset and variable, refer to Guedes et al. (2015a).

$$\eta_{ij} = \beta_{0j} + \beta_{\theta j}\theta_{ij} + \beta_{\lambda j}\lambda_{ij} + \boldsymbol{\beta}_j^* \boldsymbol{X}_i^T + \boldsymbol{\beta}_j \boldsymbol{W}_i^T + \mu_i, \tag{26.1}$$

where μ_i is a random effect quantifying personal features of individual i not captured by the covariates. In the mixed logistic regression, the probability of the individual i has high intention of adopting the protective action j is given by $\pi_{ij} = P(y_{ij} = 1|\boldsymbol{\beta}^*, \boldsymbol{\beta}, \boldsymbol{\mu}, \boldsymbol{\theta}, \boldsymbol{\lambda}, \boldsymbol{X}, \boldsymbol{W}) = \exp\{\eta_{ij}\}[1 + \exp\{\eta_{ij}\}]^{-1}$.

Assume that $y_{ij}|\boldsymbol{\beta}^*, \boldsymbol{\beta}, \boldsymbol{\mu}, \boldsymbol{\theta}, \boldsymbol{\lambda}, \boldsymbol{X}, \boldsymbol{W} \overset{ind}{\sim} Ber(\pi_{ij})$, $i = 1, \ldots, n$ and $j = 1, \ldots, J$. As a consequence, the likelihood function is

$$f(\mathbf{y}|\boldsymbol{\beta}^*, \boldsymbol{\beta}, \boldsymbol{\gamma}, \boldsymbol{X}, \boldsymbol{W}) = \prod_{i=1}^{n}\prod_{j=1}^{J}\left[\frac{\exp\{\eta_{ij}\}}{1 + \exp\{\eta_{ij}\}}\right]^{y_{ij}}\left[\frac{1}{1 + \exp\{\eta_{ij}\}}\right]^{1-y_{ij}}. \tag{26.2}$$

where $\boldsymbol{\gamma} = (\boldsymbol{\theta}, \boldsymbol{\lambda}, \boldsymbol{\mu})$, which represent the latent parameters of interest.

To complete the model specification, we assume that $\beta_{kj} \sim \text{Normal}(\mu_{kj}, \tau_{kj})$ with mean $\mu_{kj} \in \mathcal{R}$ and variance $\tau_{kj}^{-1} \in \mathcal{R}^+$ for all k and all j, $\mu_i \sim \text{Normal}(m, v)$, where $m \in \mathcal{R}$ is a known parameter and $v \sim \text{Gamma}(a, b)$, where $a > 0$ and $b > 0$ are such that $E(v) = a/b$.

Before proceeding to the inference, we need to estimate $\boldsymbol{W}$, θ_{ij} and λ_{ij}.

Obtaining the Proxies for Risk Aversion and Risk Perception (W)

Risk aversion was not directly asked, so we proxied it combining the two questions, one related to the perceived consequence of a flood event on the person's and his/her family's life integrity and another on the intention to search for information about flood consequences and evacuation routs to safe places in the neighborhood. Because the latter variable used to create the proxy for risk aversion is also a response variable (INFO), this one was excluded from our systems of equations to avoid endogeneity bias in parameters estimation.

The person was asked about the likelihood of threat to his/her own and family's life in case a flood hit his/her house, ranging from very unlikely (1) to very likely (5). Intention to seek for information was also measured using the same range. The risk aversion scale derived from combining these two questions resulted in nine different levels, from very low (threat very likely and adoption intention very unlikely) to very high (threat very unlikely and adoption intention very likely).

The question on perceived likelihood of a flood hitting the household in the future, ranging from very unlikely (1) to very likely (5), was multiplied by the mean of four perceived consequences items (damage to the city public infrastructure, damage to the house structure and personal belongings, life treath, and long term disruption of daily life) to form a single measure of risk perception ranging from

1 to 25. The same strategy was used in Terpstra and Lindell (2013), resulting in a perceived likelihood of flood risk weighted by its consequences.

Obtaining the Individual Perception the Effectiveness and Opportunity Cost $(\boldsymbol{\theta}, \boldsymbol{\lambda})$

A key point in our model is the specification of the latent traits θ_{ij} and λ_{ij}. In this first modeling stage, we consider an item response theory (IRT) model to estimate such traits. One advantage of this approach, instead of fitting the model in (26.1) considering the covariates related to these traits in their original scale, is parsimony. A substantial reduction in the parameter dimension is obtained by using this method.

IRT is a well-known psychometric theory used in educational assessments and cognitive psychology (Rasch 1960; Birnbaum 1968; Lord 1980). IRT models relate the probability of a correct response in a test to latent traits, such as the individual abilities, intelligence or language dominance, as well as to the characteristic of the items, such as difficulty and discrimination. These traits are not directly measured but the responses given by a person to a test provide valuable information from which they can be inferred.

To estimate the perception θ_{ij} of subject i about the effectiveness of protective action j, we assume a "test" with the following three items (covariates): effectiveness to protect people (EPa_{ij}), effectiveness to protect properties (EPr_{ij}) and the usefulness for other purposes (UP_{ij}). These covariates are selected in agreement to the theory related to PADM discussed in Lindell and Perry (2012).

For the sake of simplicity, was define the three different covariates as E_{ijk}, where k stands for the type of effectiveness. Assume that E_{ijk} is equal to 0 if the response of individual i in the Likert scale is up to 3, and is 1 otherwise, for $i = 1 \ldots, n$, $j = 1, \ldots, J$ and $k = 1, 2, 3$. We assume a probit IRT model as follows:

$$\begin{aligned} E_{ijk} | p_{ijk} &\overset{ind}{\sim} Bernoulli(p_{ijk}), \\ p_{ijk} &= \Phi(a_{kj}(\theta_{ij} - b_{kj})), \end{aligned} \tag{26.3}$$

in which a_k and b_k are, respectively, the discrimination and difficulty of item k and θ_{ij} denotes the perception (ability) of subject i about the effectiveness of protective action j. The link function $\Phi(m)$ is the cumulative distribution function (cdf) of the standard normal distribution evaluated at m. Under these assumptions, the likelihood function becomes

$$f(\mathbf{E}|\boldsymbol{\theta}, \boldsymbol{a}, \mathbf{b}) = \prod_{k=1}^{3} \prod_{i=1}^{n} \prod_{j=1}^{J} \left[\Phi(a_{kj}(\theta_{ij} - b_{kj}))\right]^{E_{ijk}} \left[1 - \Phi(a_{kj}(\theta_{ij} - b_{kj}))\right]^{1-E_{ijk}}.$$

The model specification is completed assuming a priori that $\boldsymbol{\theta}$, $\mathbf{a}$, and $\mathbf{b}$ are independent with $\theta_{ij} \overset{iid}{\sim}$ Normal$(\mu_\theta, \sigma_\theta^2)$, $b_{kj} \overset{iid}{\sim}$ Normal(μ_b, σ_b^2) and $a_{kj} \overset{iid}{\sim}$ Normal(μ_a, σ_a^2).

The perception λ_{ij} of individual i about the opportunity cost of protective action j is estimated using a similar probit IRT model. Following Lindell and Perry (2012), to access λ_{ij} the "test" is composed by the items (covariates): cost to implement the action ($COST_{ij}$), time to implement the action ($TIME_{ij}$), effort needed in the action implementation ($Effort_{ij}$), knowledge and skills need in the action implementation (KS_{ij}) and the necessity of cooperation to implement the action ($CoIA_{ij}$). As for effectivness, we call all the costs above as C_{ijk}, where k represents each one of the five measures. We dichotomize the covariates C_{ijk} as previously described obtaining the following probit IRT model:

$$\begin{aligned} C_{ijk}|p^*_{ijk} &\overset{ind}{\sim} Bernoulli(p^*_{ijk}), \\ p^*_{ijk} &= \Phi(a_{kj}(\lambda_{ij} - b_{kj})), \end{aligned} \tag{26.4}$$

in which a_{kj} and b_{kj} are, respectively, the discrimination and difficulty of item k and λ_{ij} denotes the perception (ability) of subject i about the opportunity cost of protective action j.

We also assume that $\boldsymbol{\lambda}$, $\mathbf{a}$, and $\mathbf{b}$ are independent with $\lambda_{ij} \overset{iid}{\sim}$ Normal$(\mu_\lambda, \sigma_\lambda^2)$, $a_{kj} \overset{iid}{\sim}$ Normal(μ_a, σ_a^2), and $b_{kj} \overset{iid}{\sim}$ Normal(μ_b, σ_b^2).

Posterior Inference

The posterior distributions in both stages of our modeling strategy is not completely known. Given their complexity, we resort to MCMC algorithm to explore the posterior distribution. We use the JAGS (Plummer 2003) software and the *Rjags* (Plummer 2016) package of the R software (R Core Team 2015) for the computational implementation of the models.

One of the goals is to evaluate if the assumptions supporting PADM are satisfied. To test the significance of each factor, that is to test the null hypothesis that the fixed effects β are equal to zero, we consider the Bayes factor. To compute it, the following approximation for the predictive distribution is assumed:

$$\hat{p}(y) = \left[\frac{1}{m} \sum_{k=1}^{m} \frac{1}{f(\mathbf{y} \mid \boldsymbol{\beta}^{*(k)}, \boldsymbol{\gamma}^{(k)}, \boldsymbol{\lambda}^{(k)}, \boldsymbol{X}, \mathbf{W})} \right]^{-1} \tag{26.5}$$

The subscript k indicates the k-th element of the posterior distribution generated by the MCMC.

Fitting Our Proposed Model to Data on Preparedness Behavior Against Flood Hazards in Governador Valadares, Brazil

To analyze the data presented in Sect. 26, we fitted the proposed model in (26.1) estimating the latent traits related to the individual perceptions about effectiveness and opportunity costs as proposed in Sect. 26. As the latent traits θ_{ij}, λ_{ij} and the difficulty parameter β_k are in the same scale, in the first stage of our modeling we follows the IRT literature and assume that $\theta_{ij} \stackrel{D}{=} \lambda_{ij} \stackrel{D}{=} b_k \stackrel{iid}{\sim}$ Normal(0, 1). We also assume a flat prior distribution for $a_k \stackrel{iid}{\sim}$ Normal(0, 1). In the second stage, flat prior distributions are elicited for all fixed effects by letting $\beta_{kj} \stackrel{iid}{\sim}$ Normal(0, 10000). For the random effects, we consider $\mu_i \stackrel{iid}{\sim}$ Normal(0, τ^{-1}) and $\tau \sim$ Gamma(0.001, 0.001).

To estimate the latent traits, in the MCMC we run 400,000 iterations, discarded the first 10,000 iterations as the burn-in and consider a lag of 400 to avoid the correlation. This resulted in a posterior sample of size 975. To analyze the probability of adopting a protective action (the logistic model in Sect. 26) we run a chain of size 100,000 with a burn-in of 20,000 and consider the lag 100.

Inferring the Individual Perceptions About Effectiveness and Opportunity Cost

Figure 26.1 presents the box-plots of the posterior means of θ_{ij} and λ_{ij} for each category (in the original scale) of the protective action j, $j = 1 \ldots, 6$. The estimates are obtained fitting the models presented in (26.4) and (26.3) and considering the prior distributions previously mentioned.

Among those individuals with high intention of adopting the protective action, their perception of effectiveness is high. This was observed for the majority of individuals in all protective actions. The medians of θ point out that, in general, the perception of effectiveness increases as the intention of adopting the protective action increases. However, for individuals with low intention (responses 1 or 2 in the graph) of acquiring insurance (INSUR) perception about effectiveness presents high variability and, for great part of the individuals, it is as large as for those individuals that have high intension of adopting it.

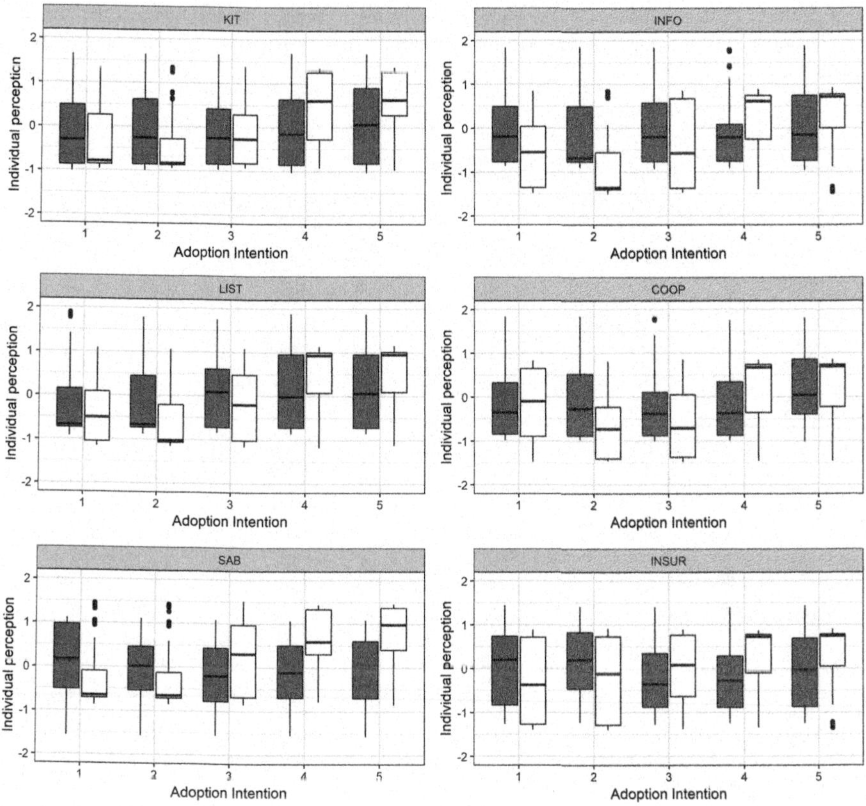

Fig. 26.1 Box-plots of the posterior means of the individual perceptions about the effectiveness θ_{ij} (white) and opportunity cost λ_{ij} (gray) within each category of protective actions KIT, INFO, LIST, COOP, SAB and INSUR, Governador Valadares Data

Inferring the Probability of Adopting the Protective Action

Table 26.1 shows the posterior mean and the highest posterior density (HPD) intervals with probability 0.95 for the effect of all covariates in the model. As expected, risk aversion and the perception of effectiveness have a positive and significant effect (See the Bayes Factor in Table 26.2) in the probability of adopting any protective action. The highest the perception of effectiveness of a protective action the highest the probability of adopting it, no matter how expensive it is (see Fig. 26.2 for the probability for an individual with particular features). The perception of opportunity cost of a protective action only affects the decision about adopting or not for flood insurance acquisition. In this case, the highest the perception of opportunity cost the smallest the probability of adopting the flood insurance. This covariate is not significant for the other protective actions. Household income positively influences the probability of adopting the flood insurance and negatively influence

Table 26.1 Posterior means and the 95% highest posterior density intervals (in brackets) for the effects of risk aversion, risk perception, gender and income, respectively. Governador Valadares data

Prot.Action	Covariate effects β_0	PEffec.	POpCost	
KIT	−3.838 (−4.839, −2.855)	1.846 (1.592, 2.114)	0.118 (−0.111, 0.342)	
LIST	−2.814 (−3.681, −1.842)	1.693 (1.461, 1.939)	0.192 (−0.038, 0.418)	
COOP	−1.857 (−2.757, −1.027)	1.399 (1.149, 1.63)	0.210 (−0.041, 0.432)	
SAB	−3.511 (−4.486, −2.397)	1.727 (1.457, 1.992)	−0.108 (−0.364, 0.135)	
INSUR	−2.594 (−3.400, −1.675)	1.122 (0.879, 1.357)	−0.463 (−0.676, −0.249)	
	Risk Aver.	Risk Percep.	Gender	log Income
KIT	0.414 (0.326, 0.511)	0.060 (0.028, 0.093)	−0.001 (−0.422, 0.381)	0.055 (−0.066, 0.180)
LIST	0.427 (0.343, 0.534)	0.033 (−0.005, 0.062)	−0.037 (−0.427, 0.375)	−0.006 (−0.120, 0.111)
COOP	0.531 (0.434, 0.624)	0.062 (0.033, 0.095)	0.135 (−0.298, 0.539)	−0.058 (−0.165, 0.056)
SAB	0.344 (0.224, 0.442)	−0.004 (−0.042, 0.031)	−0.080 (−0.500, 0.353)	−0.013 (−0.129, 0.110)
INSUR	0.307 (0.226, 0.397)	−0.032 (−0.064, −0.002)	−0.083 (−0.427, 0.292)	0.066 (−0.045, 0.18)

Table 26.2 Bayes Factor for testing the covariate effects in all protective actions

Prot.Action	PEffec.	POpCost	Risk Aver.	Risk Percep	Gender	log Income
KIT	$1.30 \cdot 10^{-40}$	$2.36 \cdot 10^{3}$	$9.96 \cdot 10^{-8}$	$2.39 \cdot 10^{2}$	$1.45 \cdot 10^{9}$	1.07
LIST	$2.58 \cdot 10^{-30}$	$3.94 \cdot 10^{5}$	$3.28 \cdot 10^{-12}$	2.60	$1.40 \cdot 10^{1}$	$6.05 \cdot 10^{5}$
COOP	$3.54 \cdot 10^{-23}$	$3.58 \cdot 10^{5}$	$8.89 \cdot 10^{-33}$	$3.82 \cdot 10^{-4}$	$1.15 \cdot 10^{4}$	$2.68 \cdot 10^{-6}$
SAB	$4.87 \cdot 10^{-20}$	$2.11 \cdot 10^{8}$	$1.50 \cdot 10^{-4}$	$4.08 \cdot 10^{2}$	$3.54 \cdot 10^{8}$	$8.33 \cdot 10^{8}$
INSUR	$6.45 \cdot 10^{-15}$	$4.10 \cdot 10^{-15}$	$1.68 \cdot 10^{-17}$	$1.98 \cdot 10^{6}$	$5.36 \cdot 10^{3}$	$2.73 \cdot 10^{-4}$

the protective action COOP but results are inconclusive about its significance to the probability of adopting the emergency kit. These results are expected once all but contractual insurance may be seen as cooperative actions where budget constraints can be relaxed by pooling external help as "income" shocks.

The effect of risk aversion (Tables 26.1 and 26.2) can also be observed in Fig. 26.2 when comparing the probability of adopting the flood insurance by levels of aversion. When perceived effectiveness is to low, the likelihood to buy insurance seems little insensitive to risk aversion. As effectiveness increases, the influence of risk aversion also increases. The influence of aversion on demand for insurance

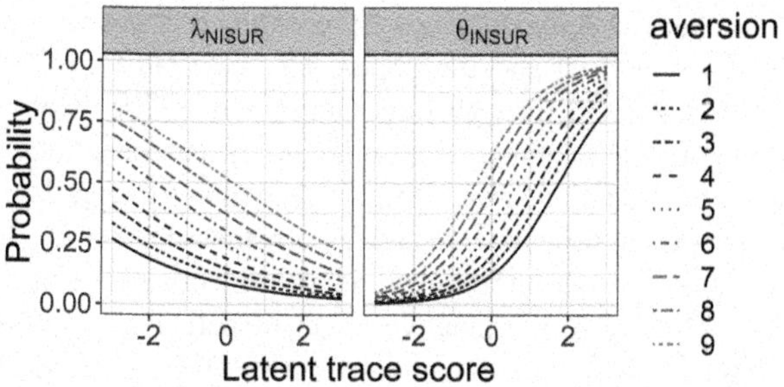

Fig. 26.2 Evolution of the probability of high intention of acquiring flood insurance for different levels of perceptions of effectiveness (right) and opportunity cost (left) for individuals with median responses for all other covariates

seems even higher for opportunity costs. Under very prohibitive costs, even highly averse individuals tend to be underinsured.

Concluding Remarks

This study investigated if the hypotheses supporting the Protective Action Decision Model (PADM) are satisfied in a study involving individuals under risk of river floods in Brazil. The PADM establishes that individuals' willingness to adopt protective measures is modified based on their perception of how effective and costly those measures are. Although PADM considers many other concurrent factors for decision making, such as sociodemographic characteristics of individuals, societal norms and values, and cues from changes in environmental conditions, it does not explicitly condition results on risk aversion. This is a key component of any decision making process in insurance economics.

Studies testing PADM main hypotheses regarding the effect of perceived effectiveness and relative prices have found mixed results, with more evidence in favor of effectiveness (Cova et al. 2009; Terpstra and Lindell 2013; Heath et al. 2018) over costs (Houts et al. 1984). We argue that these conflicting or inconclusive findings are the result of two processes: first, these studies do not take into account risk aversion explicitly (although some of them do control for individual income); second, they use very imperfect proxies for effectiveness and opportunity costs of actions instead of modeling them as latent variables that indirectly measure personal traits. Both limitations yield potential endogeneity bias in estimators, arising from omission bias and measurement errors respectively.

The statistical model we used in this paper improves previous efforts in many ways: (1) it is based on a probabilistic sample, with 1,164 individuals interviewed in a city with a large share of the population under risk of river floods; (2) it introduces a hierarchical Bayesian logistic model relating the probability of adopting protective measures against floods to covariates directly measured from individuals, as well as to latent covariates representing risk-aversion and perceptions about the effectiveness (PE) and the opportunity cost (PCO) of those measures; (3) it measures PE and PCO through Bayesian item response theory (IRT) models, appropriately quantifying the uncertainty inherent to such quantities; (4) it includes a random effect reflecting unmeasured individual features to correlate the individual responses to the different protective measures considered.

Our results show that the intention to adopt protective measures against flood hazards is more sensitive to the positive effect of perceived effectiveness than to the negative effect of the measures of relative price. This result has been found by empirical work on insurance against earthquakes and floods in Europe (Terpstra and Lindell 2013). Most strategies for self-insurance presented in this study (and similar studies elsewhere) show that the perceived effectiveness is relatively high, and the perceived opportunity costs associated are low. So why there are several still unprotected and living in areas under risk, even when income constraint, a likely cause of underprotection, should not be an issue?

The literature on preparedness behavior raises many problems that lead to a large pool of uninsured individuals. Heuristic bias is a major cause of risk misconception, as discussed by many social psychologists (Weinstein 1984) and confirmed by empirical findings (Guedes et al. 2015b). This type of bias is caused when the *interpretation* of the risk deviates from the objective risk assessment, increasing exposure to harmful situations. Kunreuther and Pauly's model of demand for insurance suggests that the overall low demand for insurance (which includes self-protection and self-insurance) may be explained by another cause: the high information costs incurred by agents (Kunreuther and Pauly 2004). Subjective errors in risk assessment is a very important area for research in social psychology, risk communication and behavioral economics, since this type of information asymmetry has important implications for individuals' wellbeing and societal welfare (Wu et al. 2015).

By modeling perceived effectiveness as a latent trait we mirror empirically the analog of subjective uncertainty in insurance transfers, as discussed in theoretical economic models of demand for insurance. Despite the positive effect of this latent trait on the probability of adoption intention, the overall low demand for self-insurance against low-probability high-loss events is probably caused by a mechanism that mimics consumers' anticipation of insurers' default in traditional contractual insurance markets. Or may be simply a result of in situ adaptation. More research is needed on this particular subject.

The perception of opportunity cost (PCO) in our study only seems to affect demand for formal, contractual insurance but not for other measures. This is expected, since many of the other measures (including seeking information about flood consequences, evacuation routes and safe/high places in the neighborhood, the creation of a list telling what to do in case of an evacuation and agreements with family/relatives, friends, and neighbors about how to help each other during an evacuation) can all be perceived as a shared cost. Again, event recurrence may explain why these protective measures are less sensitive to relative prices: the repetitive practice of group preparedness behavior works as a type of adaptation strategy against floods with declining costs.

Finally, we found a positive and significant impact of risk aversion on self-insurance against flood hazards for all types of measures used. We also found that the impact of perceived effectiveness and costs on preparedness behavior is modified depending on the level of aversion. Specially for very costly insurance contracts, the demand would be sufficiently low even among highly averse individuals.

These findings show the importance of efforts to model perceptions in self-protection and self-insurance models against natural hazards. These efforts will improve the pool of evidence that relative prices may be of second importance when it comes to private protection against hazards, with likely welfare consequences in the long run. A key research area in protective action decision-making is the optimal mix of protective strategies when individuals face multiple constraints. Survival maximization and loss minimization are simultaneous goals that may require different communication strategies depending on the characteristics of individuals under risk (Steelman and McCaffrey 2013), some level of subsidy for improving in situ adaptation in socially vulnerable communities (Bouwer and Vellinga 2005) and more research on the role of uncertainty in perception of risk, effectiveness and price of actions on preparedness behavior (Guedes et al. 2019; Raad et al. 2019).

Acknowledgments The authors like to thank CNPq, CAPES, FAPEMIG and Rede Clima/FINEP for the financial support to their research throught FAPEMIG Grant CSA-APQ-00244-12, FAPEMIG Grant CSA-PPM-00305-14, FAPEMIG Grant CSA-APQ-01553-16, CNPq Grant 4837/2012-7, CNPq Grant 472252/2014-3, CNPq Grant 431872/2016-3, CNPq Grant 314392/2018-1), and FINEP/ Rede CLIMA Grant Number 01.13.0353-00).

References

Alary, D., Gollier, C., & Treich, N. (2013). The effect of ambiguity aversion on insurance and self-protection. *The Economic Journal, 123*(573), 1188–1202.

Angus, J. E. (1994) The probability integral transform and related results. *SIAM Review, 36*(4), 652–654.

Birnbaum, A. (1968). Some latent trait models and their use in inferring an examinee's ability. In: F. M. Lord, & M. R. Novick (Eds.), *Statistical theories of mental test scores*, (pp. 379–479). Reading: Addison-Wesley.

Bouwer, L. M., & Vellinga, P. (2005). Some rationales for risk sharing and financing adaptation. *Water Science and Technology, 51*(5), 89–95.

Cova, T. J., Drews, F. A., Siebeneck, L. K., & Musters, A. (2009). Protective actions in wildfires: Evacuate or shelter-in-place? *Natural Hazards Review, 10*(4), 151–162.

Groves, R. M., et al. (2011). *Survey methodology. [S.l.]*. Wiley: Hoboken.

Guedes, G. R., Raad, R., & Raad, L. (2015a). Modeling and measuring protective action decisions under flood hazards in Brazil. In *Proceedings of the Annual Meeting of the Population Association of America*.

Guedes, G., Simao, A., Dias, C. A., & Braga, E. (2015b). Risco de adoecimento por exposição às águas do Rio Doce: um estudo sobre a percepção da população de Tumiritinga, Minas Gerais, Brasil. (in Portuguese) *Cadernos Saúde Pública, 31*(6), 1257–1268.

Guedes, G., Raad, R., & Raad, L. (2019). Welfare consequences of persistent climate prediction errors on insurance markets against natural hazards. *Estudos Econômicos, (São Paulo) 49*(2), 235–264.

Heath, R. L., Lee, J., Palenchar, M. J., & Lemon, L. L. (2018). Risk communication emergency response preparedness: Contextual assessment of the protective action decision model. *Risk Analysis, 38*(2), 333–344.

Houts, P. S., Lindell, M. K., Hu, T. W., Cleary, P. D., Tokuhata, G., & Flynn, C. B. (1984). Protective action decision model applied to evacuation during the three mile island crisis. *International Journal of Mass Emergencies and Disasters, 2*(1), 27–39.

Kellenberg, D., & Mobarak, A. M. (2011). The economics of natural disasters. *Annual Reviews, 3*, 297–312.

Kievik, M., & Gutteling, J. M. (2011). Yes, we can: Motivate Dutch citizens to engage in self-protective behavior with regard to flood risks. *Natural Hazards, 59*(3), 1475.

Kunreuther, H., & Pauly, M. (2004). Neglecting disaster: Why don't people insure against large losses? *Journal of Risk and Uncertainty, 28*(1), 5–21.

Lindell, M. K., & Perry, R. W. (1992). *Behavioral foundations of community emergency planning*. Washington, DC: Hemisphere Publishing Corporation.

Lindell, M. K., & Perry, R. W. (2012). The protective action decision model: Theoretical modifications and additional evidence. *Risk Analysis, 32*(4), 616–632.

Lord, F. M. (1980). *Applications of item response theory to practical testing problems*. London: Routledge, Taylor & Francis.

Plummer, M. (2003). JAGS: A program for analysis of Bayesian graphical models using Gibbs sampling. In K. Hornik, F. Leisch, & A. Zeileis (Eds.), *Proceedings of the 3rd International Workshop on Distributed Statistical Computing*,Vienna.

Plummer, M. (2016). *rjags: Bayesian Graphical Models using MCMC*. R package version 4-6.

Raad, R., Guedes, G., & Vaz, L. (2019). Insurance contracts under beliefs contamination. *Economics Bulletin, 39*(4), 2890–2903.

R Core Team. (2015). *R: A language and environment for statistical computing*. Vienna: R Foundation for Statistical Computing.

Rasch, G. (1960). Probabilistic models for some intelligence and attainment tests. In *Studies in mathematical psychology I* (pp. VII + 184). Denmark: Danmarks paedagogiske Institut.

Steelman, T. A., & McCaffrey, S. (2013). Best practices in risk and crisis communication: Implications for natural hazards management. *Natural Hazards, 65*(1), 683–705.

Terpstra, T., & Lindell, M. K. (2013). Citizens' perceptions of flood hazard adjustments: An application of the protective action decision model. *Environment and Behavior, 45*(8), 993–1018.

Weinstein N. D. (1984). Why it won't happen to me: Perceptions of risk factors and susceptibility. *Health Psychol, 3*, 431–457.

Wu, H. C., Lindell, M. K., & Prater, C. S. (2015). Strike probability judgments and protective action recommendations in a dynamic hurricane tracking task. *Natural Hazards, 79*(1), 355–380.

Zaalberg, R., et al. (2009). Prevention, adaptation, and threat denial: Flooding experiences in the Netherlands. *Risk Analysis: An International Journal, 29*(12), 1759–1778.

Chapter 27
Assessing Labour Market Mobility in Europe

Maria Symeonaki and Glykeria Stamatopoulou

Introduction

The theory of Markov systems is used in the literature in order to describe and model the way members of a population system move between exclusive and exhaustive categories (Vassiliou (1982), Bartholomew (1982), Bartholomew et al. (1991), Vassiliou and Symeonaki (1998, 1999), Symeonaki and Stamou (2004), Symeonaki et al. (2000, 2002), Symeonaki and Kalamatianou (2011), Symeonaki (2015), among others). Mobility within a Markov population system is commonly measured in different studies with the use of mobility indices. Well-established mobility indices are the Prais-Shorrocks mobility index (Shorrocks (1978), Prais (1955)), the Bartholomew mobility index (Bartholomew (1982)) and the Prais and Bibby mobility index (Prais (1955), Bibby (1975)). In the present paper the concept of *positive labour market mobility* is used as introduced in Symeonaki et al. (2017) and Symeonaki and Stamatopoulou (2019) and its values are estimated for European countries with the latest available raw data drawn from the European Labour Force Survey (EU-LFS) for the year 2016. In the latter studies a new labour market mobility index was introduced, where the transition probabilities included in the estimation of the new index were the transition probabilities amongst/to employment and education-training. These transitions are assumed as having equal importance. A distinction between 'good' moves and 'bad' moves in the labour market was initially suggested by Caliendo (2009). In his study movements from unemployment and inactivity to employment were distinguished as 'good', while movements from employment to unemployment and inactivity were distinguished as 'bad'. In the present study the positive labour market mobility index is estimated

M. Symeonaki (✉) · G. Stamatopoulou
Department of Social Policy, Panteion University, Athens, Greece
e-mail: msymeon@panteion.gr

C. H. Skiadas, C. Skiadas (eds.), *Demography of Population Health, Aging and Health Expenditures*, The Springer Series on Demographic Methods and Population Analysis 50, https://doi.org/10.1007/978-3-030-44695-6_27

for the year 2016 for all member states, but also for the EFTA countries Norway and Switzerland. More specifically, Sect. 27.2 provides the description of the data, the limitations of the data and the methodology that will be used. Section 27.3 presents the transition probability matrices and the positive labour market mobility index for all countries, while in Sect. 27.4 the conclusions and discussion for further research are put forward.

Data, Limitations and Methodology

Raw data drawn from the EU-LFS data will be used in order to estimate the positive labour market mobility for young individuals aged 15–29, for the year 2016. More specifically, cross-sectional data for this year will be used in order to cover the EU member states, as well as the EFTA countries Norway and Switzerland. The advantages of using the EU-LFS data lie in the facts that:

- it yields significant information on labour market participation and working conditions of Europeans,
- it allows multivariate analysis by numerous socio-demographic characteristics,
- common principles and guidelines are used to ensure cross-country comparability, and
- the sample size for each country is sufficiently large.

The variables that will be used in the present analysis are the *current labour market status at the time of the survey* (MAINSTAT) and the *labour market status one year before the survey* (WSTAT1Y). Both variables provide a self-perceived view of the individuals' labour market state at the time of the survey and one year before the survey. The main limitation regarding those variables is that MAINSTAT and WSTAT1Y are not obligatory variables, but it is up to the National Statistical Authorities of each member state to include them in their main questionnaire.

In the EU-LFS survey, for both "current labour market state" and "labour market state one year before the survey" the categories amongst which the respondent could choose are the following:

1. Carries out a job or profession, including unpaid work for a family business or holding, including an apprenticeship or paid traineeship, etc.
2. Unemployed.
3. Pupil, student, further training, unpaid work experience.
4. In retirement or early retirement or has given up business.
5. Permanently disabled.
6. In compulsory military service.
7. Fulfilling domestic tasks.
8. Other inactive person.

These categories are recoded into the following states: 1 = Employment, 2 = Education or training, 3 = Unemployment, and 4 = Inactivity and the transition probabilities are estimated, where:

$$p_{ij}(t) = prob\Big\{\text{an individual moves to state } j \text{ at time } t \mid \text{the individual was at state } i \text{ at time } t-1\Big\}. \tag{27.1}$$

Thus, the transition probability matrix that contains all the possible transitions occured inside the system is given by the following equation:

$$\mathbf{P}\,(\mathbf{t}) = \begin{pmatrix} p_{11} & p_{12} & p_{13} & p_{14} \\ p_{21} & p_{22} & p_{23} & p_{24} \\ p_{31} & p_{32} & p_{33} & p_{34} \\ p_{41} & p_{42} & p_{43} & p_{44} \end{pmatrix} \tag{27.2}$$

The transition probabilities located at the first two columns correspond to positive movements, in the sense that they show flows into employment and education or training and do not lead to exclusion from the labour market.

Accordingly, we provide the following definition of positive labour market mobility index (Symeonaki and Stamatopoulou (2019)):

Definition 1 Let a labour market system that consists of individuals stratified into the following states:

1. Employment,
2. Education or training,
3. Unemployment, and
4. Inactivity.

Then positive labour market mobility index is defined by:

$$Mob_P(t) = \frac{1}{4}\sum_{i=1}^{4}\sum_{j=1}^{2} p_{ij}(t). \tag{27.3}$$

The index takes values between 0 and 1, with 1 indicating greater positive mobility.

Results

The estimates of the positive labour market mobility index are now provided based on Definition 1, where equal weights are assumed. The respective results refer to EU member states, Norway and Switzerland for 2016 for the countries where the necessary variables (MAINSTAT and WSTAT1Y) have been included in the countries' national questionnaires.

Table 27.1 depicts the values of the positive labour market mobility index for the European countries under study, for the year 2016. It appears that the index takes lower values for all the southern European countries, with Spain and Greece scoring worse ($Mob_P(t) = 0.534$ and $Mob_P(t) = 0.540$ respectively).

Moreover, we notice that the positive mobility in Bulgaria is equal to 0.541, in Croatia 0.546, while in Romania it takes the lowest value, that of 0.525. On the other hand, Luxembourg, Sweden and Denmark are at the top when measuring positive mobility, meaning that young individuals in these countries are more likely to have

Table 27.1 Transition probability matrices and Mob_P, EU-LFS, 2016

Country	Transition Matrix	Mob_P
AT (Austria)	$\begin{pmatrix} 0.891 & 0.025 & 0.045 & 0.040 \\ 0.132 & 0.813 & 0.027 & 0.028 \\ 0.459 & 0.042 & 0.448 & 0.051 \\ 0.263 & 0.078 & 0.064 & 0.595 \end{pmatrix}$	0.675
BE (Belgium)	$\begin{pmatrix} 0.910 & 0.017 & 0.049 & 0.024 \\ 0.065 & 0.891 & 0.025 & 0.019 \\ 0.374 & 0.035 & 0.519 & 0.071 \\ 0.315 & 0.046 & 0.056 & 0.763 \end{pmatrix}$	0.618
BU (Bulgaria)	$\begin{pmatrix} 0.932 & 0.007 & 0.042 & 0.019 \\ 0.033 & 0.925 & 0.023 & 0.019 \\ 0.198 & 0.015 & 0.752 & 0.035 \\ 0.052 & 0.005 & 0.020 & 0.923 \end{pmatrix}$	0.541
CH (Switzerland)	$\begin{pmatrix} 0.838 & 0.039 & 0.083 & 0.040 \\ 0.204 & 0.732 & 0.019 & 0.045 \\ 0.479 & 0.137 & 0.297 & 0.086 \\ 0.296 & 0.190 & 0.037 & 0.504 \end{pmatrix}$	0.733
CY (Cyprus)	$\begin{pmatrix} 0.875 & 0.026 & 0.084 & 0.015 \\ 0.067 & 0.785 & 0.058 & 0.092 \\ 0.427 & 0.047 & 0.503 & 0.024 \\ 0.102 & 0.196 & 0.111 & 0.590 \end{pmatrix}$	0.630
CZ (Czech Republic)	$\begin{pmatrix} 0.939 & 0.004 & 0.054 & 0.001 \\ 0.081 & 0.895 & 0.022 & 0.001 \\ 0.653 & 0.017 & 0.331 & 0.000 \\ 0.000 & 0.000 & 0.017 & 0.983 \end{pmatrix}$	0.647
DK (Denmark)	$\begin{pmatrix} 0.529 & 0.436 & 0.021 & 0.014 \\ 0.152 & 0.776 & 0.051 & 0.020 \\ 0.405 & 0.274 & 0.271 & 0.050 \\ 0.195 & 0.216 & 0.085 & 0.503 \end{pmatrix}$	0.745
EE (Estonia)	$\begin{pmatrix} 0.846 & 0.051 & 0.037 & 0.066 \\ 0.088 & 0.843 & 0.022 & 0.048 \\ 0.462 & 0.118 & 0.278 & 0.142 \\ 0.205 & 0.058 & 0.095 & 0.641 \end{pmatrix}$	0.667

Table 27.1 (continued)

Country	Transition Matrix	*Mob_P*
ES (Spain)	$\begin{pmatrix} 0.867 & 0.017 & 0.088 & 0.028 \\ 0.086 & 0.678 & 0.109 & 0.126 \\ 0.314 & 0.028 & 0.590 & 0.068 \\ 0.071 & 0.078 & 0.117 & 0.734 \end{pmatrix}$	0.534
FI (Finland)	$\begin{pmatrix} 0.813 & 0.097 & 0.041 & 0.049 \\ 0.123 & 0.789 & 0.042 & 0.046 \\ 0.327 & 0.238 & 0.361 & 0.074 \\ 0.294 & 0.160 & 0.117 & 0.429 \end{pmatrix}$	0.710
FR (France)	$\begin{pmatrix} 0.856 & 0.028 & 0.100 & 0.015 \\ 0.094 & 0.842 & 0.048 & 0.016 \\ 0.329 & 0.044 & 0.576 & 0.051 \\ 0.131 & 0.091 & 0.189 & 0.589 \end{pmatrix}$	0.603
GR (Greece)	$\begin{pmatrix} 0.881 & 0.021 & 0.084 & 0.015 \\ 0.022 & 0.895 & 0.050 & 0.032 \\ 0.168 & 0.027 & 0.782 & 0.023 \\ 0.100 & 0.048 & 0.168 & 0.684 \end{pmatrix}$	0.540
HR (Croatia)	$\begin{pmatrix} 0.893 & 0.012 & 0.094 & 0.001 \\ 0.064 & 0.839 & 0.097 & 0.000 \\ 0.355 & 0.014 & 0.628 & 0.003 \\ 0.010 & 0.000 & 0.004 & 0.950 \end{pmatrix}$	0.546
HU (Hungary)	$\begin{pmatrix} 0.919 & 0.009 & 0.005 & 0.022 \\ 0.083 & 0.865 & 0.030 & 0.021 \\ 0.510 & 0.011 & 0.427 & 0.052 \\ 0.122 & 0.008 & 0.055 & 0.814 \end{pmatrix}$	0.631
IT (Italy)	$\begin{pmatrix} 0.888 & 0.013 & 0.088 & 0.011 \\ 0.039 & 0.893 & 0.059 & 0.008 \\ 0.242 & 0.023 & 0.713 & 0.022 \\ 0.069 & 0.043 & 0.094 & 0.794 \end{pmatrix}$	0.552
LT (Lithuania)	$\begin{pmatrix} 0.927 & 0.006 & 0.050 & 0.016 \\ 0.106 & 0.854 & 0.019 & 0.021 \\ 0.499 & 0.035 & 0.397 & 0.070 \\ 0.165 & 0.018 & 0.092 & 0.725 \end{pmatrix}$	0.652
LU (Luxembourg)	$\begin{pmatrix} 0.833 & 0.043 & 0.033 & 0.091 \\ 0.096 & 0.835 & 0.015 & 0.054 \\ 0.521 & 0.116 & 0.165 & 0.198 \\ 0.227 & 0.393 & 0.107 & 0.273 \end{pmatrix}$	0.766
LV (Latvia)	$\begin{pmatrix} 0.863 & 0.012 & 0.072 & 0.053 \\ 0.106 & 0.853 & 0.023 & 0.019 \\ 0.500 & 0.027 & 0.408 & 0.065 \\ 0.190 & 0.017 & 0.072 & 0.720 \end{pmatrix}$	0.642

Table 27.1 (continued)

MT (Malta)	$\begin{pmatrix} 0.839 & 0.119 & 0.027 & 0.014 \\ 0.187 & 0.781 & 0.025 & 0.007 \\ 0.513 & 0.032 & 0.437 & 0.019 \\ 0.182 & 0.051 & 0.020 & 0.747 \end{pmatrix}$	0.676
NL (the Netherlands)	$\begin{pmatrix} 0.817 & 0.103 & 0.054 & 0.027 \\ 0.096 & 0.859 & 0.030 & 0.015 \\ 0.235 & 0.235 & 0.471 & 0.059 \\ 0.111 & 0.067 & 0.067 & 0.756 \end{pmatrix}$	0.630
NO (Norway)	$\begin{pmatrix} 0.690 & 0.250 & 0.039 & 0.021 \\ 0.128 & 0.780 & 0.080 & 0.011 \\ 0.349 & 0.244 & 0.372 & 0.035 \\ 0.211 & 0.167 & 0.162 & 0.461 \end{pmatrix}$	0.704
PO (Poland)	$\begin{pmatrix} 0.918 & 0.008 & 0.039 & 0.035 \\ 0.067 & 0.883 & 0.033 & 0.017 \\ 0.384 & 0.014 & 0.557 & 0.045 \\ 0.113 & 0.019 & 0.037 & 0.830 \end{pmatrix}$	0.601
PT (Portugal)	$\begin{pmatrix} 0.858 & 0.029 & 0.108 & 0.005 \\ 0.065 & 0.873 & 0.056 & 0.006 \\ 0.397 & 0.040 & 0.550 & 0.012 \\ 0.055 & 0.048 & 0.065 & 0.832 \end{pmatrix}$	0.591
RO (Romania)	$\begin{pmatrix} 0.952 & 0.001 & 0.008 & 0.039 \\ 0.016 & 0.963 & 0.016 & 0.040 \\ 0.131 & 0.008 & 0.852 & 0.009 \\ 0.031 & 0.001 & 0.041 & 0.927 \end{pmatrix}$	0.525
SE (Sweden)	$\begin{pmatrix} 0.817 & 0.111 & 0.032 & 0.040 \\ 0.200 & 0.705 & 0.071 & 0.025 \\ 0.469 & 0.211 & 0.253 & 0.066 \\ 0.324 & 0.180 & 0.103 & 0.392 \end{pmatrix}$	0.753
SI (Slovenia)	$\begin{pmatrix} 0.768 & 0.176 & 0.051 & 0.005 \\ 0.069 & 0.884 & 0.043 & 0.005 \\ 0.336 & 0.021 & 0.637 & 0.006 \\ 0.064 & 0.045 & 0.089 & 0.803 \end{pmatrix}$	0.590
SK (Slovakia)	$\begin{pmatrix} 0.924 & 0.003 & 0.048 & 0.025 \\ 0.077 & 0.871 & 0.045 & 0.006 \\ 0.404 & 0.001 & 0.561 & 0.025 \\ 0.085 & 0.005 & 0.069 & 0.841 \end{pmatrix}$	0.592

Source: EU-LFS, 2016; own calculations

positive movements and not to be excluded from the labour market. Figure 27.1 gives us a better representation of the results. For the estimates of the index for the years 2008–2015 see Symeonaki and Stamatopoulou (2019).

Figure 27.2 provides the results from Symeonaki and Stamatopoulou (2019), p. 6, and show the evolution of the proposed index over the years of the economic

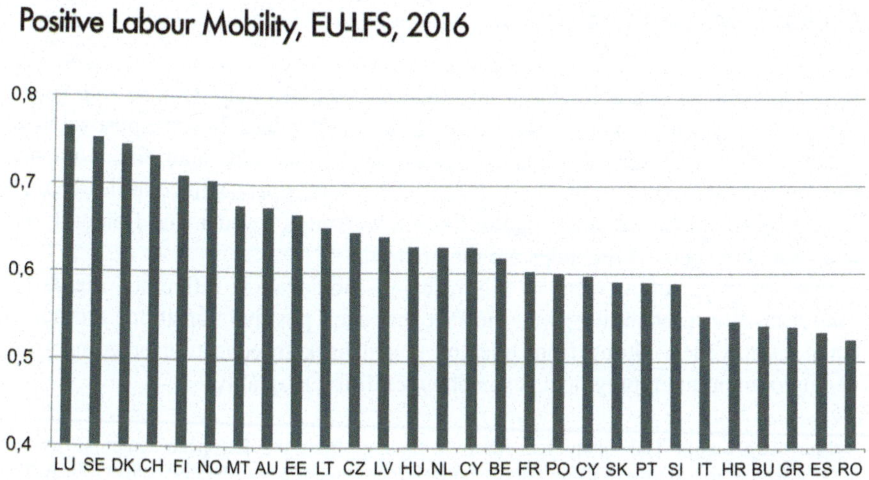

Fig. 27.1 Positive labour market mobility of individuals aged 15–29 in Europe, EU-LFS, 2016

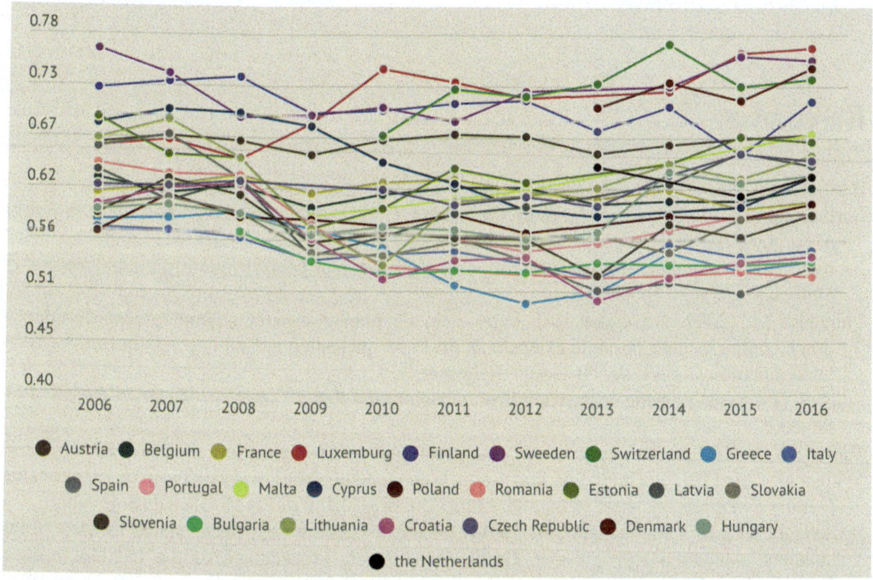

Fig. 27.2 The evolution of positive labour market mobility of individuals aged 15–29 in Europe, EU-LFS, 2006–2016

crisis. Each line corresponds to a different country and a variation amongst them is obvious. If we now add the positive mobility rates for the year 2016, we had a more complete picture of the influence that the crisis had to the positive movements of young individuals in Europe, as the index seems to decrease in the years of the crisis and gradually increased through the years in the majority of the countries.

Conclusions and Discussion

It is clear that in southern European countries (namely Italy, Greece and Spain) the positive labour market mobility index takes small values. In some post-socialist countries considered as 'developing' regarding their welfare state, like Romania, Croatia and Bulgaria, the proposed index is also low, whereas in countries characterised by advanced welfare regimes, like Luxembourg, Sweden and Denmark, for example, the values of the index are the highest.

An aspect of future work of this research should allow different weights for each transition probability given the fact that even positive transitions are not of equal significance. Going from unemployment back to education or going from unemployment to employment is somehow of different quality.

Acknowledgements This study has received funding from the European Union's Horizon 2020 research and innovation programme under grant agreement No 649395 (NE-GOTIATE – Negotiating early job-insecurity and labour market exclusion in Europe, Horizon 2020, Societal Challenge 6, H2020-YOUNG-SOCIETY-2014, Research and Innovation Action (RIA), Duration: 01 March 2015–28 February 2018).

References

Bartholomew, D. J. (1982). *Stochastic models for social processes*. London: Wiley.

Bartholomew, D. J., Forbes, A. F., & McClean, S. I. (1991). *Statistical techniques for man-power planning*. New York: Wiley.

Bibby, J. (1975). Methods of measuring mobility. *Quality and Quantity: International Journal of Methodology, 9*(2), 107–136.

Caliendo, M. (2009, November 26). Labour flows, transitions and unemployment duration – Some comments and possible extensions. In: Paper presented at Employment in Europe 2009 Dissemination Conference, Brussels, Belgium.

Prais, S. (1955). Measuring social mobility. *Journal of the Royal Statistical Society, Series A, 118*, 56–66.

Shorrocks, A. (1978). The measurement of social mobility. *Econometrica, 46*, 1013–1024.

Symeonaki, M. (2015). Theory of fuzzy non homogeneous Markov systems with fuzzy states. *Quality and Quantity, 49*, 2369–2385. https://doi.org/10.1007/s11135-014-0118-4.

Symeonaki, M., & Kalamatianou, A. (2011). Markov systems with fuzzy states for describing students' educational progress in Greek Universities, Proceedings of the ISI Conference, Dublin, Ireland, Vol. 5956–5961.

Symeonaki, M., & Stamatopoulou, G. (2019). On the measurement of positive labour market mobility, paper submitted to *Open Sage*, accepted.

Symeonaki, M., & Stamou, G. (2004). Theory of Markov systems with fuzzy states. *Fuzzy Sets and Systems, 143*, 427–445.

Symeonaki, M., Stamou, G., & Tzafestas, S. (2000). Fuzzy Markov systems for the description and control of population dynamics. In S. Tzafestas (Ed.), *Computational intelligence in systems and control design and application* (pp. 301–310). Dordrecht: Kluwer Academic Publisher.

Symeonaki, M., Stamou, G., & Tzafestas, S. (2002). Fuzzy non-homogeneous Markov systems. *Applied Intelligence, 17*(2), 203–214.

Symeonaki, M., Stamatopoulou, G., & Karamessini, M. (2017). Introducing an index of labour mobility for youth. In: Paper presented at 13th Conference of the European Sociological Association, (Un)Making Europe: Capitalism, Solidarities, Subjectivities, Athens, Greece, 29th August – 1st September, 2017.

Vassiliou, P. C. (1982). Asymptotic behaviour of Markov systems. *Journal of Applied Probability, 19*, 851–857.

Vassiliou, P. C., & Symeonaki, M. (1998). The perturbed non-homogeneous Markov system in continuous time. *Applied Stochastic Models and Data Analysis, 13*(3–4), 207–216.

Vassiliou, P. C., & Symeonaki, M. (1999). The perturbed non-homogeneous Markov system. *Linear Algebra and Applications, 289*(1–3), 319–332.

Chapter 28
The Implications of Applying Alternative-Supplementary Measures of the Unemployment Rate to Subpopulations and Regions: Evidence from the European Union Labour Force Survey for Southern Europe, 2008–2015

Aggeliki Yfanti, Catherine Michalopoulou, and Stelios Zachariou

Introduction

The unemployment rate is an important indicator with both social and economic dimensions, considered to signify a country's social and economic well-being. For its measurement well-defined concepts of the labour force, i.e. the employed and unemployed, are required. In this respect, the European Union Labour Force Survey (EU-LFS) uses a synthesized economic construct computed according to the International Labour Organization (ILO) conventional definitions of the employed, unemployed and inactive as agreed at the 13th and 14th International Conference of Labour Statisticians (ICLS) held in 1982 and 1987, respectively Hussmanns et al. (1990). The EU-LFS, by adopting these conventional definitions, satisfies one of the essential requirements permitting the cross-national and over time comparability of the data Kish (1994).

Although this definition is internationally used, considered by many to be the most objective measure, being by definition a single measure for an otherwise complex social phenomenon, it has been sharply criticized mainly in that it under estimates unemployment, raising public concern especially in recessionary times (see e.g., "Assessing labour market slack" (2017), Bartholomew et al. (1995),

A. Yfanti (✉) · C. Michalopoulou
Panteion University of Social and Political Sciences, Athens, Greece
e-mail: aggelikiyfanti@panteion.gr; kmichal@panteion.gr

S. Zachariou
European Commission, Brussels, Belgium
e-mail: Stylianos.ZACHARIOU@ec.europa.eu

C. H. Skiadas, C. Skiadas (eds.), *Demography of Population Health, Aging and Health Expenditures*, The Springer Series on Demographic Methods and Population Analysis 50, https://doi.org/10.1007/978-3-030-44695-6_28

Sorrentino 2000, 2002). In this paper, the impact of applying two broader alternative measures of unemployment to the ILO conventional definitions for the data sets of Greece, Italy, Portugal and Spain was first assessed and the social and demographic "profile" of the unemployment rate based on gender, age, marital status and level of educational attainment was inspected. Then we assessed further the impact of the application of these two alternative measures to the regions of each country as defined by Eurostat (Nomenclature des Unités Territoriales Statistiques, NUTS II).

The ILO Conventional Definitions of the Employment Status

The EU–LFS classifies the population of working age (15+ years) into three mutually exclusive and exhaustive categories: employed, unemployed and economically inactive Eurostat (2003, 2008, 2009, 2013). Figure 28.1 presents the detailed EU-LFS measurement of the employment status based on a number of variables according to the ILO conventional definitions. Note that, the definition of the unemployed applies only to respondents aged 15–74 years. As shown, a number of variables is used that define the labour status (WSTATOR), whether respondents were seeking employment (SEEKWORK), the methods for doing so (METHOD) and their availability to start work immediately within 2 weeks (AVAILABLE).

The variablle WSTATOR measures the labour status during the reference week for all respondents aged 15 years or more according to the ILO conventional definitions. This variable takes the value one (1) when respondents did any work for pay or profit for 1 h or more, including family work during the reference week. The second value (2) refers to respondents who despite of having a job or business did not work during the reference week because they were temporarily absent. The third value (3) is assigned to respondents who were not working because of lay-off. The fourth value (4) indicates the respondent who was a conscript on compulsory military service or community service. The fifth value (5) designates respondents who did not work nor had a job or business during the reference week Eurostat (2016).

The variable SEEKWORK is defined as seeking employment during the previous 4 weeks. This variable takes the value one (1) when respondents had found a job that would begin within a period of at most 3 months. The second value (2) refers to respondents who had already found a job that would begin in more than 3 months. The third value (3) is assigned to respondents who were not seeking employment and had not found any job beginning later. The fourth value (4) indicates respondents who were seeking employment. The variables METHODA to METHODM, which are briefly presented in Fig. 28.1, record the specific steps (methods) taken by respondents to find work Eurostat (2016).

The variable AVAILABLE is defined as the availability to begin working within 2 weeks, if work can be found. This variable takes the value one (1) when

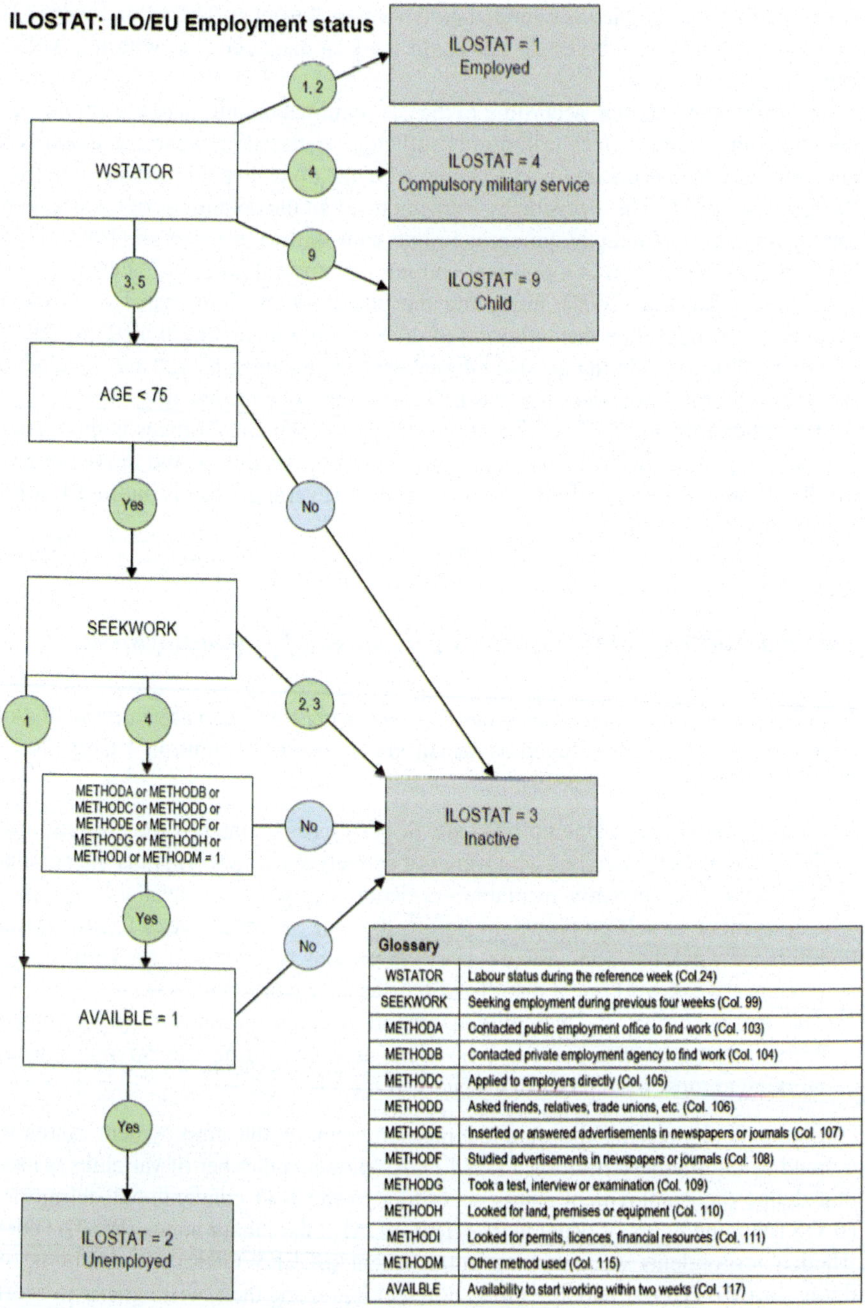

Glossary	
WSTATOR	Labour status during the reference week (Col.24)
SEEKWORK	Seeking employment during previous four weeks (Col. 99)
METHODA	Contacted public employment office to find work (Col. 103)
METHODB	Contacted private employment agency to find work (Col. 104)
METHODC	Applied to employers directly (Col. 105)
METHODD	Asked friends, relatives, trade unions, etc. (Col. 106)
METHODE	Inserted or answered advertisements in newspapers or journals (Col. 107)
METHODF	Studied advertisements in newspapers or journals (Col. 108)
METHODG	Took a test, interview or examination (Col. 109)
METHODH	Looked for land, premises or equipment (Col. 110)
METHODI	Looked for permits, licences, financial resources (Col. 111)
METHODM	Other method used (Col. 115)
AVAILBLE	Availability to start working within two weeks (Col. 117)

Fig. 28.1 The ILO conventional definitions of the employment status used in the EU-LFS. (Reproduced from "EU Labour Force Survey database user guide," by Eurostat (2016: 55))

respondents could begin work immediately (within 2 weeks). The second value (2) indicates respondents who could not begin work immediately, i.e. within 2 weeks Eurostat (2016).

It should be noted that, according to the conventional definition of the employed, the one hour's work a week criterion is sufficient to classify a person as employed and consequently, as unemployed a person who had done no paid work at all during the reference week. This decision by Eurostat to adopt the definitions of the ICLS on the measurement of unemployment is in line with Bartholomew et al. (1995: 377), who pointed out that "it is a political and not a statistical question." Furthermore, Garrido and Toharia (2004), analyzing the effects of the European Commission regulation 1897/2000 on the definition of the unemployment, questioned the nature of the regulation in deleting passive job seekers from the unemployed and concluded that the adoption of this decision was political rather than technical.

The official unemployment rate published by the statistical offices is the unemployed as a percent of the labour force (the sum of the employed and unemployed). For its computation, a combination of answers to several questions on the EU-LFS questionnaire is required.

Two Alternative-Supplementary Measures of Unemployment

In this paper, as in our previous work, the following two alternative definitions of unemployment were formulated as variations of the ILO definition Yfanti (2010, 2019), Yfanti et al. (2017, 2019):

Alternative definition 1: Unemployed are all persons as defined by the conventional definition, plus the persons who were not actively seeking work, plus the persons who were not currently available for work, i.e. were not available for paid employment or self-employment before the end of the 2 weeks following the reference week but wanted work, plus the persons who were at work during the reference week working at most for 4 h or usually worked for 4 h.

Alternative definition 2: Unemployed are all persons as defined by alternative definition 1, plus the persons who were at work during the reference week working at most for 8 h or usually worked for 8 h.

Figure 28.2 presents the detailed measurement of the employment status as defined by alternative definition 1 (ALT1). As shown, a number of variables is used that define the employment status according to the ILO conventional definitions (ILOSTAT), the hours of usual work (HWUSUAL), the labour status (WSTATOR), whether respondents were seeking employment (SEEKWORK), the methods for doing so (METHOD) and for those not seeking work their willingness to work (WANTWORK). The values of this latter variable are defined as the individuals not seeking employment but would nevertheless like to have work (value 1) or do not want to have work (value 2).

Note that, for both alternative definitions, the usual hours of work a week were used instead of the conventional 1 h of work during the reference week. Otherwise

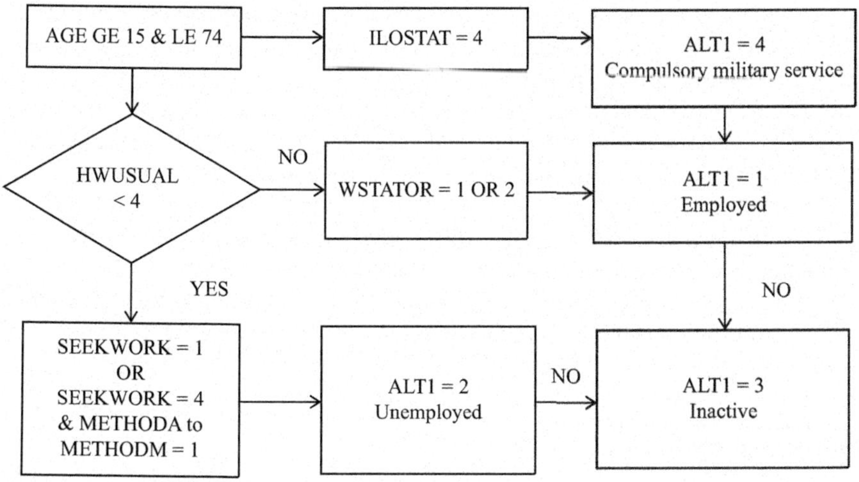

Fig. 28.2 The measurement of the employment status according to the alternative definition 1 (ALT1). *ILOSTAT* the ILO conventional definitions, *HWUSUAL* the hours of usual work, *WSTATOR* the labour status, *SEEKWORK* seeking work, *METHOD* methods of seeking work

those on leave, sickness or other reason for absence from work would have been included in the definitions.

The one hour's work a week criterion was addressed at the 14th ICLS (1987). But, while the Conference agreed to retain the criterion on the basis of its international comparability, it was also emphasized that the employment data should be classified by hours of work, thus making it possible to apply different criteria. However, the one hour's work a week criterion is considered by many as clear and unambiguous in an international context. Hussmanns (Hussmanns 2007) and the Department of Economic and Social Affairs-International Labour Office (Department of Economic and Social Affairs-International Labour Office 2009) presented a detailed account for the reasons to use the one hour's work a week criterion in the international definition. However, as Bartholomew et al. (1995: 368) pointed out, although this definition is relatively easily applied on a cross-national basis, "it does not correspond closely to what the typical person would see as the dividing line between employment and unemployment."

Method

Prerequisites for Comparability

The EU-LFS is a set of independent national multipurpose large-scale sample surveys conducted by the respective statistical offices of the member countries providing quarterly and annual results on labour participation and those outside the labour force. The survey population is defined centrally as all persons aged 15 years or more living in private households, excluding persons in compulsory military or community service and those residing in collective dwellings. Therefore, the survey population overtime within-nations and cross-national comparability is ensured Kish (1994), see also Gagliardi et al. (2009), Karanikoli (2009), Michalopoulou (2004, 2008).

Our analysis was based on the 2008–2015 annual datasets for the following Southern European countries: Greece, Italy, Portugal and Spain. Furthermore, we decided to report the results for the age group 15–74 to allow for comparability with the ILO conventional definition of unemployment (de la Fuente 2011).

Statistical Analyses

In our analysis, three releases of the EU-LFS annual datasets by Eurostat were used: the data release of 2013 for 2008–2012; the data release of 2016 for 2013–2014; and the data release of 2017 for 2015. All the results presented are based on the weighted datasets. Statistical analyses were performed using IBM SPSS Statistics Version 20.

First, the unemployment rate was computed based on the ILO conventional measurement of the employment status and the two alternative measurements to the ILO definitions for each country. These three frequency distributions were compared to assess the impact of applying different definitions to the unemployment rate. Subsequently, the demographic and social "profile" of the unemployment rate defined by the three measurements was obtained for each country based on gender (male and female), age (15–24, 25–34, 35–44, 45–54, 55–64 and 65–74), marital status (single, married and other), and level of educational attainment (primary, secondary and tertiary).

In the next step of the analysis, the regional distribution of the unemployment rate defined by the three measurements was computed for each country. According to the NUTS II classification used by Eurostat, 13, 21, 7 and 19 regions are defined for Greece, Italy, Portugal and Spain, respectively. It should be noted, that these regions are used in all the datasets without changes thus allowing for within-country over time comparability.

Specifically, the 13 regions of Greece are: Eastern Macedonia, Central Macedonia, Western Macedonia, Thessaly, Epirus, Ionian Islands, Western Greece, Central Greece, Peloponnese, Attica, North Aegean, South Aegean and Crete. The 21 regions of Italy are: Piemonte, Valle d' Aosta/Vallée d' Aoste, Liguria, Lombardia, Provincia Autonoma di Bolzano/Bozen, Provincia Autonoma di Trento, Veneto, Friuli Venezia Giulia, Emilia-Romagna, Toscana, Umbria, Marche, Lazio, Abruzzo, Molise, Campania, Puglia, Basilicata, Calabria, Sicilia, Sardegna. The 7 regions of Portugal are: Norte, Algarve, Centro (PT), Area Metropolitana de Lisboa, Alentejo, Região Autónoma dos Açores, Região Autónoma da Madeira. The 19 regions of Spain are: Galicia, Principado de Asturias, Cantabria, País Vasco, Comunidad Foral de Navarra, La Rioja, Aragón, Comunidad de Madrid, Castilla y León, Castilla-La Mancha, Extremadura, Cataluña, Comunidad Valenciana, Illes Balears, Andalucía, Región de Murcia, Ciudad Autónoma de Ceuta, Ciudad Autónoma de Melilla and Canarias.

Results

The Unemployment Rate

In Fig. 28.3, the unemployment rate according to the ILO conventional definitions and the two alternative measures to the ILO definitions is presented for Greece, Italy, Portugal and Spain.

As shown, the unemployment rate of Greece measured according to the ILO conventional definition, the first and the second alternative measurements ranged from 7.7–27.5%, 9.8–30.3% to 10.0–30.7%, respectively. The increase in the unemployment rate as measured by the ILO conventional definition when applying the first and the second alternative measures ranged from 2.1–2.9% to 2.3–3.2%, respectively.

The unemployment rate of Italy measured according to the ILO conventional definition, the first and the second alternative measurements ranged from 6.4–12.7%, 20.7–26.6% to 22.6–28.9%, respectively. The increase in the unemployment rate as measured by the ILO conventional definition when applying the first and the second alternative measures ranged from 12.9–14.0% to 14.4–16.2%, respectively.

The unemployment rate of Portugal measured according to the ILO conventional definition, the first and the second alternative measurements ranged from 7.7–16.4%, 10.5–25.7% to 12.9–28.2%, respectively. The increase in the unemployment rate as measured by the ILO conventional definition when applying the first and the second alternative measures ranged from 2.7–9.3% to 4.8–11.8%, respectively.

The unemployment rate of Spain measured according to the ILO conventional definition, the first and the second alternative measurements ranged from 11.3–26.1%, 22.9–38.3% to 22.9–38.87%, respectively. The increase in the unemployment rate as measured by the ILO conventional definition when applying the first and the second alternative measures ranged from 8.9–10.0% to 9.9–12.2%, respectively.

The most remarked percentage increase in the 2008–2015 unemployment rate measured according to the ILO conventional definitions and the two alternative measures to the ILO definitions was detected in the case of Italy.

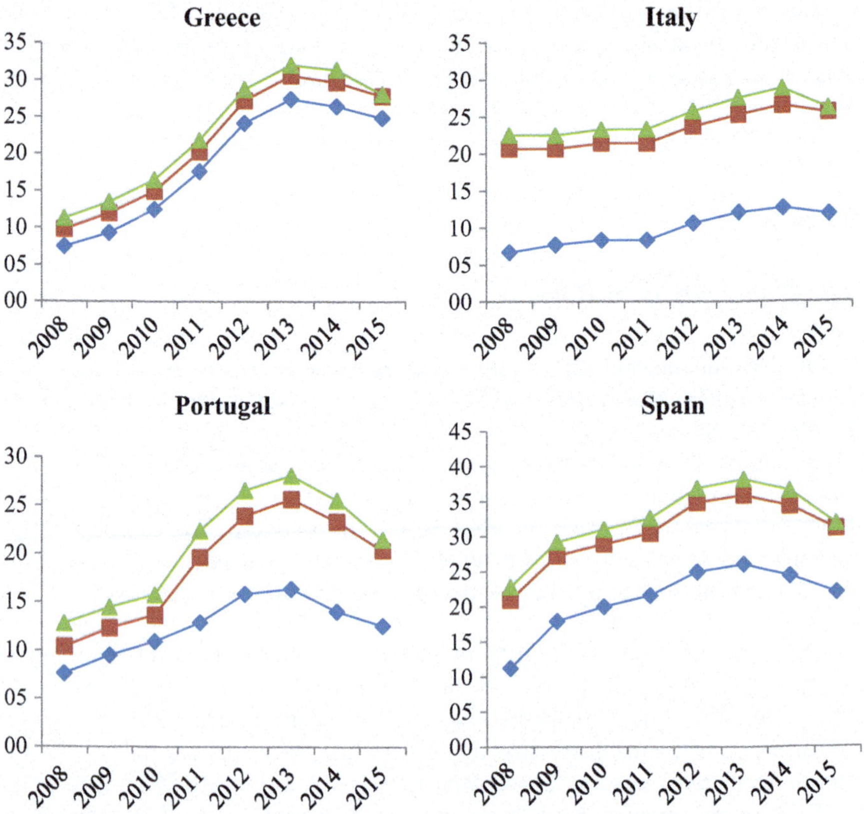

Fig. 28.3 The unemployment rate (%) measured according to the three different measurements: European Union Labour Force Survey, 2008–2015. The ILO conventional definitions and first and second alternative measurements to the ILO definitions are represented by the lines with rhombus, squares and triangles, respectively

The Demographic and Social "Profile" of the Unemployment Rate

The investigation of the demographic and social "profile" of the unemployment rate measured according to the ILO conventional definitions and the two alternative to the ILO definitions measurements showed significant results ($p < .001$) for all countries and datasets (2008–2015).

More specifically, in the case of Greece, the unemployment rate as defined by the ILO conventional definition showed that those more at risk were mainly women (11.4–31.4%), aged 15–24 years old (22.2–58.4%), single (12.7–39.6%) with secondary education (8.9–31.4%). The unemployment rate as defined by the first alternative measure to the ILO conventional definition resulted in mainly women (20.8–39.8%), 15–24 years old (29.2–63.5%), single (17.9–43.9%) with secondary education (14.0–36.9%). The unemployment rate as defined by the second alternative measure to the ILO conventional definition resulted in mainly women (15.2–35.9%) and 15–24 years old (27.0–62.6%), single (14.9–42.2%) with secondary education (11.3–34.7%).

In the case of Italy, the unemployment rate as defined by the ILO conventional definition showed that those more at risk were mainly women (8.5–13.8%), aged 15–24 years old (21.3–42.7%), single (11.2–20.7%) with primary education (9.0–19.2%). The unemployment rate as defined by the first alternative measure to the ILO conventional definition resulted in mainly women (27.8–32.4%), 15–24 years old (52.8–65.1%), single (29.0–36.7%) with primary education (31.7–43.0%). The unemployment rate as defined by the second alternative measure to the ILO conventional definition resulted in mainly women (31.0–36.2%) and 15–24 years old (54.5–67.0%), single (30.5–38.7%) with primary education (34.5–46.4%).

In the case of Portugal, the unemployment rate as defined by the ILO conventional definition showed that those more at risk were mainly women (9.0–16.6%), aged 15–24 years old (16.4–38.0%), single (13.2–24.6%) with secondary education (8.8–18.0%). The unemployment rate as defined by the first alternative measure to the ILO conventional definition resulted in mainly women (12.9–27.0%), 15–24 years old (20.7–56.9%), single (16.4–35.7%) with primary education (13.4–28.9%) from 2009 to 2015 and secondary education (11.2%) in 2008. The unemployment rate as defined by the second alternative measure to the ILO conventional definition resulted in mainly women (16.4–29.9%) and 15–24 years old (22.8–59.2%), single (18.2–37.6%) with primary education (14.2–33.4%).

In the case of Spain, the unemployment rate as defined by the ILO conventional definition showed that those more at risk were mainly women (13.0–26.7%), aged 15–24 years old (24.6–55.5%), single (15.4–33.0%) with primary education (17.7–40.8%). The unemployment rate as defined by the first alternative measure to the ILO conventional definition resulted in mainly women (25.5–38.8%), 15–24 years old (39.0–68.5%), single (25.1–42.5%) with primary education (32.1–54.8%). The unemployment rate as defined by the second alternative measure to the ILO conventional definition resulted in mainly women (28.9–42.4%) and 15–24 years old (42.2–71.7%), single (26.7–45.0%) with primary education (34.4–57.9%).

In Table 28.1, the percentage range of the increase in the 2008–2015 demographic and social "profile" of the unemployment rate measured according to the ILO conventional definitions and the two alternative measurements to the ILO definitions is presented for Greece, Italy, Portugal and Spain.

As shown, the most remarked percentage increase in the 2008–2015 demographic and social "profile" of the unemployment rate measured according to the ILO conventional definitions and the two alternative measures to the ILO definitions was detected in the case of Italy.

Table 28.1 The percentage range of the increase in the demographic and social "profile" of the unemployment rate Based on the ILO conventional definitions and the two alternative measures to the ILO definitions (A1 and A2): European Union Labour Force Survey, 2008–2015

	Percentage range of the increase in the unemployment rate							
	Women		The young (15–24)		Single		Education	
Country	A1	A2	A1	A2	A1	A2	A1	A2
Greece	4.2–10.1	3.7–4.6	5.1–7.0	4.2–6.8	2.8–5.4	2.2–3.1	3.3–5.7**	2.4–2.8**
Italy	18.1–20.3	19.3–23.6	22.4–31.5	23.8–33.2	15.3–17.8	16.7–19.3	21.8–23.8*	24.4–27.2*
Portugal	3.6–10.5	6.6–13.3	4.1–19.9	6.4–22.3	3.0–11.7	4.7–13.3	3.3–13.1*	6.9–17.5*
Spain	10.1–12.4	11.3–15.9	11.4–14.4	13.7–17.6	8.6–10.0	10.8–12.3	11.9–14.4*	12.7–17.2*

* Primary education; ** Secondary education

Note that in the case of Portugal, the results for education refer to 2009–2015. In 2008, the increase of the unemployment rate for those with secondary education as measured by the ILO conventional definition when the first and the second alternative measurement were applied ranged from 2.1–8.8% to 6.9–17.5%, respectively

The Regional Distribution of the Unemployment Rate

In the case of Greece, the regions with the highest unemployment rate based on the ILO conventional definition were Western Macedonia (12.4–31.9%) for 2008–2013, Central Macedonia and Western Greece (28.7%) for 2014 and Western Greece (30.8%) for 2015. The regional distribution of the unemployment rate changed when the first alternative measure to the ILO conventional definition was applied as the regions with the highest unemployment rate were Western Macedonia (15.2–36.1%) for 2008–2014 and Western Greece (35.2%) for 2015. A similar regional distribution was detected when the second alternative measure to the ILO conventional definition was applied. The increase in the regional unemployment rate as measured by the ILO conventional definition when the first and the second alternative measurements were applied ranged from 0.7–5.9%. The highest increase was detected in the case of Western Macedonia (5.9%) for 2014.

In the case of Italy, the regions with the highest unemployment rate based on the ILO conventional definition were Sicilia (13.8%) for 2008, Umbria (13.9–14.7%) for 2009–2010, Veneto (15.5%) for 2011, Toscana and Veneto (19.3%) for 2012 and Toscana (19.3–23.5%) for 2013–2015. The regional distribution of the unemployment rate changed when the first alternative measure to the ILO conventional definition was applied as the regions with the highest unemployment rate were Calabria (38.7%) for 2008, Veneto (39.5% and 42.3%) for 2009–2012, Toscana and Umbria (43.5–45.5%) for 2013 and 2014 and Toscana (46.2%) for 2015. A quite similar regional distribution was detected when the second alternative measure to the ILO conventional definition was applied as the regions with the highest unemployment rate were Campania (39.7%) for 2008, Veneto (39.9–42.7%) for 2009–2012, Umbria (43.9–46.0%) for 2013–2014 and Toscana (46.5%) for 2015. The increase in the regional unemployment rate of Italy as measured by the ILO conventional definition when the first and the second alternative measurements were applied ranged from 3.8–27.2% and 4.2–28.4, respectively. The highest increase was detected in the case of Toscana (27.2–28.4%) for 2010.

In the case of Portugal, the regions with the highest unemployment rate based on the ILO conventional definition were Alentejo (9.1%) for 2008, Norte (11.2%) for 2009, Algarve (13.5–18.3%) for 2010–2012, Area Metropolitana de Lisboa (18.6%) for 2013, Região Autónoma dos Açores (16.5%) for 2014 and Região Autónoma da Madeira (14.6%) for 2015. The regional distribution of the unemployment rate changed when the first alternative measure to the ILO conventional definition was applied as the regions with the highest unemployment rate were Alentejo (11.1%) for 2008, Norte (13.1%) for 2009, Algarve (15.4%) for 2010 and Região Autónoma da Madeira (21.7–30.6%) for 2011–2015. A quite similar regional distribution was detected when the second alternative measure to the ILO conventional definition was applied as the regions with the highest unemployment rate were Norte and Alentejo (11.4%) for 2008, Norte (13.7%) for 2009, Algarve (15.8%) for 2010 and Região Autónoma da Madeira (23.2–32.4%) for 2011–2015. The increase in the regional unemployment rate of Portugal as measured by the ILO conventional definition when the first and the second alternative measurements were applied ranged from 1.7–13.5% to 1.9–15.6%, respectively. The highest increase was detected in the case of Região Autónoma da Madeira (13.5–15.6%) for 2014.

In the case of Spain, the regions with the highest unemployment rate based on the ILO conventional definition were Ciudad Autónoma de Ceuta (19.4–42.4%) for 2008, 2011, 2012, 2014 and 2015, Canarias (26.2%) for 2009 and Ciudad Autónoma de Melilla (29.0–38.9%) for 2010 and 2013. The regional distribution of the unemployment rate changed when the first alternative measure to the ILO conventional definition was applied as the regions with the highest unemployment rate were Ciudad Autónoma de Ceuta (37.8–51.4%) for 2008, 2011 and 2012, Ciudad Autónoma de Melilla (40.6–41.7%) for 2009–2010, Extremadura (42.6–47.2%) for 2013 and 2015, and Andalucia (44.7%) for 2014. A similar regional distribution was detected when the second alternative measure to the ILO conventional definition was applied as the regions with the highest unemployment rate were Ciudad Autónoma de Ceuta (37.8–51.4%) for 2008, 2011 and 2012, Ciudad Autónoma de Melilla (40.6–44.4%) for 2009–2010, Extremadura (43.1–47.9%) for 2013 and 2015, and Andalucia (45.5%) for 2014. The increase in the regional unemployment rate of Spain as measured by the ILO conventional definition when the first and the second alternative measurements were applied ranged from 2.6–18.5% and 3.2–18.5%%, respectively. The highest increase was detected in the case of Ciudad Autónoma de Ceuta (18.25%) for 2008.

In Table 28.2, the regions with the highest unemployment rate as measured by the ILO conventional definition and the first and second alternative measurements are presented.

As shown, the regional distribution of the unemployment rate as measured by the ILO conventional definition alters when the two alternative measurements were applied.

Conclusions

In this paper, we investigated the impact of applying two broader alternative measures to the official unemployment rate to subpopulations and regions of interest in social policy purposes. Addressing the use of using a single all-purpose measurement of the unemployment rate based on the ILO conventional definitions for an otherwise complex social phenomenon, two broader alternative measures of the unemployment rate were formulated in the context of the EU-LFS as variations of the ILO conventional definitions. The analysis was based on the 2008–2015 annual datasets of the EU-LFS for Greece, Italy Portugal and Spain. The results are reported for the age group 15–74 to allow for comparability with the ILO conventional definition of the unemployed.

Applying these two alternative-supplementary measures to the EU-LFS data, the overall national results showed, as expected by definition, an increase of the official unemployment rate in all countries. The increase in the unemployment rate as

Table 28.2 The regions with the highest unemployment rate as measured by the ILO conventional definition and the first and second alternative measures (A1 and A2): European Union Labour Force Survey, 2008–2015

Country	2008	2009	2010	2011	2012	2013	2014	2015
Greece								
ILO	Western Macedonia						C. Macedonia, W. Greece	W. Greece
A1/A2	Western Macedonia							W. Greece
Italy								
ILO	Sicilia	Umbria		Veneto	Toscana, Veneto	Toscana		
A1	Calambria	Veneto				Toscana, Umbria		Toscana
A2	Campania	Veneto				Umbria		Toscana
Portugal								
ILO	Alentejo	Norte	Algarve			Area Metropolitana de Lisboa	Região Autónoma dos Açores	Região Autónoma da Madeira
A1	Alentejo	Norte	Algarve	Região Autónoma da Madeira				
A2	Alentejo, Norte	Norte	Algarve	Região Autónoma da Madeira				
Spain								
ILO	Ciudad Autónoma de Ceuta	Canarias	Ciudad Autónoma de Melilla	Ciudad Autónoma de Ceuta		Ciudad Autónoma de Melilla	Ciudad Autónoma de Ceuta	
A1/A2	Ciudad Autónoma de Ceuta	Ciudad Autónoma de Melilla		Ciudad Autónoma de Ceuta		Extremadura	Andalucia	Extremadura

measured by the ILO conventional definition when the first alternative measurement was applied to the datasets of Greece, Italy, Portugal and Spain ranged from 2.1–2.9%, 12.9–14.0%, 2.7–9.3% to 8.9–10.0%, respectively. The respective increase in the unemployment rate when the second alternative measurement was applied to the datasets of Greece, Italy, Portugal and Spain ranged from 2.3–3.2%,14.4–16.2%, 4.8–11.8% to 9.9–12.2%, respectively.

Investigation of the demographic and social "profile" of the unemployment rate measured according to the ILO conventional definitions and the two alternative measures to the ILO definitions revealed threatening unemployment rates for women, the young (15–24), the single and those with primary (Italy, Portugal and Spain) and secondary (Greece) education.

In all countries, the regional unemployment rate measured according to the ILO conventional definitions was increased when the two alternative measures were applied. Also, the regional distribution of the unemployment rate measured according to the ILO conventional definitions was altered when the two alternative measurements to the ILO conventional definition were applied. These findings and the resulting regional disparities indicate different priorities for social policy purposes.

Certainly, the changes in the definitions presented do not exhaust all possibilities. However, the findings indicate the need, especially in recessionary times, for implementing alternative measures of the unemployment rate to the EU-LFS in the tradition of the Current Population Survey.

References

"Assessing Labour Market Slack." (2017). *ECB economic bulletin, Issue 3, Box 3*, 31–35.

Bartholomew, D. J., Moore, P., Smith, F., & Allin, P. (1995). The measurement of unemployment in UK. *Journal of the Royal Statistical Society, Series A (Statistics in Society), 158*(3), 363–417.

de la Fuente, A. (2011). *New measures of labour market attachment: 3 new Eurostat indicators to supplement the unemployment rate (Statistics in Focus 57)*. Brussels: Eurostat, European Commission.

Department of Economic and Social Affairs-International Labour Office. (2009). *Handbook on measuring the economically active population and related characteristics in population censuses. Studies in methods, series F, No. 102*. New York: United Nations. Retrieved from http://unstats.un.org/unsd/demographic/source/census/Entire Handbook.pdf.

Eurostat. (2003). *The European Union Labour Force Survey: Methods and definitions — 2001*. Brussels: Office for Official Publications of the European Communities. Retrieved from http://epp.eurostat.ec.europa.eu/cache/ITY_OFFPUB?KS-BF-03-002- EN.PDF.

Eurostat. (2008). *Labour Force Survey revised explanatory notes (to be applied from 2008Q1 onwards)*. Brussels: European Commission.

Eurostat. (2009). *Task Force on the quality of the Labour Force Survey: Final report*. Brussels: European Commission.

Eurostat. (2013). *Labour Force Survey revised explanatory notes (to be applied from 2014Q1 onwards)*. Brussels: European Commission.

Eurostat. (2016). *EU Labour Force Survey database user guide*. Brussels: European Commission.

Gagliardi, F., Verma, V., & Ciampalini, G. (2009). *Methodology of European labour force surveys: (2) sample design and implementation* (Working paper no. 79). Retrieved from http://www.unisi.it/reicerca/dip/dmg/workingpaper/DMQ_WP_79.pdf

Garrido, L., & Toharia, L. (2004). What does it take to be (counted as) unemployed? The case of Spain. *Labour Economics, 11*, 507–523. https://doi.org/10.1016/j.labeco.2004.01.007.

Hussmanns, R. (2007). *Measurement of employment, unemployment and underemployment: Current international standards and issues in their application.* Retrieved from http://www.ilo.org/global/statistics-and-databases/WCMS_099394/lang-en/index.htm. Geneva: International Labour Office.

Hussmanns, R., Mehran, F., & Verma, V. (1990). *Surveys of economically active population, employment, unemployment and underemployment: An ILO manual on concepts and methods.* Geneva: International Labour Office.

Karanikoli, I. (2009). *Comparability issues of sample surveys: The case of the Labour Force Survey (in Greek).* M. Sc. Thesis, Department of Social Policy. Athens: Panteion University of Social and Political Sciences

Kish, L. (1994). Multi-population survey designs: Five types with seven shared aspects. *International Statistical Review, 62*(2), 167–186.

Michalopoulou, C. (2004). *In the empire of indications: The history of sampling practice in Greece (in Greek).* Athens: Papazisis.

Michalopoulou, C. (2008). *Comparability issues of the IPUMS-International microdata for Greece, 1971–2001.* In IPUMS Workshop: Panteion University of Social and Political Sciences, Athens, March 18 and 20 (17–18). Retrieved from http://www.hist.umn.edu/~rmccaa/ipums-europe/

Sorrentino, C. (2000). International unemployment rates: How comparable are they? *Monthly Labour Review, 123*(6), 3–20.

Sorrentino, C. (2002). Providing comparable international labour statistics. *Monthly Labour Review, 125*(6), 3–14.

Yfanti, A. (2010). *The definition of unemployment is a political decision and not a statistical choice* (Masters' thesis, in Greek). Panteion University of Social and Political Sciences.

Yfanti, A. (2019). *The impact of definitions to the measurement of unemployment: Evidence from the EU-LFS 2008–2015 (Doctoral dissertation).* Athens: Panteion University of Social and Political Sciences.

Yfanti, A., Michalopoulou, C., & Charalampi, A. (2017). *The impact of applying alternative definitions to the European Union Labour Force Survey measurement of the unemployment rate for Southern Europe 2008–2014.* In Paper presented at the 13th Conference of the European Sociological Association (ESA), Athens, August–September, 29-1, 2017.

Yfanti, A., Michalopoulou, C., & Zachariou, S. (2019). The impact of definitions in classifying the employed, unemployed and inactive when comparing measurements from different sources. *Communications in Statistics – Case Studies, Data Analysis and Applications, Advance Online Publication.* https://doi.org/10.1080/23737484.2019.1578703.

Chapter 29
Reverse Mortgages: Risks and Opportunities

E. Di Lorenzo, G. Piscopo, M. Sibillo, and R. Tizzano

Reverse Mortgage and the European Pension Systems Sustainability

Current insurance and pension schemes are undergoing remarkable changes due to a range of interconnected causes. First, one ought to consider demographic changes, revealing the precariousness of public pension systems, and drew governments' and international authorities' attention to the financial and welfare needs characterizing the more elderly and weaker segments of the population. At present, and in contrast to previous experience, these groups do not exclusively include marginalized or homeless people, but, also people with diverse social and personal characteristics, such as low- and very low-income elderly home-owners. Within this category one can also notice a significant gender gap, given that employment and remuneration inequalities accrued throughout one's working life are amplified during one's old age: based on EUROSTAT 2018 and 2020 data, poverty and social exclusion risk rates are higher for women than men across all age segments as of 2016, with the gender gap reaching 5.6% in over-65 populations.

Operationally speaking, it is necessary to foster the development of financial and pension products aiming at addressing needs changing in relation to socio-economic and demographic trends; a recent report by EIOPA (see EIOPA 2016) emphasized the importance of "personal voluntary pension savings", in light of national pension systems' shortcomings in terms of sustainability and adequacy, and in relation to a

E. Di Lorenzo (✉) · G. Piscopo · R. Tizzano
Department of Economic and Statistical Science, University of Naples Federico II, Naples, Italy
e-mail: emilia.dilorenzo@unina.it; gabriella.piscopo@unina.it; roberto.tizzano@unina.it

M. Sibillo
Department of Economics and Statistics, University of Salerno, Fisciano, Italy
e-mail: msibillo@unisa.it

C. H. Skiadas, C. Skiadas (eds.), *Demography of Population Health, Aging and Health Expenditures*, The Springer Series on Demographic Methods and Population Analysis 50, https://doi.org/10.1007/978-3-030-44695-6_29

context marked by financial and employment instability. At any rate, this report also underlined the high costs of personal pension products, which struggle to truly take off homogeneously in member states precisely due to this reason.

An interesting contractual solution that appears to be especially suited for new segments of the pension market revolves around re-describing home-ownership as a source of income at the end of one's working life. Via these contracts, known as Reverse Mortgage, retired people with substantial expenses (e.g. medical or welfare expenses) are granted access to credit. Simply put, Reverse Mortgages represent a form of contracts designed for house-rich-cash-poor population groups; recent surveys led in Italy show that 1.3 million elderly households receive an income lower than 20,000 € per year, but own a home that is worth at least 200,000 € (Beltrametti 2017).

Via a Reverse Mortgage contract, an elderly home-owner can receive money in a lump sum or in lifetime installments, weighted against the value of their home, in which they will be entitled to live until they die.

Alongside the clientele's new needs and changing conditions outlined thus far, the insurance world is currently facing a new challenge, linked with extraordinary data availability and their shifting usage in the wake of new technological developments. The financial and insurance sectors are coming to terms with the ever-changing macrocosm of big data, which has a massive impact on market features, in terms of creativity, competition and regulation (Malohtra and Sharma 2018). Therefore, the role that data play and will play in the future in the financial context will unquestionably be at the core of contract pricing and, even more prominently, the forecasting of assets' future value in terms of real-world evaluation (Giordano and Siciliano 2013; Gu et al. 2018). More specifically, this includes forecasting the cost of insurance contracts (Verbelen et al. 2018), as well as the possibility to consequently implement an effective contract pricing method.

These considerations link up the two themes outlined thus far, as they are intertwined in Reverse Mortgage contracts' elaboration and evaluation.

In this work, we introduce a Reverse Mortgage contract understood as an instrument aiming at supporting the expense needs of non-working people, by enabling the disinvestment of real estate assets, or, in other words, by accessing the savings represented by such real estate assets.

Moreover, we aim at employing neural networks in order to include a range of explanatory variables as part of Reverse Mortgage evaluation algorithms, in ways that may affect the product design, whose expiration is linked to the residual contractor's lifetime and whose underlying asset value depends upon the evolution of the real estate market prices within the same residual lifetime period. The existing literature has already covered the success of neural networks' application in the insurance and pension context quite extensively. Hejazi and Jackson (2016) provided an application targeting particular contract types (variable annuities), employing an approach that could be adapted to other products, too. They subsequently (Hejazi and Jackson 2017) explored neural networks' effectiveness and accuracy in the context of the implementations dictated by the Solvency II's regulatory indications; Wüthrich (2018) then demonstrated how neural networks might benefit reserving methods; Wüthrich and Merz (2019) synthetically demonstrated neural networks' application value in the insurance field; lastly, Doyle and Groendyke (2018) compared neural networks' effectiveness to that of traditional methods in the context of various Variable Aannuities guarantees' pricing.

Concerning the Reverse Mortgage contracts, the prediction of future values is a function of the perspective trends in real estate value; it is therefore necessary to discard evaluations based on risk-neutral probabilities and rather resort to real-world evaluations, which would enable the best estimate of the product, on the basis of careful assessment of the many variables and risk sources featured in the contract profile. Such considerations therefore encouraged us to use neural networks as part of the evaluation procedure.

The paper presents the contractual structure of the Reverse Mortgage and proposes some formulas in order to obtain the cash flows due to the contractor. The main variables impacting the final cash flows are described by means of a neural network algorithm and a numerical example provides some empirical evidences.

The paper is organized as follows. Section 28.2 presents the mathematical description of the contract. In Sect. 28.3 the empirical results are provided.

Reverse Mortgage and Life Annuity: A Balancing Approach

The starting point of the construction of a reverse mortgage contract is the determination of the real estate market value that must be projected over a time period that covers the duration of the contract; according to the literature, such a projection may be rather made through a real world probability approach, than through a risk neutral approach. This is why simulation methods, neural networks and machine learning techniques have been used to obtain the future tendencies of the real estate market values and to quantify the lump sum that a lender can deliver to a borrower while hedging the risk of a capital loss at the end of the contract period: the amount of the loan should not exceed the estimated market value of the asset at the borrower's death. In Di Lorenzo et al. 2019, the authors propose a model for calculating the lump sum by considering the three main risks that impact its determination: the risk of house prices, the interest rate risk and the longevity risk of the owner/borrower [on the other hand, any excesses in the sale price of the asset over the amount due to the owner/borrower at his death will be transferred to the heirs].

Since Reverse Mortgage contracts may take flexible forms, in which the delivery of the lump sum to the borrower can be replaced by a commitment by the lender to pay recurring sums, we will consider the following two cases:

the lump sum is replaced by a financial annuity of limited and determined duration, or,

the lump sum is replaced by a life annuity on the owner/borrower

We will do so in a framework where we recognize (as in Di Lorenzo et al. 2019), the volatility of the real estate market, that is the *house price risk*, the volatility of the financial market, that is the *interest rate risk* and the uncertainty of the property owner's/borrower's lifetime, i.e. the *longevity risk*. We represent the first two by means of the following rates estimated in $t = 0$:

- the random interest rates $r_0^{HV}(t-1,t)$, that is the *house price appreciation rate* for the period $t-1, t$ (that will be derived by the application of a neural network model) and
- $r_0^{RM}(t-1,t)$, that is the *reverse mortgage interest rate* for the period $t-1, t$.

Such rates operate dynamically in the opposite direction, giving rise to the following summary rate:

$$r_0(t-1,t) = \frac{r_0^{HV}(t-1,t) - r_0^{RM}(t-1,t)}{1 + r_0^{RM}(t-1,t)} \tag{29.1}$$

The non-negative equity condition requires that the estimated market value of underlying asset (the house whose liquidation ~~that~~ will reimburse the loan) at the inception of the contract is carried out in such a way that the guarantee is verified. This implies that the dynamics of the valuation are carried out by means of the minimum between the following two values:

$$f_0(0,t) = min\left\{\prod_{k=1}^{t}[1 + r_0(k-1,k)]\,;\,\prod_{k=1}^{t}[1 + i_0(k-1,k)]\right\} \tag{29.2}$$

where, giving the interest rate that the borrower pays for raising funds on the financial market (which, in turn, may depend upon its own financial standing) $i(t-1,t)$, we have that:

$$i_0(t-1,t) = \frac{i(t-1,t) - r_0^{RM}(t-1,t)}{1 + r_0^{RM}(t-1,t)} \tag{29.3}$$

The amount K_0, forecasted in $t = 0$, of the sum equivalent to the value that the lender recognizes to the borrower, is given by the formula:

$$K^{(0)} = \sum_{t=1}^{\omega} {}_{t-1/}q_x \,\alpha\, HV_0\, f_0(0,t) \tag{29.4}$$

where.

- $0 \leq \alpha \leq 1$ is a portion of the house value (HV_0) estimated in $t = 0$ which the lender is willing to entrust the borrower according to his *risk appetite* and
- ${}_{t-1/}q_x$ is the probability that the borrower aged x at the issue time, dies between the ages $x+t-1, x+t$.

The following expression declines the Reverse Mortgage in a contract in which the borrower/house owner is granted a series of recurring level payments P due for a predetermined m number of years:

$$P\ddot{a}_{\overline{m}|} = K^{(0)}{}_{m}E_x \tag{29.5}$$

or a series of recurring level payments R due to the borrower/house owner throughout his or her life:

$$R\ddot{a}_x = K^{(0)}{}_{w-x}E_x. \tag{29.6}$$

Since the result of the implementation of (29.4), (29.5) and (29.6) is influenced — through (29.2) — by (29.1) and (29.3), either amount of the lump sum, or the installments, will depend, in addition to the tendencies of the real estate market value, upon the level of the interest rate that the lender pays for raising funds on the financial market (that determines the level of $i_0(t-1,t)$) and its risk appetite (that determines the level of α); both of those are elements of the competitiveness of the lender in the RM market.

Quantitative Evidence

Let us consider an RM contract offered to a retired male aged x = 80 who resides in London. He owns a house value HV_0= 200,000 at the inception of the contract. Let's assume the following rates:

$$\mathrm{r}_0^{\mathrm{RM}}\,(\mathrm{t}-1,\mathrm{t}) = \mathrm{r}_0^{\mathrm{RM}} = 0.05$$

$$\mathrm{i}\,(\mathrm{t}-1,\mathrm{t}) = \mathrm{i} = 0,015.$$

The r_0^{HV} will be determined by analyzing the real estate market whose data have been collected by the Greater London Authority City Hall (www.london.gov.uk). They are described in the survey "Housing in London 2018" from which we have extracted the following input variables from 1997 up to 2016, that we considered explicating of the house prices (cf. Di Lorenzo et al. 2019):

POP: historic London population
OO: annual trend in household tenure — Owned outright
OM: annual trend in household tenure — Owned with mortgage
NB: New build homes
NS: Net housing supply
NM: Net migration (domestic and international)
TJ: Trend of jobs in London

The Output Variable is the Average House Price (AHP) in the same period.

Then we have implemented a Neural Network (from here on NN) based on a back-propagation algorithm with the free software R and the Neuralnet package. The training is supervised; the number of the hidden nodes has been chosen empirically to minimize the error in the fitting. In order to catch the non-linear relations between features and outputs, we have chosen as activation function a sigmoid logistic one, due to the fact that its slope insures strong influence on the output for small changes in the input variables. For this reason it is suitable for prediction purposes (cf. Gan 2013; Hejazi and Jackson 2017).

As a first step the lump sum A is calculated in correspondence to the house price forecasted through NN and ARIMA time series (TS). As a second step the lump sum is converted into an annuity with installment equal to R (formula 29.6).

The actuarial computation is derived using a projected life table. It has been produced implementing the Lee Carter model (1992) on the UK male mortality dataset downloaded from the Human Mortality Database (HMD).

Tables 29.1 and 29.2 offer some results for studying the sensitivity of the installments to increasing values of α as well as to increasing values of the interest rate i.

Fixing the values as shown in the Table above, we observe that the differences in the installments calculated with NN and TS are due to the presence of the non-equity guarantee. In Table 29.1 we report evidence of the effects of different values of α on R calculated by NN and TS procedures. As α increases, the lump sum and consequently the installment due to the home-owner increases, but the installments calculated by the NN procedure are lower than those calculated by the TS procedure. This is due to the fact that the valuations through NN, including the possibility of negative returns to the forecasted house market prices, appear more prudential than those obtained through Arima Time Series. Such differences are emphasized when the lender accepts a higher risk exposure (for values of α next to or equal to 1) and are nullified if he takes a conservative attitude (for values of α close to 0.5 od 0.6).

In Table 29.2, fixing the values as shown, we analyze the sensitivity of the installments to the increase of i: on one hand, the final payment due at borrower's

Table 29.1 Installments calculated through *NN* and *TS* for different values of α

HV_0	200000
X	80
r_0^{RM}	0.05
i	0.015

α	*NN*	*TS*
0,5	4149,55	4149,55
0,6	4979,46	4979,46
0,7	5788,94	5809,37
0,8	6389,41	6639,28
0,9	6864,91	7469,19
1	7074,27	8299,09

Table 29.2 Installments calculated through *NN* and *TS* for different values of the loan interest rate

HV_0	200000	
X	80	
r_0^{RM}	0.05	
α	1	
i	*NN*	*TS*
0,01	6677,29	7581,16
0,015	7074,27	8299,09
0,02	7458,05	9078,82
0,025	7822,73	9925,18
0,03	8196,94	10789,10

death increases when the house market value is higher than the accumulated amount; on the other hand, there is the effect of the interest on the conversion of the lump sum into an annuity: since the discounted cash flow decreases when the interest rate increases, in order to verify the financial equivalence with the lump sum, the installments have to increase. Consequently, as shown in Table 29.2, the installments increase as the interest increases; however, in the case of NN the increment is less evident because in some years a depreciation of the house certainly occurs and the cash flows to be discounted will always be equal to the disposable value of the house.

References

Beltrametti, L. (2017). *House rich, cash poor. Come rendere liquida la ricchezza rappresentata dalla casa di abitazione*. Fondazione Cariplo, Quaderni dell'Osservatorio n. 26.

Di Lorenzo, E., Piscopo, G., Sibillo, M., & Tizzano, R. (2019). *Pricing of Reverse Mortgage through Machine Learning: new opportunities for the actuaries*. http://www.actuary-isoa.com/downloads/afir-erm-2019/slides/may-24/AM/2-Parallele/4A_Michelangelo/3_Piscopo_Pricing%20of%20Reverse%20Mortagages%20through.pdf

Doyle, D., & Groendyke, C. (December 2018). Using neural networks to price and hedge variable annuity guarantees. *Risks, 7*(1), 1–19.

EIOPA. (2016). *EIOPA's advice on the development of an EU Single Market for personal pension*. EIOPA-16/457.

EUROSTAT: *Living conditions in Europe*. 2018 Edition.

EUROSTAT. (2020). *Europe 2019 indicators – Poverty and social exclusion*. https://ec.europa.eu/eurostat/statistics-explained/index.php/Main_Page

Gan, G. (2013). Application of data clustering and machine learning in variable annuity valuation. *Insurance: Mathematics and Economics, 53*, 795–801.

Giordano, L., & Siciliano, G. (2013). *Real-world and risk-neutral probabilities in the regulation on the transparency of structured products*. Quaderni di Finanza, CONSOB 74.

Gu, S., Kelly, B., & Xiu, D. (2018). *Empirical asset pricing via machine learning*. Chicago Booth School of Business Research Paper Series. 31st Australasian Finance and Banking Conference 2018. https://papers.ssrn.com/sol3/papers.cfm?abstract_id=3159577

Hejazi, S. A., & Jackson, K. R. (2016). A neural network approach to efficient valuation of large portfolios of variable annuities. *Insurance: Mathematics and Economics, 70*, 169–181.

Hejazi, S. A., & Jackson, K. R. (2017). Efficient evaluation of SCR via a neural network approach. *Journal of Computational and Applied Mathematics, 313*, 427–439.

Lee, R. D., & Carter, L. R. (1992). Modelling and forecasting U.S. mortality. *Journal of the American Statistical Association, 87*, 659–671.

Malohtra, R., & Sharma, S. (2018). *Machine learning in insurance: Enabling insurers to become AI- driven enterprise powered by automated machine learning.* https://www.accenture.com/_acnmedia/PDF-84/Accenture-Machine-Leaning-Insurance.pdf

Verbelen, R., Antonio, K., & Claeskens, G. (2018). Unravelling the predictive power of telematics data in car insurance pricing. *Journal of the Royal Statistics Society, Applied Statistic, 67*(5), 1275–1304.

Wüthrich, M. V. (2018). Neural networks applied to chain–ladder reserving. *European Actuarial Journal, 8*(2), 407–436.

Wüthrich, M. V., & Merz, M. (2019). Editorial: Yes, we Cann! *ASTIN Bulletin, 49*(1), 1–3.

Chapter 30
Estimating the Health State at Retirement: A Stochastic Modeling Approach

Christos H. Skiadas and Charilaos Skiadas

Introduction

After introducing and applying the theory of the first exit time of a stochastic process from a barrier and published it in several papers and books, we use the findings to estimate the optimal retirement age of the population. Important results are in the two books published recently in volumes 45(2018a) and 46(2018b) in The Springer Series on Demographic Methods and Population Analysis by Skiadas and Skiadas while other publications from the authors (2010, 2014, 2015) support the theoretical and applied material. Other related publications are due to Janssen and Skiadas (1995) and Skiadas and Zafeiris (2015), Torrance (1976) and Weitz and Fraser (2001).

Related Theory

Following the first exit time theory we assume that the health state of an individual is expressed by a stochastic function denoted by S_x. The associated stochastic paths over time t or age x are estimated after integrating the stochastic differential equation

$$dS_x = h_x dx + \sigma \, dW_x, \tag{30.1}$$

C. H. Skiadas (✉)
ISAST, Athens, Attiki, Greece
e-mail: skiadas@cmsim.net

C. Skiadas
Department of Mathematics and Computer Science, Hanover College, Hanover, IN, USA
e-mail: skiadas@hanover.edu

C. H. Skiadas, C. Skiadas (eds.), *Demography of Population Health, Aging and Health Expenditures*, The Springer Series on Demographic Methods and Population Analysis 50, https://doi.org/10.1007/978-3-030-44695-6_30

with drift h_x and determining the formula for the stochastic paths S_x

$$S_x = \int_0^x h_s ds + \sigma \int_0^x dW_s = H_x + \sigma\ W_x, \tag{30.2}$$

where W_x is the Wiener process and the Health State H_x is provided as the integral of the instantaneous change h_x

$$h_x = \frac{dH_x}{dx}. \tag{30.3}$$

The death occurs when $S_x = 0$ and from (30.2) follows that

$$H_x + \sigma\ dW_x = 0. \tag{30.4}$$

The simpler form for the Health State H_x should be a decreasing process of the form (see related bibliography in Janssen and Skiadas 1995 and Skiadas and Skiadas 2007, 2010, 2014, 2015):

$$H_x = l - (bx)^c, \tag{30.5}$$

where *l*, *b and c* are parameters. The form of (30.4) becomes

$$l - (bx)^c + \sigma\ W_x = 0. \tag{30.6}$$

This is a very important relation providing different ways of simulation of the stochastic process.

This is demonstrated by observing the new form of (30.2) that is

$$S_x = l - (bx)^c + \sigma\ W_x. \tag{30.7a}$$

This form is important for constructing stochastic paths for the health state S_x.

The next form arises from (30.7a) by a simple transformation

$$S_x + (bx)^c = l + \sigma\ W_x. \tag{30.7b}$$

A third form given by

$$S_x - l + (bx)^c = \sigma\ W_x. \tag{30.7c}$$

The three forms (30.7a), (30.7b) and (30.7c) are mathematically the same. However, they provide three distinct simulation opportunities very important to explain the development of the health status and the development of the death probability density function.

According to the theory developed (Janssen and Skiadas 1995) first the associated Fokker-Planck equation for the appropriate boundary conditions was solved in order to find the transition probability density function, and then we could determine the formula for the first exit time probability density function $g(x)$ for the health state stochastic process crossing or hitting for the first time a barrier set to zero.

$$g(x) = \frac{\lfloor H_x - xH'_x \rfloor}{\sigma\sqrt{2\pi x^3}} e^{-\frac{[H_x]^2}{2\sigma^2 x}}. \quad (30.8)$$

Stochastic Simulations

According to the theory developed earlier we recall eq. (30.5) by setting $l = 1$ that is

$$H_x = 1 - (bx)^c. \quad (30.9)$$

This simpler form is very important for the applications and simulations that follow.

It is assumed that the health state H_x is one at zero ($H_0 = 1$) and then the health state is decreasing slowly in the first period of the lifetime and faster in the middle and later stages as far as the deterioration term $(bx)^c$ is getting larger and larger growing exponentially. We call as the deterioration function (*Det*) the following relation

$$Det = (bx)^c. \quad (30.10)$$

Now by using Eq. (30.9), eq. (30.8) takes the form

$$g(x) = \frac{1 + (c-1)(bx)^c}{\sigma\sqrt{2\pi x^3}} e^{-\frac{[1-(bx)^c]^2}{2\sigma^2 x}}. \quad (30.11)$$

Figure 30.1 illustrates the Health State for Sweden female in 2016. The mean health state $H(x)$ is presented by the heavy magenta curve. The health state of the individuals $S(x)$ is expressed by light dashed lines, whereas the confidence intervals are represented by the two dashed black curves. The left skewed death distribution function $g(x)$ is expressed by the yellow dashed curve with a perfect coverage with the blue fit curve. We have also reproduced the death distribution function by using the health state $H(x)$, the variance \sigma and the standard Wiener function to account for the stochastic character of the health state of the individuals

$$S(t) = H(t) + \sigma\,[W(t) - W(0)].$$

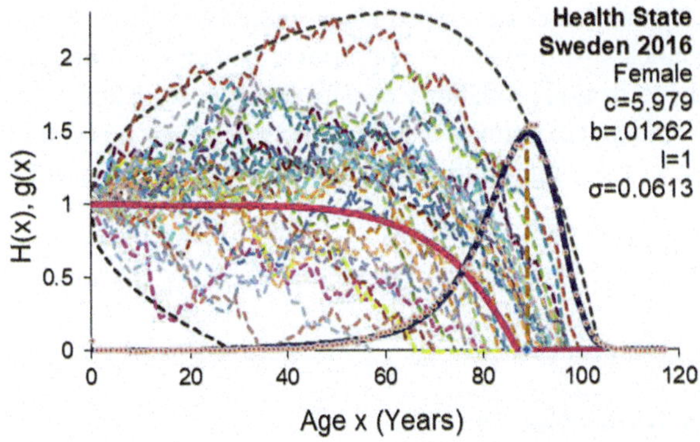

Fig. 30.1 Health state and stochastic paths

We assume that the health state of every individual starts at level 1 at birth and then follows a stochastic process at higher or lower values until the end which takes place when the stochastic path of an individual reaches the zero health level represented by the horizontal axis X at a point at age x. By collecting the number of the first exit time individual health states at every point x of the X axis, the death distribution function is derived. This is a first exit time probability density function $g(x)$. This function is found in the yearly data collected by the bureau of the census in every country. Following the theory already developed we can find the Health State of the population $H(x)$ from $g(x)$ and also reproduce $g(x)$ from $H(x)$ by stochastic simulation. This first exit time or hitting time theory was applied earlier in applications in physics when observing stochastic processes including the Brownian motion phenomena. What is completely new is the derivation of the appropriate first exit time probability density function in a form expressing the human health state problem so that it extracts the health state function from the death probability density function. As we already have shown in our earlier publications cited above, the derivation of the health state function $H(x)$ was very important for expressing the aging process from another point of view. The health state is a declining process over time whereas mortality is a growing process over time.

In Fig. 30.2 we present how the two methods of stochastic estimation and simulation are equal. There are two main lines to simulate and reproduce the death distribution function used to estimate the health state function and the deterioration function. In the first line presented in Fig. 30.1 the health state function is used as a guide to the stochastic paths with barrier the horizontal axis, whereas in the second line presented in Fig. 30.3 the deterioration function is used as a barrier to stop the stochastic paths. Both methods produce identical results as is presented in Fig. 30.2 where two stochastic paths based on the first and the second approach stop at the same age the $f1(x)$ at the X axis barrier and the $f2(x)$ at the deterioration barrier. Clearly the deterioration approach is simpler as it avoids passing to negative values as for the health state function. The latter characterizing a critical and supercritical health range for the ages included in negative part. Note that the zero health level

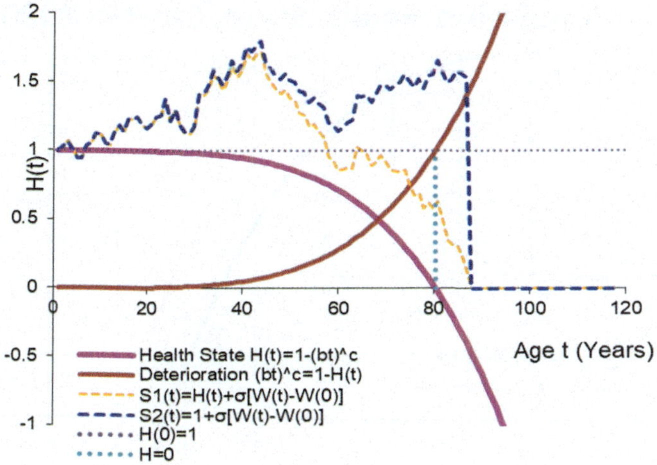

Fig. 30.2 Health state, deterioration and stochastic paths

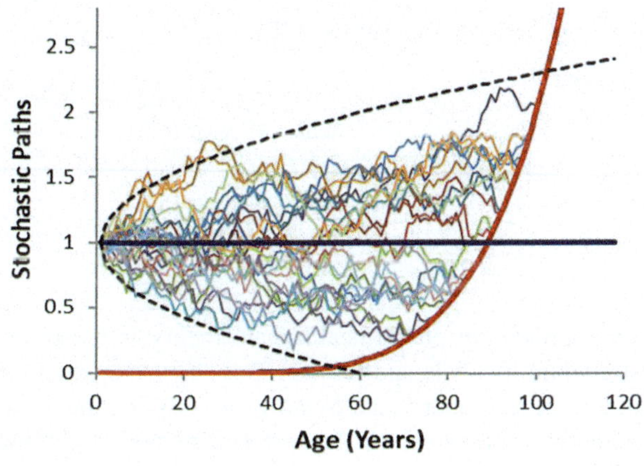

Fig. 30.3 Deterioration function and stochastic paths

corresponds to ages just below the "modal" age at death (this is the age with the maximum of the death distribution).

The Deterioration function expressing how the mean deterioration of the population is growing is presented in Fig. 30.3. It is an exponential like function of the form $Det = (bt)^c$. In the first stages it is slowly varying while it is fast growing in the last stages at high age levels. It thus sets a barrier for the health state of the individuals expressed in the graph by light lines. The two dashed curves express the confidence intervals of the stochastic paths of the individuals. The vast majority of the health states (99%) are included within these intervals. The exceptional cases of people living longer are stopped by the deterioration curve.

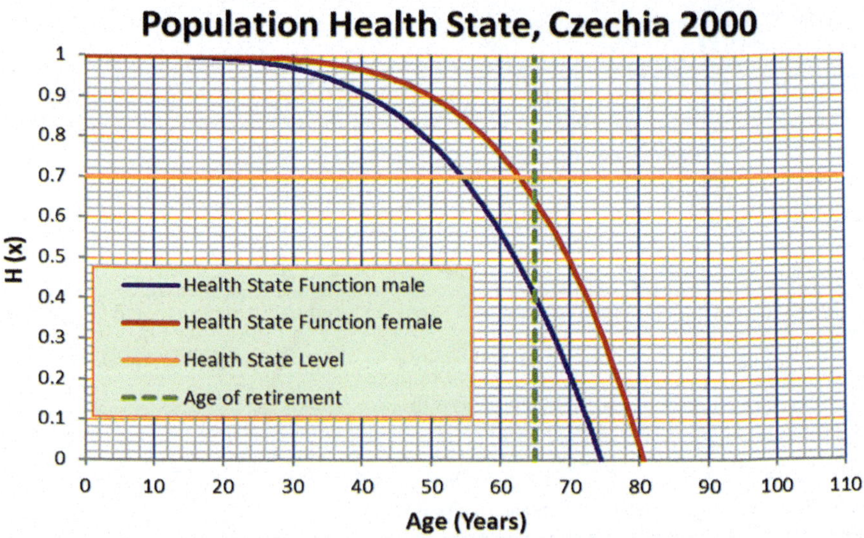

Fig. 30.4 Health state in the Czech Republic for 2000

Table 30.1 Age at 70% health level

Year	Age at health level 70%		Age difference
	Female	Male	Years
1990	60.2	49.7	10.5
2000	62.8	54.4	8.4
2016	68.1	58.7	9.4

An interesting application of the health state theory is presented in Fig. 30.4 for the Czech Republic for male and female. This graph form is very useful when deciding the optimum retirement age. What is very important is to keep people working with a relatively high health state. In the presented example we set a barrier at the 70% level of the total health state for the optimal retirement age. The Health state for female is higher than male. Other estimates can be found by moving the barrier to higher or lower places to account for the scenarios of decision makers.

Table 30.1 includes the years of age at 70% health state along with the differences between female-male for the various periods investigated. It is important to have as small as possible a difference between sexes in the case of decisions of equal age of retirement. The difference was rather high for all three time periods with 10.5 years in 1990, dropping to 8.4 in 2000 and growing to 9.4 in 2016. In all cases the difference is quite large and needs further measures to reduce the gap.

As one of the standard barriers for retirement is 65 years of age we have set this barrier to include more information. The Health Level (%) at retirement at age 65 is presented in Table 30.2. The health state differences are very large in this case especially for the year 1990 where the health differences between female – male are

Table 30.2 Health level (%) at retirement at age 65

Year	Health level (%) at retirement at age 65		Health difference (%)
	Female	Male	Years
1990	57.4	25.8	31.6
2000	64.7	40.8	23.9
2016	77.2	54.4	22.8

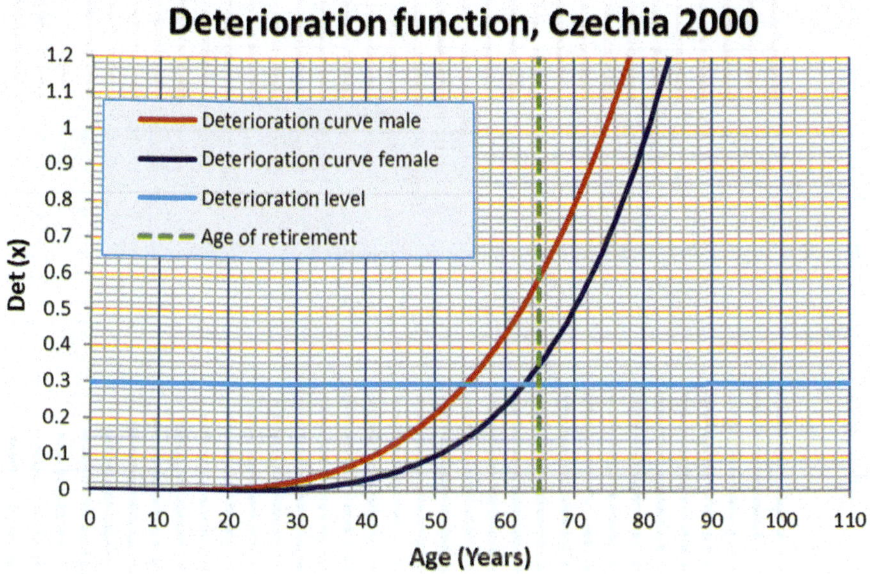

Fig. 30.5 Deterioration function in the Czech Republic for 2000

31.6% with a clear improvement in the year 2000 at 23.9% and a small improvement at 22.8% the year 2016.

Figure 30.5 illustrates the deterioration function for the Czech population in 2000. Female show also a higher level of survival than male. The deterioration level corresponding to 70% health state level is 0.30.

The age at 70% health state level for female in various countries is illustrated in Fig. 30.6. Japan, France, Switzerland, Spain and Australia are at the top 5 places whereas Latvia, Slovakia, Hungary, Belarus and Russia are in the last places. The Czech Republic is ranked in the 22 place just before USA.

The age at 70% health state level for male in various countries is illustrated in Fig. 30.7. Iceland, Switzerland, Australia, France and Italy are at the top 5 places whereas Hungary, Lithuania, Latvia, Belarus and Russia are in the last places. The Czech Republic is ranked in the 24 place just before Croatia.

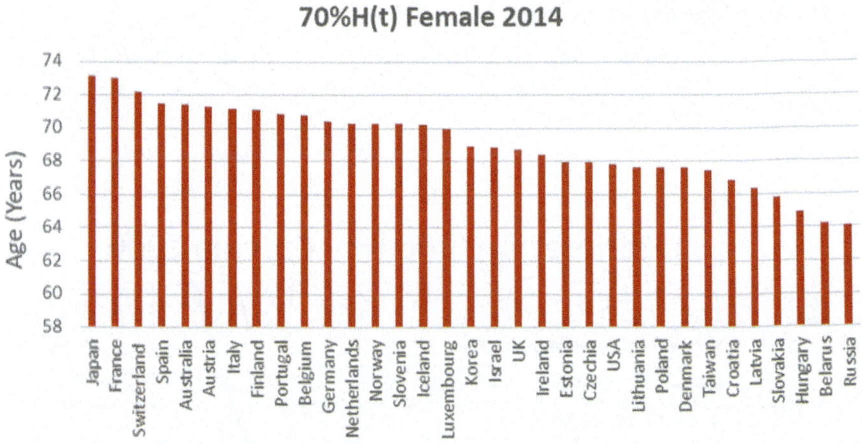

Fig. 30.6 Age at 70% health state level for various countries

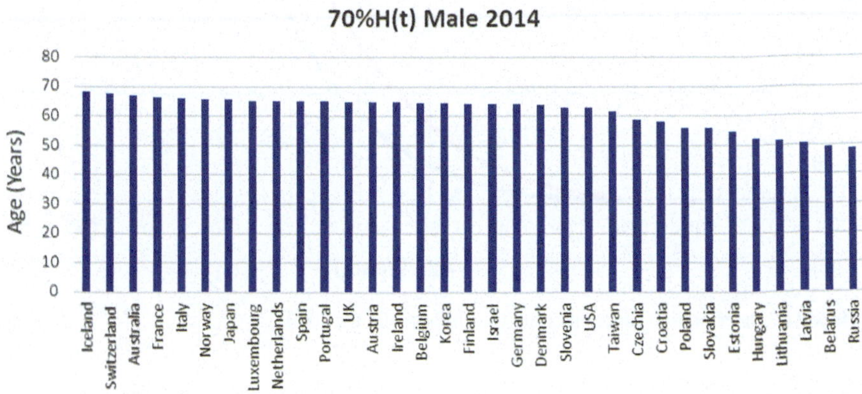

Fig. 30.7 Age at 70% health state level for various countries

Optimal Retirement Strategy

Retirement age and related decisions and applications are the main issue in governmental decisions in almost every country in the World. Especially after the huge improvements in the health systems and the advancements in life expectancy leading to an "aging society", the question is what is the optimal retirement strategy. Many discussions have to do with an equal retirement age for male and female. The quest for equality is a classical approach and tends to be a leading opinion for many people especially from countries with high levels in life expectancy. However, it looks like that closing the gap between life expectancy of females and males is also a very important task. Our approach of estimating the health state of the population along with the deterioration function gives important tools for assessing the health

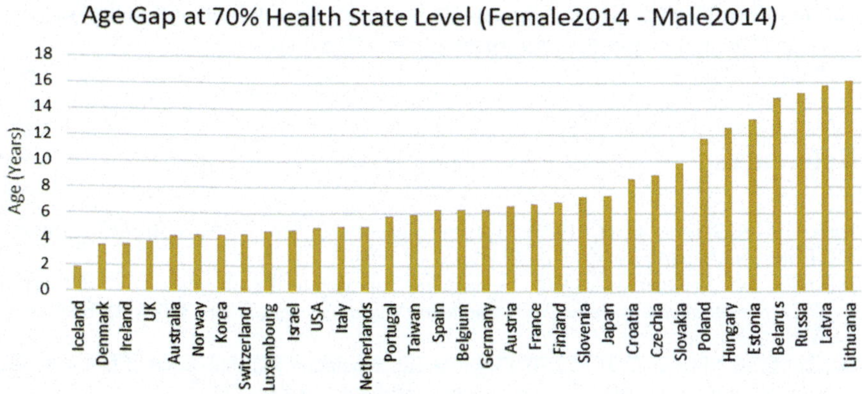

Fig. 30.8 The gap between females and males

state of the population along with determining the optimal retirement strategy. The important new finding with the new method is that we have a quantitative method for estimating the health state of the population or of the deterioration function in the various age levels thus providing a simple and powerful tool when dealing with the optimal retirement strategy. Retirement has not only to do with age levels and economic decisions but mainly to succeed in giving the opportunity to people after retirement to start their new life at a relatively high level as to enjoy the remaining years of life. Another point is also to select the retirement age level after reducing the health level gap between female and male. This gap is presented in Fig. 30.8. The smallest gap with 1.8 years of age is due to Iceland. Denmark, Ireland and UK follow with the gap of over 3 years and less than 4 years. Less than 5 years gap but larger than 4 years are followed by UK, Australia, Norway, Korea, Switzerland, Luxembourg, Israel, USA and Italy. The last 7 countries in the rank are Poland, Hungary, Estonia, Belarus, Russia, Latvia and Lithuania showing a gap larger than 10 years of age. The Czech Republic with 8.9 years of age gap is ranked 25 at a higher place than Slovakia and at a lower than Croatia and 2 places below Japan.

Clearly the countries with a gap higher than 5 years of age have to take strong measures to reduce the gap. This in an important action along with the equalization of the retirement age for both male and female.

Conclusion

We have applied a stochastic theory of the health state function to develop appropriate graphs useful to estimating the health state of a population or, alternatively, the deterioration function at several periods of time and age. The related methodology is used to determine the health state at the retirement age or the retirement age at an optimum health level. The results should be useful for decision and policy-makers in ~~to~~ developing the optimal retirement strategy in a country.

References

Janssen, J., & Skiadas, C. H. (1995). Dynamic modelling of life-table data. *Applied Stochastic Models and Data Analysis, 11*(1), 35–49.

Skiadas, C. H., & Skiadas, C. (2007). A modeling approach to life table data. In C. H. Skiadas (Ed.), *Recent advances in stochastic modeling and data analysis* (pp. 350–359). Singapore: World Scientific.

Skiadas, C., & Skiadas, C. H. (2010). Development, simulation and application of first exit time densities to life table data. *Communications in Statistics, 39*, 444–451.

Skiadas, C. H., & Skiadas, C. (2014). The first exit time theory applied to life table data: The health state function of a population and other characteristics. *Communications in Statistics-Theory and Methods, 43*, 1585–1600.

Skiadas, C. H., & Skiadas, C. (2015). Exploring the state of a stochastic system via stochastic simulations: An interesting inversion problem and the health state function. *Methodology and Computing in Applied Probability, 17*(4), 973–982.

Skiadas, C. H., & Skiadas, C. (2018a). *Exploring the health state of a population by dynamic modeling methods*. Cham: Springer. https://doi.org/10.1007/978-3-319-65142-2.

Skiadas, C. H., & Skiadas, C. (2018b). *Demography and health issues — Population aging, mortality and data analysis*. Cham: Springer. https://doi.org/10.1007/978-3-319-76002-5.

Skiadas, C. H., & Zafeiris, K. N. (2015). Population aging and healthy life: Lessons from the related studies. In *RELIK*, 12–13 November 2015.

Torrance, G. W. (1976). Health status index models: A unified mathematical view. *Management Science, 22*(9), 990–1001.

Weitz, J. S., & Fraser, H. B. (2001). Explaining mortality rate plateaus. *Proceedings of the National Academy of Sciences of the United States of America, 98*(26), 15383–15386.

Correction to: Births by Order and Childlessness in the Post-socialist Countries

Filip Hon and Jitka Langhamrová

Correction to:
Chapter 15 in: C. H. Skiadas, C. Skiadas (eds.), *Demography of Population Health, Aging and Health Expenditures*, The Springer Series on Demographic Methods and Population Analysis 50, https://doi.org/10.1007/978-3-030-44695-6_15

The chapter "Births by Order and Childlessness in the Post-socialist Countries" was previously published with a spelling mistake to the first author's name. It has now been updated to "Filip".

The updated online version of this chapter can be found at
https://doi.org/10.1007/978-3-030-44695-6_15

C. H. Skiadas, C. Skiadas (eds.), *Demography of Population Health, Aging and Health Expenditures*, The Springer Series on Demographic Methods and Population Analysis 50, https://doi.org/10.1007/978-3-030-44695-6_31

The manufacturer's authorised representative in the EU is Springer Nature Customer Service Centre GmbH, Europaplatz 3, 69115 Heidelberg, Germany. If you have any concerns regarding our products, please contact ProductSafety@springernature.com

Printed and bound by CPI Group (UK) Ltd, Croydon, CR0 4YY
22/09/2026
02220980-0004